Nutrient Symposium 2015

San Jose, California, USA
26 – 28 July 2015

ISBN: 978-1-5108-7156-4

Printed from e-media with permission by:

Curran Associates, Inc.
57 Morehouse Lane
Red Hook, NY 12571

Some format issues inherent in the e-media version may also appear in this print version.

Printed by Curran Associates, Inc. (2018)

For permission requests, please contact Water Environment Federation (WEF)
at the address below.

Water Environment Federation (WEF)
601 Wythe Street
Alexandria, Virginia 22314
USA

Phone: 1-800-666-0206
Fax: 1-703-684-1545

csc@wef.org

Additional copies of this publication are available from:

Curran Associates, Inc.
57 Morehouse Lane
Red Hook, NY 12571 USA
Phone: 845-758-0400
Fax: 845-758-2633
Email: curran@proceedings.com
Web: www.proceedings.com

Water Environment Federation
the water quality people™

Nutrient Symposium 2015

July 26-28, 2015
Doubletree San Jose
San Jose, California, USA

WERF
Water Environment Research Foundation
Collaboration. Innovation. Results.

California Water Environment Association

BACWA
BAY AREA CLEAN WATER AGENCIES

WERF
Water Environment Research Foundation
Collaboration. Innovation. Results.

This conference is held by the Water Environment Federation in cooperation with the California Water Environment Association, The Bay Area Clean Water Agencies and The Water Environment Research Foundation.

July 26, 2015

Dear Colleagues,

Welcome to sunny San Jose, California! The Water Environment Federation, the Water Environment Research Foundation, and the California Water Environment Association are honored to have you join us for the Nutrient Symposium 2015. On their behalf, we invite you to fully participate in the exceptional opportunity for education and collaboration!

This event continues WEF's tradition of addressing the challenge of sustainable nutrient management through education and sharing of best practices, and is dedicated to serving the practitioners who are addressing nutrient issues in today's world. The focus for this event will be on the *practical* design and process operations of dealing with nutrient issues across the nation and around the globe.

The speakers and attendees for this event come from a wide variety of backgrounds including regulatory, research, design, implementation, and plant operations. The conference technical program is composed of 15-20 minute presentations, short briefings, and interactive facilitated discussion in every session; session topics include:

- Nutrient Removal and Recovery;
- BNR Optimization and Process Demonstrations;
- Research and New Technologies;
- Operations;
- Process Control Optimization;
- Nutrient Management Lessons Learned; and
- Compliance Strategies.

While you network with attendees, make a point to learn from them and share your experiences. We look forward to interacting with each of you, and to the many interesting exchanges that will take place in the next few days – and beyond.

Enjoy this informative event and the innovative culture of Silicon Valley and the San Francisco Bay Area.

Sincerely,

Thor Young
Conference Co-Chair

Michael Connor
Conference Co-Chair

CONFERENCE COMMITTEE

Thor Young
GHD
Conference Co-Chair

Michael Connor
East Bay Dischargers Authority
Conference Co-Chair

NUTRIENT SYMPOSIUM STEERING COMMITTEE

Charles Bott
Hampton Roads Sanitation District

Jeanette Brown
Manhattan College

Leon Downing
CH2M

Sarina Ergas
USF

Aren Hansen
Brown and Caldwell

Joe Husband
ARCADIS

Sam Jeyanayagam
CH2M

Jose Jimenez
Brown and Caldwell

Wendell Khunjar
Hazen and Sawyer

Jim McQuarrie
MWRD

Kristi Perri
GHD

Tina Pham
City of San Jose, CA

Robert Sharp
Manhattan College

Kumar Upendrakumar
Veolia North America

Jianfeng Wen
American Water5

NUTRIENT SYMPOSIUM PROGRAM COMMITTEE

REGISTRATION

All events are held in the Doubletree Hotel.

The Registration Desk is located in the **Gateway Foyer** and will be open at the following times:

Hours:

Sunday, July 26	11:00am – 5:00pm
Monday, July 27	7:30am – 5:00pm
Tuesday July 28	7:30am – 5:00pm

PRESENTER AND MODERATOR INFORMATION

All presenters should sign-in at Conference Registration before presenting and must attend the assigned briefing.

Presenters, including technical briefs and session moderators participating Monday and Tuesday should attend the briefing on their respective mornings. The Speaker Briefing and room assignment is as follows:

Monday, July 27 – **Cedar Ballroom** *Sessions 1 through 4	7:15am - 8:00am
Tuesday, July 28- **Cedar Ballroom** *Sessions 5 through 9	7:15am - 8:00am

If you are not able to attend the assigned briefing, please contact your session moderator immediately and arrive early for your technical session to make up for any missed discussions.

Networking Reception – Monday, July 26 **Sierra/Cascade Ballroom**
5:00pm – 6:30pm

The Networking Reception will be held on the exhibit floor. It is your opportunity to view the exhibits, renew acquaintances, and make new contacts. Light hors d'oeuvres and refreshments will be served.

Luncheon

Lunch will be provided for all full conference, daily registrants, and exhibitors. Please advise staff as soon as possible if you have any special dietary requirements. The luncheons will be held:

Monday, July 27 – Siskiyou Ballroom
12:00pm - 1:30pm

Tuesday, July 28 – Siskiyou Ballroom
12:00pm - 1:30pm

Networking Breaks
The Networking Breaks will be held:

Monday, July 27
Sierra/Cascade Ballroom
10:15am – 11:00am and 3:00pm – 3:45pm

Tuesday, July 28
Sierra/Cascade Ballroom
10:00am – 10:45am and 3:00pm – 3:45pm

CONTINUING EDUCATION

How Do I Receive Credit For this Conference?
To receive credit for all technical sessions, pre-conference workshops, and facility tours please fill out a Continuing Education Request Form and be sure to have a room monitor initial at each entry and exit from a session. Attendees will have to submit their CE Request Form to the Registration Desk at the end of the conference. Please check your bag for a copy of the form or visit the Registration Desk if you need a copy or if you have further questions.

Pre-Conference Workshops:
WEF offers Continuing Education Units (CEUs) for participation in workshops. **One CEU is the equivalent to 10 hours** of training or formal instruction. These are distributed for structured, relevant professional training above and beyond that of initial certification or employment in a particular field.

Technical Sessions:
WEF offers Professional Development Hours (PDHs) for participation in technical sessions. **A PDH is defined as one hour** spent engaged in an activity that contributes to the advancement or enhancement of professional skills or scientific knowledge of a professional engineer or operator.

Off-Site Facility Tours:
General Contact hours are given for participating in Facility Tours. **A General Contact Hour is defined as one hour** spent engaged in an activity that contributes to the professional skills of the participant.

When Will I Receive Credits For this Conference?
Certificates and transcripts will be emailed to participants within 8 weeks after this conference. Please keep in mind that although WEF does provide these files, most states will require the individual licensee to report continuing education credits.

Note: *Educational Credits will not be recorded and documentation will not be distributed unless the attendee is a confirmed registrant of this event and the proper form is completed as indicated in the directions provided.*

Are WEF Continuing Education Credits Approved in My State?
WEF applies for approval in many states, and will be happy to work with individuals and Member Associations for additional state or agency approvals upon request. In addition, WEF has been approved as a Training Provider through the following:

CONTINUING EDUCATION

The Florida Board of Professional Engineers, the New York State Department of Education, and the Ohio EPA. Most states accept WEF PDH credits and WEF CEU credits without issue. For example: California (CWEA), Michigan, Nevada, and New Jersey.

What Else Do I Need To Know?
WEF follows the International Association of Continuing Education and Training (IACET) guidelines along with state-specific regulations to achieve strict policies and procedures regarding its Continuing Education Program. WEF calculates education credits following a standardized method that is the most widely accepted by certification and licensing agencies. However, many states differ in the type and/or number of credits they will approve for educational events. Because of this, participants are responsible for exploring their state requirements and for ensuring that WEF conference credits are accepted.

Service and Support...
In keeping with IACET guidelines, WEF maintains a database of all continuing education files for a minimum of 7 years. You may contact WEF's Customer Service Team between the hours of 8:30am and 5:00pm EST, Monday through Friday to request these files. Please call 1-800-666-0206 or submit an email request to csc@wef.org.

CEU & PDH credits are available for Workshops to Professional Engineers licensed in the state of New York (NYSED). Please visit www.wef.org/CECs for more details related to WEF's Continuing Education Program

Market Efficiencies
for Nutrient Sensors and Services

www.nutrients-challenge.org

Please join us for a one hour meeting on Monday, July 27th from 12:30pm – 1:30pm in the Cedar Ballroom where EPA, USGS, and WEF will lead an interactive discussion on how we can improve market efficiencies for nutrient sensors.

Purpose:
To explore and catalyze opportunities for market efficiencies for acquisition, services and data sharing associated with nutrient sensors.

Background:
Nutrient pollution is one of America's most significant and costly environmental problems, and requires monitoring to assess conditions, evaluate progress, and target investments. New sensors are emerging that allow for nutrient measurement at desired temporal and spatial scales; however, organizations and agencies have a limited capacity to purchase and maintain them. Water Environment Federation (WEF), with the support of the Association of Clean Water Administrators (ACWA) and the federal agency partners of the Challenging Nutrients Coalition are inviting a range of stakeholders that include users, suppliers, regulatory groups, trade organizations, laboratories and system integrators to propose and discuss strategies that will facilitate effective deployment and services of nutrient sensors and enabling technologies.

As food is not allowed in meeting room the hotel kindly asks that you eat before the meeting. Thank you for your cooperation!

USB PROCEEDINGS

Conference proceedings, consisting of manuscripts for each presentation, have been produced on a USB drive. Full-conference and daily registrants will receive the proceedings at the registration desk at the time of badge pickup or onsite registration.

Copies of proceedings may be ordered after the conference at the member rate of $75 USD/nonmember rate of $115 USD, plus shipping and handling. All orders will be processed immediately after conclusion of the conference. You may call 1-800-666-0206. Ask for Stock Number CPNR1504.

We would like to thank the following sponsoring companies for their contributions to the conference and program.

www.cdmsmith.com
Opening General Session

www.hdrinc.com
Conference Lanyards

www.suez-environnement.com
Conference Tote Bags
Booth 100

San Jose-Santa Clara Regional Wastewater Facility Tour

Monday, July 27, 2015 **Contact Hours: 3.0**
Time: 1:30pm – 4:30pm
$35.00

Visit the largest BNR facility in Northern California. Observe an intricate 8 to 10 hour treatment process that cleans wastewater from 1.4 million residents in the Silicon Valley. Over 75% of nitrogen and 90% of phosphorus is removed using no significant chemical addition; a miraculous process that simulates the way nature cleans water.

Participants should gather no later than 1:15pm at the lobby level, convention entrance of the Doubletree Hotel.

OPENING GENERAL SESSION

Opening General Session
Monday, July 27, 2015
8:30am – 12:00pm

Room: Oak Ballroom
PDHS: 2.4

8:30am	**Welcome and Introduction** Ed McCormick, *WEF President*	
8:50am	**Nutrient Treatment as a part of Urban Wastewater Resource Recovery Facilities in 25 year** Craig Criddle, *Stanford University*	N/A
9:10am	**Pragmatic Utility Sustainability through Rapid Implementation of Emerging Technology** Charles Bott, *HRSD*	N/A
10:00am	**Networking Break**	
10:45am	**The San Francisco Bay Area Nutrient Watershed Permit: Balancing Innovative Technology Uncertainty and Environmental Prediction Uncertainty** Michael Connor, *East Bay Dischargers Authority*	1
11:05am	**The San Francisco Bay Area Nutrient Watershed Permit: A Streamlined Approach to Identify Cost Effective Strategies for Nutrient Load Reductions** Michael Falk, *HDR Engineering Inc*	N/A
11:25am	**Applying 30 years of Knowledge from the Chesapeake Bay Program to Nutrient Removal Upgrades in the San Francisco Bay Area** Thor Young, *GHD*	N/A
11:45am	**Questions and Discussion**	
12:00pm	**Session Adjourns**	

TECHNICAL PROGRAM

Session 1: Nutrient Removal and Recovery
Monday, July 27, 2015 **Room: Oak Ballroom**
1:30pm – 5:00pm **PDHS: 2.75**

Moderator: Jim McQuarrie, *Metro Water Reclamation District, Denver CO*

1A 1:30pm **Mainstream Nitrite-Shunt with Biological Phosphorus Removal Using a Low Dissolved Oxygen Aeration Control Strategy** 18
Jose Jimenez, Pusker Regmi, *Brown And Caldwell;* GeorgeWise, *City of St. Petersburg FL.*

1B 1:50pm **Demonstration of a Novel Separate Centrate Deammonification Process at the 26th Ward Wastewater Treatment Plant** 24
Sarah Galst, Wendell Khunjar, *Hazen and Sawyer, P.C.;* Robert Sharp, *Manhattan College*; Kathleen O'Connor; *NYSERDA;* Allen Deur; *NYCDEP*; Paul Pitt, *Hazen and Sawyer, P.C.*

1C 2:10pm **Recommendations and Lessons Learned from the Startup of the First Two Full-Scale Sidestream Deammonification Processes in North America: DEMON and ANITA Mox** 30
Stephanie Klaus, *Virginia Tech;* Andrea Nifong, *World Water Works, Inc.*; Andy Nelson, Bob Rutherford, Rick Baumler, Charles Bott, *Hampton Roads Sanitation District*

2:30pm **Facilitated Discussion**

3:00pm **Networking Break**

1D 3:45pm **Reclaiming Ammonia from Anaerobic Digestate as a Profitable Product** 35
Dennis Burke, *Environmental Energy & Engineering Co*

1E 4:05pm **Implementing EBPR at an Existing WRRF: Operational Adjustments to Low pH, Alkalinity Limitations, and Future Biosolids Resource Recovery** 47
Leon Downing, *CH2M;* Mike Young, Bill Tatum, Trinity Hill, *River Authority of Texas*

TB1 4:25pm **Generating Class A Struvite at the West Boise WWTP** 49
Jordi Figueras, *Pharmer Engineering*

4:30pm **Facilitated Discussion**

5:00pm **Session Adjouns**

TECHNICAL PROGRAM

Session 2: BNR Troubleshooting and Optimization

Monday, July 27, 2015
1:30pm – 5:00pm
Moderator: Joe Husband, *Arcadis*

Room: Fir Ballroom
PDHS: 2.75

Session 3: Compliance Strategies

Monday, July 27, 2015 **Room: Pine Ballroom**
1:30pm – 5:00pm **PDHS: 2.75**
Moderator: Aren Hansen, *Brown and Caldwell*

Session 4: New Technologies

Tuesday, July 28, 2015 **Room: Oak Ballroom**
8:30am – 12:00pm **PDHS: 2.75**
Moderator: Wendell Khunjar, *Hanzen and Sawyer*

4A 8:30am **Redefining Being Green: Extended Piloting of an Algae-Based Technology for Low-Level Nutrient Removal and Recovery** 171
Alexandra Doody, *CDM Smith; Karla Sangrey,* Mark Johnson, *Upper Blackstone Water Pollution Abatement District*; Jane Madden, *CDM Smith;* Rick Johnson, Will Holm, *Clearas Water Recovery*

4B 8:50am **Pilot Testing Of Struvite Recovery from Dewatering Centrate of a German Municipal WWTP through Electrochemical Precipitation (ePhos Technology)** 189
Iosif Mariakakis, Jennifer Bilbao, Siegfried Egner, Thomas, Hirth, *Fraunhofer*

4C 9:10am **Aerobic Granular Sludge Bench-Scale Startup and Testing to Full-Scale Conceptual Design** 194
Roland Jezek, Ed Kobylinski, Mark, Steichen, James Barnard, *Black & Veatch;* Belinda Sturm, *University of Kansas*

9:30am **Facilitated Discussion**

10:00am **Networking Break**

4D 10:45am **Hybrid Adsorption Biological Treatment Systems (HABiTS) for Improved Nitrogen Removal in Onsite Wastewater Treatment** 216
Sarina Ergas, *University of South Florida;* Lucie Krayzelova, *Institute of Chemical Technology Prague, Czech Republic*; Laura, Rodriguez-Gonzalez, Karl Payne, Maya Trotz, *University of South Florida*

4E 11:05am **Industry Update for WASSTRIP® Impacts On Nutrient Flows, Digestion and Dewatering of EBPR Solids** 221
Ahren Britton, Steve Wirtel, Ram Prasad, Nate Turner, *Ostara Nutrient Recovery Technologies*

11:30am **Facilitated Discussion**

12:00pm **Session Adjourns**

TECHNICAL PROGRAM

Session 5: Process Control Optimization

Tuesday, July 28, 2015
8:30 AM – 12:00 PM
Moderator: Leon Downing, *Ch2M*

Room: Fir Ballroom
PDHS: 2.75

5A	**8:30am**	**Implementation of Ammonia-Based Aeration Control (ABAC) at Full-Scale Wastewater Treatment Plants** Kshitiz Uprety, *HRSD/ODU;* Amanda Kennedy, Bill Balzer, Rick Baumler, Ryan Duke, *Charles Bott, HRSD*	231
5B	**8:50am**	**Seeking the Optimum Approach for Generating VFA/rbCOD To Ensure Reliable EBPR at the Robert W. Hite Treatment Facility** Kimberly Cowan, Kurt, Carson, Cynthia, Lynch, James McQuarrie, *Metro Wastewater Reclamation District*	241
5C	**9:10 am**	**Aeration Control Strategies for Nitrogen Removal: A Pilot and model Based Investigation** Pusker Regmi, *Brown and Caldwell;* Mike Sadowski, *Virginia Tech;* Jose Jimenez, *Brown and Caldwell*; Charles Bott, *HRSD*	259
	9:30am	**Facilitated Discussion**	
	10:00am	**Networking Break**	
5D	**10:45am**	**Using Step-Feed to Improve Secondary Effluent AmmoniaControl** Bruce Johnson, Jennifer Phillips, *CH2M;* Greg Smith, John Sherlock, *Pima County Regional Wastewater Reclamation Department*	264
5E	**11:05am**	**Better Process Control Enables Performance Exceeding Limits of Technology in Nutrient Removal - Truckee Meadows Water Reclamation Facility Experience** Kishora Panda, Michael Drinkwater, Todd Saxberg, *Truckee Meadows Water Reclamation Facility*	273
TB	**11:25am**	**Aeration Considerations in the Design and Operation Of Biological Nutrient Removal Processes** Joe Rohrbacher, Paul Pitt, Jim Cramer, *Hazen and Sawyer;* Diego Rosso, *University of California, Irvine*	288
	11:30am	**Facilitated Discussion**	
	12:00pm	**Session Adjourns**	

Session 6: Nutrient Management Lessons Learned

Tuesday, July 28, 2015 **Room: Pine Ballroom**
8:30am – 12:00pm **PDHS: 2.75**
Moderator: Tina Pham, *City of San Jose Environmental Services Dept.*

6A	8:30am	**Nutrient Recovery At the F. Wayne Hill Water Resources Center: Design and Full Scale Start Up** Ron Latimer, *Hazen and Sawyer*	300
6B	8:50am	**Prove It! Demonstrating Performance of a Combined Carbon and Ammonia Removal BAF** Jon Pearson, *AECOM*; Terry Desmarais, Paula Anania, *City of Portsmouth, NH.;* Jon Pearson, *AECOM*; Erik Grotton, *Blueleaf, Inc.*	303
6C	9:10am	**Startup of a Low-Phosphorus Nitrifying Tertiary Membrane System** Mario Benisch, *HDR Engineering Inc;* Casey Fisher, *City of Coeur d'Alene*	328
	9:30am	**Facilitated Disucssion**	
	10:00am	**Networking Break**	
6D	10:45am	**Planning, Design, and Operation Of A BNR Plant With Air Mixed Selectors** Donavan Esping, *Brown and Caldwell;* Carol Mordorski, George Sprouse, Christine Voigt, *MCES*	333
6E	11:05am	**Low Cost Biological Nutrient Removal for Treatment Plants in the Upper Long Island Sound Watershed** Jeanette Brown, *JJ Environmental, LLC;* Emily Bird, *New England Interstate Water Pollution Control Commission*; Stacy Passaro, *Passaro Engineering*	343
TB	11:25am	**IFAS Systems with Structured Sheet Media for Simultaneous Carbon Oxidation, Nitrification, Denitrification and Implications** Hua Li, *Brentwood Industries, Inc.;* Jia Zhu, Michael Faeth, Bryan Rothermel, *Brentwood Industries*	349
	11:30am	**Facilitated Discussion**	
	12:00pm	**Session Adjourns**	

Session 7: Leading Edge Nutrient Removal Research and its Practical Application for Front Line Practitioners

Tuesday, July 28, 2015
1:30 PM – 5:00 PM
Moderator: Daniel Dair, *World Water Works*

Room: Oak Ballroom
PDHS: 2.75

Session 8: Research and New Technologies

Tuesday, July 28, 2015
1:30 PM – 5:00 PM
Moderator: Sarina Ergas, *University of South Florida*

Room: Fir Ballroom
PDHS: 2.75

8A	**1:30pm**	**Phototrophic Membrane Bioreactor (PMBR) for the Management of Nutrients in Domestic Wastewater within the Context of Algal Biorefinery and Biofactory Concepts** Onur Ozcan, Herby Jean, Robert Bair, Ana Prieto, Daniel Yeh, *University of South Florida*	419
8B	**1:50pm**	**A New Kind of Nutrient Standard: Strengthening Linkages Between Nutrients, Algal Blooms, and Aquatic Life** Clifton Bell, *Brown and Caldwell*	421
8C	**2:10pm**	**Real Time Process Control Algorithms Combining Process Modeling and Instrumentation** Joseph Rohrbacher, Katya Bilyk, David Wankmuller, Paul Pitt, *Hazen and Sawyer*	427
	2:30pm	**Facilitated Discussion**	
	3:00pm	**Networking Break**	
8D	**3:45pm**	**Demonstration of an Up-flow Biological Aerated Filter System for Nitrogen Removal from Non-nitrified Secondary Effluent** Liyan Jin, Young Eum, Joon Min, Hongkeun Park, Alice Robinson, *BKT United*	436
8E	**4:05pm**	**Nutrient Recovery as a Sustainable Nutrient Management Tool** Wendell Khunjar, *Hazen and Sawyer P.C.*	441
	4:30pm	**Facilitated Discussion**	
	5:00pm	**Conference Adjourns**	

Session 9: BNR Process Demonstrations

Tuesday, July 28, 2015
1:30 PM – 5:00 PM
Moderator: Kam Law, *Greeley and Hansen*

Room: Pine Ballroom
PDHS: 2.75

Session	Session Title	CE Credit Total/Type	Room
Workshop	Nutrient Management: How to Get Started	**0.3 CEUs**	Fir Ballroom
OGS	Opening General Session	**2.4 PDHs**	Oak Ballroom
Session 1	Nutrient Removal and Recovery	**2.75 PDHs**	Oak Ballroom
Session 2	BNR Troubleshooting and Optimization	**2.75 PDHs**	Fir Ballroom
Session 3	Compliance Strategies	**2.75 PDHs**	Pine Ballroom
Facility Tour	San Jose-Santa Clara Regional Wastewater Facility Tour	**3.0 GCHs**	Off-site
Session 4	New Technologies	**2.75 PDHs**	Oak Ballroom
Session 5	Process Control Optimization	**2.75 PDHs**	Fir Ballroom
Session 6	Nutrient Management Lessons Learned	**2.75 PDHs**	Pine Ballroom
Session 7	Leading Edge Nutrient Removal Research and its Practical Application for Front Line Practitioners	**2.75 PDHs**	Oak Ballroom
Session 8	Research and New Technologies	**2.75 PDHs**	Fir Ballroom
Session 9	BNR Process Demonstrations	**2.75 PDHs**	Pine Ballroom

EXHIBITION GENERAL INFORMATION

Located in Sierra/Cascade ballroom of the Doubletree San Jose, the exhibition hall hosts morning and afternoon refreshment breaks, luncheons, and on Monday, the Networking Reception.

The exhibition offers networking opportunities with companies displaying the latest product and service solutions directly related to the conference's technical program.

Exhibit Hall Schedule

Monday, July 27	**10:00am – 6:30pm**
10:00 am –10:45 am	Coffee Break
12:00 pm – 1:30 pm	Luncheon
3:00 pm – 3:45 pm	Refreshment Break
5:00 pm – 6:30 pm	Networking Reception
Tuesday, July 28	**10:00am – 3:45pm**
10:00 am –10:45 am	Coffee Break
12:00 pm – 1:30 pm	Luncheon
3:00 pm – 3:45 pm	Refreshment Break

EXHIBITION FLOOR PLAN

BAYSHORE BALLROOM

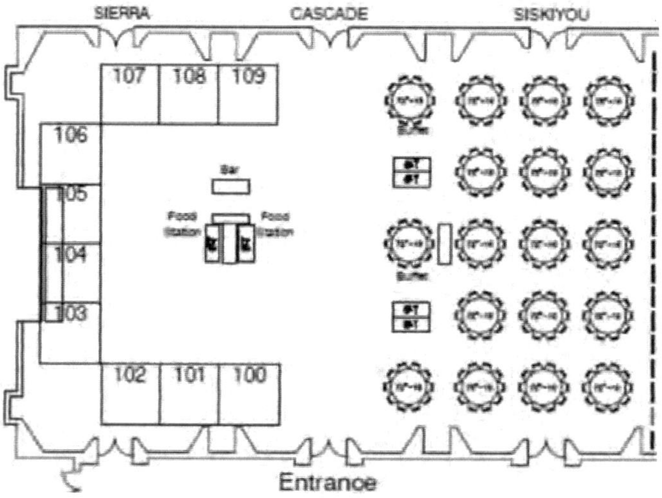

Children under 18 entering the Exhibition must be accompanied by a parent or guardian at all times. The parent or guardian assumes all risk and responsibility for the child's safety.

Photographing or video recording in the Exhibition is only allowed with exhibitor's permission

EXHIBITION DIRECTORY

As of July 14, 2015

ASA Analytics.. Booth 104
www.asaanalytics.com
Phone: (262) 893-4155
2325 Parklawn Dr Ste I
Waukesha, WI 53186
ChemScan Process Analyzers are individual instruments and complete systems for automatic analysis of nitrate, nitrite, ammonia, phosphate, oxidants, organics, metals and dissolved chemicals at single or multiple process wastewater locations.

Hach Company .... Booth 102
www.hach.com
Phone: (262) 717-9500
P.O. Box 389
Loveland, CO 80539
We make water analysis better—faster, simpler, greener and more informative—via unsurpassed customer partnerships, the most knowledgeable experts, and reliable, easy-to-use products.

Ostara Nutrient Recovery Technologies. .... Booth 109
www.ostara.com
Phone: (604) 408-6697
1199 W Pender St., Ste. 690
Vancouver, BC V6E 2R1
Canada
Ostara helps protect water resources by changing the way cities around the world manage excess nutrients in wastewater streams. The company's technology recovers phosphorus and nitrogen at wastewater treatment plants and upcycles them into a high-value, eco-friendly fertilizer, Crystal Green®.

s::can Measuring Systems LLC. .... Booth 103
www.s-can.us
Phone: (888) 694-3230
PO Box 36402
Cincinnati, OH 45236
s::can - the world technology leader for submersible online UV-Vis spectrometer probes - provides a full range of sensors, complete monitoring stations and software for drinking, environmental, waste, and industrial water applications.

EXHIBITION DIRECTORY

Schwing Bioset, Inc.. Booth 108
www.schwingbioset.com
Phone: (715) 247-3433
350 SMC Dr.
Somerset, WI 54025
Screw press dewatering, class A hauling and residuals marketing, class A bioset and drying technology, sliding frames, piston pumps, live bottoms, conveyors, turn-key biosolids systems.

Suez Environnement.. Booth 100
www.suez-environnement.com
Phone: (804) 756-8423
8007 Discovery Dr.
Richmond, VA 23229
As an industrial services and solutions company specializing in securing and recovering resources, Suez Environnement provides its customers (local authorities, industry and consumers) with concrete solutions to address new resource management challenges.

World Water Works, Inc. .... Booth 106
www.worldwaterworks.com
Phone: (405) 943-9000
PO Box 892050
Oklahoma City, OK 73189
World Water Works is a leading manufacturer of specialized process and wastewater treatment technologies. With a focus on cost effective performance, flexibility, and durability, the company works alongside clients to create the most appropriate treatment solutions.

YSI, a Xylem Brand.. Booth 107
www.ysi.com
Phone: (937) 767-7241
1725 Brannum Ln.
Yellow Springs, OH 45387
YSI designs and manufactures innovative water quality, water velocity, and flow measurement sensors and instrumentation for sampling and long term monitoring applications online, in the field and laboratory.

PRESENTER AND MODERATOR DIRECTOR

Clifton Bell, *Brown and Caldwell,* **8B**

Mario Benisch,*HDR Enginerring Inc.,* **6C**

Ahren Britton, *Ostra Nutrient Recovery Technologies,* **4E**

Jeanette Brown, *Manhattan College,***6E, Workshop A**

Dennis Burke, *Environmental Energy & Engineering Co.,* **1D**

David Clark, *HDR Engineering Inc.,**7C***

Luke Cole,*ORISE*, **3D**

Timothy Constantine, *CH2M,* **3E**

Kimberly Cowan, *Metro Wastewater Reclamation District,* **5B**

Daniel Dair, *World Water Works,* **Moderator Session 7**

Alexandra Doody, *CDM Smith,* **4A**

Leon Downing, *CH2M,* **1E, Moderator Session 5**

Zeynep Erdal,*CH2M,***Techial Brief Session 9, Workshop**

Sarina Ergas, *University of South Florida,* **4D, Moderator Session 8**

James Ervin, *City of San Jose,* **2E**

Donavan Esping, *Brown and Caldwell,* **6D**

Michael Falk, *HDR Enginerring Inc.,* **OGS**

Jordi Figueras,*Pharmer Enginerring,* **Technical Brief Session 1**

Jim Fitzpatrick, *Black & Veatch Corporation,* **Workshop**

John Gallegos, *CDM Smith,* **2B**

Sarah Galst, *Hazen & Sawyer,* **1B**

Alonso Griborio, *Hazen & Sawyer,* **2C**

Thomas Grovhoug, *Larry Walker Associates,* **3C**

Aren Hansen, *Brown and Caldwell,* **Moderator Session 3**

Seppi Henneman, *Brown and Caldwell,* **9D**

Joseph Husband, *Arcadis,* **Moderator Session 2**

Roland Jezek, *Black & Veatch*, **4C**

Sam Jeyanayagam, *CH2M,* **Workshop A**

Jose Jimenez, *Brown and Caldwell,* **1A**

PRESENTER AND MODERATOR DIRECTORY

Alex Johnson, *The Freshwater Trust,* **3B**

Bruce Johnson, *CH2M,* **5D**

Wendell Khunjar, *Hazen and Sawyer,* **8E, Moderator Session 4**

Stephanie Klaus, *Virginia Tech,* **1C**

Ron Latimer, *Hazen and Sawyer*, **6A**

Kam Law, *Greeley and Hansen,* **Moderator Session 9**

Samuel Ledwell, *Environmental Operating Solutions,* **2D**

Hua Li, *Brentwood Industries Inc.,* **Technical Brief Session 6**

Jane Madden, *CDM Smith,* **Workshop**

Iosif Mariakakis, *Fraunhofer,* **4B**

Jim McQuarrie, *Metro Water Reclamation District, Denver, Co.,* **Moderator Session 1**

Rion Merlo, *Brown & Caldwell,***9B**

Megan Moody, *CDM Smith,***9E**

J Neethling, *HDR Engineering Inc.,* ***7A***

Onur Ozcan, *University of South Florida,***8A**

Kishora Panda,*Truckee Meadows Water Reclamation Facility,* **5E**

Jon Pearson, *AECOM,* **6B**

Tina Pham, *City of San Jose,* **Moderator Session 6**

Paul Pitt, *Hazen & Sawyer,* **2A**

Pusker Regmi, *Brown and Caldwell,* **5C**

Rod Reardon,*Carollo Engineer,* **Workshop**

Joseph Rohrbacher, *Hazen & Sawyer,***Technical Brief Session 5, 8C**

Julian Sandino, *CH2M,* ***7D***

Linda Sawyer, *Brown and Caldwell,* **3A**

Jeffrey Sober, *Garver,* **Technical Bried Session 2**

H Stensel, *University of Washington,* **7B**

Gerry Stevens, *AECOM,* **9A**

Ryujiro Tsuchihashi, *AECOM,* ***7E***

Heng Zhang, *MWRD,* **9C**

The San Francisco Bay Area Nutrient Watershed Permit: A Streamlined Approach to Identify Cost Effective Strategies for Nutrient Load Reductions

Michael Falk, PhD, PE[1]*; JB Neethling, PhD, PE[1]; Holly Kennedy, PE[2]; Rion Merlo, PhD, PE[3]; Jim Graydon, PE[3]; Michael Connor, PhD[4]; David Williams, PE[5]

1-HDR Engineering, 2365 Iron Point Road, Suite 300, Folsom, CA95630-8709
2-HDR Engineering, 2121 N. California Boulevard, Suite 475, Walnut Creek, CA 94596
3-Brown and Caldwell, 201 North Civic Drive, Walnut Creek, CA 94596
4-East Bay Dischargers Authority, 2651 Grant Avenue, San Lorenzo, CA 94580
5-Bay Area Clean Water Agencies, PO Box 24055, MS 59, Oakland, CA 94623
*-Corresponding Author: Michael.Falk@hdrinc.com

ABSTRACT

The San Francisco Regional Water Quality Control Board (Water Board) issued a Regional Watershed Permit for 37 Water Resource Recovery Facilities (WRRFs) discharging into five identified hydrologic sub-embayments in April, 2014 (Order No. R2-2014-0014). The permit requires that Bay dischargers assess their nutrient loads and evaluate strategies for nutrient load reduction, should reductions be required in the future. The permit includes four key elements for evaluating nutrient load reduction opportunities at WRRFs (if supported by sound science):

1. Plant optimization
2. Sidestream treatment
3. Plant upgrades
4. Nutrient reduction by other means (including source control, natural treatment systems, diversion of effluent to water recycling, and others)

The 37 WRRFs are working collectively under the joint powers agency, Bay Area Clean Water Agencies (BACWA), to submit one coordinated study. This paper will outline the studies approach to reduce nutrient loads to San Francisco Bay (SF Bay).

KEYWORDS: San Francisco Bay Area, Nutrient Regulations, Estuary, Sidestream Treatment, Nitrogen Removal, Phosphorus Removal

INTRODUCTION

Nutrients in the SF Bay are a growing concern for the Bay Area water quality community. Historically, the SF Bay has not been adversely impacted by nutrient loading, although there are indications that its historic resilience to the effects of nutrient enrichment may be weakening (Cloern and Jassby, 2012; SFEI, 2013). While the definition of impairment has not been reached, there is concern that the SF Bay has reached a tipping point that might lead to impairment. Numerous scientific studies are being conducted, such as the Nutrient Numeric Endpoints

(NNE), to understand the impact of nutrients on the SF Bay. Initial findings have indicated that an estimated 63 percent of the annual average total nitrogen load to the SF Bay is from publically owned WRRFs. As a result, it may be necessary to limit the availability of essential nutrients, by implementing some form of nutrient removal, to address three potential challenges:

1. Ammonia toxicity and/or inhibition of phytoplankton growth. Full or partial nitrification may be required.
2. Eutrophication. Full or partial denitrification may be required where total inorganic nitrogen is the limiting nutrient.
3. Undesirable phytoplankton assemblage changes due to the nitrogen to phosphorus ratio. Phosphorus reduction may be required.

While various studies are ongoing, BACWA members have been proactively collaborating with the San Francisco Regional Water Quality Control Board (Water Board) to ensure that future nutrient limits are based on the documented assimilative capacity of the SF Bay. BACWA members are in a unique circumstance because the nutrient endpoints for the SF Bay have not yet been established, and because of that, wastewater agencies can help inform the discussion and shape potential future limits, targeting the operational sweet spot for nutrient removal that is practical - technically, sustainably, and economically. Rather than having to react to unachievable endpoints driven by unrealistic Total Maximum Daily Load (TMDL), as has been the case in so many other areas of the country, with this study, BACWA can help set the stage for future limits that are reasonable.

On April 9, 2014, the Water Board issued Order No. R2-2014-0014, *Waste Discharge Requirements for Nutrients from Municipal Wastewater Discharges to San Francisco Bay* (Watershed Permit). The Watershed Permit sets forth a regional framework to facilitate collaboration on studies that will inform future management decisions and regulatory strategies. A component of the permit is to conduct treatment plant optimization and upgrade studies for nutrient removal (due in July, 2018). The permit includes four key elements for evaluating nutrient load reduction opportunities at WRRFs (if supported by sound science):

1. Plant optimization
2. Sidestream treatment
3. Plant upgrades
4. Nutrient reduction by other means (including source control, natural treatment systems, diversion of effluent to water recycling, and others)

The sidestream refers to the nutrient laden return stream from biosolids dewatering for WRRFs that have solids stabilization and dewatering. The benefits of removing nutrients in the sidestream are well documented:

- Warm water (favorable kinetics; small footprint)
- Concentrated nutrients (favorable kinetics; small footprint)
- Low flows (ability to equalize)

- More cost-effective as $/lb nutrient removed than complete liquid stream treatment conversion
- Less aeration and chemicals than liquid stream treatment (limited to nitrogen removal)
- Easier to phase construction than liquid stream treatment
- The sidestream process can remain operational to provide additional reliability and reduce the overall nutrient removal cost if Levels 2 and 3 are required in the future.

The "by other means" refers to opportunities to reduce nutrient loads outside the plant fence.

A map showing the 37 participating WRRF locations is presented in Figure 1. These studies will increase the understanding of external nutrient loads, improve the accuracy of the inputs used in load response models, and identify potential load reductions and costs for different dischargers to the Bay.

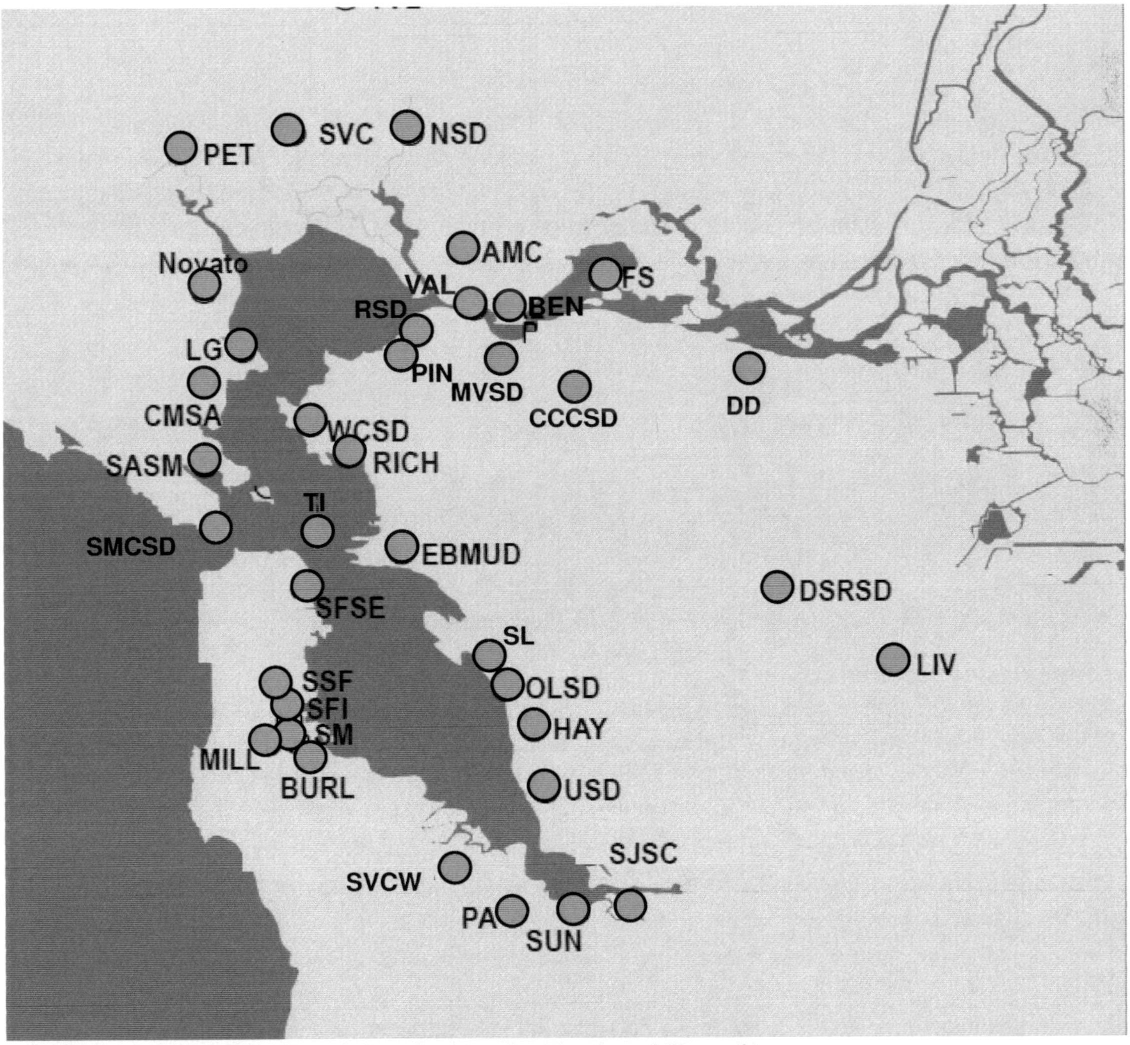

Figure 1. 37 WRRFs Participating in the Regional Permit

The 37 WRRFs tasked with satisfying the Watershed Permit requirements opted to collectively submit one coordinated study under the joint powers agency, BACWA. The Watershed Permit requires a Scoping and Evaluation Plan that describes the approach and schedule for completing the nutrient reduction studies by plant optimization and plant upgrade, as well as by other means.

The objectives of this manuscript are as follows:

1. Provide an overview of the study approach
2. Present the Scoping and Evaluation Plan
3. Present initial results from the questionnaire
4. Identify the proportion of WRRFs that are candidates for sidestream treatment and discuss the potential nutrient load reductions
5. Discuss the on-going efforts, in particular the site visits and reports
6. Discuss future work

APPROACH

The BACWA Consultant Management Group (CMG) selected the HDR and Brown and Caldwell Team (HDR/BC) in the summer of 2014 to perform the collective Watershed Permit study under BACWA's direction. A systematic approach was developed by the HDR/BC team as shown in Figure 2 for satisfying the Watershed Permit requirements. The sub-sections below discuss the approach for the steps depicted in Figure 2.

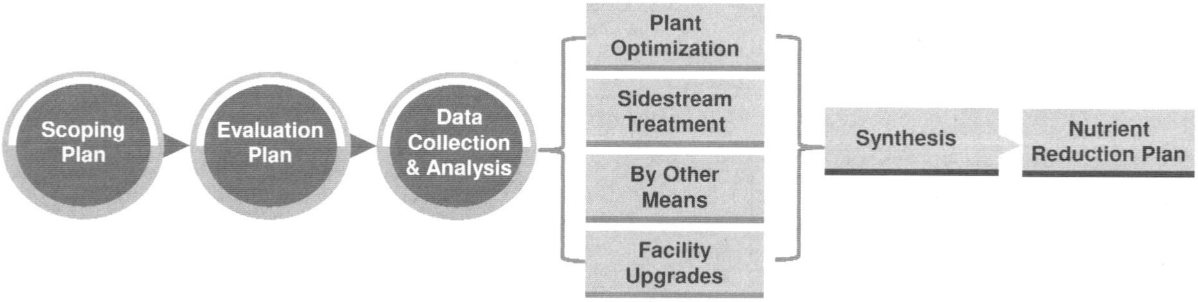

Figure 2. Approach to Evaluating Nutrient Load Reductions

Scoping and Evaluation Plan

The Watershed Permit requires a Scoping and Evaluation Plan that describes the approach and schedule for completing the nutrient reduction studies by plant optimization and plant upgrade, as well as by other means (BACWA, 2015). Nutrients of interest are ammonia, total nitrogen, and total phosphorus. The evaluation considers flows projected forward to year 2025 for plant optimization/sidestream treatment and the permitted design capacity flows for plant upgrades. The scoping and evaluation plan lays out the following steps for the study, which must be completed by July 1, 2018:

- Establish a range of nutrient removal levels
- Collect data for each plant and conduct a preliminary assessment based on this data

- Evaluate nutrient reductions achievable through plant optimization, sidestream treatment, and upgrades for each WRRF
- Compile existing information to identify options for reducing nutrient loads by other means, such as water recycling, wetlands, etc.

Nutrient Levels

The Watershed Permit does not explicitly state nutrient removal goals. As a result, nutrient removal target levels were developed for the purposes of this study. As shown in Table 1, three seasonal nutrient levels were identified.

Table 1. Nutrient Removal Targets for Seasonal Averaging Periods*

Treatment Level	Study	Ammonia	Total Nitrogen	Total Phosphorus	Comment
Level 1	Optimization	--	--	--	Removal potential to be determined
Level 2	Upgrades	2 mg N/L	15 mg N/L	1.0 mg P/L	Without filters and external carbon**
Level 3	Upgrades	2 mg N/L	6 mg N/L	0.3 mg P/L	Filters and external carbon source required***

* The seasonal impacts will be considered for all three treatment levels.

** Assumed to be achievable by conventional nutrient removal processes without effluent filtration and without adding an external carbon source. Certain participating plant configurations and technologies will require chemicals.

*** An external carbon source will not be required for certain plant configurations and technologies

Level 1 does not include numeric targets because each plant will have a unique potential for optimization. Level 2 was developed to reflect numeric targets that are generally achievable without effluent filtration and without adding an external carbon source. Level 3, the most stringent level, has numeric targets that would generally require an external carbon source and effluent filtration.

Treatment Levels 2 and 3 were selected based on the typical tipping point for treatment technologies to achieve the respective effluent levels. For most plant configurations, the less stringent Level 2 can be achieved with conventional nutrient removal processes without adding an external carbon source (e.g., methanol) and without adding effluent filtration. The more stringent Level 3 requires an external carbon source for nitrogen removal and metal salt addition with filtration for most plant configurations. These factors contribute to a tipping point due to the well documented increase in cost, operational and safety burdens, energy demand, and greenhouse gas (GHG) emissions (Falk et al., 2013). Ammonia levels are selected to provide stable ammonia reduction (typically nitrification). By using this clearly defined tipping point, the results for both Treatment Levels are beneficial for making informed future management decisions.

Nitrogen and phosphorus typically have seasonal impacts on receiving waters. Thus, targets for total nitrogen and phosphorus removal should be based on long averaging periods linked to the specific waterbody response to nutrient enrichment. Short averaging periods based on guidance applicable to toxics constituents (Brown and Caldwell, 2014) will would result in overly conservative designs for nutrient removal facilities in order to provide the required reliability to meet the targets, but would provide little, or no, additional water quality benefit. As a result, seasonal averaging periods for total nitrogen and phosphorus discharges are proposed.

In order to capture seasonality variations, both wet season (September 30 to April 30) and dry season (May 1 to September 30) discharges are considered. A dry season average was considered because it excludes sizing treatment facilities for peak wet weather events and low temperatures. Biological process kinetics are more rapid at warmer temperatures and thus result in a reduced footprint if sized for the dry season. During a significant precipitation event, plants are subjected to peak flows with subsequently less hydraulic residence time within the plant. Wet and dry season nutrient impacts on the estuary may differ as well.

Data Collection and Analysis

A data information request, referred to as the questionnaire herein, was disseminated to all 37 WRRFs in the fall of 2014. The questionnaire included the following requests:

1. General WRRF background (e.g., location, service area description, permit limits, etc.)
2. Historical plant flows and loads (last three years of data)
3. Historical unit process data (last three years of data; performance, energy and chemical data)
4. List of treatment assets and sizing information for major unit processes
5. Series of questions to identify candidate WRRFs for sidestream treatment
6. Information on any on-going bench or pilot scale studies related to nutrient removal, such as sidestream treatment to remove nitrogen
7. Submit any reports related to nutrient removal (e.g., Master Plan, CIP, etc. and nutrient removal by other means (e.g., water recycling, wetlands, etc.))

Screen captures of the questionnaire workbook distributed to the WRRFs are shown in Figure 3.

Based on the information received from the questionnaire, the team performed a preliminary assessment to identify potential optimization strategies and plant upgrades for each plant. For example, unit processes identified as having additional capacity or unused tanks were tagged as locations for potential nutrient load reduction opportunities. The capacity criteria used were based accepted industry values (e.g., WEF MOP8) and engineering judgment.

Sidestream Treatment

A series of questions were included in the questionnaire to identify candidate WRRFs for sidestream treatment using a structured approach. A graphic depicting the questions as a decision tree is shown in Figure 4.

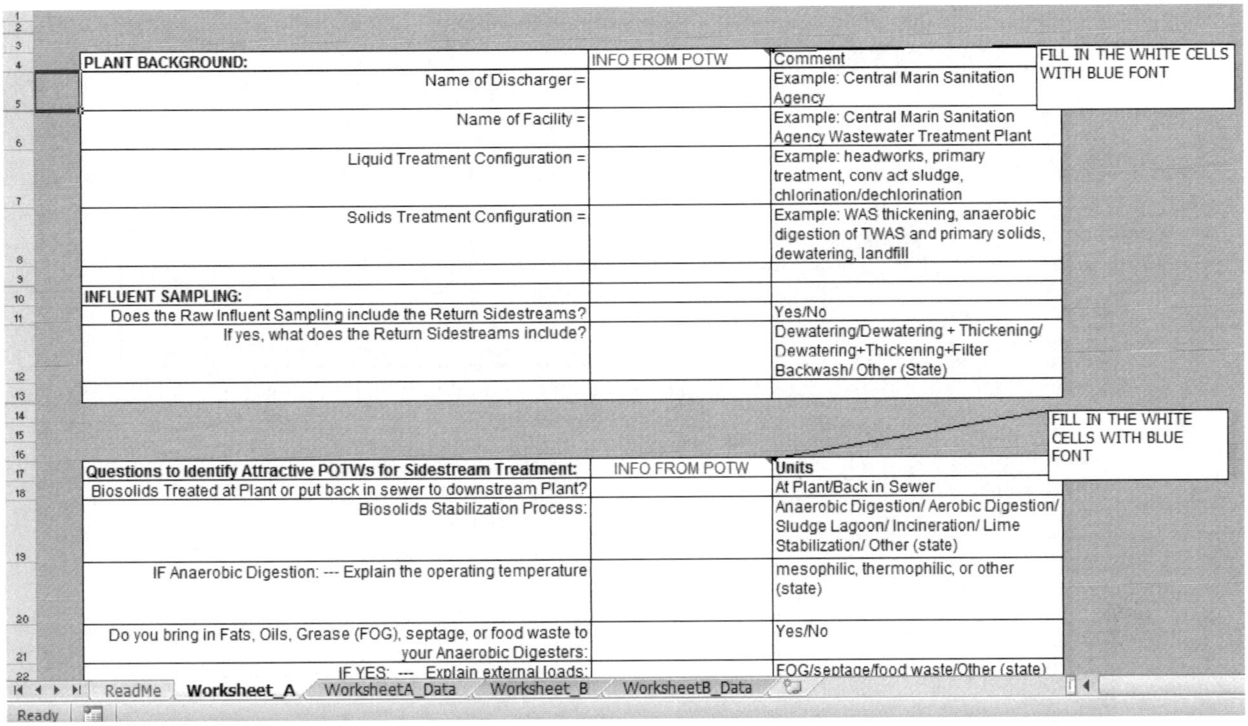

PLANT BACKGROUND:	INFO FROM POTW	Comment	FILL IN THE WHITE CELLS WITH BLUE FONT
Name of Discharger =		Example: Central Marin Sanitation Agency	
Name of Facility =		Example: Central Marin Sanitation Agency Wastewater Treatment Plant	
Liquid Treatment Configuration =		Example: headworks, primary treatment, conv act sludge, chlorination/dechlorination	
Solids Treatment Configuration =		Example: WAS thickening, anaerobic digestion of TWAS and primary solids, dewatering, landfill	
INFLUENT SAMPLING:			
Does the Raw Influent Sampling include the Return Sidestreams?		Yes/No	
If yes, what does the Return Sidestreams include?		Dewatering/Dewatering + Thickening/ Dewatering+Thickening+Filter Backwash/ Other (State)	

FILL IN THE WHITE CELLS WITH BLUE FONT

Questions to Identify Attractive POTWs for Sidestream Treatment:	INFO FROM POTW	**Units**
Biosolids Treated at Plant or put back in sewer to downstream Plant?		At Plant/Back in Sewer
Biosolids Stabilization Process:		Anaerobic Digestion/ Aerobic Digestion/ Sludge Lagoon/ Incineration/ Lime Stabilization/ Other (state)
IF Anaerobic Digestion: --- Explain the operating temperature		mesophilic, thermophilic, or other (state)
Do you bring in Fats, Oils, Grease (FOG), septage, or food waste to your Anaerobic Digesters:		Yes/No
IF YES: --- Explain external loads:		FOG/septage/food waste/Other (state)

ReadMe | **Worksheet_A** | WorksheetA_Data | Worksheet_B | WorksheetB_Data

Ready

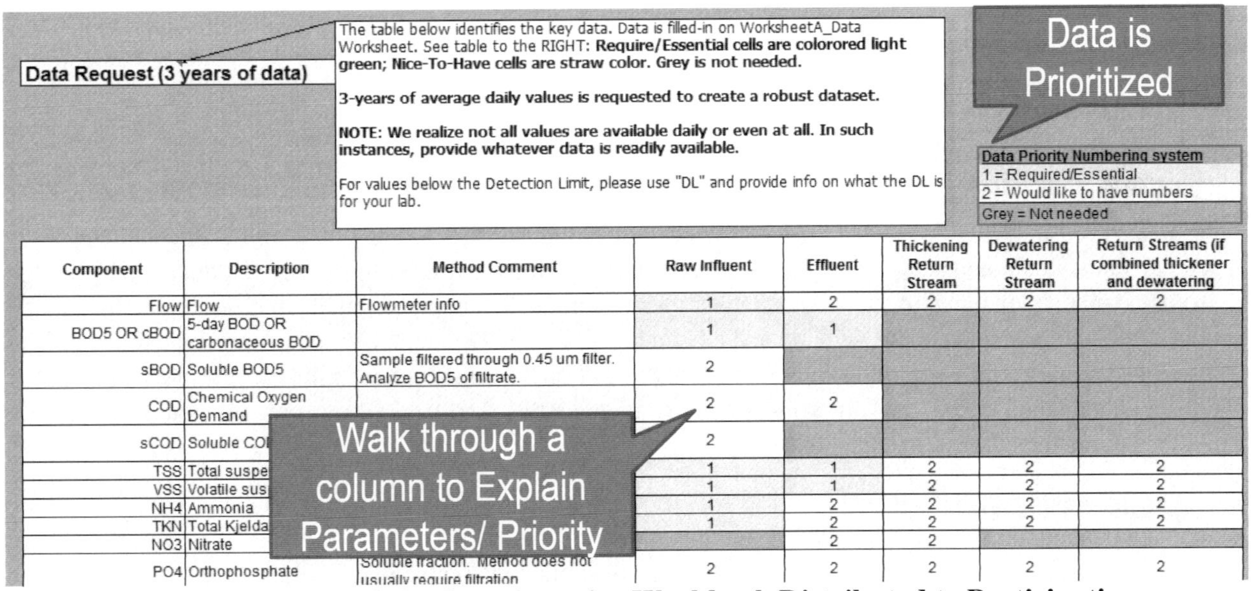

Data Request (3 years of data)

The table below identifies the key data. Data is filled-in on WorksheetA_Data Worksheet. See table to the RIGHT: **Require/Essential cells are colorored light green; Nice-To-Have cells are straw color. Grey is not needed.**

3-years of average daily values is requested to create a robust dataset.

NOTE: We realize not all values are available daily or even at all. In such instances, provide whatever data is readily available.

For values below the Detection Limit, please use "DL" and provide info on what the DL is for your lab.

Data is Prioritized

Data Priority Numbering system
1 = Required/Essential
2 = Would like to have numbers
Grey = Not needed

Component	Description	Method Comment	Raw Influent	Effluent	Thickening Return Stream	Dewatering Return Stream	Return Streams (if combined thickener and dewatering
Flow	Flow	Flowmeter info	1	2	2	2	2
BOD5 OR cBOD	5-day BOD OR carbonaceous BOD		1	1			
sBOD	Soluble BOD5	Sample filtered through 0.45 um filter. Analyze BOD5 of filtrate.	2				
COD	Chemical Oxygen Demand		2	2			
sCOD	Soluble CO		2				
TSS	Total suspe		1	1	2	2	2
VSS	Volatile sus		1	1	2	2	2
NH4	Ammonia		1	2	2	2	2
TKN	Total Kjelda		1	2	2	2	2
NO3	Nitrate			2	2		
PO4	Orthophosphate	Soluble fraction. Method does not usually require filtration	2	2	2	2	2

Walk through a column to Explain Parameters/ Priority

Figure 3. Screen Captures of the Questionnaire Workbook Distributed to Participating WRRFs

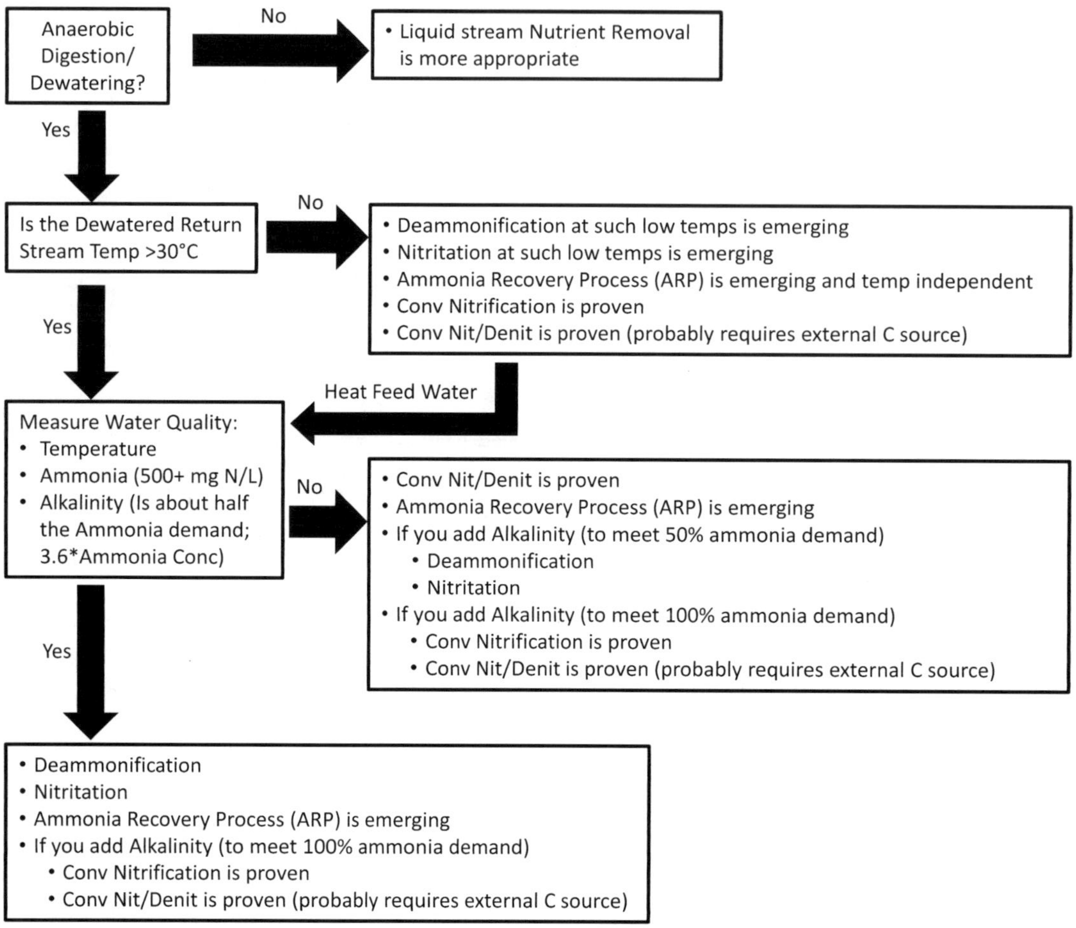

Figure 4. Decision Tree to Identify Candidate WRRFs for Sidestream Treatment

WRRFs deemed candidates for sidestream treatment are subjected to preliminary engineering to develop facility needs and quantify the corresponding nutrient load reduction potential. Similar to the capacity criteria, the sidestream treatment facility needs were developed using accepted industry values (e.g., WEF MOP8) and engineering judgment. Site visits are being used to verify the preliminary engineering results and identify any additional opportunities (e.g., unused tanks that can serve as a reactor), potential for nutrient recovery and beneficial reuse, or site constraints for refining the facility needs.

Optimization Strategies

Optimization of existing facilities is a potential first step toward nutrient reduction. Nutrient removal might be possible at existing facilities operating below design load and thus existing unused "capacity" might be devoted for nutrient reduction on an interim basis. Optimization strategies would take advantage of unused tankage, new process approaches, instrumentation improvements, and, without a permit limit with potential enforcement penalties, achieves as much nutrient reductions as possible in the short term.

Any proposed optimization strategies are viewed as interim solutions as most strategies will take advantage of unused capacity (i.e., facilities not needed to meet the current load but may be required to treat the design load). The unused capacity was typically constructed using fees to accommodate future growth so it may not be available for nutrient reduction in the future as that growth occurs or as stepping stones for either Level 2 or Level 3 technology changes in the plant.

The plant optimization strategies are based on each individual plant's documented plans for future growth and recorded existing flows and BOD/nutrient loadings in 2015 and also what they are projected to be in 2025. For plants where no documentation exists, a 15% increase in BOD/nutrient loadings will be assumed for the 10 year period, and no increase in flows.

It is important to stress that implementing some of the strategies will likely impact overall treatment capacity and operational complexity in the long term. The plant might need to revert back to the prior mode of operation or add new facilities as flows and loads increase over time.

A list of the most common optimization strategies for each treatment category will be generated during the preliminary assessment effort. For example, a plant could implement chemically enhanced primary treatment (CEPT) as a means to remove total phosphorus and unlock downstream aeration basin capacity for ammonia removal. The following strategies are being considered for each plant:

- Use offline tankage to provide additional treatment
- Modify operational mode, such as raising the solids residence time
- Modify blower operating set points
- Operate in split treatment mode
- Change to simultaneous nitrification/denitrification operation
- Shut down aeration to create anoxic zones
- Add instruments for nutrient removal in ammonia based aeration control mode
- Add chemicals for phosphorus removal
- Add chemicals to reduce load, unlock capacity
- Add anoxic and/or anaerobic zones for biological nutrient removal
- Add internal recycle for denitrification
- Add mixers for unaerated zones

During the site visits, the optimization strategies from the preliminary assessment are confirmed.

Upgrade Options

Each WRRF is in the process of being evaluated to determine capital improvements necessary to provide nutrient removal to meet the Level 2 and Level 3 target levels listed in Table 1. Established treatment technologies will be used to determine cost estimates (both capital and operating) and to determine site footprint requirements. However, innovative and/or emerging technologies will be identified for future consideration where they may be appropriate at individual facilities. As part of the evaluation, both beneficial and adverse ancillary impacts associated with plant upgrades will be identified for each facility and will be incorporated into the cost estimates.

The general approach is to consider Level 3 nutrient removal as a potential endpoint for all facilities. The intent is to avoid situations where a Level 2 scenario requires the construction of facilities that would be stranded in a Level 3 scenario.

In determining upgrade requirements, each facility will be evaluated based on existing infrastructure and space constraints. Space constraints will determine which technologies will be considered for implementation. For instance, a facility with limited footprint may consider a compact technology (e.g., membrane bioreactor) and a facility with ample footprint could consider a 5-stage Bardenpho process for meeting Level 3 requirements. In cases of severely constrained sites, removal and replacement of existing facilities may be required.

Figure 5 shows a progression of how technologies could be selected to meet nitrification and Level 2 and Level 3 nitrogen removal targets. This approach illustrates the progression of unit processes to meet Level 2 and later Level 3 requirements. For instance, Figure 5 shows that if a facility were upgraded to a membrane bioreactor (MBR) facility to meet Level 3 nitrogen targets, then a MBR process would also be used for Level 2 nitrogen removal and nitrification. A similar progression was developed for phosphorus as presented in Figure 6.

Once a representative technology that will comply with Level 3 nitrogen and phosphorus requirements has been selected for each facility, conceptual cost estimates will be prepared to determine capital and operating costs for the most attractive option. Operating costs will represent the change in cost due to nutrient removal. For instance, upgrading from a conventional activated sludge process to a membrane bioreactor will increase electrical, chemical, and labor costs and only that increase will be quantified. Cost estimates will be presented so that unit processes are line items that can be removed to evaluate other scenarios. For instance, change from a Level 2 nitrogen removal scenario to a nitrification-only scenario by eliminating anoxic zones.

Changes in greenhouse gas (GHG) emissions from additional energy and chemical demands will be estimated. Expected changes in sludge production will be identified where appropriate. A qualitative estimate of changes in pharmaceuticals removal will be provided.

Nutrient Reduction By Other Means

Strategies that reduce nutrient loadings to SF Bay are not limited to options inside the WRRF fence. The optimization and plant upgrades sections focus on concepts that would be implemented inside the WRRF fence. This section serves as a first step in considering outside the WRRF fence concepts by compiling previous reports for each plant. Potential nutrient reduction by other means options are listed below:

- Effluent Management: Nutrient trading, water recycling and reuse
- Effluent Polishing: Wetlands treatment (e.g., Hayward Marsh)
- Solids Management: Biosolids export (un-stabilized) to a joint facility
- Source Control: Septic source abatement, urine separation, etc.
- Non-Point Sources: Non-point source reduction program

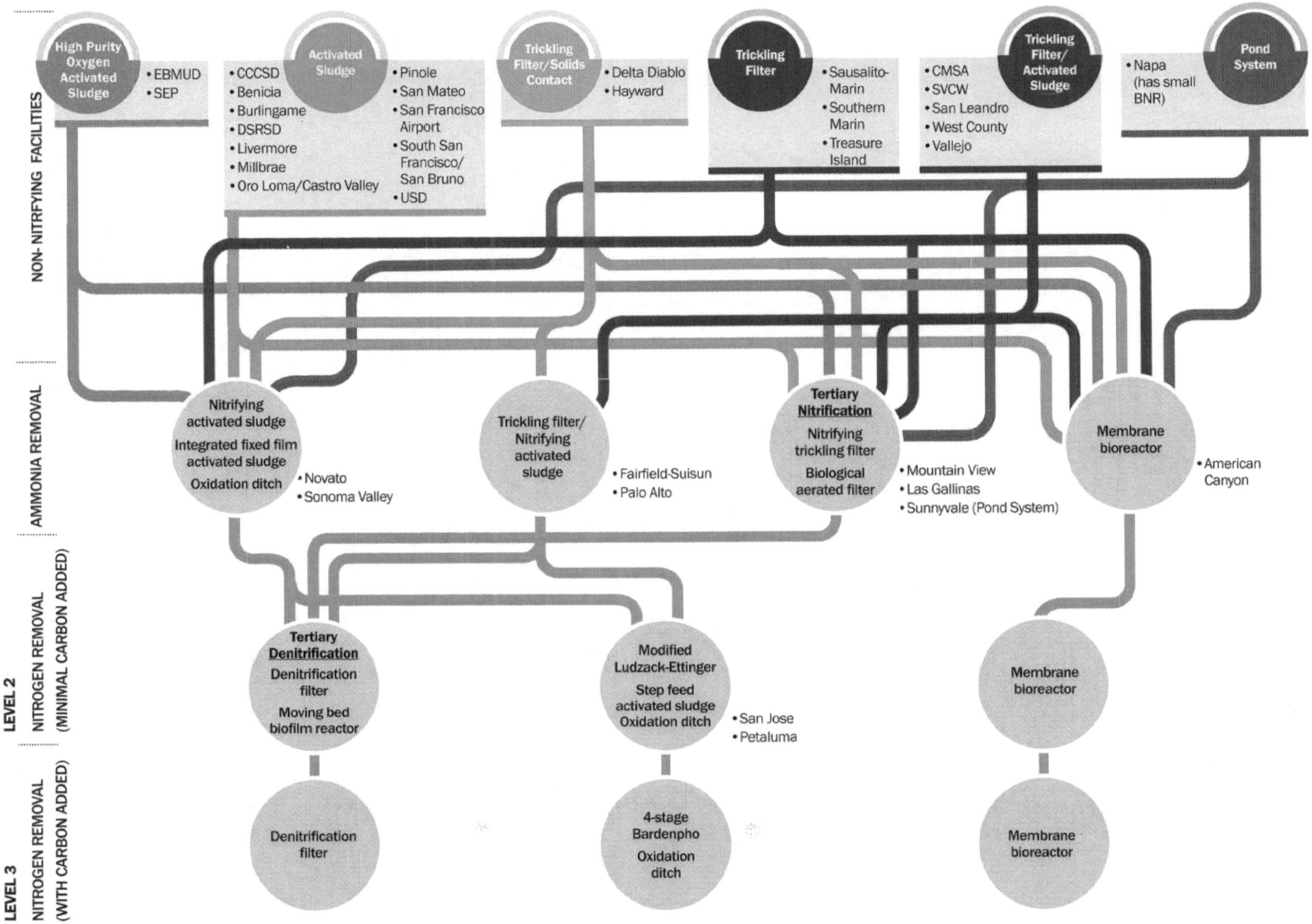

Figure 5. Existing Treatment Categories and the Nitrogen Removal Technologies Progression

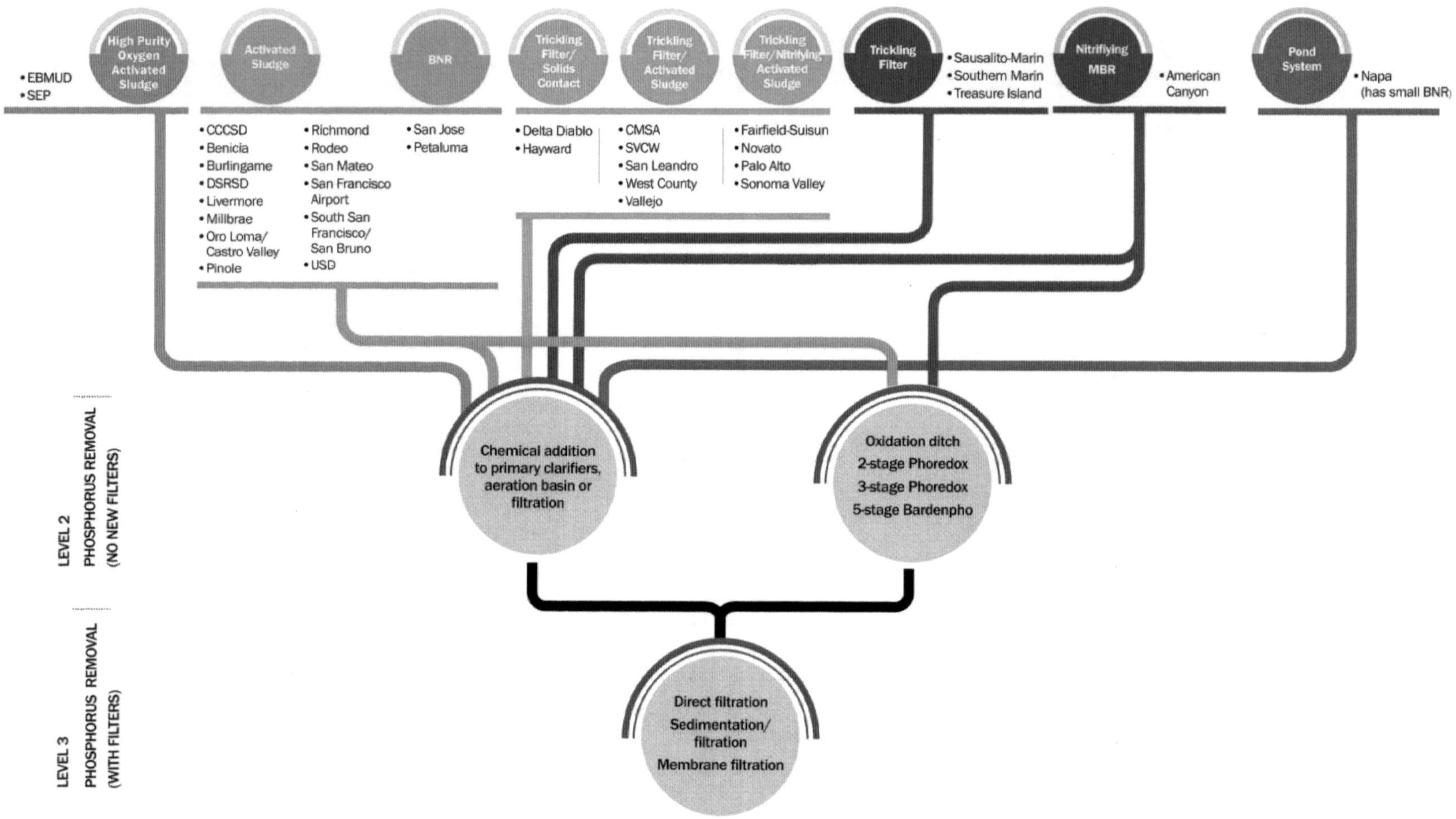

Figure 6. Existing Treatment Categories and the Phosphorus Removal Technologies Progression

This task compiles previous reports or documents prepared for each WRRF that addresses nutrient reduction by other means. Inclusion of such information into the evaluation might identify cost-effective and innovative solutions for nutrient load reductions. The ancillary benefits and adverse impacts for those identified strategies will be discussed. Additionally, the compiled report will identify institutional barriers to water recycling along with approaches for overcoming such barriers.

Impacts of Sea Level Rise

Participating agencies that are vulnerable to the impacts of sea level rise will be identified. The analysis is based on publically available data from the United States Army Corps of Engineers, the Federal Emergency Management Agency, and publically available topography data. Participating agencies will provide key plant elevation data in the data collection template.

The impacts of sea level rise with respect to potential for inundation of facilities needed to achieve nutrient reduction will be determined for each of those identified agencies. Results will be presented in a map format, illustrating location of the participating plants and areas of inundation. The costs associated with sea level rise mitigation will not be determined as these additional costs are highly site specific.

Site Visits

Following the preliminary assessment, a site visit for each WRRF is carried out to confirm the preliminary assessment and identify additional nutrient reduction strategies. The HDR/BC site visit teams include a two-person tactical team (one process engineer and one operations expert). The benefit is that each expert has their own perspective on how a plant operates and they connect effectively with plant engineering and operations staff.

The site visits will confirm our understanding of how the plant operates, validate chemical use, identify "low cost" and "medium cost" optimization strategies, and identify upgrade options. For example, they may look for any unused tanks for additional treatment, or examine operational practices such as the dissolved oxygen set-point. An example list of information that will be generated during the site visit is as follows:

- Validate and confirm facility mode of operation
- Validate and confirm whether the plant is a candidate for sidestream treatment
- Validate and confirm the historical performance trends, number of units in service, etc.
- Generate a list of optimization and upgrade strategies and their implications, such as:
    - Flow routing
    - Chemical dosing strategy
    - Pumping strategy
    - Aeration strategy
    - Impact to plant capacity
    - Non-economic impacts (e.g., biosolids yield)
    - Impacts on sustainability (e.g., energy demand and GHG emissions)

- Confirm the on-going optimization/upgrade projects and summarize their potential impacts on nutrient discharge loads

At the conclusion of each site visit, a site visit report is submitted to the respective WRRF for their review. These reports are intended to capture the information generated from each site visit, such as site constraints, potential treatment concepts and the implications on plant operation/performance. Each report will serve as an Appendix in the Watershed Permit submittal.

Watershed Permit Submittal

Information gathered from the questionnaire and site visits will be synthesized and captured as a nutrient reduction plan for the watershed permit submittal. The submittal will include an overall findings summary and individual reports for each WRRF. The optimization, sidestream, and upgrades sections under the memo produced for each WRRF will include the following:

- Capital and operations and maintenance (O&M) cost estimates for the recommended strategy (if pertinent). The O&M cost will discuss the impacts on energy, chemicals, and labor.
- Summary of adverse and ancillary impacts (e.g., greenhouse gas impacts)
- Estimates of nutrient reduction and unit costs for the recommended strategy (e.g., $/lb nutrient; lb GHG/lb nutrient)
- Footprint estimate for strategies involving construction (limited to optimization and upgrades)
- Discussion of seasonal nutrient reduction as some of the optimization strategies might only apply during the dry season and vice versa
- Discuss reduced capacity, process residuals, operational complexity and/or potential regulatory compliance issues that would be created as result of these modifications

RESULTS AND DISCUSSION

The results and discussion will include a discussion on the questionnaire and the site visits. About two-thirds of the site visits have been concluded by the end of June 2015 with plans to conclude the site visits by early September 2015.

Questionnaire Findings

All 37 participating WRRFs submitted their questionnaires in 2015. In addition to these submittals, BACWA provided previous nutrient influent/effluent nutrient loading information for each WRRF.

Preliminary evaluations suggest that 32 out of 37 WRRFs are candidates for sidestream treatment. The sidestream is relevant due to the nutrient load contribution. Despite their small flows (typically about 1 percent of raw plant flow), the sidestream represents about 15 to 40 percent of the discharge nutrient load as shown in Figure 7.

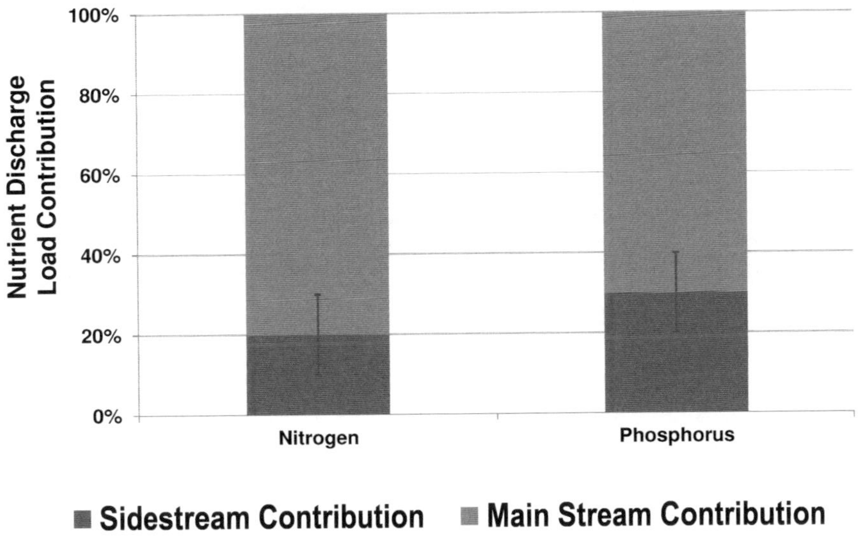

Figure 7. Nutrient Discharge Load Contribution

The majority of WRRFs do not have sufficient sidestream data to support developing facility needs. As a result, additional sampling on the sidestream (3 composite daily samples) will be conducted in July 2015. Once collected, facility needs will be refined for all 32 candidate WRRFs based on the permitted flow capacity for established (conventional nitrifying) or innovative (deammonification) sidestream technologies.

The preliminary sampling results will be discussed at the WEF Nutrient Symposium 2015.

Site Visits

As previously stated, about two-thirds of the site visits have been concluded through the end of June 2015 with plans to conclude the site visits by early September 2015. A picture of the site visit team at Oro Loma Sanitary District is provided in Figure 8. In this picture, the site visit team is investigating Oro Loma's sidestream treatment demonstration facility that is evaluating the zeolite/anammox nitrogen removal technology.

Figure 8. Site Visit Picture: Zeolite/Anammox Demonstration Facility at Oro Loma Sanitation District (Left) and the Zeolite Media (Right)

The site visits have been instrumental in developing innovative concepts that would not have been considered otherwise. For example, the site visits have identified tankage or space previously thought to be unavailable. One of the facilities has secondary clarifiers that were abandoned as part of a previous secondary clarifier upgrade. It was determined during the site visit that with minor modifications, the abandoned secondary clarifiers could be used for denitrification during the dry season and for contact stabilization during wet weather events. Concepts such as this are viable with the long seasonal averaging periods and separate dry and wet seasons. As previously stated, shorter averaging periods result in more conservative designs for nutrient removal facilities in order to provide the required reliability to meet the targets.

Water Board Submittal

Following completion of the site visits in September 2015, a draft of the Water Board Submittal will be developed for distribution to participating WRRFs by the end of 2015. The vision for the submittal is to develop unit costs ($/lb nutrient removed) for optimization, sidestream treatment, and upgrades at each WRRF. A hypothetical unit cost plot for nitrogen and phosphorus removal/recovery at WRRFs is captured in Figure 9. In addition to unit costs, the Water Board submittal will include a discussion on the plant wide implications for each recommended nutrient load reduction concept, such as ancillary impacts (e.g., energy, chemicals, GHG emissions, biosolids production, etc.), footprint estimate for technologies involving construction), and capacity and redundancy impacts.

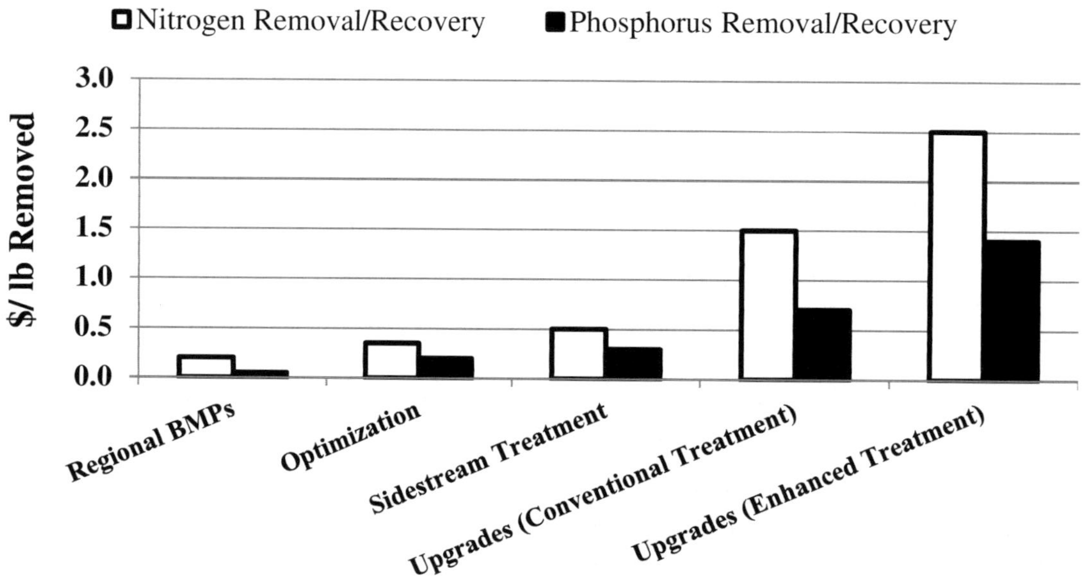

Figure 9. Hypothetical Unit Cost for Nutrient Load Reduction Strategies at each WRRF

CONCLUSIONS

Coordinating the nutrient reduction assessment effort for all 37 WRRFs provides a comprehensive assessment of strategies and unit costs for nutrient reduction to SF Bay. These results will assist with making informed rationale decisions on strategies for maintaining the historical documented assimilative capacity of SF Bay.

More than 85% of the WRRFs are candidates for sidestream nutrient reduction. An estimated reduction of 10-30% in nitrogen load and 20-40% in phosphorus load (see Figure 7) can be achieved for these WRRFs. Evaluation of additional nutrient reduction strategies such as optimizing existing infrastructure and unit processes, plant upgrades, reclamation, and others will provide detailed cost and nutrient reduction information.

In addition to this study, other scientific studies are underway in the SF Bay to determine nutrient load impacts on the water quality. All these efforts will be combined to develop a strategy to preserve all assigned beneficial uses in the SF Bay.

The approach from this effort to use science to identify the health of the SF Bay coupled with strategies for nutrient reduction by WRRFs around the SF Bay may have the potential to serve as a template for other watersheds considering nutrient load reduction strategies.

ACKNOWLEDGEMENTS

This study was funded by the joint powers agency, BACWA as part of the Regional Watershed Permit for 37 WRRFs (Order No. R2-2014-0014). The authors would like to thank the 37 WRRFs for their inputs throughout the process, as well as the BACWA CMG.

REFERENCES

Brown and Caldwell (2014) Review of USEPA Methods for Setting Water Quality-Based Effluent Limits for Nutrients. Prepared for the National Association of Clean Water Agencies, Washington D.C.

Cloern, J.E. and Jassby, A.D. (2012) Drivers of change in estuarine-coastal ecosystems: Discoveries from four decades of study in San Francisco Bay. Reviews of Geophysics, 50, RG4001, page 21.

Falk, M.W.; Reardon, D.J.; Neethling, J.B.; Clark, D.L.; and Pramanik, A. (2013) Striking the Balance between Nutrient Removal, Greenhouse Gas Emissions, Receiving Water Quality, and Costs. Water Environment and Research, 85(12):2307-2316.

HDR and Brown and Caldwell (2014) BACWA: Potential Nutrient Reduction by Treatment Optimization and Treatment Upgrades. Scoping and Evaluation Plan submitted to the San Francisco Regional Water Quality Control Board.

Novick, E. and Senn, D. (2013) DRAFT – External Nutrient Loads to San Francisco Bay. San Francisco Estuary Institute, Richmond, CA.

San Francisco Estuary Institute (SFEI) (2013) Nutrient Conceptual Model Draft, May 1, 2013, page 14. San Francisco Estuary Institute, Richmond, CA.

Water Environment Association (2009) Design of Municipal Wastewater Treatment Plants, WEF Manual of Practice 8, 4th ed.; ASCE - Manuals and Reports on Engineering Practice No. 76, Volume 1. Water Environment Federation: Alexandria, Virginia.

MAINSTREAM NITRITE-SHUNT WITH BIOLOGICAL PHOSPHORUS REMOVAL USING A LOW DISSOLVED OXYGEN AERATION CONTROL STRATEGY

Jose Jimenez[1*], Pusker Regmi[1], George Wise[2]

[1]Brown and Caldwell, Maitland, Florida USA
[2] City of St. Petersburg, Florida USA
* Corresponding author: jjimenez@brwncald.com

ABSTRACT
The City of St. Petersburg (Florida) Southwest Water Reclamation Facility achieves significant levels of nitrogen (N) and biological phosphorus (P) removal. N removal is via SND from operating the aerated zones at low DO. Despite the low DO operation, very good biological P removal performance is achieved. P removal is sustained at mixed liquor summer temperatures reaching 30°C. Full-scale performance data and detailed bench-scale testing were conducted to assess the N and P removal at low DO. Full-scale results showed that the plant achieves effluent total inorganic nitrogen (TIN) and total phosphorus (TP) concentrations of approximately 2.0 mgN/L and 0.5 mgP/L, respectively, at an average influent C: N ratio of 7:1. Specific nitrification rate (SNR) and specific denitrification rate (SDR) testing indicated that the population of nitrite oxidizing bacteria (NOB) is repressed significantly due to the low DO operation and that nitrite-shunt occurs at the plant. Phosphorus release and uptake (PRU) tests indicated that the low DO operation did not hinder the uptake of phosphate and biological P removal was achieved at the low DO conditions.

KEYWORDS: Nitrogen removal, nitritation, denitritation, nitrite-shunt, phosphorus removal, low DO operation, sustainability.

INTRODUCTION
The traditional biological *nitrogen* removal process employed in wastewater treatment plants (WWTPs) involves the autotrophic oxidation of ammonia (NH_3-N) to nitrate (NO_3^-) (nitrification) and subsequent heterotrophic reduction to nitrogen gas (N_2) with an organic carbon source (denitrification). Nitrite (NO_2^-) is an intermediate in both nitrification and denitrification. Therefore, if nitrification is stopped at NO_2^- (nitritation - NH_3-N conversion to NO_2^-), then denitritation from NO_2^- to N_2 can be achieved. The coupled process is known as nitritation-denitritation or nitrite-shunt. Nitrite-shunt may yield a reduction in aeration of up to 25% (if an external carbon source was required for denitrification) and a reduction in COD requirements of up to 40% when compared to conventional nitrification-denitrification. Implementing nitrite-shunt requires repression of the nitrite oxidizing bacteria (NOB; those that oxidize NO_2^- to NO_3^-) while retaining activity of the NH_3 oxidizing bacteria (AOB; those that oxidize NH_3-N to NO_2^-).

Simultaneous nitrification and denitrification (SND) at low DO has been demonstrated in many full-scale plants (Jimenez et al., 2010); however, the widespread application has not been extensive due to a lack of complete understanding of underlying mechanisms and proper design, control and operational guidelines. Further, mainstream nitrite-shunt remains a topic of research since the underlying mechanisms of NOB out-selection are still not completely understood. There are conflicting reports regarding the role of DO on NOB out-selection for nitrite-shunt. Low DO was shown to repress NOB (Blackburne et al., 2008) while high DO was claimed to favor AOB over NOB in other mainstream studies (Cao et al., 2013, De Clipplier et al., 2013, Wett et al., 2013, Regmi et al., 2014).

Biological *phosphorus* removal (BPR) relies on anaerobic release of phosphate (with simultaneous sequestration and storage of acetate as poly-hydroxy-alkanoates, PHAs) followed by aerobic phosphate uptake by polyphosphate accumulating organisms (PAOs). Typically BPR treatment plants are designed with dedicated anaerobic zones followed by aeration zones operated at high DO levels (i.e. 2 mg/L). Low DO operation is considered unfavorable for biological aerobic uptake of phosphorus by phosphorus accumulation bacteria (PAO) and is avoided in biological P removal plants (Barnard, 1975 and 1992). Several studies on combined N and P removal have evaluated phosphorus uptake under aerobic and anoxic conditions (Applegate et al., 1980, Kerrn-Jespersen and Henze, 1993; Wachtmeister et. al., 1997; Meinhold et. al., 1999). These studies showed lower phosphorus uptake under anoxic conditions compared to aerobic conditions, but did indicate that nitrate can be used as an electron acceptor by so-called denitrifying PAOs (DPAOs).

This study was initiated to assess aspects of this plant's operation. Specific objectives included: (a) monitoring and evaluating the full-scale performance of the Southwest Water Reclamation Facility (WRF); (b) establishing whether N removal occurs via nitrite-shunt or through full nitrification-denitrification; (c) verifying biological phosphorus activity at low DO conditions.

METHODOLOGY

Southwest Water Reclamation Facility

The Southwest WRF is owned and operated by the City of St. Petersburg, Florida. The facility is permitted to treat an annual average daily wastewater flow of 75,000 m^3/day from the southwest section of the City. Current flow is approximately 30,000 m^3/day. Final effluent is distributed in the City's reclaimed water system or disposed through deep injection wells located on the plant property. Figure 1 presents a schematic flow diagram for the facility that includes screening, grit removal, activated sludge process, secondary clarification, deep-bed filtration, and disinfection. Residuals treatment includes gravity belt thickening, anaerobic digestion, and dewatering by belt filter presses.

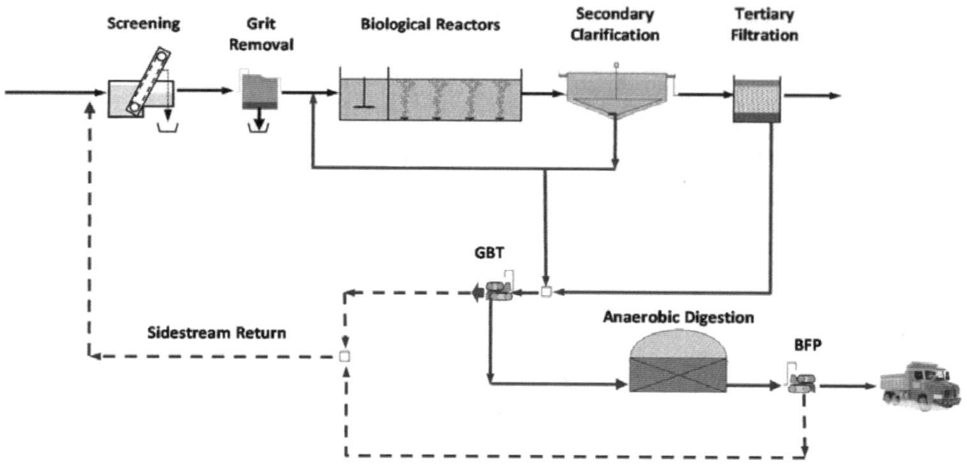

Figure 1 Process flow diagram for the Southwest WRF

The biological process at the plant is an activated sludge system using an anaerobic-aerobic (A/O) process. The system currently operates at total and aerobic sludge retention times (SRT) of approximately 5 days and 3.5 days, respectively. No supplemental carbon or chemicals are added for nutrient removal or pH control. The biological process currently is operated with a simple control strategy where the average dissolved oxygen (DO) in the reactors is low (e.g. 0.1 – 0.4 mg/L) to minimize the accumulation of NO_3^- in the effluent. Effluent NH_3-N, NO_2-N and NO_3-N are monitored regularly at the treatment plant and used as indicators to change SRT and DO set points. The wastewater temperature ranges from 23 to 30°C.

Special Sampling and Concentration Profiling

A detailed sampling campaign was performed at the plant to understand the biological conversions of nitrogen, COD and phosphorus through the activated sludge process. Concentration profiles of important parameters at the inlets and outlets of each zone were collected to understand the conversion pathways of targeted parameters. Parameters such as mixed liquor concentration, DO, waste flows, recycle stream, temperature, pH and alkalinity samples were monitored for the process control and to close capture actual operating conditions during the tested period.

Bench Scale Testing

A series of bench-scale tests were conducted at the plant including specific nitrification rate (SNR), and phosphorus release and uptake (PRU) tests. Two 10-L reactors were operated in parallel, allowing two tests to be run simultaneously. One reactor was equipped with a Hach LDO probe in combination with a Hach sc200 DO controller to maintain target DO setpoint concentration in the reactors. A Jenco membrane DO probe (ID-900-3-DO) connected to a Jenco analyzer (Model 6309PDT) was used to measure the DO concentration in the reactor run alongside the reactor with the Hach LDO probe and controller. Aeration was provided using aquarium air pumps and air stones, and a stand mixer provided mixing.

The ratio NO_3PR/NO_XPR [i.e. $(\Delta NO_3/\Delta t) / (\Delta NO_X/\Delta t)$] is linked directly to the ratio NOB/AOB (but not equal to NOB/AOB because NO_3PR/NO_XPR incorporates the maximum growth rates of AOBs and NOBs). For example, in a fully nitrifying plant the ratio typically will be close to 0.8.

However, if a lower value is measured in a fully-aerated SNR test on plant mixed liquor, then it is likely that the NOB population is suppressed and nitrite shunt is occurring in the plant.

RESULTS AND DISCUSSION

Special Sampling and Concentration Profiling Results

Influent samples were collected daily during the sampling campaign as presented in Table 1. Based on these results, an average C:N ratio of approximately 7:1 (total COD and TKN) and 2.8:1 (based on rbCOD and NH_3-N) were observed. An influent TP-to-COD ratio of approximately 0.013:1 was observed.

Table 1 Summary of Influent Characteristics

Parameters (mg/L)	Value (Standard Deviation)
COD	300 (±65)
Soluble COD	120 (±20)
Readily Biodegradable COD	65 (±15)
Unbiodegradable COD	20 (±8.5)
VFAs	13 (±5)
TSS	140 (±35)
TKN	30 (±5.6)
NH_3-N	23 (±4.5)
TP	3.9 (±0.75)
PO_4-P	2.4 (±0.30)
Alkalinity (mg/L $CaCO_3$)	210 (±20)
pH (SU)	7.1 (±0.22)

Table 2 shows the concentration profile for DO, NH_3^-, NO_2^-, NO_3^- and PO_4^- at the end of the anaerobic zone and each aerobic zone of the activated sludge reactor. The average total and aerobic SRT and temperature were approximately 5.2 and 3.9 days and 29.2°C, respectively. As shown in Table 2 with considering the NH_3-N dilution and NO_2^- and NO_3^- brought by the recycled RAS, there was essentially no NH_3-N removal or accumulation of NO_2^- or NO_3^- in the anaerobic zone. In the aerobic zones, NH_3-N was removed from 10.5 ±3.3 to 1.4± 0.32 mgN/L at the low DO condition. A small accumulation of NO_2^- up to about 1.4 mgN/L was observed at the end of the first aerobic zone with no significant accumulation of NO_3^-. After the first aerobic zone, no observed accumulations of NO_2^- or NO_3^- were detected despite NH_3-N being oxidized by AOB. The profile results demonstrated the good nitrogen removal performance of the plant. Bench tests were conducted to verify the nitrogen removal was the result of nitritation with minimum NOB activity and denitritation from NO_2^- to N_2 and the detailed results were presented in the next section. In addition to the nitrogen removal down to a level of 2 mgN/L, low phosphate concentration was also achieved through biological phosphorus removal activities. Significant PO_4^- release was observed through the anaerobic zone and phosphorus uptake was observed in the aerobic zones despite the low DO condition in the activated sludge reactors. Biological activities were also verified through bench tests.

Table 2 Dissolved oxygen, ammonia, nitrate and nitrite concentration profiles

Zone	Profiles (mg/L)					
	DO	COD	NH_3-N	NO_2^-	NO_3^-	PO_4^-
Influent	--	65 (±15)	23.0 (±4.5)	0.1 (±0.02)	--	2.4 (±0.3)
Anaerobic	0.02 (±0.01)	41 (±17)	10.5 (±3.3)	0.2 (±0.01)	0.1 (±0.07)	12.3 (±0.23)
Aerobic 1	0.42 (±0.12)	11.1 (±5.4)	4.2 (±2.1)	1.4 (±0.43)	0.3 (±0.12)	3.2 (±0.70)
Aerobic 2	0.12 (±0.10)	2.4 (±3.2)	2.5 (±1.2)	0.3 (±0.14)	0.2 (±0.08)	0.7 (±0.12)
Aerobic 3	0.08 (±0.12)	--	1.4 (±0.32)	0.20 (±0.14)	0.15 (±0.05)	0.21 (±0.11)

Figure 2 shows the effluent inorganic nitrogen concentrations during the sampling campaign. The effluent NH_4^+, NO_2^- and NO_3^- averaged approximately 1.8, 0.41 and 0.23 mg N/L, respectively. Figure 3 shows the influent and effluent total phosphorus (TP) concentrations with averages of 3.5 and 0.5 mg P/L, respectively, indicating good biological phosphorus removal even at very low DO conditions. Based on the monthly operating reports for the first quarter of 2014, the effluent NH_4^+, NO_2^-, NO_3^- and PO_4^- averaged 3 mgN/L, 0.4 mgN/L, 0.2 mgN/L and 0.1 mgP/L, respectively.

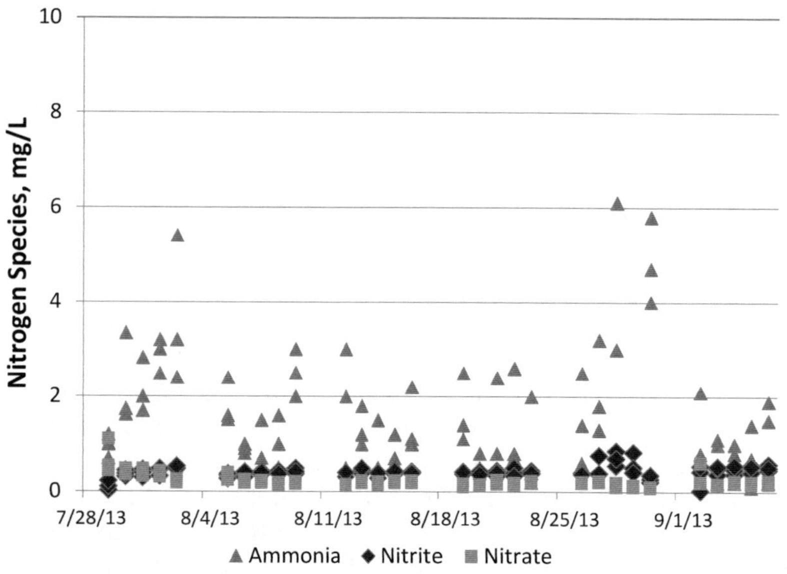

Figure 2 Final effluent ammonia, nitrate and nitrite concentrations

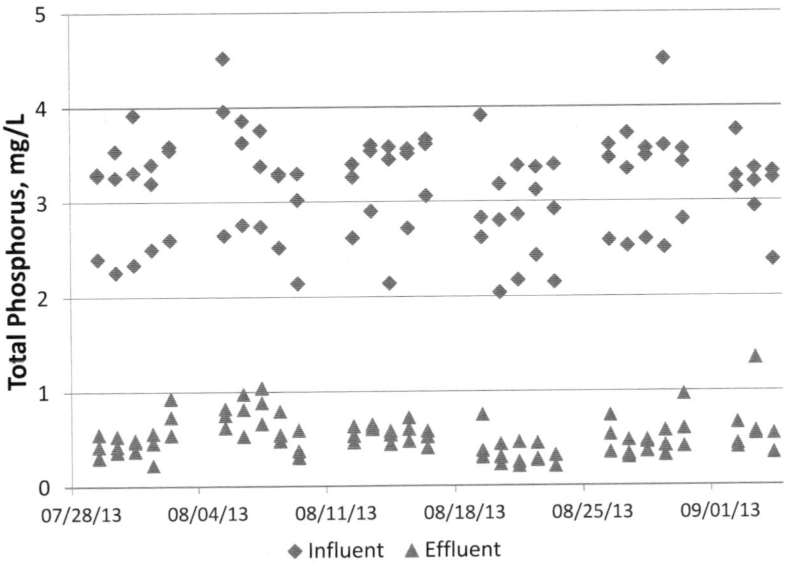

Figure 3 Influent and final effluent total phosphorus concentrations

Bench Scale Testing Results

Specific Nitrification Rate (SNR) Tests

SNR tests were conducted to measure specific ammonia removal and nitrite/nitrate production rates at specific operating conditions. SNR tests were seeded with a grab sample of 4 L of RAS and 4 L of influent collected at the plant. Ammonium chloride was added such that the initial NH_3 concentration was approximately 25 mgN/L. A small amount of supplemental phosphorus (in the form of potassium dihydrogen phosphate) was added to ensure nutrient limitations would not impact the test. Additional alkalinity in the form of sodium bicarbonate was added as needed to ensure stable pH throughout the test.

Two sets of SNR tests were conducted. The first SNR test was fully aerated to achieve a DO concentration of at least 5 mg/L whereas the second SNR test was operated at a DO setpoint of 0.3 mg/L, which is close to the upper limit of the DO range at which the aeration basins at the plant are operated. The fully aerated SNR Test was used to determine the NO_3PR/NO_2PR ratio and hence the NOB/AOB ratio to determine whether the NOB population is suppressed and nitrite-shunt is occurring in the plant. The results from the SNR Test operated at low DO levels were used to determine the impact of low DO on the nitrification rate.

Linear regression analysis was used to estimate the removal rate of NH_3 and production rates of NO_2, NO_3 and NO_X. Dividing these rates by the batch volatile suspended solids concentration yields the specific rates, as shown in Table 3. The specific rates were corrected to 20°C. An Arrhenius value of 1.072 was used for the SNH_3RR, SNO_3PR and SNO_XPR.

Demonstration of a Novel Separate Centrate Deammonification Process at the 26th Ward Wastewater Treatment Plant

Authors: Sarah Galst[1]*, Wendell Khunjar[1], Paul Pitt[1], Robert Sharp[2], Kathleen O'Connor[3], Allen Deur[4]

1 Hazen and Sawyer, P.C.
2 Manhattan College
3 New York State Energy Research and Development Authority
4 New York City Department of Environmental Protection
*sgalst@hazenandsawyer.com

ABSTRACT

Increasingly stringent nutrient discharge limits require utilities to employ energetically and cost intensive biological nutrient removal (BNR) processes. This project, "Demonstration of a Novel Separate Centrate Deammonification Process," will explore the application of anaerobic ammonia oxidation (anammox) for reducing the energy demand and operating costs associated with biological nutrient removal. In this hybrid approach, facilities can reduce theoretical oxygen and alkalinity demand requirements by up to 50%, as well as reduce supplemental carbon costs between 60 to 80% versus the conventional approach. These savings can significantly reduce plant wide energy costs and the over-all carbon footprint of plant operations.

INTRODUCTION

Recently, treatment of sidestream process flows generated from solids handing processes (e.g., belt filter press (BFP) filtrate, gravity belt thickener filtrate, centrate, and digester supernatant), has emerged as an economical complement to mainstream BNR due to the relatively low volume and high concentration of nutrients present in these sidestream flows. In performing sidestream treatment focused on nutrient mitigation, utilities can effectively provide for a higher factor of safety in the mainstream process as well as reduce the total nutrient load needs to be treated at the head of the plant.

Conventional Sidestream Treatment

Sidestream treatment of nitrogen can be accomplished using the conventional approach of nitrification and denitrification (**Figure 1**). In this strategy, ammonia is oxidized to nitrite and then nitrate by aerobic ammonia oxidizing bacteria (AOB) and aerobic nitrite oxidizing bacteria (NOB) respectively. This nitrate is then denitrified to nitrogen gas using heterotrophic bacteria. This process requires 4.57 pounds of oxygen (O_2) per pound of ammonia (NH_3-N) processed, 7.14 pounds of alkalinity per pound of ammonia (NH_3-N) processed and 6 pounds of supplemental carbon (COD) per pound of nitrate (NO_3^--N) removed through heterotrophic denitrification.

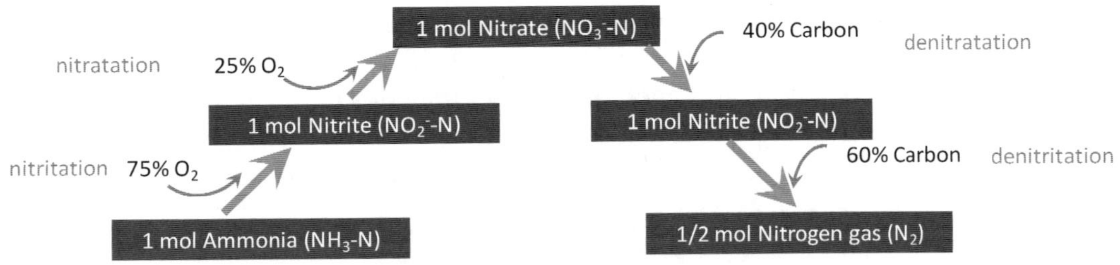

Figure 1: Nitrification and Denitrification pathway currently employed at 26th Ward SCT

Sidestream Deammonification

An alternative approach for removing nitrogen in sidestream flows is the use of the deammonification process (**Figure 2**). The deammonification process requires conversion of approximately 50% of the influent ammonia into nitrite by AOB followed by the simultaneous removal of ammonia and nitrite by anammox bacteria. For deammonification processes to be successful, stable production of nitrite is needed. Further, since anammox bacteria have extremely low growth rates, deammonification processes must provide sufficiently long solids retention times (SRT) for anammox bacteria growth.

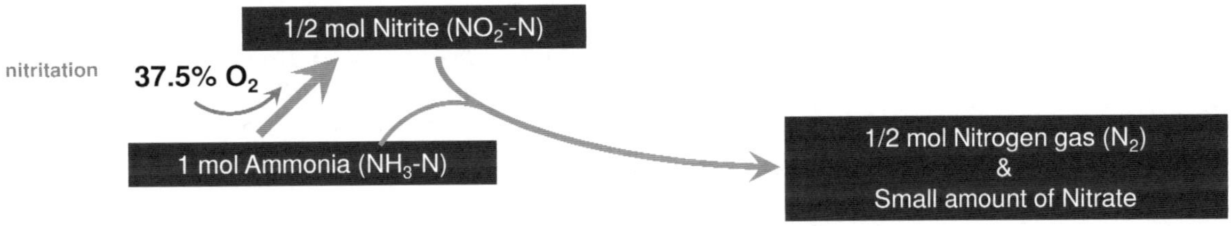

Figure 2: Overview of the Deammonification Process

There are existing commercial deammonification processes that are available: DEMON™, AnitaMOX™, ANAMMOX™, CLEARGREEN™. In each of these existing processes, nitrite production is achieved using patented approaches for repressing NOB. These approaches employ some combination of free ammonia, dissolved oxygen concentration gradients and selective wasting. These processes also utilize sequential batch reactor configurations (DEMON™, CLEARGREEN™), fixed film media (AnitaMOX™) or completely stirred reactors (ANAMMOX™) with solids/media separating devices to provide sufficiently long SRTs for anammox bacteria retention. In all cases, the technology providers also utilize seed bacteria from elsewhere to startup and maintain the process. Given the existing configuration of SCT processes at NYC WWTPs, conversion to any of these processes would require substantial capital upgrades and modifications to operating practices that would be unfamiliar to operations staff. Additionally, NYCDEP would be required to import anammox bacteria as part of a paid license agreement with the technology providers representing these existing technologies.

Separate Centrate Deammonification (SCAD) Process

In this project, we are working to develop a novel strategy that exploits existing infrastructure to circumvent these challenges. In our proposed separate centrate deammonification process (SCAD), approximately 50% of the influent ammonia in the centrate will be converted nitrite and then nitrate by AOB and NOB. This nitrate will then be subjected to denitratation (conversion of nitrate to nitrite) using glycerol based denitrification that is powered by specialist heterotrophic bacteria. This will be followed by the simultaneous removal of ammonia and nitrite via anaerobic ammonia oxidation **(Figure 3)**.

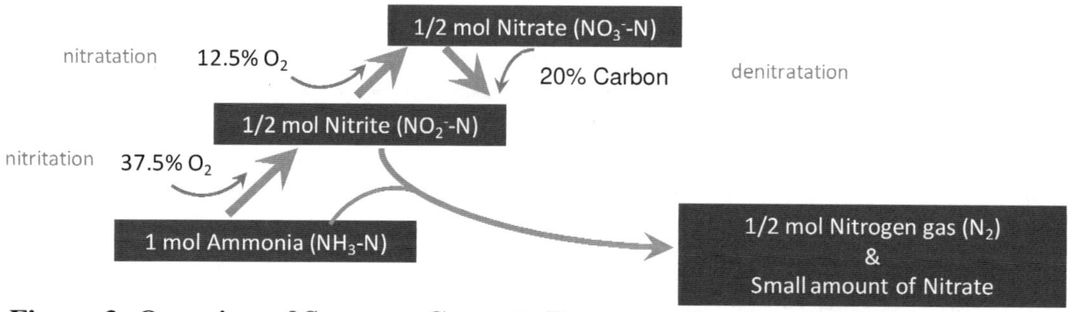

Figure 3: Overview of Separate Centrate Deammonification (SCAD) pathway

In this proposed SCAD option (**Figure 3**), theoretical savings of 50% and 60 to 80% in oxygen supply and supplemental carbon respectively can be achieved, respectively, relative to the existing conventional nitrification/denitrification approach employed. Further, the SCAD process is novel in that it does not seek to repress NOB activity, which can be challenging, but instead uses glycerol to stimulate nitrite production and thereby exploits the natural predisposition of the bacteria to produce nitrite. Previous work performed by Hazen and Sawyer with SCT has demonstrated that this approach is successful in growing the anammox bacteria, which are notoriously slow growing organisms.

For a full-scale application, enhanced retention of the anammox bacteria will need to be achieved. In the full-scale scenario, the retained anammox bacteria would be recycled within the SCAD process, while the discharged solids, containing mostly AOB and NOB, would be used to bioaugment and improve the mainstream process nitrification performance. Attempts can be made to bioagument the mainstream process with the anammox bacteria to stimulate mainstream deammonification, which would yield further energy and cost savings.

APPROACH

For this project's pilot demonstration of the SCAD sidestream treatment, the existing pilot facilities at the New York City 26th Ward WWTP, **Figure 4**, were leveraged and reconfigured. An existing plate settler was re-purposed to be both the reactor and plate settler. The system is alternated between periods of aeration and anoxia using a Program Logic Control (PLC), and aeration is provided via a fine bubble diffuser in the reactor. pH, TSS, and DO probes installed in the reactor provide continuous monitoring of important process control parameters. Ammonia, nitrite, nitrate, soluble COD and orthophosphorus are monitored utilizing Hach TNT reagents.

Figure 4: 26th Ward WWTP

RESULTS AND DICUSSION

Bench-scale Proof of Principles

Bench-scale work has demonstrated the proof of principles for the SCAD process (**Figure 5**). Oxidation of ammonia and nitrite resulted in full nitrification to nitrate by the end of the aerobic cycle. After glycerol feed, nitrite production was observed. At the start of the unaerated anammox cycle, ammonia from centrate combined with nitrite from denitratation allowed for nitrogen removal via anaerobic ammonia oxidation.

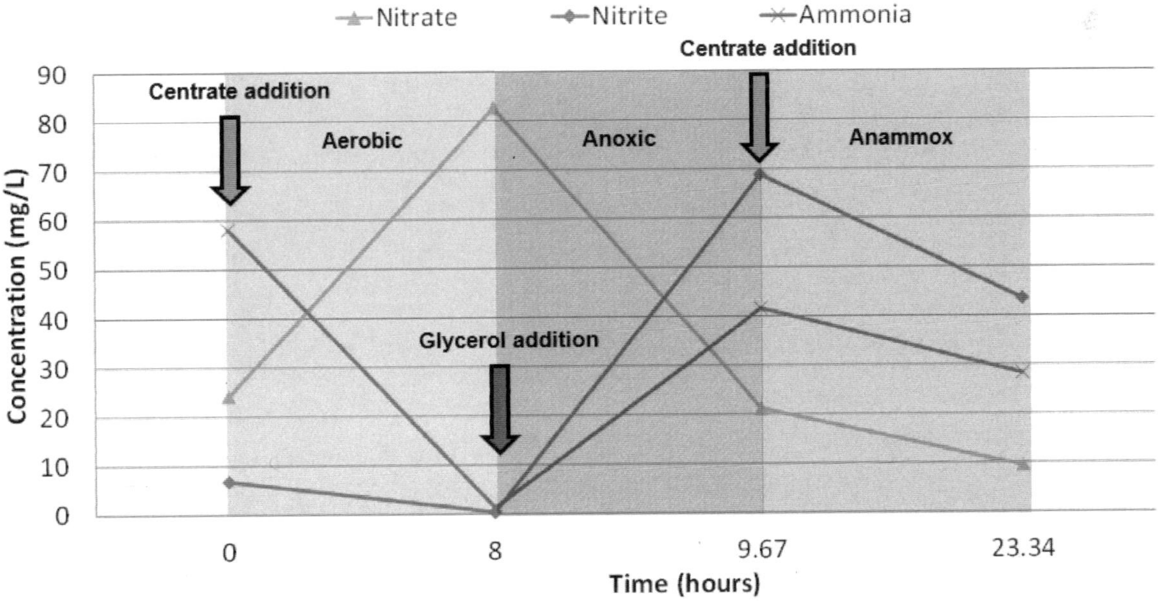

Figure 5: Bench Scale Testing Results Demonstrating SCAD

Pilot Status

The SCAD pilot was installed at the 26th Ward WWTP. Centrate addition and aeration are controlled to mimic hydraulic retention times and operating DO concentrations that would be achieved a full-scale installation.

Enrichment of anammox bacteria in the SCAD process is currently underway, using a combination of high SRT (using plate settlers) and supply of equal amounts of nitrite and ammonia. Nitrite is produced by adding glycerol under anoxic conditions to stimulate the rapid production of nitrite, while ammonia is utilized directly from centrate.

During the March and April 2015 enrichment period, the volumetric TIN loading rate was 0.05 kg N/m^3-day. Results indicated that the TIN removal rate increased from 0.41 to 0.71 mg TIN/g TSS-hr. The corresponding TIN removal efficiency increased from 14 to 34%. During this period of operation, between 20 and 30 mg NO_2-N/L was produced via nitritation while approximately 10 mg NO_3-N/L was produced via complete nitrification. The occurrence of nitritation in the system is expected due to the temperature and free ammonia concentrations present during the aerobic cycle. It should be noted that greater than 90% of the nitrite produced in the aerobic period was consumed in the anoxic period. Nitrite removal during the anoxic phase was accompanied by ammonia removal confirming the presence and activity of anammox bacteria. Further, almost complete removal of nitrite indicated that the system anammox capacity was not limiting under these conditions.

Between April 10 and April 18, 2015, the TIN removal rate increased from 0.71 to 1.21mg TIN/g TSS-hr, as shown in **Figure 7**. TIN removal efficiency increased moderately from 34 to 38%. During this period, nitrite concentrations during the aerobic period increased up to 60 mg N/L. Greater than 95% of the nitrite produced via nitritation was removed in the anoxic phase. This consumption of nitrite was coupled to ammonia removal confirming anammox activity. It is important to note that a portion of the nitrite removal was also associated with heterotrophic denitrification. This denitrification was likely fueled by the presence of soluble COD in the centrate and through endogenous denitrification. As noted previously, the intent of this mode of operation was to enrich anammox biomass. Therefore a high overall removal efficiency was not the primary driver. Instead, the strategy focused on increasing the specific removal rate capacity (a proxy of anammox bacteria enrichment).

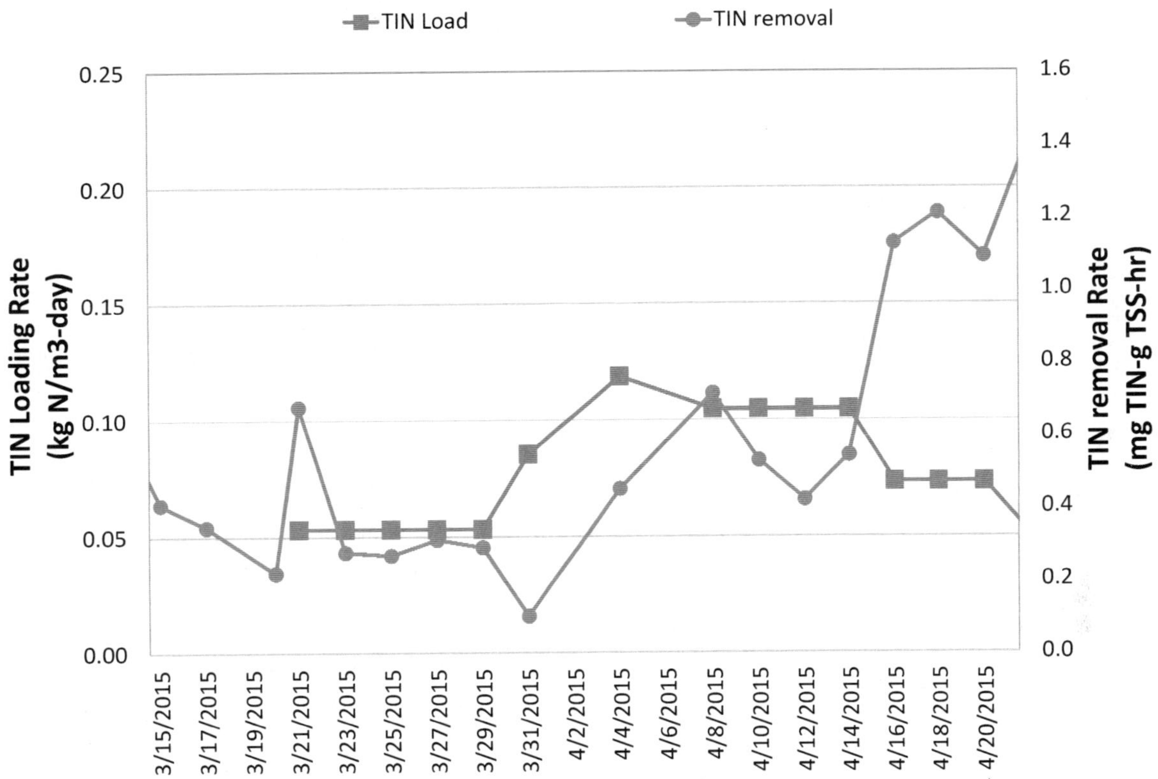

Figure 6: TIN Loading and Removal for March and April 2015

FUTURE WORK

After enrichment of anammox bacteria is achieved, the goal will be to manipulate SCAD operation to achieve approximately 80% total inorganic nitrogen removal. This will include manipulating three main variables: Aerobic and anoxic reactions times, Glycerol dose, and Internal recycle rates

Recommendations and Lessons Learned from the Startup of the First Two Full-Scale Sidestream Deammonification Processes in North America: DEMON® and ANITA™ Mox

Stephanie Klaus[1,3], Andrea Nifong[2], Andy Nelson[3], Bob Rutherford[3], Rick Baumler[3], Charles Bott[3]

[1]Civil and Environment Engineering Department, Virginia Tech, Blacksburg, VA 24060
[2]World Water Works, Inc., 4000 SW 113th Street, Oklahoma City, OK 73173
[3]Hampton Roads Sanitation District, 1436 Air Rail Ave., Virginia Beach, VA 23455

Keywords: Sidestream Treatment, Nutrient Removal, Operations and Maintenance

Abstract

Deammonification is an economical option for sidestream treatment due to decreased aeration energy requirements, no required external carbon or alkalinity, and decreased sludge production over traditional nitrification/denitrification. While this technology has been proven at multiple locations in Europe since 2004, the installation of the DEMON® process at the Hampton Roads Sanitation District (HRSD) York River Wastewater Treatment Plant in 2012 and ANITA™ Mox at the HRSD James River Wastewater Treatment Plant in 2013 represent the first two operating sidestream deammonification installations in North America. Lessons learned from the startup of DEMON at York River Plant were applied to the startup of ANITA Mox at James River Plant. The goal of this paper is to share lessons learned and recommendations to aid in future startups of sidestream deammonification processes.

Introduction

Deammonification is a two-step process which combines partial nitritation and anaerobic ammonium oxidation (anammox). In the first step aerobic ammonium oxidizing bacteria (AOB) convert a portion of the incoming ammonia to nitrite and in the second step anaerobic ammonium oxidizing bacteria (AMX) convert the remaining ammonium and nitrite to nitrogen gas and a small amount of nitrate.

The DEMON® process is a sequencing batch reactor (SBR) which is intermittently fed and intermittently aerated. In this process partial nitritation occurs during aerated intervals while anammox occurs during anoxic intervals. This process utilizes a biomass separation device (hydrocyclone) to control the solids retention time (SRT) of the bacterial populations in the reactor and effectively decouple the SRT of AOB from AMX. The ability to separate the SRT of AOB/NOB/heterotrophs/debris and AMX allows for the flexibility to control the NOB population mechanically as well as with the pH-DO based control strategy.

The ANITA™ Mox process is a continuously aerated single-stage moving bed biofilm reactor (MBBR) in which partial nitritation and anammox occur simultaneously within a biofilm attached to plastic carriers. A key component of the process is controlling dissolved oxygen (DO), again using pH, to maximize ammonia removal (and alkalinity consumption) and to limit the growth of nitrite oxidizing bacteria (NOB) which compete with anammox bacteria for substrate and to limit oxygen penetration into the biofilm which allows anammox growth in this location.

Operation Objectives

Operating objectives for a sidestream deammonification reactor are:

1. Ammonia concentration below a level that could cause free ammonia (FA) inhibition, while staying above a level that would cause substrate limitation (50-100 mg/L).
2. Nitrite below a level that could cause inhibition of AMX (10-40 mg/L).
3. Alkalinity above a level that would cause inorganic carbon limitation, low pH inhibition, and/or unstable pH conditions (150-250 mg/L).
4. Target ammonia removal around 80-90% (70-80% TIN removal).
5. Maintain nitrate production below 12-15% NO3 produced/NH4 removed. If all of the nitrate production in the reactor is due to AMX activity then the nitrate production ratio will be around 11% (or less with some heterotrophic denitrification due to influent COD). If the nitrate production ratio is any higher than 11% then it can be assumed that the excess nitrate production is due to NOB activity.

DEMON Control Strategy

Online sensors were used to monitor pH, DO, specific conductivity, and temperature in the DEMON® reactor. Centrate feed pumps, the submersible mixer, blowers, cyclone feed pumps, online sensors, and the SBR decanter are all controlled from the dedicated DEMON PLC that is accessible both at the facility and via an internet connection which is password protected..

The typical DEMON SBR cycle time is 6 or 8 hours. The last hour of this cycle the biomass is settled and the treated effluent is decanted. The remaining cycle time is referred to as the react period and is composed of multiple mini-cycles typically 10-30 minutes in length each. During each mini-cycle the following occur sequentially:

- Aerate
- Feed/Mix
- Mix

pH-based aeration control determines the intermittent aeration duration. When aerating, a DO controller is used to maintain 0.2 to 0.5 mg/L DO through typical proportional-integral-derivative (PID) controller or a constant airflow mode.
SRT is controlled by hydrocyclones (used to separate AMX from AOB and to waste NOB from the system). Set operation time during the react period achieves the desired floc-phase SRT. If the percentage of NO_3 produced/NH_4 removed increases above 12-15%, the hydrocyclone runtime is increased to decrease the floc-phase SRT and waste NOB from the system.

Anita MOX Control Strategy

Online sensors were used to monitor NH_4-N, NO_3-N, pH, DO, specific conductivity, and temperature in the ANITA™ Mox tank. NH_4-N and temperature were monitored in the EQ basin. The sensors, centrate pump, air flow control valve, and mixers are all connected to the plant distributed control system (DCS). Centrate pump speed is controlled by a variable frequency drive and is measured by a flow meter to meet a flow setpoint. Air flow rate to the reactor is controlled and measured by a modulating, motor actuated control valve. Air flow can be either operated continuously or intermittently. Intermittent aeration is utilized during the beginning of startup to encourage the growth of AMX by providing a distinct anoxic period and to control the production of nitrite.

Airflow can be controlled to meet an airflow setpoint (measured by a flow meter upstream of the valve), a DO setpoint (measured by a process probe), or a pH setpoint (measured by a process probe). In all three modes a low pH setpoint safeguards against running out of alkalinity. When the low pH setpoint is reached the airflow shuts off while centrate feed continues to allow the system to recover. If NOB growth occurs, resulting in an increase in effluent nitrate, airflow is decreased at the expense of ammonia removal until the nitrate production ratio is less than the value that would be expected to be produced by AMX alone (12-15%). Mechanical mixers keep the media in suspension and keep the tank completely mixed during periods of non-aeration.

Sampling and Monitoring

For both systems, grab samples from the influent and effluent were analyzed at the plant for NH_4-N, NO_3-N, and NO_2-N using Hach TNT kits and a Hach DR-2800 spectrophotometer. Grab samples were also analyzed at the HRSD Central Environmental Laboratory for the following parameters: NH_3-N, NO_x-N, NO_2-N, TKN, TSS, VSS, COD, sCOD, TP, OP, and alkalinity. Sampling was initially performed daily, gradually reducing the sampling schedule to two days per week. Two to three operators were trained to perform this analysis.

DEMON Startup

A full-scale DEMON® process was installed at HRSD's 57 ML/d (15 MGD) York River Treatment Plant (YR) to treat anaerobically digested sludge liquor (centrate). The process was anticipated to decrease alkalinity (sodium hydroxide), methanol, and aeration energy usage at YR, which currently includes a fully aerobic nitrifying activated sludge system with post denitrification filters. The DEMON® process was installed in existing SBR tanks that were part of a decommissioned reuse system. Design criteria are listed in Table 1.

Table 1: YR Design Criteria

Centrate Feed	
TKN-N (average month)	0.35 kgN/ m^3/day
Flow (7 day equalized)	270 m^3 (71,500 gpd)
Removal	
TN	80%
NH_3	90%
SRT	
Floc Phase SRT	~4-6 days
Volumes	
EQ Tank	1,517 m^3 (400,830 gal)
DEMON Tank (minimum)	329 m^3 (86,930 gal)
DEMON Tank (maximum)	363 m^3 (95,920 gal)

The DEMON® system at YR was seeded October 10th, 2012 with 18 m^3 (5,000 gal.) of concentrated (5% solids) biomass. On October 12th the SBR cycling was initiated, feeding the system at 20% total capacity and reaching 50% capacity within the first four weeks of operation. During this period, specific load was increased no more than 10% per week, and aerobic/anoxic time intervals adjusted as needed. The anoxic time length was slowly decreased throughout startup while the aerobic time length was increased (Figure 1). This trend held true except for the initial startup when aerobic time was purposely increased to promote the growth of AOB. During startup, a couple of instances of NO_2 spikes occurred causing no detrimental effects to the system.

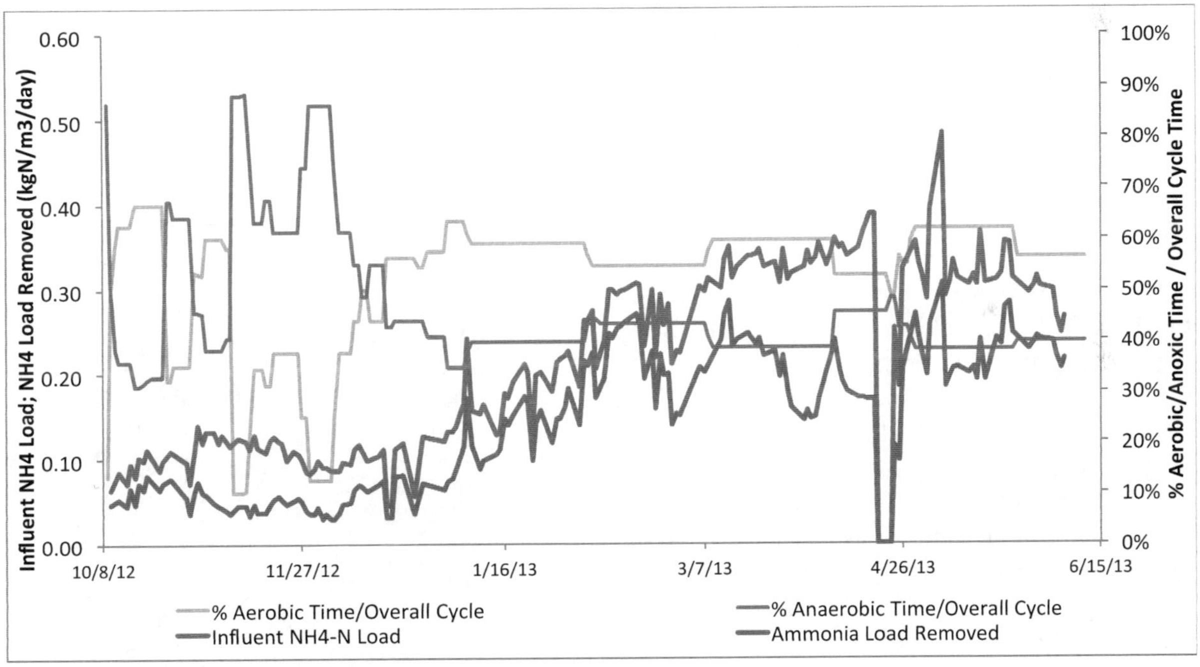

Figure 1: DEMON Aerobic/Anoxic times per ammonia loading

The DEMON® reactor achieved 100% treatment within 3 months of operation at a loading rate of 0.50 kgN/m^3/day, but it should be recognized that this design load is less than that which can be achieved by a DEMON® process and was a function of the size of the already existing tank that was available at the York River plant and retrofitted to serve as the DEMON® reactor. NH4-N removal varied during startup between 40% (prior to insulation of the reactors causing severe temperature loss) and 85% and TIN removal ranged from 30% to 75% (Figure 2). The maximum NH_4-N removal that the DEMON system has achieved is 95.5 % (78% TIN) and a maximum loading rate of 0.58 kgN/m^3/day.

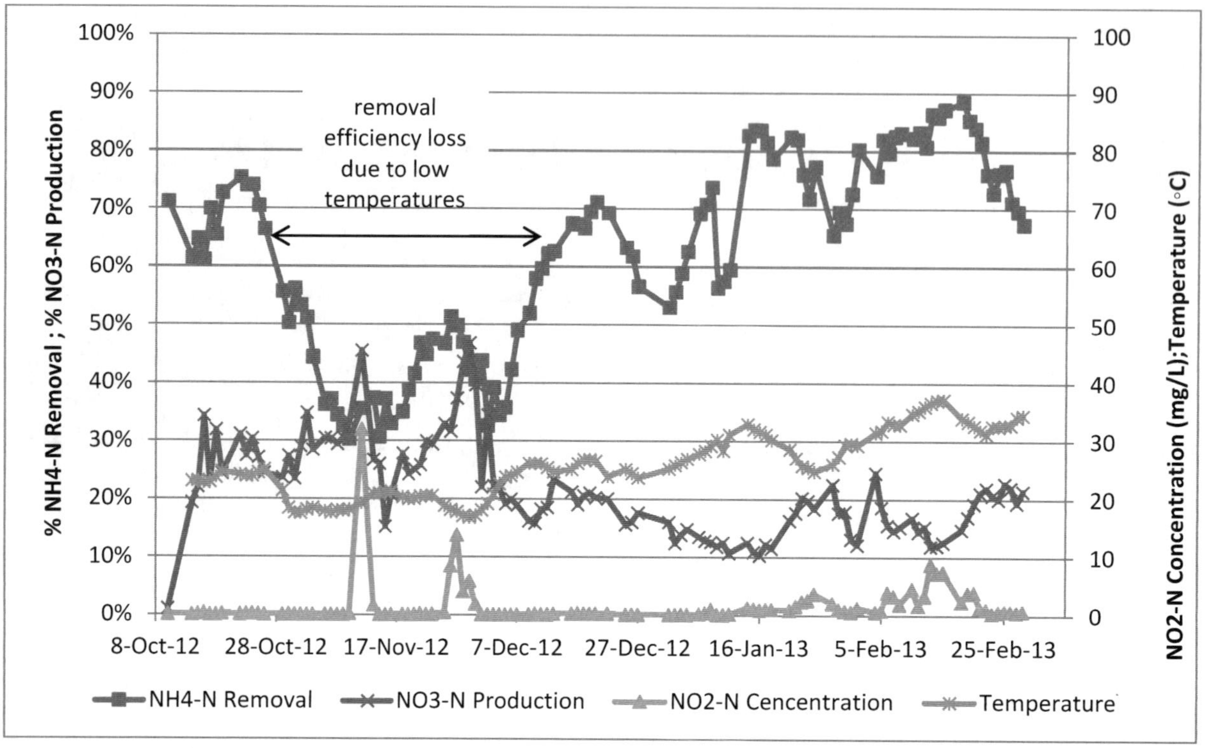

Figure 2: DEMON Startup Performance

ANITA Mox Startup

The James River Wastewater Treatment Plant is a 76 ML/d (20 MGD) plant located in Newport News, Virginia. Eight IFAS aeration basins are operated in a 4-stage Bardenpho configuration. Anaerobically digested waste activated sludge and primary sludge is dewatered using centrifuges and the centrate is sent to an equalization (EQ) basin. The goal of sidestream treatment at HRSD James River Plant was to decrease final effluent total nitrogen to help meet the "bubble" permit for nitrogen loading to the James River watershed. An existing pre-aeration basin was retrofitted to install the ANITA Mox process. Design parameters (Table 2) were based on average centrate characteristics shown in Table 3.

Reclaiming Ammonia From Anaerobic Digestate As A Profitable Product

Dennis A. Burke P.E.*
Environmental Energy & Engineering Company, (E[3])Olympia, WA
Email waeng@me.com

ABSTRACT

Most physical/chemical processes utilized to recover nutrients from anaerobic digestate are uneconomical since they produce nutrient products that have little value. On the other hand, numerous processes have been developed to economically destroy ammonia nitrogen by converting it to N_2 gas. Those processes also produce NO_x ,a precursor to ground level ozone, and N_2O a powerful greenhouse gas at 310 times the impact of CO_2. This paper presents the initial investigations of a process that strips and recovers ammonia from anaerobic digestate as a valuable product while recovering most of the phosphate. The concentrated ammonia recovered by the process has value since it can be used in: (1) lignocellulosic pretreatment for biofuel production, (2) stripping CO_2 from biogas through the "chilled ammonia process" to produce biomethane, a transportation quality fuel, and ammonium bicarbonate solids for use as a renewable fertilizer, and (3) diesel exhaust fluid used for the destruction of NOx from combustion sources through selective catalytic reduction (SCR). Those products have sufficient value to justify the use of a nutrient recovery process as opposed to nutrient destruction processes. Nitrogen is recovered with low energy inputs and without the use of chemicals, by stripping ammonia from the digestate and condensing it as an ammonium hydroxide product. The process uses a fixed film of phototrophic organisms to increase the pH of the digestate while consuming bicarbonate, a prerequisite for the economic recovery of phosphate and ammonium hydroxide. The process can be designed to operate using several different configurations. The basic configuration produces a consistent 98% plus removal and recovery of ammonia at relatively low loading rate (0.2± Kg N/ m3/d). The second configuration produces a variable (75% to 85%) ammonia recovery at significantly higher loading rates. Over 90% of the phosphate is recovered under both operating configurations utilizing calcium chloride as a calcium source if sufficient calcium is not available in the substrate. This paper describes a four month pilot investigation that demonstrated the recovery of 98% of the influent ammonia as a concentrated (6%±) ammonium hydroxide solution, without the use of chemicals, and 90% of the phosphate with the use of calcium chloride if sufficient calcium is not present in the substrate. The pilot operation concentrated 98%+ of the influent ammonia to 3% of the influent flow. Over 90% of the influent phosphate was recovered as a rapidly settling concentrated amorphous tricalcium phosphate precipitate.

KEYWORDS: Anaerobic digestate, nutrient recovery, photo bioreactor, ammonia stripping, phosphate precipitation

INTRODUCTION

The National Academy of Engineering of the National Academies recently published the "Grand Challenges for Engineering" http://www.engineeringchallenges.org/cms/challenges.aspx. One of the 14 challenges was "Manage the Nitrogen Cycle". The "nitrogen issue" is the result of twice as much nitrogen being introduced into the world through anthropogenic sources than natural sources. The ability to produce reactive nitrogen (Nr) through the Haber Bosch process and enhanced agricultural nitrogen fixation, has enabled man to feed the world. However, those engineering miracles have totally distorted the nitrogen cycle leading to significant adverse environmental and public health consequences. The nitrogen cycle is further distorted by the combustion of a wide variety of fuels that produce NOx. In the US alone during the year 2002, 6.4 Tg of reactive nitrogen (Nr) was added from natural sources, 15.3 Tg was added through Haber Bosch fertilizer and non-fertilizer production, 7.7 Tg through enhanced agricultural nitrogen fixation, and 5.7 Tg as NOx from stationary and transportation fossil fuel combustion. Natural nitrogen fixation in the US represented less than 19% (6.4/35.1) of the total reactive nitrogen added to the environment i(SAB, n.d.).

Reactive nitrogen can only escape the environment through natural nitrification-denitrification or ammonia oxidation-anammox processes that convert the Nr to N_2 gas. In the ocean approximately 20% to 30% of the reactive nitrogen is removed by anammox organisms, depending on the amount of dissolved carbon present (Ward, 2013). However, in the terrestrial environment, where organic carbon is plentiful, the primary removal pathway is through denitrification a process that converts 1 to 10 percent of the input nitrogen to nitrous oxide (N_2O) (Ni, Ruscalleda, Pellicer-Nacher, & Smets, 2011) a powerful, stratospheric ozone depleting, green house gas (GHG) at 310 times the impact of CO_2. The third reactive nitrogen removal pathway is Selective Catalytic Reduction (SCR) commonly used for diesel exhaust or combustion NOx removal as shown by the following equation.

$$(2NO_2 + 4NH_3 + O_2 \Rightarrow 3N_2 + 6\ H_2O) \qquad \text{Eq 1}$$

SCR requires the use of additional reactive nitrogen (NH_3) produced through the Haber - Bosch process in order to destroy the NOx emission from combustion sources (Stephanie Kato, n.d.). The production of Haber Bosch ammonia consumes 3 to 5% of the worlds annual natural gas production or 1 to 2% of the world's energy supply resulting in significant annual GHG emissions. Those emissions would be substantially reduced if recycled ammonia could be used in SCR.

If the reactive nitrogen is not removed from the environment it accumulates, producing a wide variety of environmental and public health consequences. The accumulated reactive nitrogen is stored in the soils, groundwaters, rivers, estuaries, and oceans modifying those terrestrial and aquatic environments. The adverse impacts of nitrogen include the production of fine particulate matter ($PM_{2.5}$) that is responsible for atmospheric haze and increased human mortality, increased nitrate levels in groundwater, acidification of surface water, harmful toxin producing algae blooms, hypoxia in coastal waters, forestry decline, crop yield loss, and loss of terrestrial biodiversity (Sutton, 2011), (SAB, n.d.) .

Solutions to the "nitrogen problem" have primarily been through the use of engineered anammox or denitrification systems that increase the N_2O greenhouse gas emissions to the atmosphere. Some processes have attempted to recover and reuse ammonia from waste streams and thereby reduce the industrial production of ammonia through the Haber-Bosch process. Ion exchange, membrane draw solution separation, and stripping technologies have all been used. The recovery processes invariably sequester the ammonia as dilute acidic solutions of ammonium sulfate or ammonium nitrate that are not profitable technologies due to the low value of the products produced and the hazardous chemical reagents (H_2SO_4, HNO_3) required. Other technologies have used selective catalytic reduction (SCR) to convert combustion NOx emissions to N_2 gas through the use of ammonia nitrogen primarily in the form of dissolved urea. NOx emissions are reduced at the expense of producing more Haber Bosch ammonia nitrogen, some of which is lost to the environment as "ammonia slip". Recent attention has focused on methods to increase "nitrogen use efficiency" (NUE) and thereby reduce the demand for Haber-Bosch nitrogen fertilizers. Various chemical inhibitors have also been introduced to inhibit the nitrification and denitrification of ammonia and thereby reduce N_2O emissions. New trading schemes have also been introduced to reduce N_2O emissions (EPRI). Those schemes pay farmers to reduce fertilizer application and thereby achieve reductions of N_2O from nitrogen fertilizer calculated at 1% of the applied nitrogen (Millar & Adam, n.d.; University, n.d.). However, the primary focus of the waste treatment industry has been on the development of the Anammox process; a process that can achieve loadings as high as 1.5 $KgN.m^3d^{-1}$. That process however reclaims neither nitrogen nor phosphorus. The state of the art in phosphorus reclamation is the MAP or Struvite crystallization process that can recover approximately 12% of the nitrogen and 95% of the phosphorus in domestic anaerobic digestate through the addition of soluble magnesium while increasing the pH to 9.0±. The requirement for chemicals that increase the pH and provide sufficient magnesium results in considerable expense. Typically magnesium hydroxide or hydroxide is used with or without sodium hydroxide.

Reducing Reactive Nitrogen In The Environment

The RPB process if adopted is expected to reduce the reactive nitrogen (Nr) in the environment by 25%. The process reclaims nitrogen, in a liquid or solid form, from manure, human waste, or biomass renewable energy production facilities. If sufficiently concentrated, the ammonia nitrogen can then be used in selected catalytic reduction (SCR) to remove NOx from stationary and transportation combustion sources. The process will eliminate production of new nitrogen for use in catalytic converters and replace that nitrogen with nitrogen recovered from biomass or waste. Currently 7.5 Tg of nitrogen from manure and human waste is discharged to the environment each year in the US. That quantity represents 18% of the total reactive nitrogen discharged each year to the environment. In addition, 5.4 Tg of reactive nitrogen is introduced into the environment from NOx combustion sources in the US, 12% of the total Nr. The proposed E^3 process will use the 7.5±Tg of recovered nitrogen to destroy the 3.75±Tg of NOx nitrogen thereby achieving a 11.25±Tg annual reduction or 25% of the total reactive nitrogen introduced into the environment each year. The process assumes that reactive nitrogen is recovered from waste residuals in a liquid or solid form that can be used in selective catalytic reduction.

Diesel exhaust fluid (DEF) or "AdBlue" is a commonly used fluid for SCR. It consists of a solution of 32.5 wt% urea and distilled water. The solution contains 11.8 moles per liter of ammonia resulting in a 20% ammonia solution. Concentrated aqua ammonia or solid ammonia bicarbonate, carbonate, or carbamate crystals can also be used.

Recovery Of Nitrogen As Valuable Products

Burke received US patent 7,153,427 that recovers ammonia as an enriched organic compost. Burke also received US patent 7,811,455 that recovers ammonia as an ammonium bicarbonate/ carbonate solid. Burke's current patent 8,637,304 recovers the ammonia as a concentrated aqua ammonia condensate that can be used to produce ammonium bicarbonate / carbonate without the use of chemicals while cleaning biogas to produce biomethane transportation quality fuel. The process described in the 2011 US patent 8,637,304 provides an economical solution to the management of the "Nitrogen Cycle" by reclaiming or recycling ammonia nitrogen in a form that can be used in selective catalytic reduction, namely aqua ammonia, ammonia carbonate, ammonium bicarbonate, or urea. Using both reclaimed fugitive ammonia and fugitive NOx the process will cause the destruction of both while producing N_2 gas without the generation of N_2O. If the technology is adopted it has the potential to reduce reactive nitrogen discharged to the environment by 25%, partially fulfilling the National Academy of Engineering "Grand Challenges for Engineering".

Patent 8,637,304 describes the Rotating Photo Bioreactor, (RPB), a simple low kW energy process for recovering nutrients from aquatic environments or ammonia from anaerobic digestate without the use of chemicals. It is a low temperature, low pressure process that is carried out in a single vessel. The RPB was originally developed to scavenge nutrients from lakes, rivers, streams as a fixed film of autotrophic organisms that can be economically harvested by vacuum from the disk surface. The reactor has been described elsewhere http://www.extension.org/67578, (Burke, 2013). A video of its operation can be viewed at (http://www.youtube.com/watch?v=pD5MTx88FJE). Applications of the Rotating Photo Bioreactor are described in US Patent 8,895,279 and Patent Pending 14/504,078.

Ammonia Stripping

Stripping of ammonia from anaerobic digestate is carried out by first converting the ammonium (NH_4+) to ammonia gas (NH_3) followed by passing gas, deficient in ammonia, at a high rate through the liquid. The shifting of ammonium to ammonia gas is a function of turbidity, pH, and temperature as shown in Figure 1.

The cost of shifting ammonium to ammonia gas is a function of the cost required to increase the pH. Typically the shift is accomplished through the use of caustic chemicals, and / or heating the liquid substrate to high temperatures (> 70°C). Stripping ammonia is a function of temperature and relatively high gas to liquid flow rates, typically in the range of 500 to 1000. The cost of stripping ammonia gas is primarily a function of the energy required for the high volume stripping gas and the pressure required for high flow through packed media. The RPB is a high flow rate, low pressure gas exchange unit that operates at moderate temperatures (< 40°C). The process increases the pH of the liquid biologically through the use of autotrophic organisms that consume bicarbonate in the liquid thereby increasing the pH. Low pressure ammonia stripping occurs on the non submerged portion of the rotating plates as shown in Figure 2.

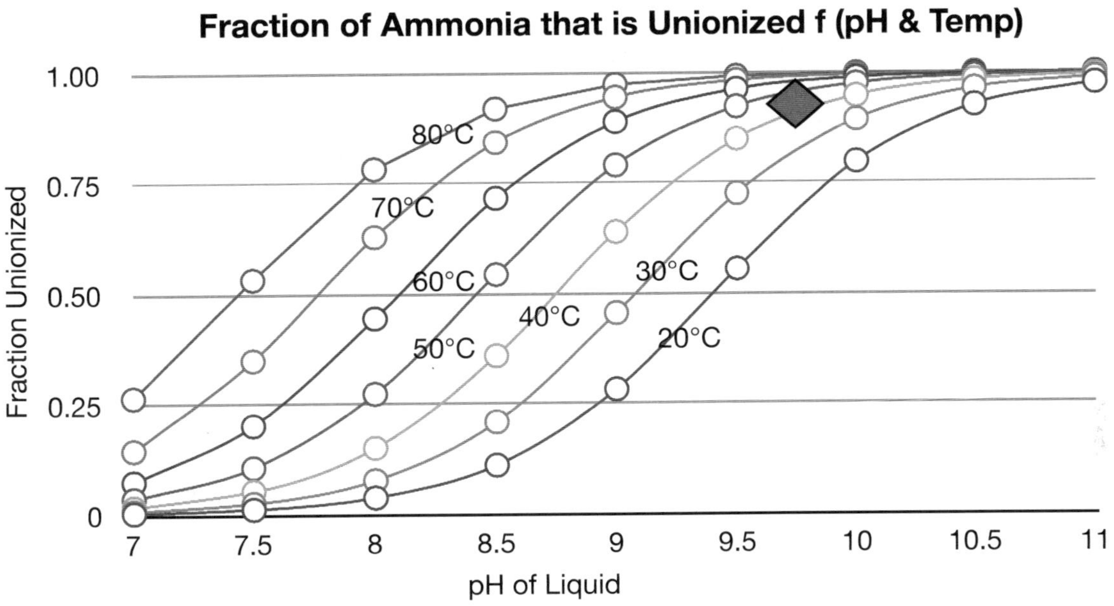

Figure 1 Fraction As Ammonia Gas (Unionized)

Stripping lowers the NH3 gas solubility while removing both ammonia and water vapor. The water vapor carries a considerable amount of heat from the stripping unit while diluting the concentration of the final product. Typically the final product of ammonia stripping is a dilute solution of ammonium sulfate or ammonium nitrite. The stripped ammonia and water vapor are condensed by cooling the gas stream. Minimizing the stripping temperature and maximizing the condensation temperature will minimize dilution of the condensate. The stripped water vapor and ammonia gas must be concentrated to produce a highly concentrated condensate for SCR. A 20% solution of aqua ammonia condensate can be used in existing SCR devices in lieu of the conventional urea solution (33% urea plus distilled water). The process concentrates the condensate through the use of several specified concentration techniques that conserve energy while producing a highly enriched condensate.

The aqua ammonia condensate can also be used to produce solid ammonium bicarbonate / carbonate as described in US patent 8,637,304 without the use of chemicals at low temperatures and pressures (Eq 2). The ammonium bicarbonate / carbonate can be used in the solid form in SCR devices as described by Fulks (Fulks et al., 2009). US patent 8,637,304 presents a method for reclaiming the ammonia nitrogen as solid ammonium bicarbonate / carbonate crystals while cleaning biogas to produce biomethane, a transportation quality fuel produced at little or no additional expense.

$$NH_3 + (H_2O + CH_4 + CO_2)_{biogas} \rightarrow CH_4 + (NH_4HCO_3)_s \qquad \text{Eq 2}$$

Reclaiming ammonia that would otherwise be discharged to the environment and using that concentrated ammonia for SCR NOx removal eliminates two significant environmental problems. The above described technologies are excellent methods of reclaiming ammonia nitrogen and producing biomethane gas and ammonium bicarbonate fertilizer.

The RPB essentially uses fixed film phototrophic organisms to consume the bicarbonate and by that consumption, increase the pH. Removal of phosphate from digestate requires a higher pH (8.5 to 10.0) and reduced bicarbonate in order to facilitate the precipitation of calcium phosphate or magnesium ammonium phosphate, commonly referred to as MAP or struvite. Struvite recovery has been promoted as the best option to recover a portion (<20%) of the nitrogen and most of the phosphate in a waste stream. However, the cost of magnesium and pH increasing chemicals such as NaOH, MgO, MgOH make the process difficult to adopt. Using a RPB will provide the means of eliminating chemicals used to increase the pH, while precipitating phosphate and reclaiming substantially all of the ammonia nitrogen.

The Rotating Photo Bioreactor Process

The Rotating Photo Bioreactor is a retained biomass reactor similar to the traditional Rotating Biological Contactor (RBC) that also incorporates a gas exchange system and source of light for phototroph organism growth. The phototrophic biofilm increase the pH by consuming bicarbonate resulting in a shift of NH4+ to NH3 that is stripped in the same reactor without the use of chemicals or heat energy. The high pH, low bicarbonate environment causes the precipitation of phosphate as a sandlike granular Tricalcium-Phosphate precipitate. Substantially all of the phosphate can be reclaimed by adding soluble calcium if sufficient calcium is not available in the substrate.

Because of the digestate's high turbidity and low light transmission the phototrophic organisms are cultured and retained on circulating plates, similar to a rotating biological contactor (RBC). The organisms are first exposed to light and then the nutrient laden substrate. At the higher pH the ammonium is converted to ammonia gas that is immediately removed, because of its high toxicity, by circulating a CO_2 deficient stripping gas between the plates in an enclosed reactor. The stripping gas continually lowers the partial pressure of ammonia over the liquid surface, thereby driving ammonia from the liquid to the stripping gas stream. As the total ammonia in the substrate decreases more ammonium is converted to ammonia gas. Over time all of the ammonium will be converted to ammonia gas and stripped from the solution.

A graphic representation of the RPB stripping unit is presented in Figure 2. As shown an ammonium rich substrate (e.g. digestate) enters the reactor, and as it flows through the reactor the pH increases from approximately 8.3 (water without CO_2) to a maximum of 10.0 causing the conversion of ammonium to ammonia gas. Stripping gas flows between the plates to remove the ammonia gas. The products are a gas rich in ammonia and an effluent containing a reduced concentration of ammonia.

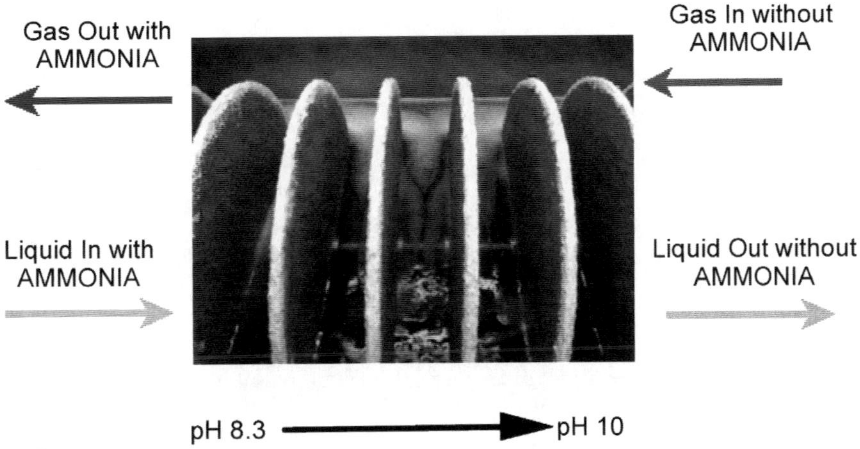

Figure 2 RPB Stripping Unit

The RPB can be operated in several configuration. The first configuration, shown in Figure 3, passes the substrate directly through the reactor. The phototrophic organisms (cyanobacteria) receive the full load of toxic ammonia which can be as high as 4,000 mg/L.

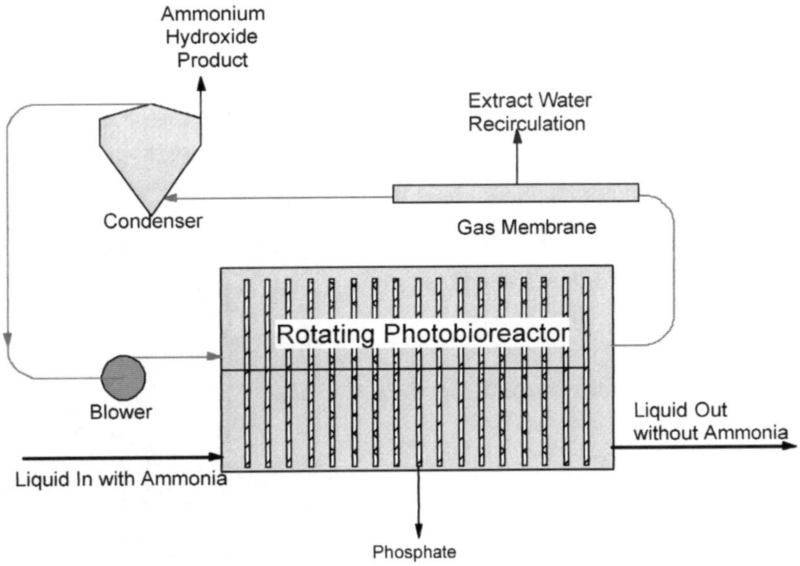

Figure 3 RPB Operating Configuration A

The reactor must rapidly remove the ammonia and thereby reduce its impact on the bicarbonate consuming organisms. Consequently, in order to function, the reactor must produce an effluent having very little ammonia. Maximum removal rates must be achieved. Maintaining the pH above 9.0 is necessary to maintain the bicarbonate consumption and process stability. In practice once the pH falls below 9.2 the ammonia load should be reduced. Removal rates at low loadings typically exceed 99%.

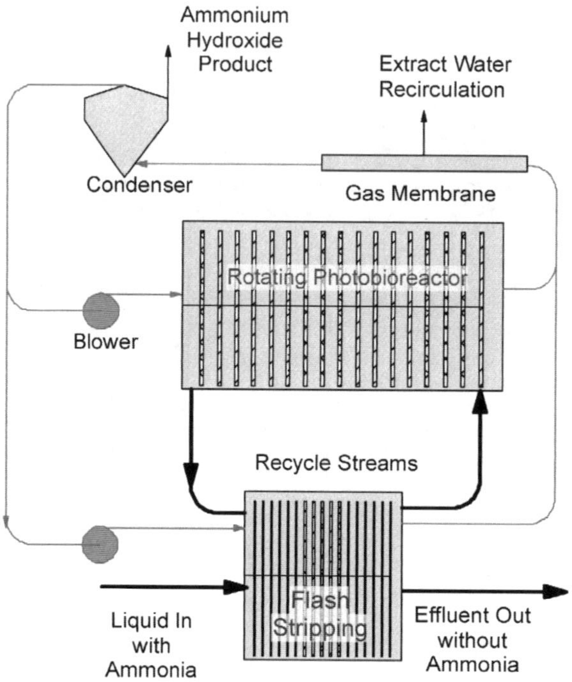

Figure 4 RPB Operating Configuration B

The second configuration uses the RPB to produce a high pH liquid stream that is recycled to a low detention time flash stripping unit that removes a bulk of the ammonia prior to recycle to the photo bioreactor. The RPB organisms are insulated from highly concentration influent stream. The recycle ratio will establish the dilution required to produce a desired effluent quality.

Figure 5 Pilot Scale operating RPB's

Nitrogen removal from the liquid substrate is related to pH, carbon dioxide concentration in the headspace gas, ammonia gas solubility, detention time, temperature, air movement over the surface of the substrate, substrate turbidity and gas humidity. The operation is a function of many variables including temperature, the Rotating Photo Bioreactor (RPB) rotation rate (rpm), surface area, reactor volume, submergence, light exposure (umole/m^2/s), hours of light exposure per day i.e, operating period (h), substrate flow rate, CO_2 content, pH, alkalinity, total ammonia and calcium concentration; recirculating gas flow rate, CO_2 content, humidity, ammonia concentration; open or covered reactor (external

CO_2). Humidity is important because as ammonia is striped water is also evaporated which establishes the maximum concentration (mg NH_3 / L water) of the product obtained. A higher humidity reduces the evaporation rate. After stripping, the gas passes through a membrane that extracts warm water thereby producing a higher concentration product that is ultimately condensed before returning to the process.

PILOT INVESTIGATIONS

Laboratory pilot investigations were conducted to verify many of the variables presented above and ultimately develop a predictive model of the RPB operating configuration A. Three RPB reactors were constructed of Plexiglas. Each reactor had a volume of 5 L. Each reactor contained four 21.6 cm disc's providing 2,930 cm^2 of surface area. The surface area to volume ratio was 0.58.

Each disk was covered with a different growth surface. Various patterns of fiberglass, carbon fibers, and rubber were used. The membrane was a silicone SSP-M823 gas permeable membrane. Light to each reactor was provided with a 5 Watt, Par 38, aquarium light with blue and white diodes. The intensity of the lights could be varied from 10% to 100% in 10% increments. The process was normally operated at 50 to 70% resulting in a light intensity of 300 to 600 umole/m^2/s. The operating temperature varied from 16° to 28° C. The RPB operated at 3.5 RPM at a 40% to 50% submergence. Operation varied from 8 to 12 hours per day. The substrate was centrate from a local municipal WWTP (LOTT, Olympia, WA). The substrate flow rate varied from 200 mL to 1.0 L /d. Average influent substrate pH and ammonia concentration were 8.12 and 2,700 mg/l as N with some values as high as 4,000 mg/L.

Figure 6 Biomass on Disk Surface
(Blue color from blue diodes)

The system was operated for four months during which time a rich cyanobacteria biofilm developed on each plate that periodically sloughed off leaving a patchy surface of old and new biofilm (Figure 6). The system was continuously operated at a pH above 9. When the pH began to fall the load was reduced.

While operating the system it was observed that a granular precipitate, having the appearance of sand, began to build up between the disk's on the bottom of each reactor (See Figure 7). At the time phosphate removal was averaging 30± mg/L as P whereas the total influent phosphate averaged 256 mg/L as P. However, the effluent was totally deficient in calcium. Further investigations established that the high pH, low bicarbonate, concentrations found in the reactor's effluent provided an ideal solution for precipitating calcium phosphate. When small quantities of calcium chloride were added to the effluent rapid precipitation of amorphous calcium phosphate occurred as shown in Figure 8. Over 90% pf the phosphate was precipitated using small doses of calcium chloride.

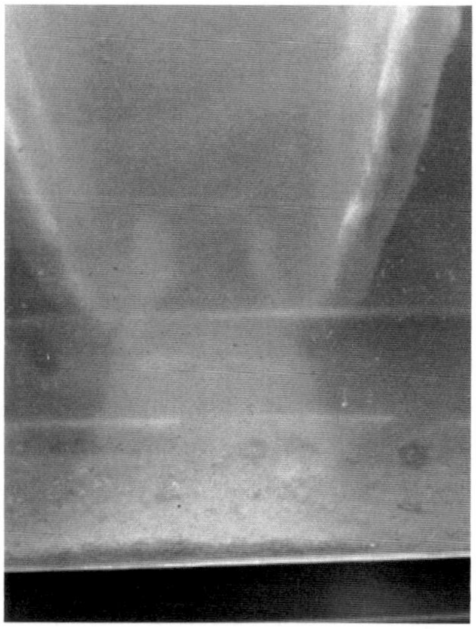

Figure 7 Phosphate Buildup Between the Disks

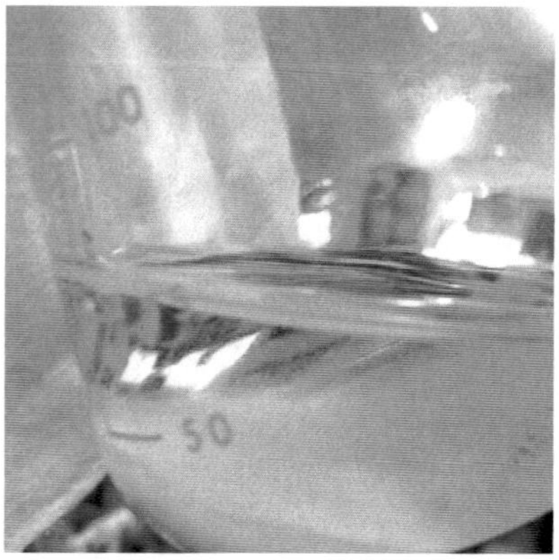

Figure 8 Decanted Calcium Phosphate Precipitate

Average results to date for 1 L /day influent flow rate using Operating Condition A (Figure 3) under low temperature conditions are presented in Figure 9. However, influent flow rates above 0.4 L/d could not be sustained because of inhibited growth that resulted in lower pH values that approached 9.0, at which point influent flow volume was reduced. Consequently, the maximum sustained loading rate achieved by the low temperature (16° to 17.5° C) operation to date was 0.16 Kg/m^3/d.

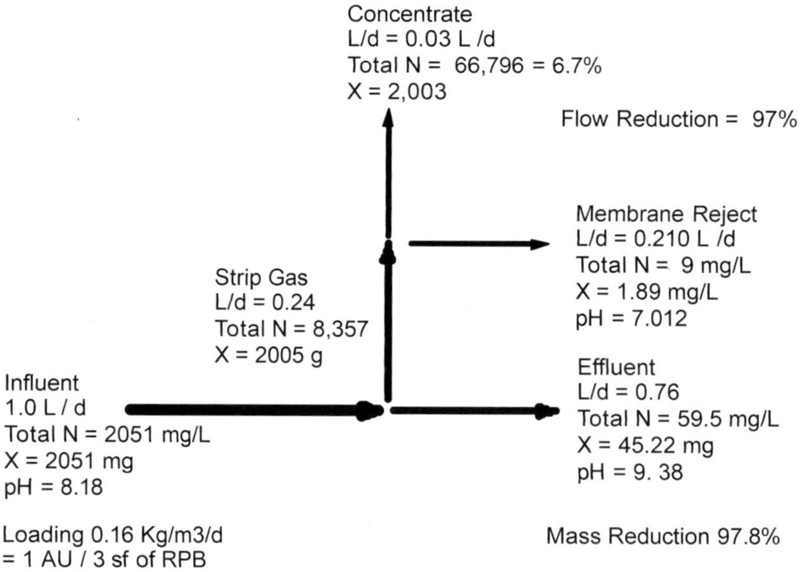

FIGURE 9 Process Operating Summary Assuming a 1 Liter Per Day Feed Rate

The high temperature operating configuration B will be presented curing the conference.

CONCLUSION

This work demonstrated the removal of substantially all of the ammonia and phosphate from a high strength anaerobic digestate stream. With the addition of calcium chloride, over 90% of the phosphate was recovered as a rapidly settling amorphous calcium phosphate. Over 98% of the ammonia was concentrated to a 6% solution without the use of chemicals in the low energy consuming process. The 20% solution concentration required for use as a diesel exhaust fluid has not been achieved to date. Several options are available to achieve the required 20% concentrations. The loading rates were lower than desired primarily due to the low temperature operation. Temperature has a significant impact on the process since the specific growth rate, bicarbonate consumption, dissociation constant, Henry's constant etc. are all strongly influenced by temperature. High ammonia gas solubilities have a significant adverse impact on the cyanobacteria used. Very low solubilities can be achieved by operating at higher temperatures with gas flow rates sufficient to significantly reduce the partial pressure of NH_3 gas. The reactors are now operating at 20°C, 30°C and 35°C using both process configuration A and process configuration B. Those operating conditions are expected to substantially increase the loading rates. This research will ultimately develop a process model of the various operating configurations over the next 12 months as well as an assessment of the economic value of the various products produced such as ammonium hydroxide concentrations, biomethane, solid ammonium bicarbonate / carbonate, calcium phosphate and MAP, aka struvite). It is anticipated that construction of a pilot scale facility will be initiated within 12 months. Figure 10 presents the overall process and the optional products produced.

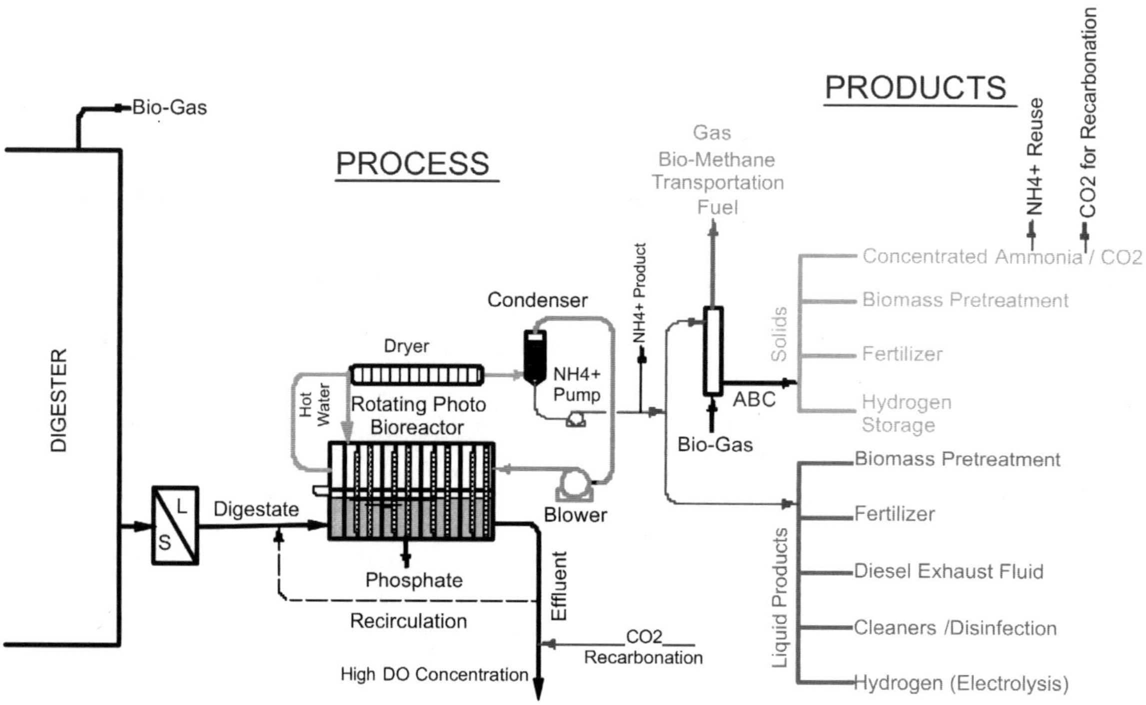

FIGURE 10 Process And Products

BIBLIOGRAPHY

Burke, D. A. (2013). The rotating photo-bioreactor as a method for removing soluble nutrients from waters. In *The rotating photo-bioreactor as a method for removing soluble nutrients from waters.Pdf* (2013 ASABE Annual International Meeting ed., Vol. 2013 ASABE Annual International Meeting). Kansas City, Missouri.

Fulks, G., Fisher, G. B., Rahmoeller, K., Wu, M. -C., DHerde, E., & Tan, J. (2009). A review of solid materials as alternative ammonia sources for lean nox reduction with SCR. *Diesel Exhaust Emission Control*, 2009-2001.

Millar, D. N., & Adam, D. D. (n.d.). *Background information MSU-EPRI N2O offset.Pdf.*

Ni, B. -J., Ruscalleda, M. L., Pellicer-Nacher, C., & Smets, B. F. (2011). Modeling nitrous oxide production during biological nitrogen removal via nitrification and denitrification: Extensions to the general ASM models. *Environmental Science & Technology*, *45*(18), 7768-7776.

SAB, E. P. A. (n.d.). *Reactive nitrogen in the united states: An analysis of inputs, flows, consequences, and management options*.

Stephanie Kato, A. R. E. (n.d.). *GAS-FIRED POWER PLANT nox EMISSION CONTROLS AND RE.Pdf.*

Sutton, M. A. (2011). *The european nitrogen assessment.* Cambridge University Press.

University, M. S. (n.d.). *Quantifying N2O emissions reductions in agricultu.Pdf.*

Ward, B. B. (2013). Oceans. How nitrogen is lost. *Science (New York, N.Y.)*, *341*(6144), 352-3. doi: 10.1126/science.1240314

Implementing EBPR at an Existing WRRF: Operational Adjustments to Low pH, Alkalinity Limitations, and Future Biosolids Resource Recovery

Leon Downing[1], Mike Young[2], Bill Tatum[2]
[1]CH2M Hill, Milwaukee, WI, leon.downing@ch2m.com
[2]Trinity River Authority of Texas, Arlington, TX

INTRODUCTION

The Trinity River Authority of Texas (TRA) Central Regional Wastewater System (CRWS) operates a 620,000 m^3/day (162 million gallon per day (mgd)) municipal wastewater treatment plant (WWTP) just outside of Dallas, Texas, USA. The activated sludge portion of the facility consists of 12 parallel aeration basins with 12 parallel rectangular secondary clarifiers. Each aeration basin is a single pass nitrified activated sludge tank, with no selector zones and an overall length to width ratio of 3:1. Past studies have focused on the quantification of simultaneous nitrification and denitrification (SND) and simultaneous nutrient removal (SNR) using process modeling (Downing and Young, 2010) and field sampling (Downing et al, 2011). Starting in 2012, TRA began full scale testing on an enhanced biological nutrient removal (EBPR) full scale demonstration basin. The investment in the full scale demonstration basin helped to confirm the feasibility of the SND focused EBPR process. In addition to confirming the base performance capabilities, the demonstration basin enabled the TRA staff to further investigate operational parameters associated with the SND and SNR processes. One of the main areas of concern was associated with operation at low pH conditions, and how alkalinity addition may have plant wide impacts associated with an upcoming biosolids improvements project.

DEMONSTRATION BASIN TESTING

In 2012, one of the 12 aeration basins at the TRA CRWS Treatment Plant was retrofitted with a single selector zone. This demonstration basin provides the ability to test nutrient removal and operational changes in a single basin receiving flow rates between 12 and 20 mgd (45 and 76 mega liters per day). Investing in this demonstration basin has provided TRA the ability to test at a full scale an innovative process configuration that relies on SND in the aeration basins as well as denitrification in the secondary clarifier sludge blankets. From a phosphorus removal perspective, effluent phosphorus has averaged 0.6 mg/L during demonstration testing. After confirming the ability to meet an effluent phosphorus concentration of less than 1 mg/L was completed, the operations staff at the TRA CRWS Treatment Plant continued to utilize the investment in the demonstration basin to test a range of operational improvements, including online sensors, alkalinity control, and low DO operation.

ALKALINITY AND LOW PH

The TRA CRWS Treatment Plant has historically operated at relatively low pH values in the aeration basins, with the activated sludge operating at a pH of between 6.5 and 6.6 on a regular basis. During 2013 and 2014, there were several periods of time where the aeration basin pH dropped below 6.5. Low pH values have been shown in impact the stability of EBPR via the selection of glycogen accumulating organisms (GOAs) rather than the beneficial phosphate accumulating organisms (PAOs)(Oehmen et al, 2005).

As part of the demonstration testing, microbial ecology was evaluated using pyrosequencing (Bibby et al, 2010). DNA was extracted from samples using a modified MoBio Powersoil (Carlsbad, CA) protocol. For 16S community analysis, PCR was performed in duplicate using 16S rRNA primers 515F and 806R (>99.9% and 98.6% bacterial coverage) including sequencing and barcoding adapters as previously described. Independent negative controls were run for each sample and primer set. PCR products were pooled, purified, and quantified prior to sequencing on an Illumina MiSeq. Sequences were analyzed within the QIIME workflow.

Previous results of this sampling related to the nitrifying community structure have been reported (Downing et al, 2014). Two sampling periods were completed relative to low pH operation. The first sampling occurred after a two week period where the pH averaged 6.61, while the second sampling event occurred after a two week period where the pH averaged 6.48. The performance of EBPR was shown to be impacted by low pH, with a higher average effluent phosphorus concentration and more instability (Figure 1a). A larger fraction of GAOs was also shown during this operational period (Figure 1B).

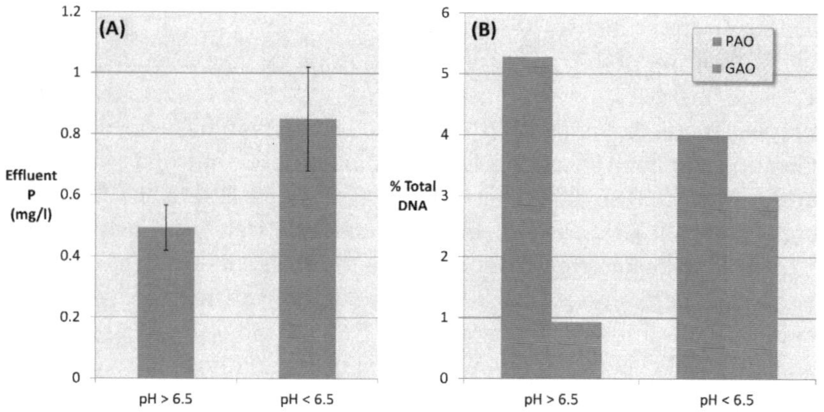

Figure 1: Impact of aeration basin pH on (A) effluent phosphorus and (B) PAO/GAO levels

ALKALINITY ADDITION AND BIOSOLIDS RESOURCE RECOVERY

In addition to implementing EBPR, the TRA CRWS Treatment Plant is also in the process of moving from a lime stabilization based biosolids processing facility to an anaerobic digestion facility with thermal hydrolysis. As the TRA staff evaluated the liquids process impacts of low pH, magnesium hydroxide addition was tested to determine the operability of a magnesium hydroxide based alkalinity control system. It was shown that addition of magnesium hydroxide to the influent increased the aeration basin alkalinity and pH, as anticipated. However, the TRA staff also questioned the potential impact of this magnesium addition to Struvite production in the future anaerobic digestion process. Additional magnesium sampling was completed with and without magnesium hydroxide addition. Simulations were completed in Biowin to examine the relative impact of the magnesium addition on available magnesium concentrations in the future anaerobic digestion complex. The available magnesium in the anaerobic digesters was dominated by the magnesium released during bio-accumulation as part of the EBPR process, and the magnesium hydroxide addition to the influent was shown to have a relatively minor impact on the Struvite production.

DISCUSSION

The TRA CRWS Treatment Plant staff has taken full advantage of their investment in a demonstration basin prior to full nutrient removal implementation. Full scale testing was able to identify low pH operation as a major impact to the reliability of biological phosphorus removal. In addition, investigation of the microbial ecology confirmed that low pH condition contributed to the selection of GAO bacteria within the system.

References

Bibby, Viau, and Peccia (2010) Pyrosequencing of the 16S rRNA gene to reveal bacterial pathogen diversity in biosolids. Water Research 44(14): 4252-4260

Downing and Young (2010) "Accounting for Nitrogen Loses in a Large-Scale Municipal WWTP" WEF/IWA Modeling Seminar 2010, St. Anne, Quebec, Canada 2010

Downing, Boksiner, Jackson, Juarez, and Young (2011) "Designing for EBPR in the Presence of SND" WEF Nutrient Recovery and Management 2011, Miami, Florida, USA 2011

Downing, Young, Cramer, Nerenberg, Bruce (2014) "Low Level DO Operation: Impacts on Energy, Nutrients, and Ecology" WEFTEC 2014, New Orleans, LA, USA 2014

Oehmen, Vives, Lu, Yuan, and Keller (2005) "The Effect of pH on the Competition Between Polyphosphate-Accumulating Organisms and Glycogen-Accumulating Organisms" Water Research 39: 3727-3737

Generating Class A Struvite at the West Boise WWTP (ID)

Jordi Figueras, P.E.
Pharmer Engineering, LLC, Boise, Idaho.
Email: jfigueras@pharmereng.com

ABSTRACT

In Idaho, NPDES permits and biosolids are regulated by USEPA Region 10. The Federal EPA has not yet identified specific definitions or regulations on struvite. The City of Boise established a goal that struvite would be treated according to the guidelines in 40 CFR 503 to be considered Class A for unrestricted use. The City defined this goal following the definitions in 40 CFR 503 part 9(w) which includes "material derived from sewage sludge". This definition may be interpreted to include struvite that is produced from the digested filtrate and/or undigested filtrate from waste activated sludge.

Laboratory and pilot tests results reveal that heating up struvite to 71-74°C (160-165°F) and maintain temperature consistently is difficult. The product temperature is not stable at this range and exposure to high temperature will cause loss of hydration and ammonia release. It is recommended to maintain heating temperature below 100°C (212°F). Multiple pilot units were tested to heat struvite to 68-74°C (155-165°F). Shear and agitation due to the configuration of the unit in addition to water vapor caused a change in material consistency from sand to a paste that is difficult to handle. A drum mixer provides the proper conditions to dry the product, maintain crystal integrity and reach the time-temperature requirements to meet Class A status.

KEYWORDS: Struvite, biosolids, 503 regulations, Class A, Multiform Harvest.

INTRODUCTION

The City of Boise is currently implementing a series of upgrades to their wastewater treatment infrastructure to meet future TP limits. As part of these upgrades, the City owns and operates a struvite production facility at their West Boise WWTP which includes five (5) Multiform Harvest Reactors designed to crystalize and harvest struvite. Granular struvite is harvested from the reactors periodically, dewatered, and distributed as a fertilizer. Struvite provides a slow release source of phosphorus, magnesium and nitrogen beneficial for plant growth.

The West Boise Struvite Production Facility is designed to receive two streams: digested filtrate and filtrate from an enhanced biological phosphorus removal treatment system (EBPR). Streams are treated in separate reactors. Digester filtrate is rich in phosphorus and ammonia. Additional chemicals are injected in the fluidized bed reactor to crystallize struvite in a controlled environment. The reactors achieve 90% phosphorus removal reducing the phosphorus loading that is recycled back to the front of the plant. Struvite is harvested from the reactors, dewatered and dried.

Struvite does not look like sewage sludge or 'biosolids.' Wastewater biosolids appear dark with a rich organic content and further degrade over time, even after being stabilized at the wastewater treatment facility. In contrast, when produced in a crystallization process, struvite looks like clean sand or semi-spherical pebbles with almost no organic material. The definition of sewage sludge, however, in 40 CFR 503 part 9(w) includes “material derived from sewage sludge”. This definition may be interpreted to include struvite that is produced from the digested filtrate that is a Class B compliant biosolids stream and an undigested filtrate from waste activated sludge. Both filtrate streams are subject to potential pathogens.

APPROACH

There is not a known existing facility currently operating in the U.S. that processes struvite to meet Class A 503 regulations. Pharmer Engineering performed an evaluation of available technologies that could potentially be used to treat the struvite produced at the West Boise Struvite Production Facility to meet 40 CFR 503 Class A requirements.

To meet Class A requirements and have unrestricted use, sludge-derived products must comply with pathogen reduction requirements in Section 503.32 and vector attraction requirements in Section 503.33. There are a number of methods to comply with Section 503.32. The West Boise Class A facility is intended to meet the requirements set forth in 503.32 - Alternative 1 - Regime A where struvite is heated in accordance with the following time–temperature equation 1:

$$D = \frac{131{,}700{,}000}{10^{0.1400t}}$$

D = time in days
t = temperature in degrees Celsius

A struvite crystal includes six water molecules. Water and ammonia will be released as the crystal is heated. If exposed to high heat the crystal can potentially turn to magnesium phosphate, a highly insoluble compound thus defeating the purpose of generating a slow release fertilizer. Therefore, during the evaluation process pilot units were limited to a maximum heat input temperature of 100°C (212°F) to maintain crystal integrity.

Plow Shear Mixer

The first unit tested was a 22-liter jacketed plow shear mixer. The unit is completely sealed and operates at 240 rpm. This unit is rated at 6.9 bar (100 psi). Approximately 9 kg (20 lbs) of struvite were used in each trial and hot water was supplied to the heating jacket at 88 °C (190 °F). Shear and moisture caused the product to change consistency from a wet sandy material to a paste similar to spackle. Struvite reached a peak temperature of 74 °C (165 °F) before it started to fluctuate. The unit was considered to not be a viable option due to the temperature fluctuation in the product and the difficulties associated with handling the treated product.

Figure 1: Plow Shear Mixer. Consistency of struvite changed from a wet sandy material to a paste similar to spackle.

Paddle Mixer

The second unit tested was a paddle mixer. This pilot unit is a horizontal cylindrical vessel with flat paddles connected to a center shaft and a thermo fluid jacket installed around the outer wall of the vessel. The unit has a capacity of 283 L (10 ft^3), operates between 6-75 rpm and it includes a 50 hp TEFC/1760 rpm/230/460V 3phase 60Hz motor. Heating fluid used is oil. The hot oil electric heater is rated at 48 kW output and operates at 9100 Lph (40 gpm). The maximum operating temperature of the oil heater is 260°C (500°F). The unit includes a temperature gauge installed through the endplate to monitor product temperature. The gauge is in contact with the product providing continuous temperature measurements. The unit includes seals rated at 5 psi.

The initial test was performed with the pilot unit operating at low rpm to reduce shearing and agitation. Struvite temperature increased steadily and ultimately it started to decline. It was observed that, similar to the plow mixer, the consistency of struvite changed from a wet sandy material to a paste similar to spackle (Figure 2), thus decreasing heat transfer efficiency and generating material handling concerns.

Figure 2: Paddle Mixer. Similar to the Plow Shear Mixer, the consistency of struvite changed from a wet sandy material to a paste similar to spackle.

The following test was an immersed approach. Struvite was placed in the vessel with a water bath at 1:1 mass ratio. The water bath was heated. The heat load associated with this method is higher than the dry heating approach given the thermal load of the water, however, heat transfer is more efficient. The struvite slurry ultimately reached 74 °C (165 °F). Class A requirements were met. However, material handling was challenging. Slurry was discharged into a drum were struvite was allowed to settle for 15 minutes. The drum was decanted and the settled material was dewatered utilizing a dewatering bag. The result was a cemented material difficult handle.

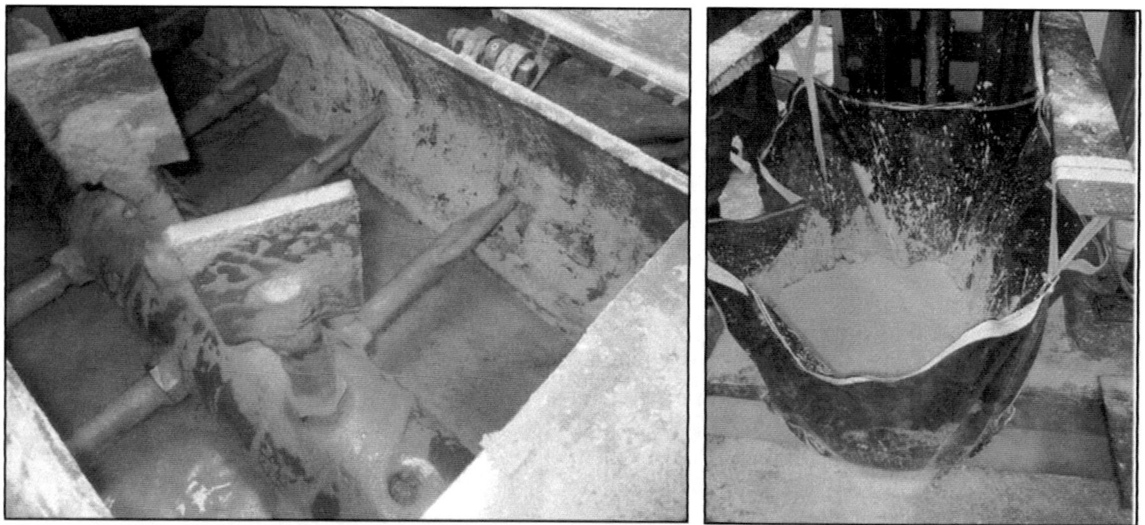

Figure 3: Struvite reached the target temperature when submerged in water. Water provided efficient heat transfer. However, material handling was challenging after meeting the time-temperature requirements.

Drum Mixer

The third pilot was a drum mixer. The pilot unit operates as a batch process in which a batch of struvite is continuously mixed in a rotating drum. The internals of the drum mixer generate a low rpm free fall type mixing providing efficient mixing an exposure of the surface area of the product to the heat source. The product is heated with direct and indirect hot air. Hot air is generated in an indirect heater with hot water coils. The heater can also use natural gas as a fuel source to heat up the air. The mixer includes load cells to automatically detect when a pre-selected batch size is loaded.

During each test, the pilot unit was initially vented to exhaust water vapor and dry the product. After a drying period the vent was closed and struvite temperature started to increase steadily while maintaining a material consistency similar to sand. Sample size varied in each test between 135 and 208 kg (300-460 lb). Results are shown in Figure 5.

Figure 4: Class A struvite is discharged from the Drum Mixer Pilot. The final product is an easy to handle clean sandy material.

PILOT RESULTS

Figure 5 shows the data obtained from four tests performed with the drum mixer pilot. It was noted in test 1 and confirmed in tests 2, 3 and 4 that beginning the test with a drying step would decrease the length of time required to reach 68°C (155 °F). Struvite drying is achieved by venting the unit during the first hour. When struvite reaches 66°C (150 °F), the rate at which temperature increases begins to decline. Heat input remains constant throughout the test. Temperature will continue to increase slowly and struvite may reach 71°C (160°F). Class A can be met at different time-temperatures based on Equation 1 (e.g. 68 °C (155°F) for 51 minutes or 71 °C (160°F) for 21 minutes). The time in which all four samples met Class A ranged from 2 hours and 17 minutes (test 2) to 3 hours and 46 minutes (test 4). The heat source in test 2 was set at 93°C (200°F) and the batch weight was 150 kg (335 lb), 60% of the rated capacity of the pilot unit. The heat source in test 4 was set at 82°C (180°F) and the batch weight was 460 lb (85% of the rated capacity of the unit).

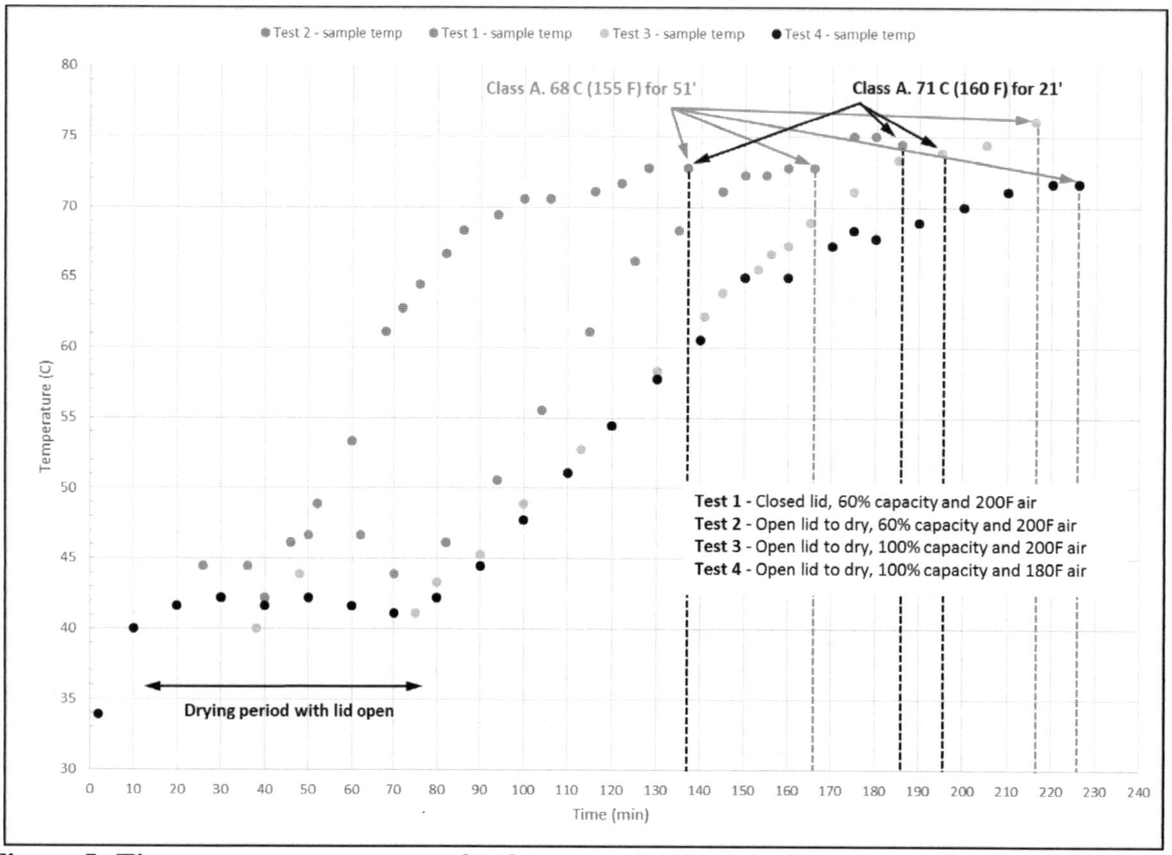

Figure 5: Time–temperature curves for four samples tested with a drum dryer. All four tests met EPA 503 Class A requirements.

FACILITY DESIGN

The Class A Struvite facility includes one 5,500 lb rated drum mixer decoupled from the struvite production facility. A 2,800 L (110 ft^3) hopper downstream of the existing struvite drying conveyors allows the operator to continue to harvest struvite from the Multiform Harvest reactors at steady state and either load a batch into the drum dryer to treat for Class A or bypass and store the harvested product. A bi-directional cross conveyor installed below the existing drying conveyors sends dried struvite to the hopper or the bypass. A second screw conveyor loads the struvite from the hopper into the drum mixer. Instrumentation built into the hopper and drum mixer provides full automation of the treatment process.

The facility design criteria is as follows:

Parameter	Value
Struvite production	8,000 lb/day
Struvite target temperature	155 °F
Expected batch treatment time	4-6 hours
Batch size	≤4,000 lb
Drum dryer capacity	5,500 lb
Heat source	Hot water, dual coil
Hot air temperature	200 °F
# of units	1
Storage prior to heat treatment	80-110 ft^3 hopper
Conveyance to drum dryer	Shafted screw conveyor
Class A Struvite storage method	2,000 supersack

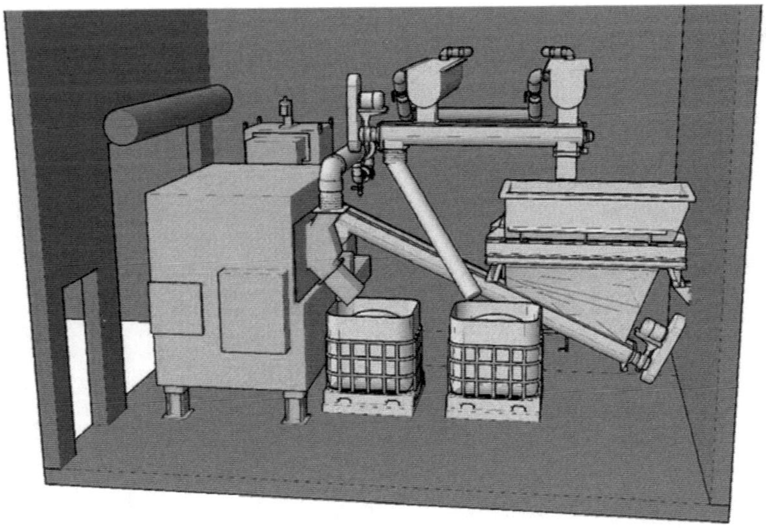

Figure 6: Class A Struvite Facility Layout

SUMMARY AND CONCLUSIONS

Pilot results show a drum mixer is a successful application to treat harvested struvite from Multiform Harvest reactors to meet 40 CFR 503 Class A requirements. Struvite crystals are sensitive to heat and exposure to high temperature will cause loss of hydration and ammonia release, eventually converting to magnesium phosphate, a highly insoluble compound. Implementing a heating approach to meet EPA 503.32 Class A requirements requires careful consideration when selecting the heat input temperature. Previous pilot systems induced a high level of agitation into the product, and combined with water vapor caused a change in material consistency from sand to a paste that is difficult to handle. A drum mixer provides the proper conditions to dry the product, maintain crystal integrity and reach the time-temperature requirements to meet Class A. A full scale unit is currently under construction at the West Boise WWTP and will be operational in July 2015.

Troubleshooting Activated Sludge Performance through Microscopic Observations: Five Case Studies

Authors: Paul Pitt[1], Sarah Galst[2], Mike Lynch[2], Bryan Atieh[2], Alonso Griborio[3]
[1] Hazen & Sawyer, 201 Mission Street, Floor 12, San Francisco, CA 94105
[2] Hazen & Sawyer, 498 Seventh Avenue, New York, NY 10018
[3] Hazen & Sawyer, 4000 Hollywood Blvd., Hollywood, FL 33021

ABSTRACT

Plant operators rely on many tools to observe and evaluate the Wastewater Treatment Plant (WWTP) process. For example, a laboratory analysis showing increasing effluent ammonia may lead an operator to increase the aerobic volume used in treatment. Or a hand-held Dissolved Oxygen (DO) meter showing low DO concentrations may inform an operator to increase the aeration output to the aeration tanks (ATs).

Another valuable tool for operational staff is microscopic analysis of the activated sludge. Routine microscopic analysis allows a continuous database of the "health" of the activated sludge to be determined and acts as an early warning of any potential settling (bulking) and/or foaming problems.

Filamentous bulking is due to the dominance of filamentous organisms, which interfere with the settling properties of the activated sludge. Activated sludge foaming is often caused by the presence of filamentous bacteria, like *Nocardioforms,* which are hydrophobic in nature and interact with gas bubbles, promoting free-form filament attachment to these bubbles in the bulk solution, which rise to the surface as foam.

Five case studies are presented, demonstrating examples where microscopic analysis provided identification of filamentous microorganism populations and associated corrective action.

INTRODUCTION

Activated sludge bulking and foaming is caused by excessive amounts of filamentous organisms in the mixed liquor. Operational conditions that can promote the growth of filamentous organisms include nutrient deficient conditions, high unaerated aeration basin fractions, long solids residence time (SRT), low dissolved oxygen (DO) concentrations, septicity and low pH. Nuisance foaming events are often related to long SRT's, physical trapping and recycling of surface foam as well as the presence of surfactants (natural or synthetic).

Continuous monitoring of the activated sludge for filamentous organisms can help with detection and mitigation of operational issues. Microscopic analysis consists of activated sludge mixed liquor (MLSS) and foam sample collection, followed by microscopic analyses. Microscopic analyses include staining techniques along with cell morphology to distinguish between the different classes of filamentous organisms. Semi-quantitative measurements of filamentous organism abundance are determined. Protocols for microscopic methods are outlined in the "Manual on the Causes and Controls of Activated Sludge Bulking, Foaming and Other Solids Separation Problems" (Jenkins, et al, 2004).

Molecular techniques like fluorescence in situ hybridization and/or quantitative polymerase chain reaction techniques can also be used for resolution of the specific species of organisms as well as quantitative measures of the abundance of the different species; however, these molecular methods often require specialized equipment and protocols for monitoring and can be expensive to perform on a routine basis. Conversely, microscopic analyses are relatively simple to implement and have proven to be a powerful tool for diagnosing the health of activated sludge systems.

Five case studies using microscopic analysis of the process activated sludge are presented.

RESULTS AND DISCUSSION

Plant A

Plant A is a 25 million gallons per day (mgd) plant located in the Northeast, operating a step-feed Biological Nitrogen Removal (BNR) process with effluent Total Nitrogen (TN) concentrations around 3-4 mgN/L, with the use of supplemental carbon in the form of methanol. Microscopic analysis of the MLSS was requested by operational staff when a decline in settling was observed in April/May 2013. Sludge Volume Indices (SVIs) showed a dramatic increase as shown in **Figure 1**.

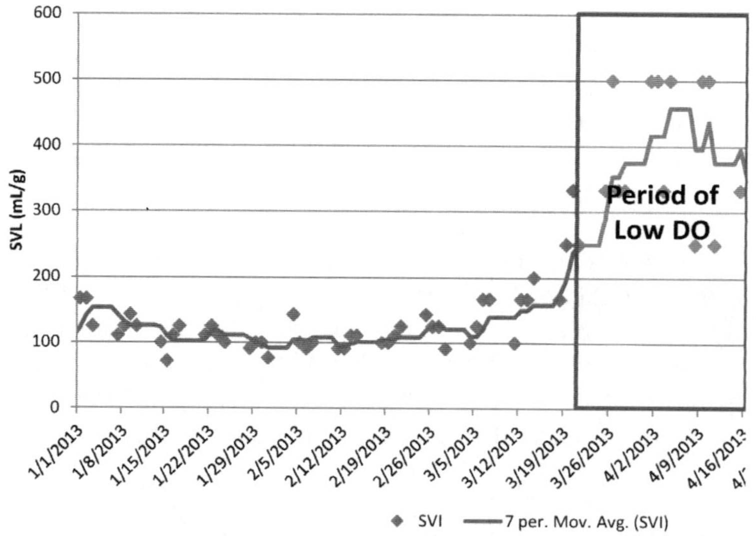

Figure 1: Increased SVIs Observed at Low DO Conditions

Microscopic analysis showed excessive amounts of *Microthrix Parvicella*, as shown in **Figure 2**. *Microthrix* was dominant both inside and outside the activated sludge floc.

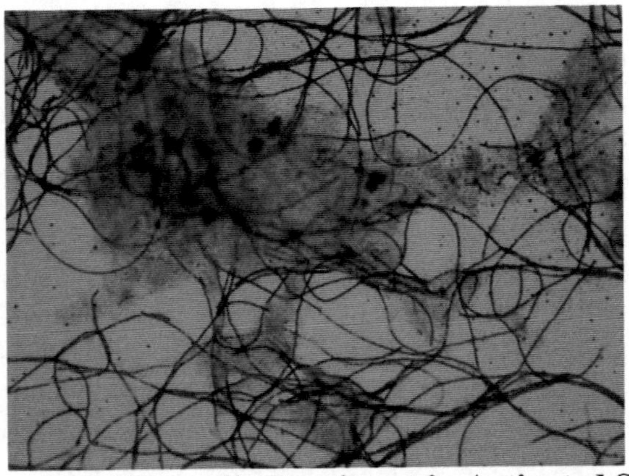

Figure 2: *Microthrix Parvicella* Dominant in Activated Sludge Sample

Recommendations for process changes to reduce *Microthrix* levels were provided and included operating at higher DO concentrations, lowering SRT, reducing anoxic volume, and initiating RAS chlorination. During this period it was determined that very low DO conditions had occurred in the aeration basin – increased unaerated volume will favor the growth of *Microthrix Parvicella*. Plant staff initiated RAS chlorination and increased aeration, resulting in an immediate increase in DO concentration with a corresponding decrease in SVI, and improvement in settling as shown in **Figure 3**.

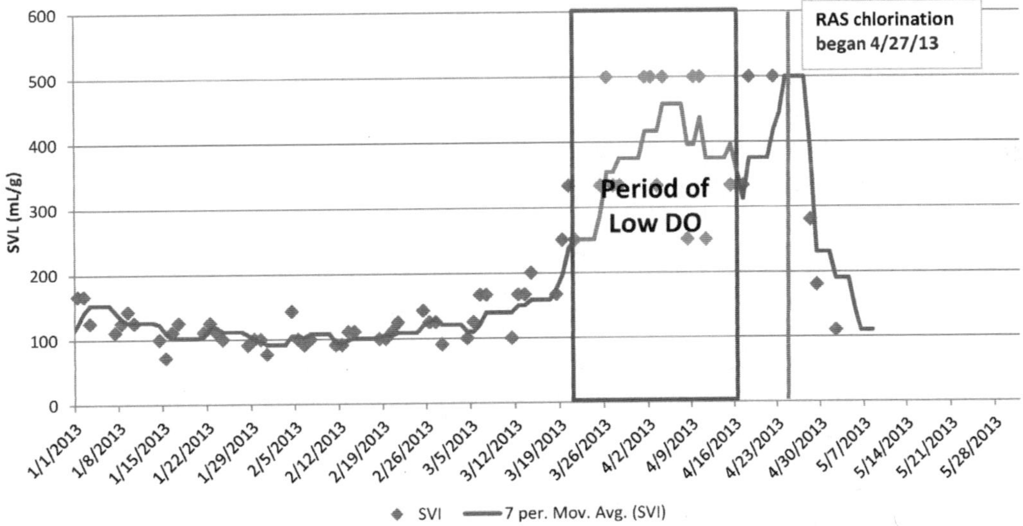

Figure 3: Improvement in SVI with the use of RAS Chlorination and Increased Aeration

Plant B

Plant B is a 120 mgd plant located in the Northeast, operating a step-feed secondary treatment process with typical effluent TSS and CBOD concentrations less than 15 mg/L. An increase in effluent TSS concentrations was observed in late 2013/early 2014, triggering an analysis of various plant parameters, including microscopic analysis of the activated sludge.

Microscopy revealed that dominant levels of the filamentous organisms *Sphaerotilus natans* and *Nostocoida Limicola II* were present in the MLSS, as shown in **Figure 4**. At the levels observed, significant interference with the settling properties of the activated sludge were occurring, with SVI values in the 300-600 ml/gm range. These high SVI values were directly correlated to the poorer effluent quality (**Figure 5**).

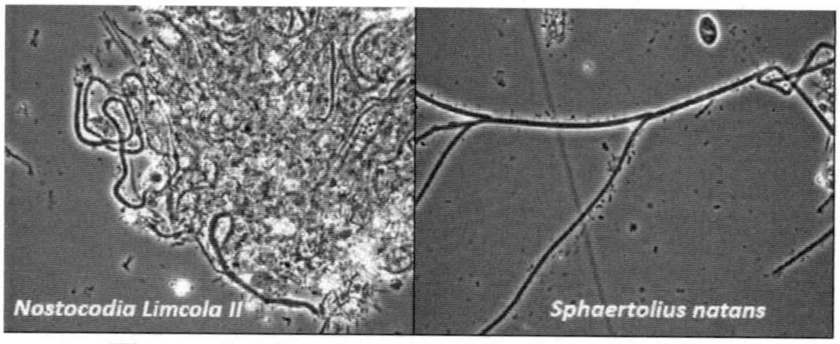

Figure 4: Microscopic Analysis of Plant B

The filaments identified represented a combination of low DO filaments (*Sphaerotilus natans*) and filaments present due to soluble organic matter in the influent wastewater (*Nostocoida Limicola II*). Control strategies recommended included increased DO levels, RAS chlorination and polymer addition. **Figure 5** shows the impact of the presence of *Sphaerotilus natans* and *Nostocoida Limicola II* on effluent quality and the immediate improvement once DO concentrations were increased.

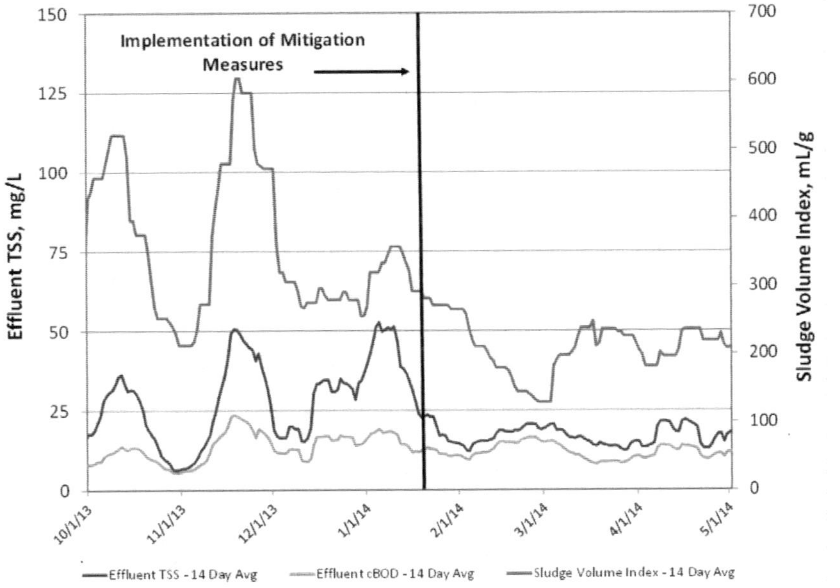

Figure 5: Improvement in Effluent Quality with Increased DO Concentrations

Plant C

Plant C is a 26 mgd plant located in the Southeast, which was originally designed for secondary standards consisting of 3 distinct activated sludge trains using diffused air, jet aeration, and mechanical aeration. The plant was faced with new limits for total nitrogen and phosphorus. The limits for total nitrogen required a step down over time, (including during construction) from 12 mg/L to 3 mg/L TN. **Figure 6** shows an aerial of the plant. **Figures 7** through **9** show the mechanical, jet and diffused aeration, respectively.

To address the near term standard TN of 12 mg/L and 8 mg/L, the plant implemented cycling of the mechanical and jet aerators to achieve simultaneous nitrification/denitrification (SNDN). Highly variable and high SVI conditions occurred on the jet aeration and mechanical aeration trains. Microscopy was used at this plant to troubleshoot the high SVI issues and monitor improvements.

Figure 6: Plant C aerial prior to construction

Figure 7: Plant C Mechanical Aeration Train

Figure 8: Plant C Jet Aeration Train

Figure 9: Plant C MLE Diffuse Aeration Train

Microscopic analysis of both the jet aeration train and mechanical aeration train indicated low DO filaments as the primary cause of the poor SVI **(Figures 10** and **11),** so strategies to improve the DO conditions in the activated sludge were undertaken. An alternative approach to cycling the aerators, utilizing the calibrated BioWin process model, for SNDN was developed to reduce extended low DO periods at the front of the mechanical aeration train. In addition, supplemental aerators were installed on the jet aeration train to increase DO levels and allow cycling of the aerators to promote SNDN. Finally, the calibrated BioWin process model was used to optimize the flow split to each train. The results for the SVI before and after the modifications are shown in **Figure 12**.

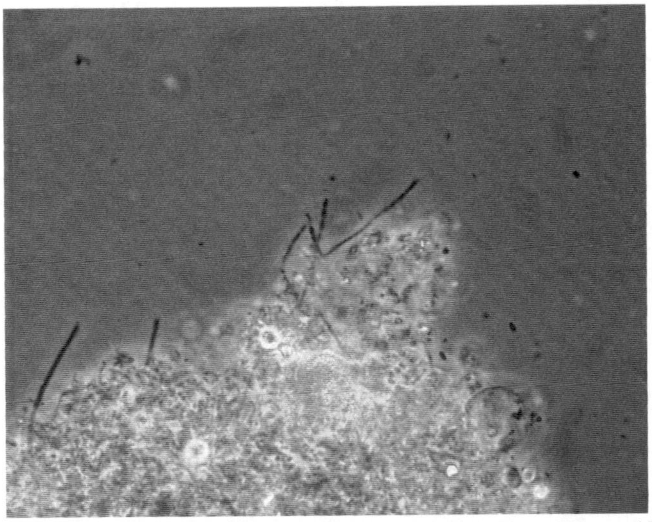

Figure 10: Microscopic Analysis of Plant C – Type 1701

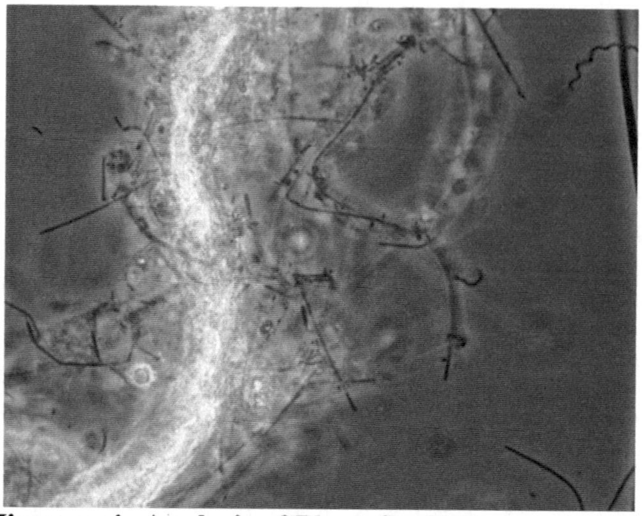

Figure 11: Microscopic Analysis of Plant C – H Hydrossis and Spirochetes

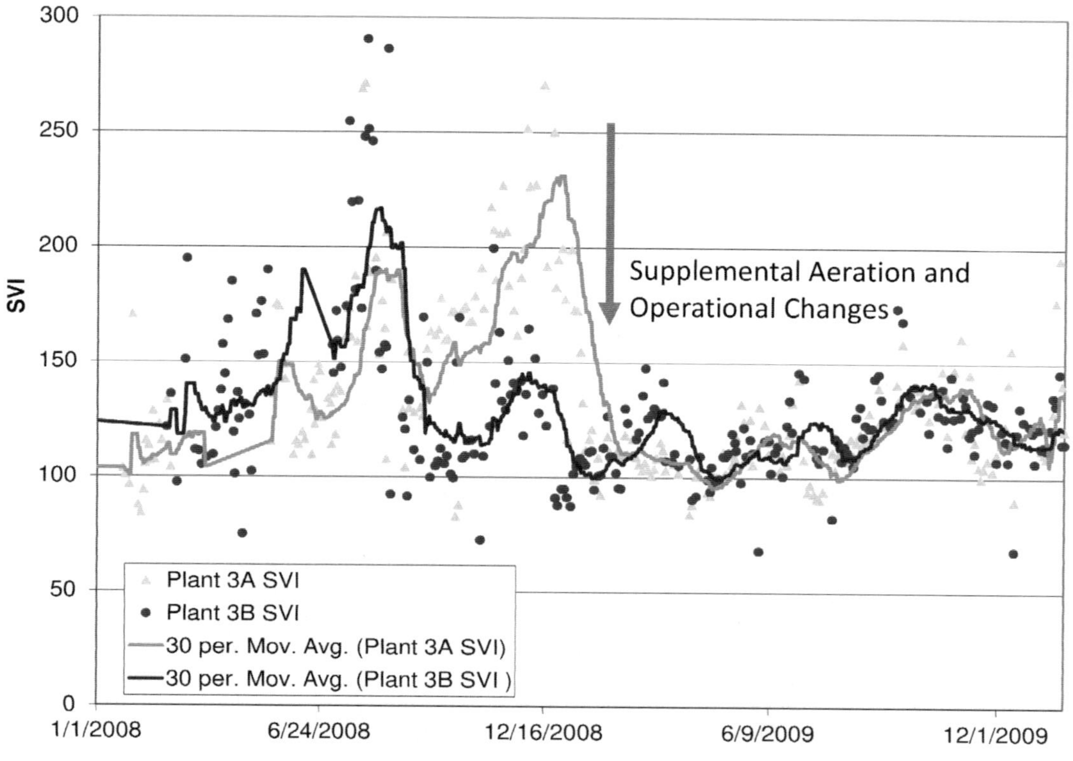

Figure 12: Plant C SVI Before and After Operational Improvements

The upgrades are now completed with all trains converted to 4-stage BNR with diffused aeration. The new basins included staged anoxic zones and aerobic zones with zone DO control. The SVI results for the new train have been excellent as shown in **Figure 13**.

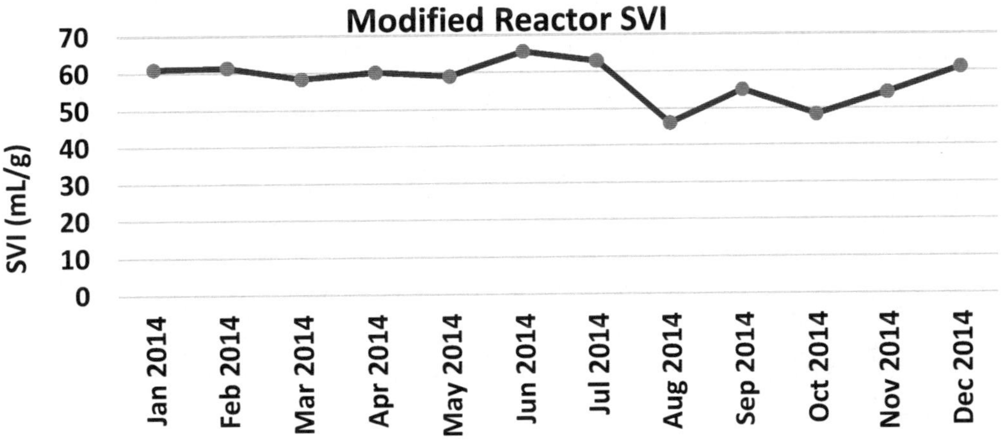

Figure 13: Plant C SVI After Completion of BNR Improvements

Plant D
Plant D is a facility composed of five individual activated sludge modules and rated for 95 mgd. Modules A, B and D have mechanical surface aerators while Modules C and E utilize fine bubble aeration. Module E has an anaerobic selector; the implementation of an Anoxic Selector was tested during a BNR pilot study in Module C. **Figure 14** shows an aerial view of the facility, and **Figure 15** shows a closer look at Module E.

Figure 14: Plant D

Figure 15: Plant D Module E (selector zone shown in red rectangle)

Table 1 shows the SVI values measured for all modules during the course of the pilot study (3/25/08 to 7/22/08). The data presented indicates considerably better settling for Modules C and E compared to Modules A, B and D. The average SVIs for Module C and E are 75 and 76 mL/g, respectively, while SVIs for the rest of the Modules are over 200 mL/g. **Figure 16** shows the SVI data during the BNR pilot study (for all modules).

Table 1. Plant D SVI Data.

Statistical Parameter	SVI (mL/g)				
	Mod A	Mod B	Mod C	Mod D	Mod E
Average	212	227	75	239	76
Max	489	891	137	538	113
Min	83	71	43	99	28
90th Perc.	311	320	97	316	91
95th Perc.	328	345	110	374	97

Modules A, B and D typically contained type 1701 (low DO organism) and type 021N and *Thiothrix* (presence of soluble influent organic matter) **Figure 18**.

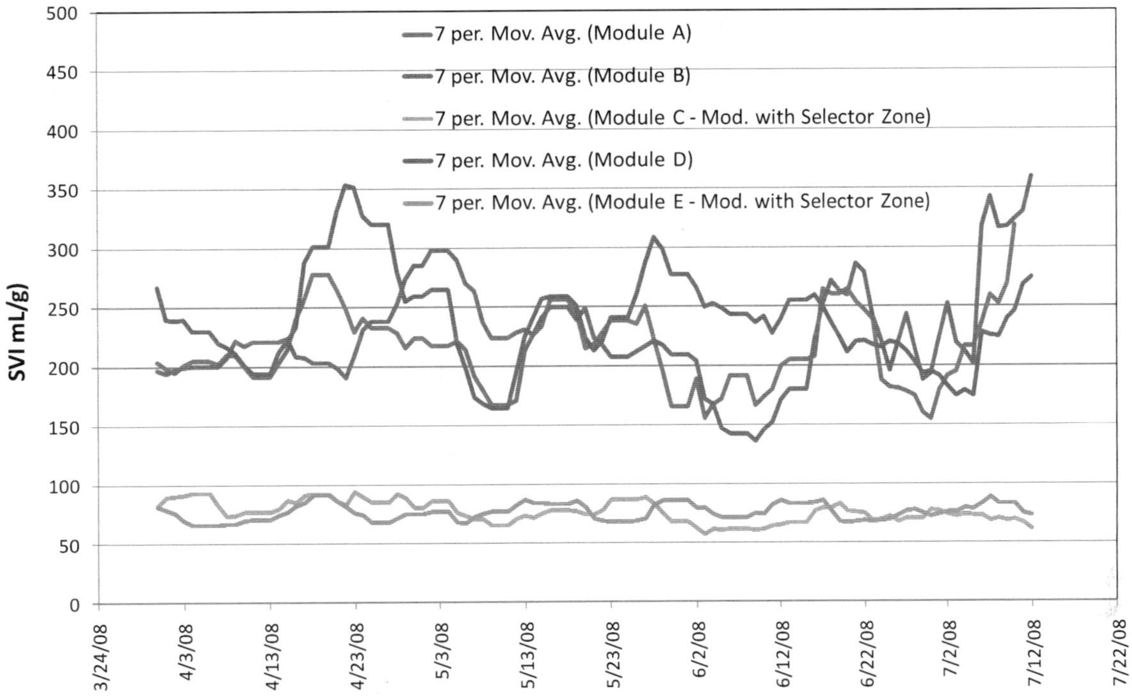

Figure 16: SVI Data during Plant D BNR Pilot Study

During the course of the pilot study, samples of the MLSS from the modules were analyzed with modules C and E containing strong, firm flocs (**Figure 17**) and no significant filamentous organism population. Both modules also revealed the presence of Poly-phosphate accumulating organisms (PAO's), which help produce a dense well-settling activated sludge floc.

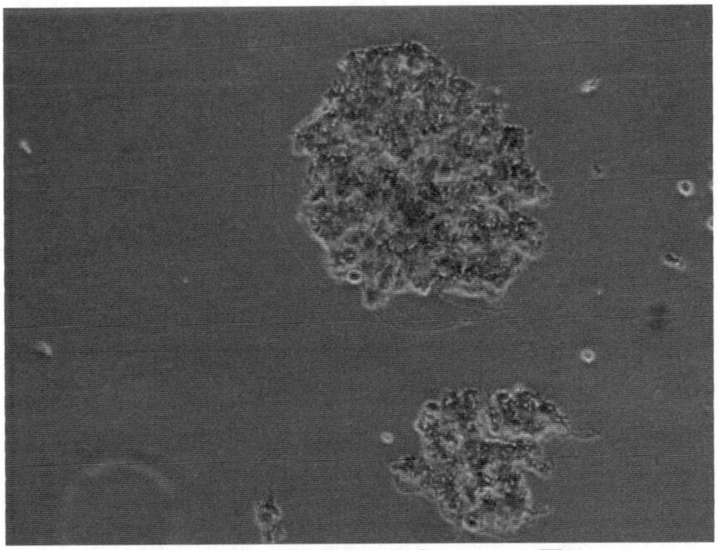

Figure 17: Module C Compact Floc

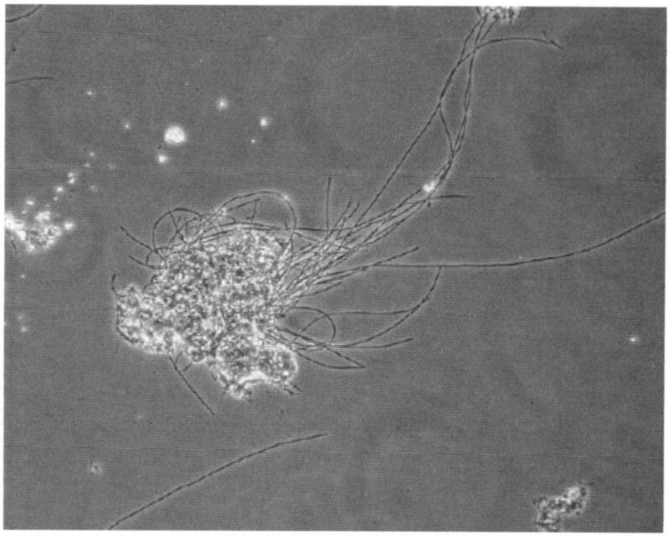

Figure 18: Module B Floc with Filamentous Bacteria

Table 2 compares the SVI values for Module C before and after the implementation of the anoxic selector. While the difference between Modules C and E compared to Modules A, B and D was associated with the use of mechanical surface aerators versus the use of fine bubble diffused aeration, the SVI values presented in **Table 2** and **Figure 19** confirm that the anoxic selector zone considerably improved the settleability of the activated sludge in Module C.

Table 2. Module C SVI Data.

Statistical Parameter	**SVI (mL/g)**	
	Module C Historical Without Selector Zone	**Module C With Selector Zone**
Period of Record	8/1/07 to 2/29/08	3/25/08 to 7/22/08
Average	141	75
Maximum	635	137
Minimum	53	43
90th Percentile	221	97
95th Percentile	270	110

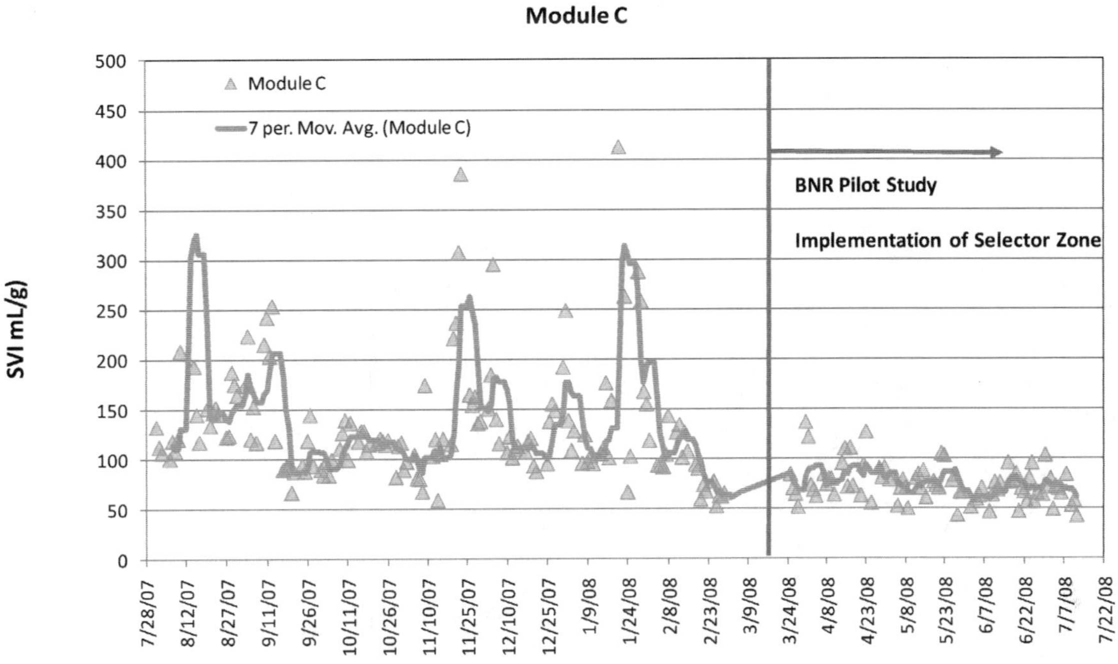

Figure 19: Module C SVI Data with and without Anoxic Selector

Plant E

Plant E is a 6 mgd advanced wastewater treatment plant that incorporates four Eimco Carrousel® A2C™ systems, which consist of an oxidation ditch preceded by anaerobic and anoxic zones. Nitrified recycle is returned from the oxidation ditch to the anoxic zone via an adjustable gate. There are no primary clarifiers at this facility.

Effluent total phosphorus (TP) concentrations are limited by Total Maximum Daily Loads (TMDLs) corresponding to a maximum 12-month rolling average effluent TP concentration of 0.42 mg/L and a maximum month effluent TP concentration of 0.83 mg/L at design flow.

The plant historically met the required TMDLs through 2004. However, the monthly and 12-month rolling average TMDLs were exceeded during most of 2005 and 2006. **Figure 20** presents the monthly effluent TP concentration from the past several years.

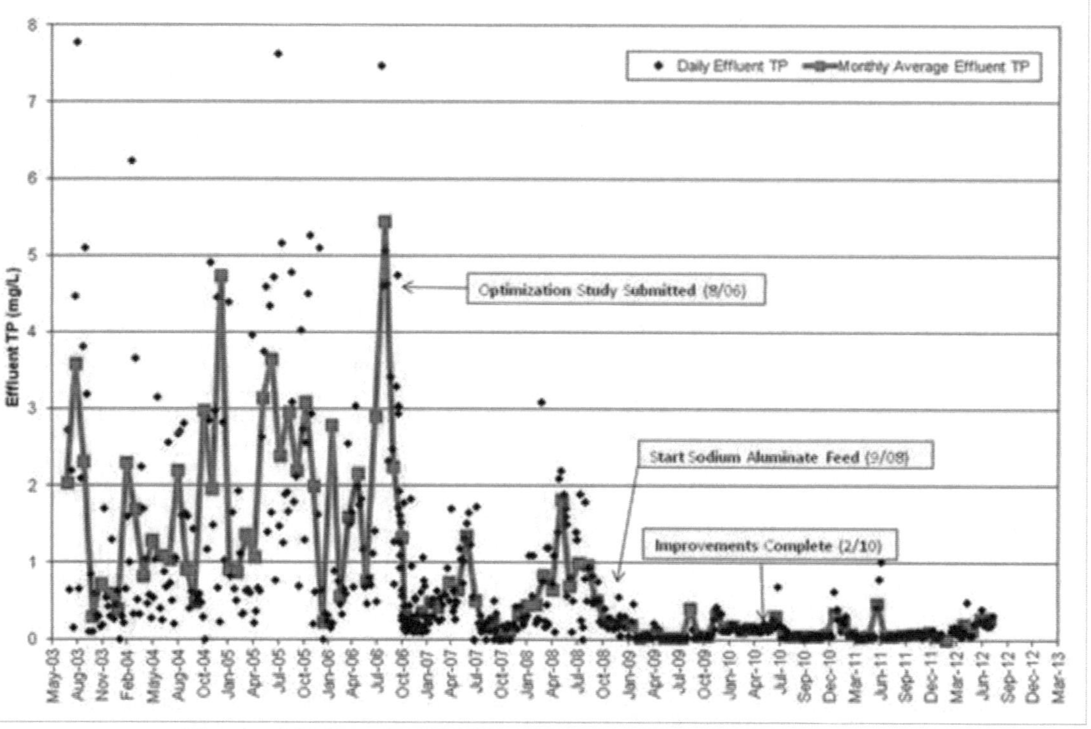

Figure 20 Plant E WWTP Effluent TP Concentrations

Multiple investigations were performed to identify the causes for the poor Biological Phosphorus Removal (BPR) performance. Microscopic analysis of the MLSS was used to identify microbial populations and indicated many small, Neisser positive cell clumps of PAOs. Filamentous organisms associated with long SRT and low DO concentrations were also observed. Glycogen accumulating organisms (GAOs) and *fungi* were observed during a period of BPR upset.

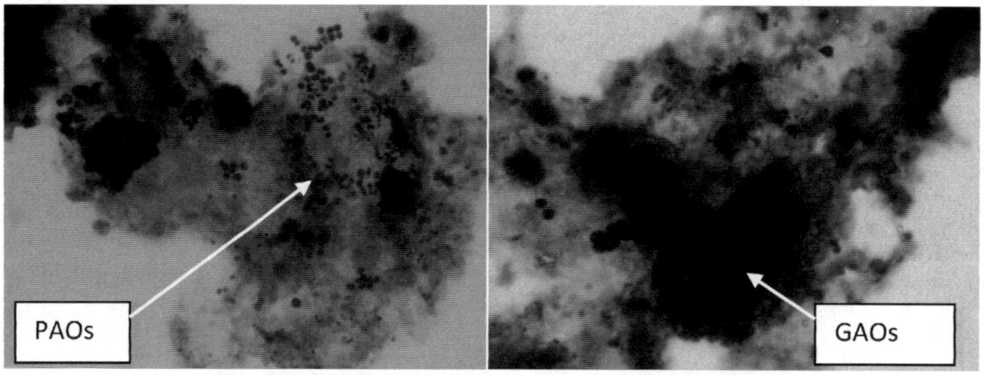

Figure 21 Plant E Microscopic Evaluation

Fungi are typically associated with operation at low pH (less than 6.5). The plant effluent data indicated effluent pH between 7.0 to 7.3, and low pH was not originally identified as a cause of reduced BPR reliability. Further investigation revealed that effluent and MLSS samples were typically collected and taken back to the laboratory for pH analysis, where there was an opportunity for the samples to re-equilibrate and increase pH. In-situ pH monitoring was performed once *fungi* was identified in the MLSS, and aerobic basin pH was measured at approximately 6.0. The low aeration basin pH was due, in part, to the feed of alum to provide chemical trim of phosphorus. The facility switched from using alum to sodium aluminate, which adds alkalinity. The pH increased through the treatment process after this change, and BPR operations stabilized, as indicated in **Figure 20**.

Biological phosphorus removal efficiency is limited at low pH , and operation at values less than pH 7 reduces phosphorus removal efficiency. GAOs were found to take up acetate faster than PAOs when anaerobic zone pH was less than 7.3, resulting in reduced PAO proliferation and GAO predominance at lower pH values. PAOs were found to consume acetate at a greater rate than GAOs at anaerobic zone pH greater than 7.5 (Filipe, et al. 2001a). Phosphate uptake is also pH dependent. It is necessary to maintain a pH greater than 7.0 in the aerobic zone since PAOs are sensitive to the pH of the aerobic zones, but GAOs are not (Filipe, et al 2001b).

Summary

The use of microscopic techniques for the analysis of activated sludge MLSS is an important tool for the efficient operation of the activated sludge process. Five examples of common operational problems due to the proliferation of activated sludge filamentous organisms are presented along with the causative organisms and the remedial actions taken.

References

Filipe, C.D.M.; Daigger, G.T.; and Grady, L., Jr. (2001a) pH as a Key Factor in the Competition Between Glycogen-Accumulating Organisms and Phosphorus-Accumulating Organisms. *Water Environ. Res.*, **73**, 223-232

Filipe, C.D.M.; Daigger, G.T.; and Grady, L., Jr. (2001b) Effects of pH on the Rates of Aerobic Metabolism of Phosphate-Accumulating and Glycogen-Accumulating Organisms. *Water Environ. Res.*, **73**, 213-222

Jenkins, D; Richard, M.G.: and Daigger, G.T. Manual on the Causes and Control of Activated Sludge Bulking, Foaming and Other Solids Separation Problems, 3rd Edition. Lewis Publishers 2004.

FINE TUNING YOUR BIOLOGICAL NUTRIENT REMOVAL SYSTEM

John Gallegos[1*]

[1]CDM Smith
[*]gallegosjn@cdmsmith.com

ABSTRACT
As reclaimed water is being utilized and planned to be utilized as an important commodity to assist with California's drought conditions, wastewater treatment plants, particularly in Southern California, are being faced with present and future stringent effluent requirements. Of particular notice, nitrogen reductions in reclaimed water effluent. This paper will explain how CDM Smith met these challenges at one of the wastewater treatment plants that was designed, constructed and eventually operated by CDM Smith under a design, build contract with the eventual operation and maintenance component that was later added by the client.

KEYWORDS: Total nitrogen, ammonia (NH_3), nitrate (NO_3), nitrite (NO_2), total kjeldahl nitrogen (TKN), biological nutrient removal and sequencing batch reactor.

INTRODUCTION
The South Regional Tertiary Treatment Plant (SRTTP) at the Marine Corp Base at Camp Pendleton (MCBCP) has been providing reclaimed water under very stringent effluent requirements with a maximum instantaneous total nitrogen limit of 5 mg/L that was further reduced to a maximum instantaneous TN limit of 4.2 mg/L. The SRTTP's Master Reclaimation Permit (MRP) defined TN as $NH_3 + NO_3 + NO_2 +$ organic nitrogen (organic nitrogen = NH_3 + TKN). For the past 24 months the SRTTP has been averaging a TN concentration of 2.0 mg/L using sequencing batch reactor (SBR) technology followed by tertiary filtration. The first part of this abstract will discuss how the SRTTP successfully attained effluent compliance with these restrictive TN effluent limits without an external carbon source. Staff eventually reduced the TN concentrations to an average of 1.7 mg/L after making process changes and introducing methanol as an external carbon source after a recent plant expansion. The SRTTP team, consisting of operations and maintenance staff and process engineers, has fine-tuned the biological nutrient removal system to provide an effluent quality that is below traditional advanced wastewater treatment plant standards of 3 mg/L TN with an external carbon source. This presentation will discuss the steps taken by the SRTTP team to accomplish the operational control strategy to provide uninterrupted reclaimed water through optimization of the treatment process with on-line instrumentation monitoring, laboratory analysis and system tracking of key process components and staff training. This operational control strategy led to significant process control adjustments by the operations staff and fine tuning the Biological Nutrient Removal System (BNR) for maximum performance.

ON LINE INSTRUMENT MONITORING

Extensive on-line instrumentation units were installed to improve the existing performance of the plant's nitrogen removal. The system tracks the concentrations of dissolved oxygen, ammonia, and nitrate during all stages of the SBR cycles, which include anoxic, aerobic, settle and decant phases in each Sequencing Batch Reactor's (SBR) timed cycle.

The monitoring program included the use of online ammonia and nitrate analyzers, monitored in conjunction with dissolved oxygen (DO) concentrations in the cycles, to determine cycle times for the optimum nitrogen removal. The DO meters also provide feedback signal for oxygen control of the aeration blowers to an operator adjusted DO set point that starts and stops the blowers and adjusts the blower motor speed. Air flow meters are also provided for monitoring of the aeration cycle. Figure 1 shows a typical DO profile with cycles indicated for one SBR. Figure 2, on the following page, shows the ammonia and nitrate response to changes in the DO throughout an SBR cycle. The use of these trend charts was a key first step in providing real time process analysis that helped the SRTTP staff with optimizing cycle times to improve total nitrogen removal.

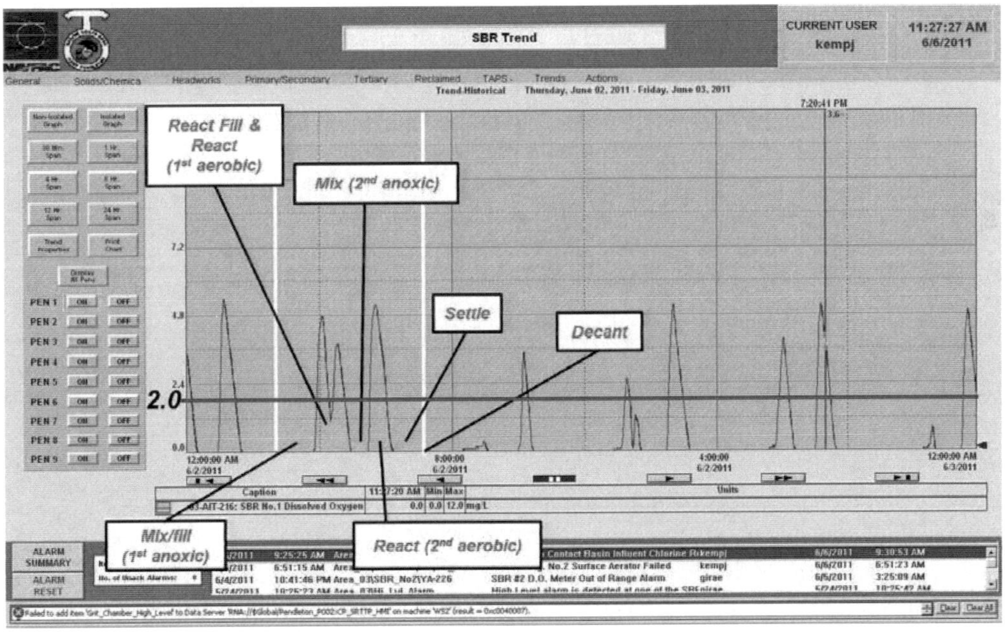

Figure 1 – DO levels (blue pen) in SBR cycle illustrating anoxic, aerobic, settling and decant stages. Note the air peaks, above the DO set points (horizontal red line). Over shooting was later corrected by staff through air control optimization of the computer controls.

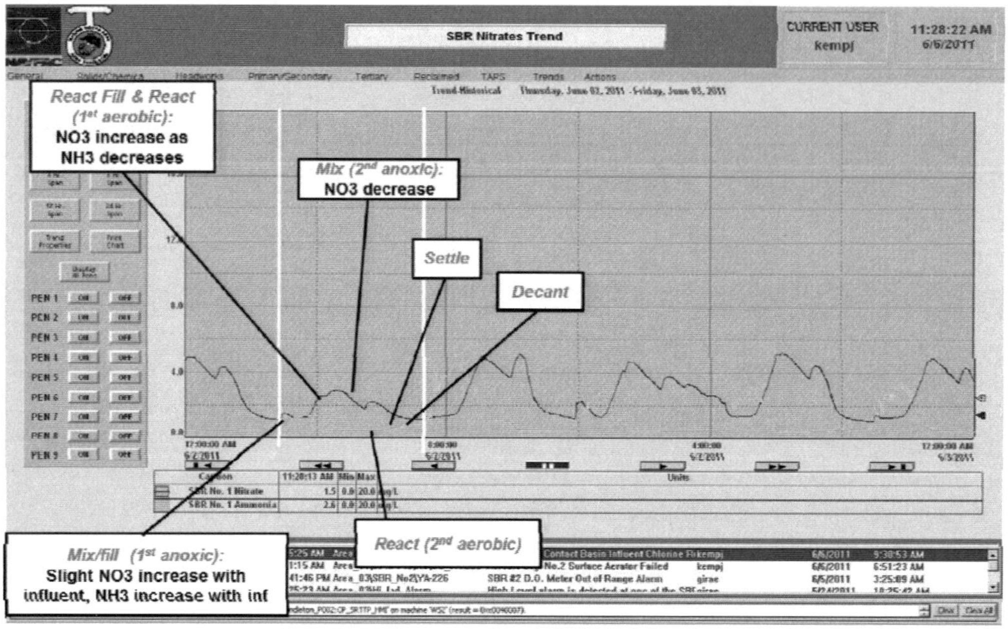

Figure 2 – Ammonia and nitrate response to DO changes in SBR cycle. Ammonia blue pen and nitrate green pen.

Secondary meters were also installed at the tertiary filter effluent to monitor ammonia and nitrate. These meter readings are sent to the SRTTP SCADA system and are used to provide a running average to ensure that the reclaimed effluent is in compliance prior to reclaiming. Although the SRTTP compliance limit is reported as TN, the ammonia and nitrate analyzers help operations stay within compliance by only allowing reclaiming when the ammonia/nitrate average is below 3.0 mg/L. This is based on the historical laboratory data of a TKN value of 0.7 mg/L (normal ammonia effluent is a non-detect based on historical laboratory data). This allows operations with a comfortable buffer of 1.3 mg/L (assuming a negligible amount of NO_2 in the effluent based on historical laboratory data).

LABORATORY ANALYSIS

This step was broken down into two areas; in-house process analysis and outside analysis conducted by a certified laboratory. The ammonia and nitrate analyzers were cross checked with in-house laboratory instruments to insure the accuracy of the field instruments on a weekly basis. The accuracy of the field meters was within 10% of the in-house laboratory instruments, on average. Additional in-house laboratory tests are conducted four to five times per week on each SBR for wasting control. These tests include mixed liquor suspended solids (MLSS), mixed liquor volatile suspended solids (MLVSS) and sludge volume index (SVI). These results were used to optimize the mean cell residence time (MCRT) for the system and verify settling cycle timing. Figure 3 shows a trend chart of TN compliance sampling results over a one year period of the SRTTP's reclaimed effluent prior to distribution. Note again that this was accomplished without the use of an external carbon source.

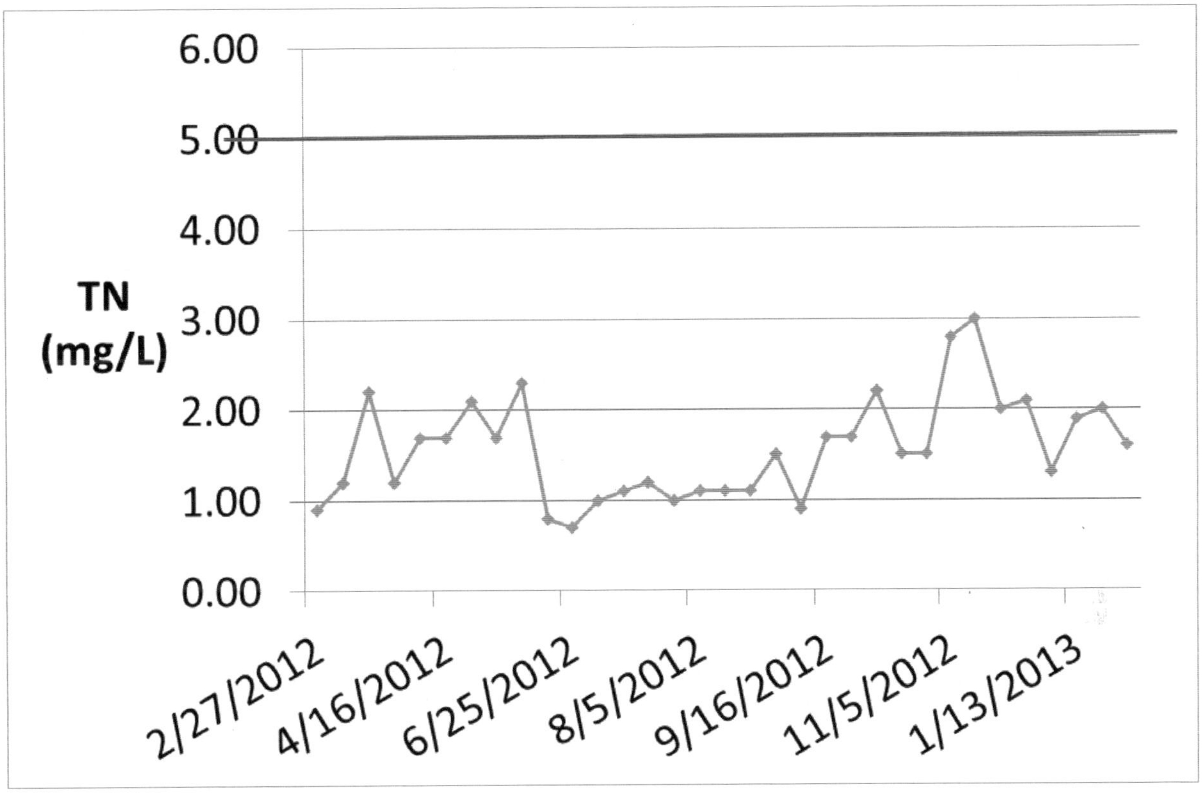

Figure 3 – TN reportable compliance trend chart for one year without an external carbon source. TN achieved through a single aeration stage of 53 minutes prior to a 75 minute anoxic stage. MLSS averaged 2500 during this period to time.

SYSTEM TRACKING OF KEY PROCESS COMPONENTS

The SRTTP operations staff use extensive plant round sheets that track key data and also serve as training aids. This provides the operators with expected normal status such as flow, typical pressure, color of the mixed liquor, air distribution of the aeration cycles inside each SBR, scum buildup and clarity of the decant cycles. The air distribution monitoring provided a key element in the revised placement of a SBR DO meter that led to further cycle time adjustments. Microscopic analysis of MLSS biological health was also performed on a weekly basis as another tool used by staff.

STAFF TRAINING

Staff training started with the previously mentioned development of in-depth SRTTP plant round sheets that are used as teaching aids to instill a proactive approach with the operations and maintenance staff. Weekly process control meetings are conducted with all staff members that cover any changes to process parameters and status of all process equipment. A schedule of reviewing individual standard operating procedures is conducted to insure that all staff members are on the same page with operational changes and to provide a chance for reviewing and editing these documents as needed. Example subjects covered with these procedures include adjusting sludge wasting rates, beginning of shift operations, daily compliance reports and continuous compliance monitoring instruments.

PROCESS CONTROL ADJUSTMENTS

By utilizing operations input as a result of staff training, refer to figures 4 and 5, staff noted less than optimum air distribution in one of the on line SBR's. After performing a DO profile, it was noted that the original placement of the DO probe was less than optimum and was located in an oxygen dead zone. A spot was selected for the DO probe that provided a more accurate DO reading that led to an overall nitrate reduction of 0.5 mg/L with a corresponding total nitrogen reduction of 0.5 mg/L as well. Note attached figure 6.

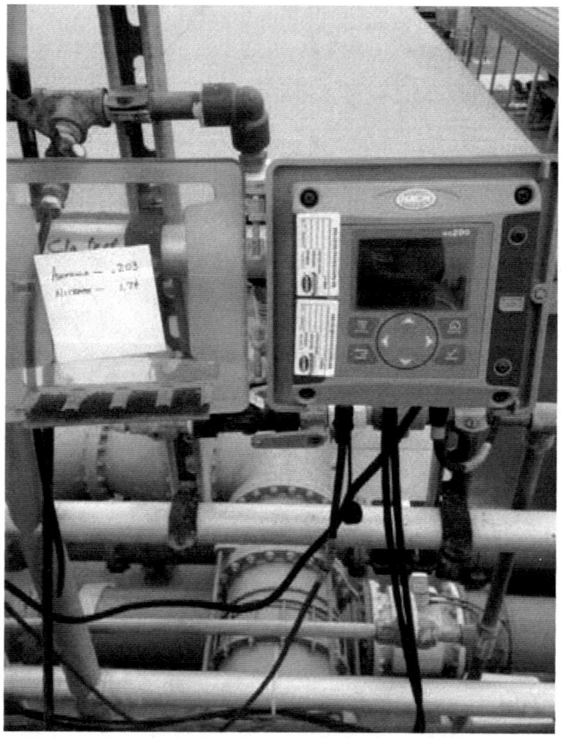

Figures 4 and 5 – Air observations during aerobic stage of SBR cycle, above left. Effluent ammonia and nitrate analyzer are checked during daily rounds, above right. Combined analyzer readings of ammonia and nitrate provides TN reading for compliance checks on a 24 hour trend chart. Staff assumes a TKN of 0.7 mg/L based on historical laboratory readings. Once combined ammonia and nitrate readings reach an inputted operator set point, reclaiming operation will shutdown automatically. A shutdown set point of 3.0 mg/L has been chosen to allow for 0.7 TKN amount and 1.3 mg/L safety factor.

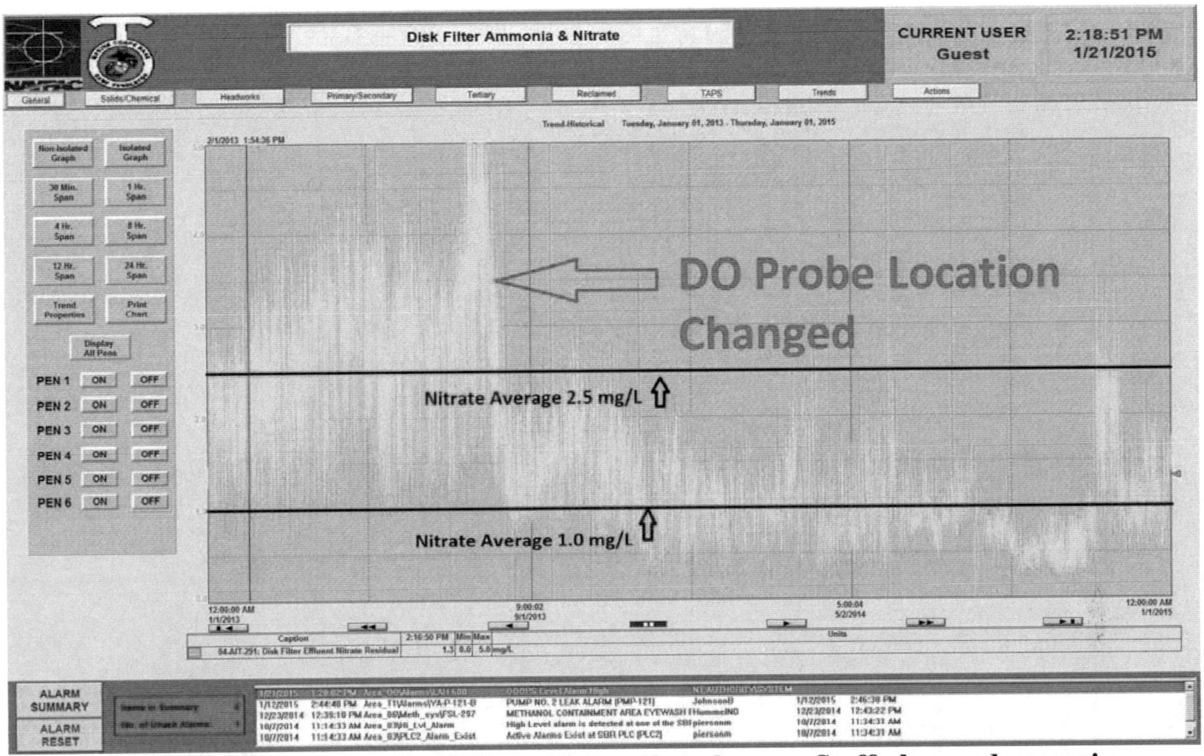

Figure 6 - Nitrate reduction after DO probe location change. Staff observed poor air circulation where DO probe was originally placed. Lower than actual DO readings led to over aeration and oxygen carryover into the anoxic stage thus inhibiting denitrification. DO profile done by staff which led to an optimal relocation of the DO probe.

ADDITION OF METHANOL AND FURTHER PROCESS ADJUSTMENTS

The last process control adjustment was made as a response to:

- Reducing filamentous growth
- Improving MLSS settling characteristics
- Reducing SBR blanket levels and floc carryover
- Maintain or improve the total nitrogen of the reclaimed effluent
- Introducing methanol as a supplemental carbon source

After discussions with the operations, maintenance and process engineers, the team decided that the best course of action was to; change the single SBR air cycle approach and introduce methanol as a supplemental carbon source. In the original plant design, based on the client's original request for proposal (RFP), methanol was not included at the plant. After the plant was designed and into construction, the previously mentioned MRP was issued that presented the TN limit of an instantaneous 5 mg/L with a further reduction to 4.2 mg/L at a later date. This led to the original SBR cycle times, that provided effluent compliance in particular with the TN limit. However, this approach required long anoxic detention times, needed for nitrate removal (denitrification) and a MLSS concentration of 2500 mg/L. This in turn encouraged filament growth, poor settling of the MLSS and endogenous respiration from the lack of a carbon food source. A multi-stage air cycle approach was decided upon, refer to figure 7 below. Several cycle iterations were done with complete laboratory and process analysis before settling on the optimum multi-stage air cycle times. Refer to figure 8 on the following page for an optimum DO profile and correlating ammonia/nitrate reductions.

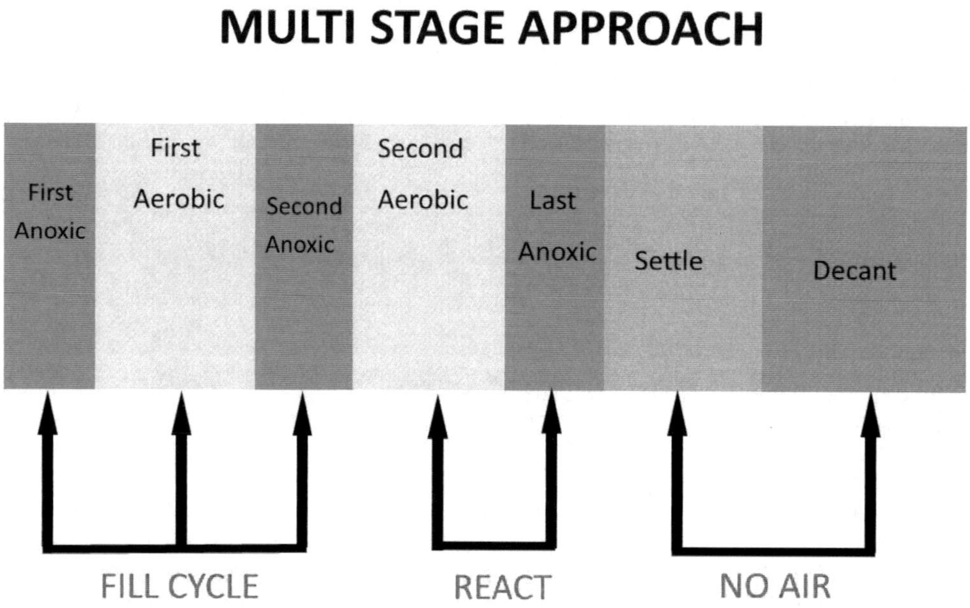

Figure 7 – Multi-stage SBR cycle approach. 3 iterations were tested before settling on the optimum stage times. Methanol added after DO level in second anoxic stage reaches 0.5 mg/L or 10 minutes into the second anoxic time. Timer starts after blower stop signal is initiated.

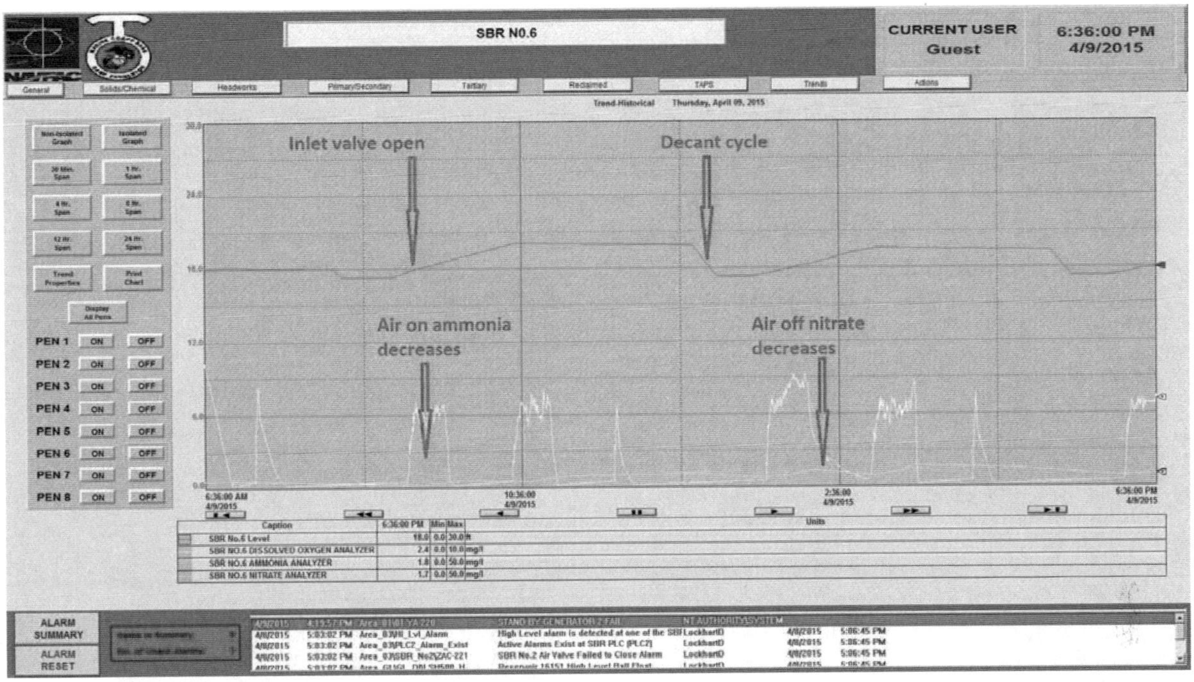

Figure 8 – DO profile and corresponding ammonia/nitrate reductions in a typical SBR. Purple pen SBR level, white pen DO, Blue pen ammonia and yellow pen nitrate. Note steady DO set point control after optimization of the DO control logic. Methanol addition starts after second aeration period/beginning of second anoxic period. (Note, inlet raw sewage valve closed and steady state SBR level.) Program logic calls for methanol addition once DO level reaches 0.5 mg/L or 10 minutes after blower shutdown command.

As a side benefit for the multi-stage cycle changes, staff noted the need to carry less MLSS in the SBR's and an improvement in stalked ciliates during normal microscopic examinations, indicating an improvement in biological activity, refer to figures 9 and 10 on the following page. Currently, MLSS concentrations of the SBR's are at 1800 mg/L. Figure 11, following page bottom, provides total nitrogen trend results (ammonia and nitrate combined) from the field instruments after each cycle time iteration.

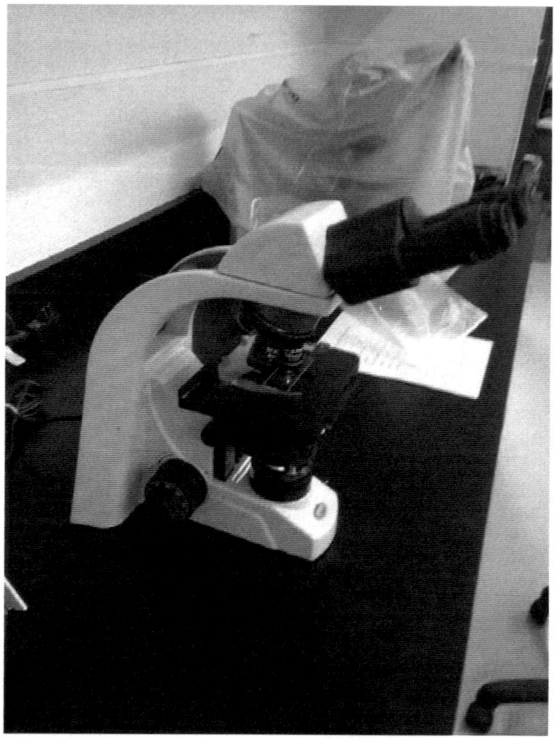

Figures 9 and 10 – Settling observations and microscope analysis. Process changes led to improvement of SBR sludge volume index (SVI) and decreased scum during decant stages. Increased stalked ciliates observed as well indicating good biological activity.

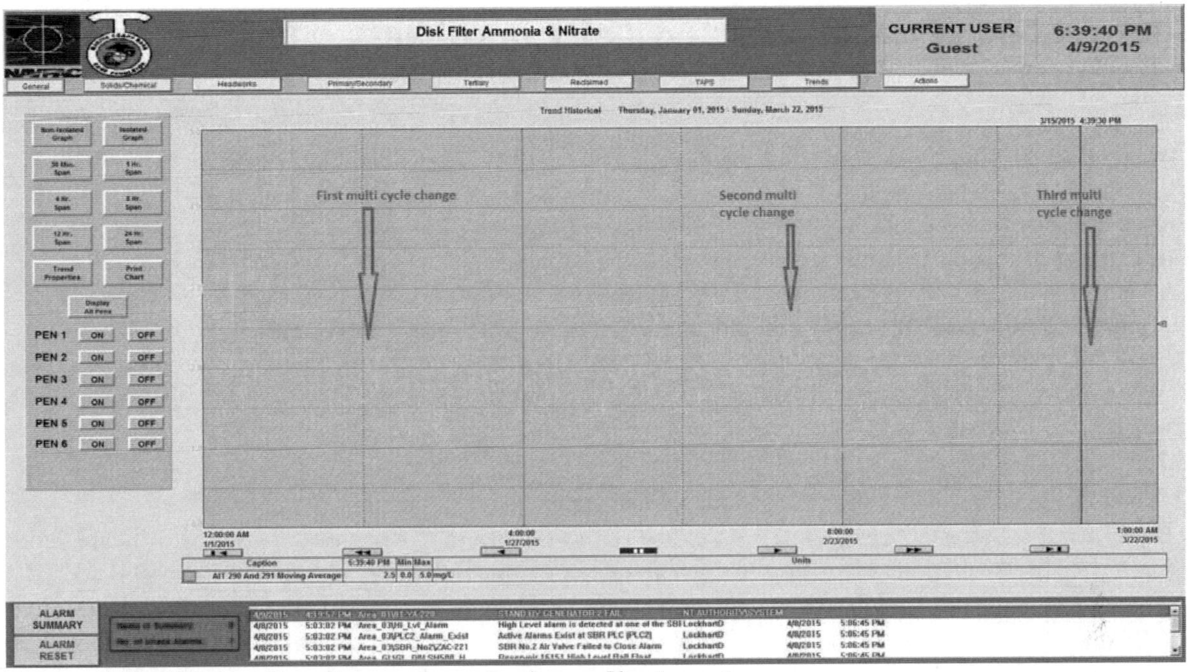

Figure 11 – TN results after each cycle time adjustment.

CONCLUSIONS

Optimizing nitrogen removal at the SRTTP required a multi-step approach involving team work that includes operators, maintenance and process engineers. Initial process conference calls were conducted on a weekly basis with key process engineers and operations staff when process changes are planned. Following these weekly conference calls, information was shared to the entire operations and maintenance staff so all team members were made aware of any changes or ideas that were discussed and to include their input as well. The SRTTP has shown that following these steps has led to the optimization of the SRTTP to meet the stringent TN requirements. It should be noted that recent changes in the MRP have lessened the TN requirements to an annual average of 5 mg/L. This has provided the SRTTP staff with the opportunity to explore future process changes. These changes include; optimizing and or eliminating the addition of methanol as an external carbon source and operating the SBR's based on seasonal reclaimed demand approach in an effort to minimize starts and stops on the aeration blowers. These proposed changes will reduce operating chemical usage (methanol) and electrical power consumption for further operational cost savings while still maintaining effluent compliance.

Optimization of clarifier performance to enhance biological nutrient removal

Chu, Irene[1]; Griborio, Alonso[1*]; Pitt, Paul[1]; M. Ahmad[2]; G. Chiu[2]; J. Desai[2]; J. Wang[3]; D. Freitas[3];
[1]Hazen and Sawyer, PC, 4000 Hollywood Blvd., Hollywood, FL, 33021, USA
[2]San Francisco Public Utilities Commission, 525 Golden Gate Ave, SF, CA 94102
[3]East Bay Municipal Utility District, 2020 Wake Ave, Oakland, CA 94607, USA
*To whom correspondence should be addressed. Email: agriborio@hazenandsawyer.com

ABSTRACT

KEYWORDS: CFD Modeling, Secondary Clarifiers
The performance of secondary clarifiers is critical to ensure that stringent effluent limits are met, particularly in Biological Nutrient Removal (BNR) and Enhanced Nutrient Removal (ENR) treatment plants. Additionally, these plants rely in slow growth microorganism populations that are difficult to restore in the event of solids washout that could happen in wet-weather storm flows. Secondary clarifiers are a key unit to ensure retention of the solids inventory. This paper illustrates the application of computational fluid dynamics (CFD) modeling at two different BNR/ENR plants and two different carbon removal facilities local to the nutrient conference (San Jose, Ca). These examples cover a wide variety of clarifier geometries, internal configurations and loading conditions.

Case study 1, the San Francisco Public Utilities Commission (SFPUC) Southeast Plant (SEP) is a carbon removal facility where CFD modeling was used to identify retrofits of their center feed circular secondary clarifiers. Case study 2, the East Bay Municipal Utility District (EBMUD) Main Wastewater Treatment Plant (MWWTP) is a carbon removal facility where CFD modeling was used to identify upgrades of their peripheral feed peripheral overflow (PFPO) secondary clarifiers. In Case study 3, the model was use to evaluate the impact that the tank depth can have on clarifier performance and capacity. Case study 4 the model was used for the validation of step feed as a wet weather strategy to prevent solids washout during storm events. In all cases, the model was used to identify improvements and operating strategies that result in improved reliability, increased capacity and reduced capital cost.

INTRODUCTION

Biological Nutrient Removal (BNR) and Enhanced Nutrient Removal (ENR) treatment plants rely on several microbial populations (heterotrophs, autotrophs, poly-phosphate accumulating organisms, etc.) to achieve Nitrogen and Phosphorus removal. Some of these populations, e.g. aerobic autotrophic bacteria responsible for nitrification, have lower growth rates and therefore are more difficult to restore in the event of solids washout that could happen in wet-weather storm flows. The retention of the solids inventory and maintenance of the microbial population in a healthy and viable state is therefore critical in ensuring adequate treatment after the storm event has passed.

Secondary clarifiers are a key unit process to ensure retention of the solids inventory and permit compliance. The performance of clarifiers depends on several interrelated processes and factors including: hydrodynamics, settling, turbulence, flocculation, sludge rheology, atmospheric conditions, tank geometry, internal features and loading conditions. Despite being a key unit process to ensure successful treatment, clarifiers have traditionally been designed following simplistic approaches. Historically, process design of secondary clarifiers was based strictly on overflow and solids loading rates, with sludge collection mechanism design based on client preference or manufacturer recommendation. State point analysis is typically the next level of design analysis complexity. State point analysis takes into account sludge volume indices, return activated sludge pumping capacity and mixed liquor concentration. While state point analysis may provide a more rigorous analysis of clarifier capacity than the historical solids and overflow rate analysis, the most reliable predictor of clarifier capacity and the only tool that allows for consideration of clarifier geometry is the computational fluid dynamics modeling.

This paper will illustrate the application of CFD modeling to four different clarifier optimization case studies (two at BNR/ENR facilities and two local carbon removal facilities) covering a wide variety of clarifier geometries, internal configurations and loading conditions. The CFD model used in these cases studies is known as 2Dc. The 2Dc model, developed at the University of New Orleans (Griborio, 2004; McCorquodale et al. 2005; Griborio and McCorquodale, 2006), is a state-of-the-art quasi-three-dimensional software that can be used for circular or rectangular clarifiers. The model is superior to traditional clarifier analysis tools, in that it:

- Accounts for the behavior of all settling mechanisms, which include discrete, zone and compression settling, as well as flocculation and floating particles.
- Accounts for temperature and/or heat exchange effects.
- Models the specific clarifier geometry (floor slope, centerwell baffles, peripheral baffles).
- Provides an estimate of the effluent suspended solids.
- Provides a graphical representation for both the hydrodynamics within the tank and the sludge blanket interface.

METHODOLOGY

As indicated before, the following sections present the application of the 2Dc model to the evaluation of four case studies. In all cases, the model was previously calibrated and validated. The calibration and validation of the model are not presented herein. The reader is referred to Griborio (2004) for a full description of the calibration and validation procedures.

CASE STUDY 1: SFPUC SEP, SF CALIFORINIA

The SFPUC SEP is a 57 mgd high purity oxygen activated sludge facility treating wastewater for carbon removal in California. The sixteen clarifiers at SEP, are 120-ft circular center-feed peripheral overflow with side water depth (SWD) of 15 ft. SFPUC is planning to upgrade these 30+ year old clarifiers as they are reaching the end of their useful life. **Table 1** summarizes the clarifier dimensions.

Table 1. SEP Secondary Clarifier Dimensions

Parameter	Dimension/Characteristic
Diameter	120-ft
Side Water Depth (SWD)	15-ft
Center Well Diameter	16-ft
Center Well Submerged Depth	8-ft
Center Column Diameter	36-inch
Bottom Slope	1/8 to 12
Inlet Ports	4 ports 18 ¾ inch x 10 ¾ inch
Energy Dissipating Inlet	None – Inlet Ports are provided with Target Baffles
Effluent Launder	Outboard Launder
RAS Withdrawal	Towbro Suction Header

A CFD model was developed to determine if improvements could be incorporated into the clarifier retro-fit design (over replacing the clarifier internals in kind) to enhance effluent quality and gain clarifier capacity. Modeling was performed for normal operation as summarized in **Table 2**. Extensive field testing was performed (including stress testing) to calibrate and validate the CFD model.

Table2. SEP Operational Characteristics

Parameter	Unit	Normal operation	Stressed condition
SVI	ml/g	85	85
Surface Overflow Rate (SOR)	gpd/sf	520	900
MLSS Concentration	mg/L	3,690	2,830
RAS Rate	%	25	25

Field testing and the baseline CFD model revealed strong density currents, high inlet port velocities and limited time for flocculation in the center well. **Figure 1** shows clarifier performance under stressed conditions. Modeling scenarios focused on mitigation of these processes that can detrimentally effect solids separation in a clarifier. Modeling scenarios are described in **Table 3.**

Table 3. Modeling Scenarios for Clarifier optimization

Modification	Inlet Modification	Center Well Diameter ft (% of diameter)	Stamford Baffle
Modification 1	Yes	24-ft (20%)	Yes
Modification 2	Yes	30-ft (25%)	Yes
Modification 2-1	Yes	30-ft (25%)	No
Modification 3	Yes	36-ft (30%)	Yes

Modeling showed that under the stressed condition, scenario 2 and 2-1 had the best performance. **Figure 1** shows modeling results for modifications 1, 2, and 3. **Figure 2** shows effluent quality for all scenarios graphically.

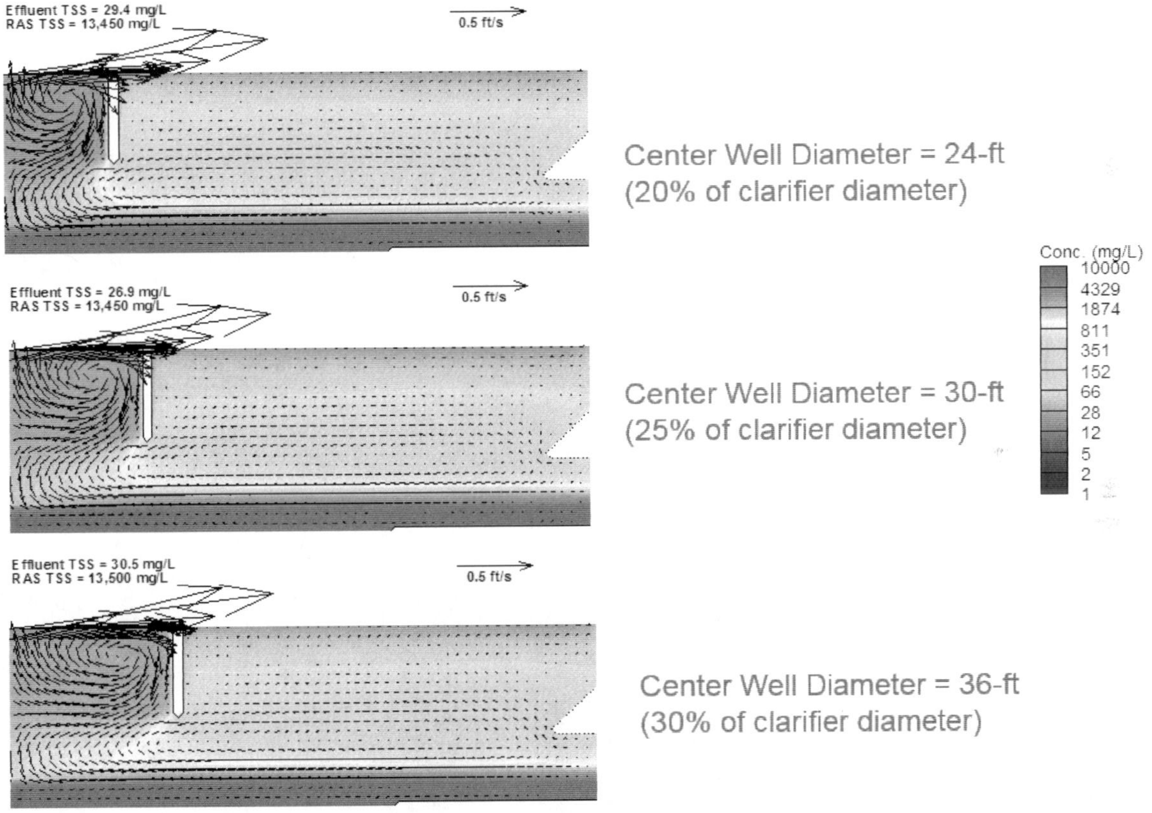

Figure 1. CFD modeling of optimization scenarios under stressed conditions

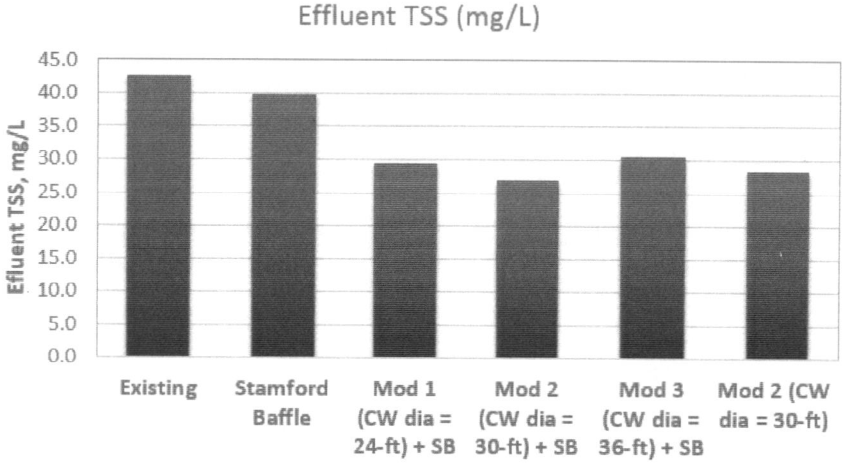

Figure 2. Effluent Quality under modeling scenarios

Figure 1 and **2** show that the optimized center well diameter was 30-ft or approximately 25% of the clarifier diameter. **Figure 2** showed minimal improvement between modification scenario 2 and 2.1; with and without the stamford baffle.

Adjustments to the stamford baffle location were modeled. **Figure 3** shows the modeling results of different stamford baffle heights above the clarifier bottom. Results showed that under high load low the optimized height for the stamford baffle waas 7 ft above the clarifier floor.

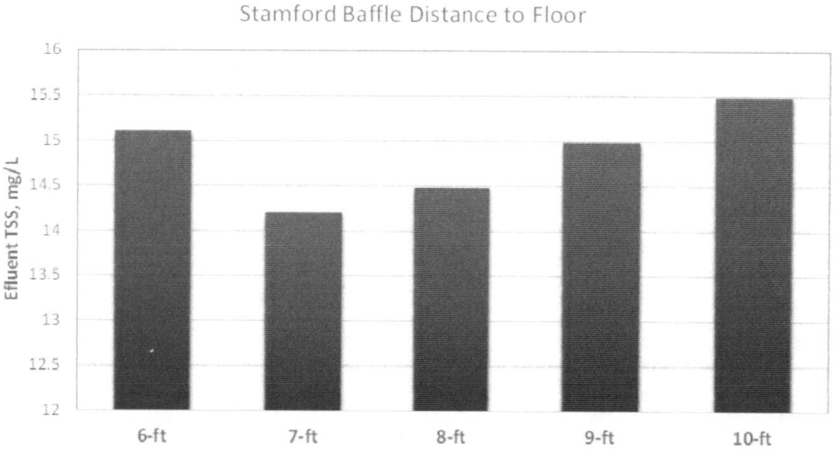

Figure 3. Modeling results for various stamford baffle locations

Ultimately the recommended changes included energy disipating inlets, enlarged centerwell, enlarged influent ports, and a stamford baffle. While these changes increased clarifier capacity at Plant 1 as shown in **Figure 4**, the secondary system capacity was ultimately limited by other upstream processes. Plant staff may employ the added capacity by leaving a secondary clarifier out of service during storm events, saving time and increasing opertional flexibility.

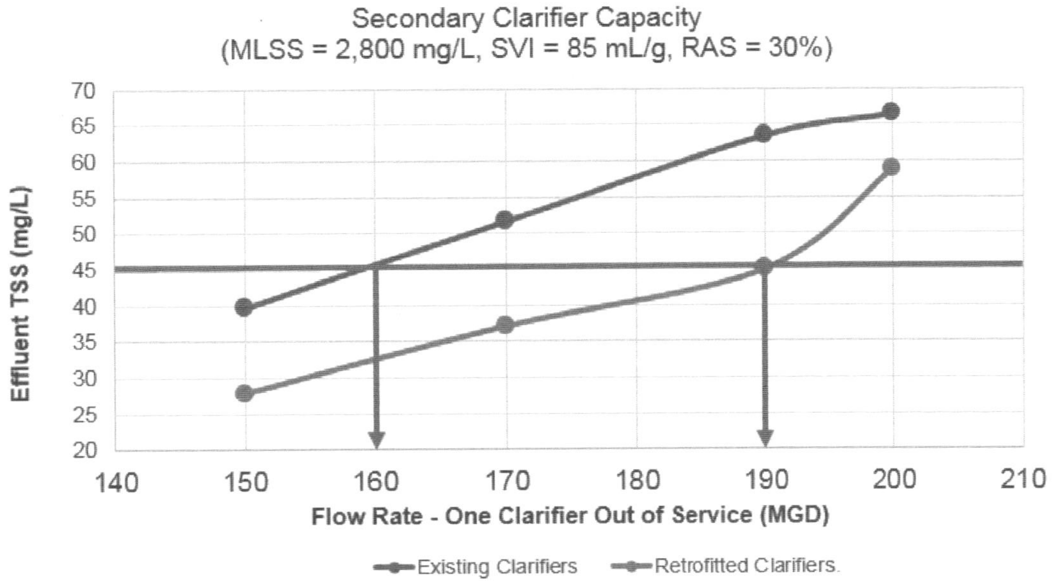

Figure 4. SEP existing and retorfitted clarifier capacity

CASE STUDY 2: PLANT 2 EVALUATION OF CIRCULAR PERIPHERAL-FEED PERIPHERAL OVERFLOW (PFPO) CLARIFIERS

The EBMUD MWWTP is a 63 mgd high purity oxygen activated sludge facility treating wastewater for carbon removal in Alameda County, CA. The twelve clarifiers at the MWWTP are 120-ft circular PFPO clarifiers with side water depth (SWD) of 14 ft. EBMUD is planning to upgrade these 30+ year old clarifiers as they are reaching the end of their useful life. **Table 4** summarizes the clarifier dimensions.

Table 4. EBMUD MWWTP Secondary clarifiers – Basic Dimensions and Characteristics

Parameter	Dimension/Characteristic
Diameter	140-ft
Side Water Depth (SWD)	14-ft
Bottom Slope	1/16 to 12
RAS Withdrawal	Towbro Suction Header

A CFD model was developed to determine if improvements could be incorporated into the clarifier retro-fit design (over replacing the clarifier internals in kind) to enhance effluent quality and gain clarifier capacity. Extensive field testing was performed (including stress testing) to calibrate and validate the CFD model.

Field testing and the baseline CFD model revealed strong density currents. The baseline model also showed that the small inlet skirts were ineffective at segregating influent and effluent in the clarifiers. **Figure 5** shows the calibrated baseline model and velocity vectors.

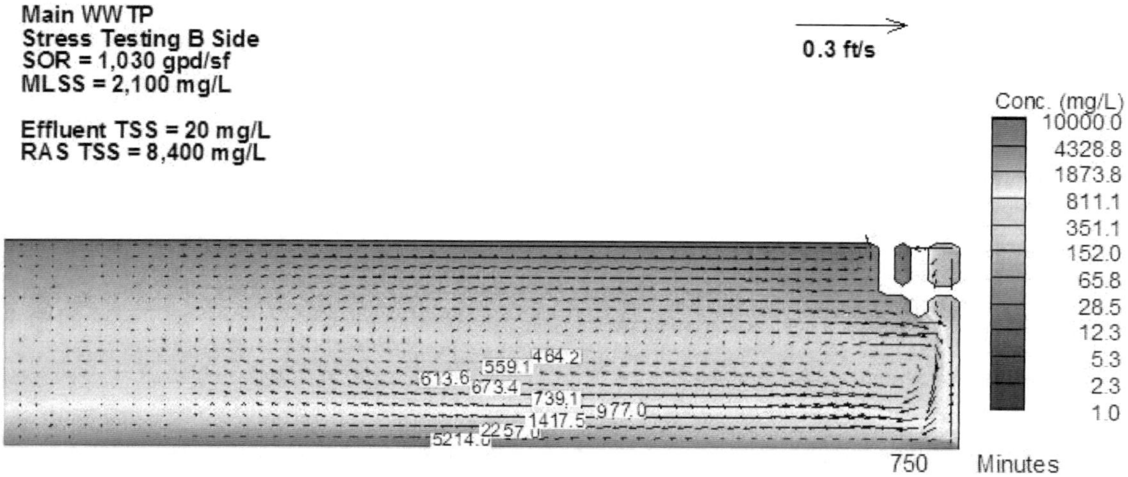

Figure 5. EBMUD MWWTP Calibrated baseline model

As the initial modeling showed that the inlet skirt baffle was ineffective, modeling was conducted to optimize the skirt baffle. **Figure 6** shows modeling of various skirt baffle lengths while **Figure 7** shows the simulated effluent quality for these lengths. Results showed that the optimal skirt baffle length was 30-inch skirt baffle.

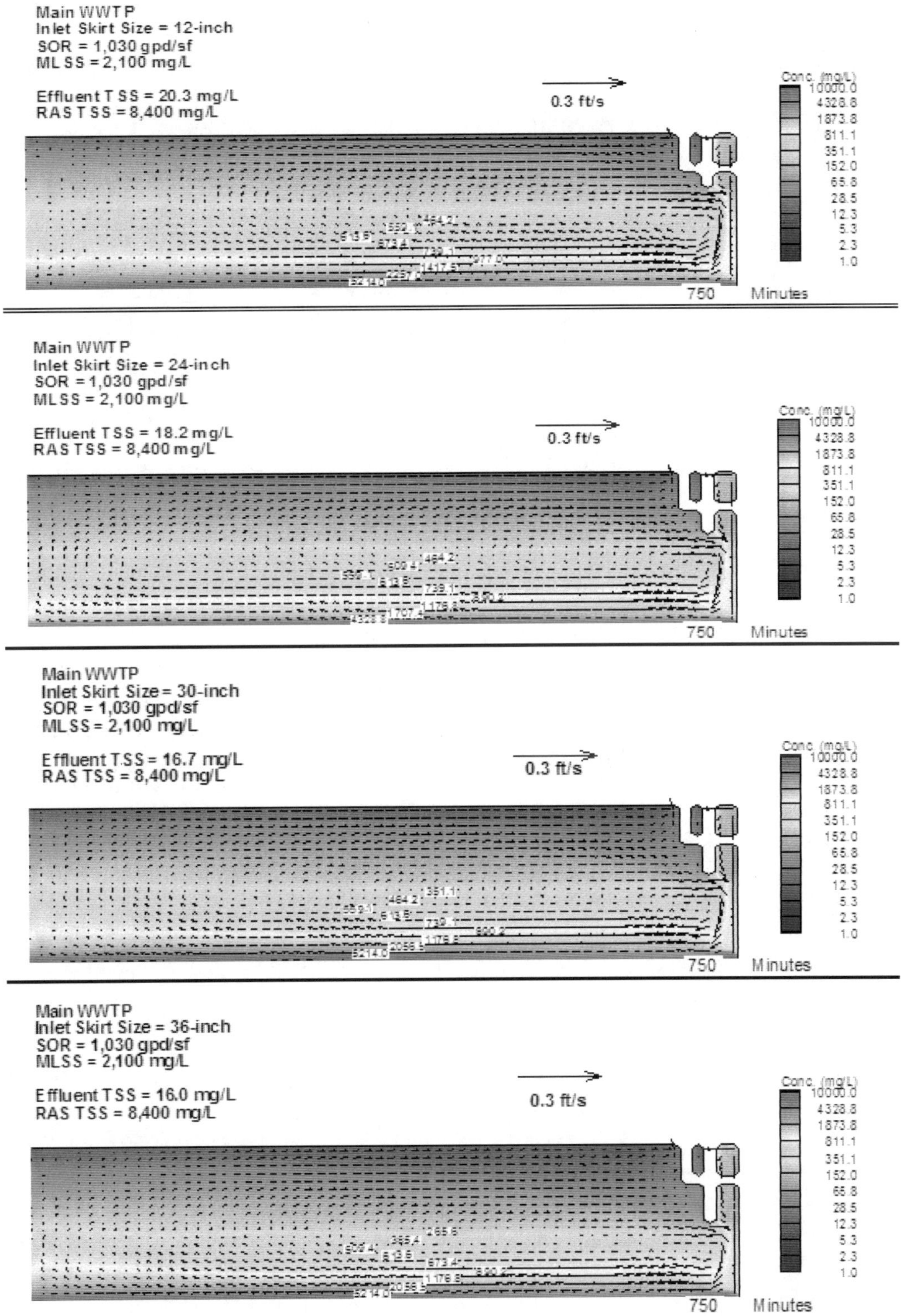

Figure 6. EBMUD MWWTP Model results for various skirt baffle lengths

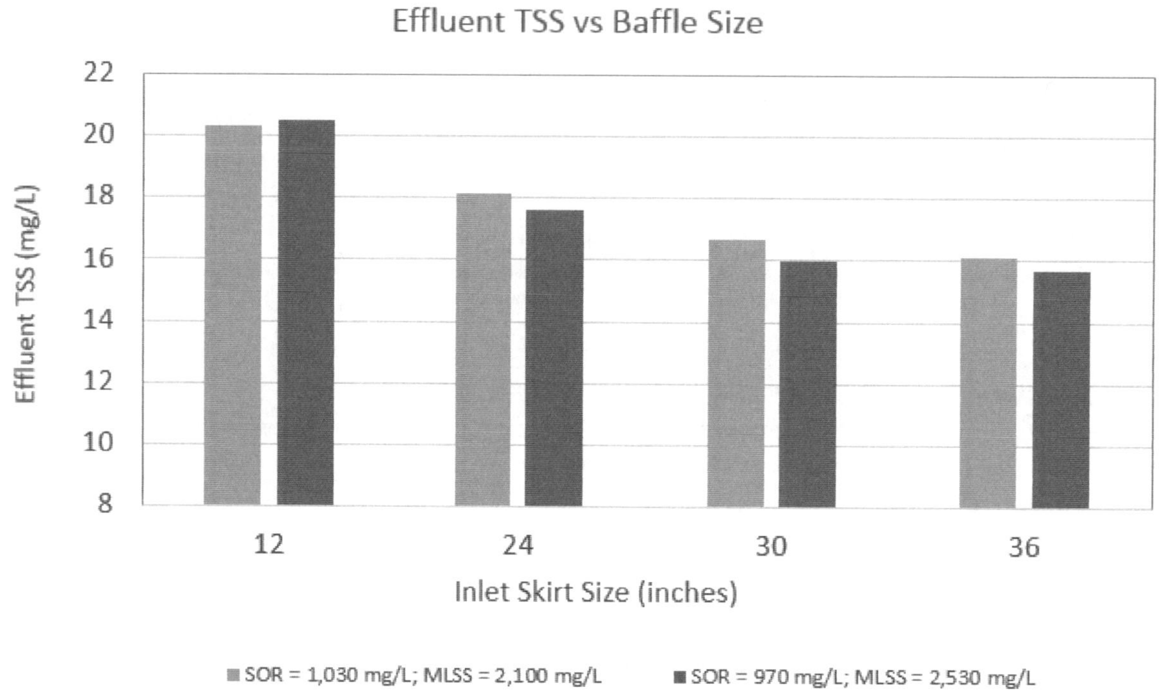

Figure 7. Simulated effluent Quality for various skirt baffle lengths

Simulationf for the optimized clarifier improved effluent quality and ultimately resulted in an increased secondary clarifier capacity for the MWWTP. **Figure 8** shows the increased capacity for the poposed retrofit vs the existing configuration. Under an excellent SVI of 84 mL/g and a slightly bulking SVI of 108 mL/g the retrofitted clarifiers show a remarkable increase in cpacity. The clarifiers do not reach the permited limit of 30 mg/L until SOR greater than 1500 gpd/sf.

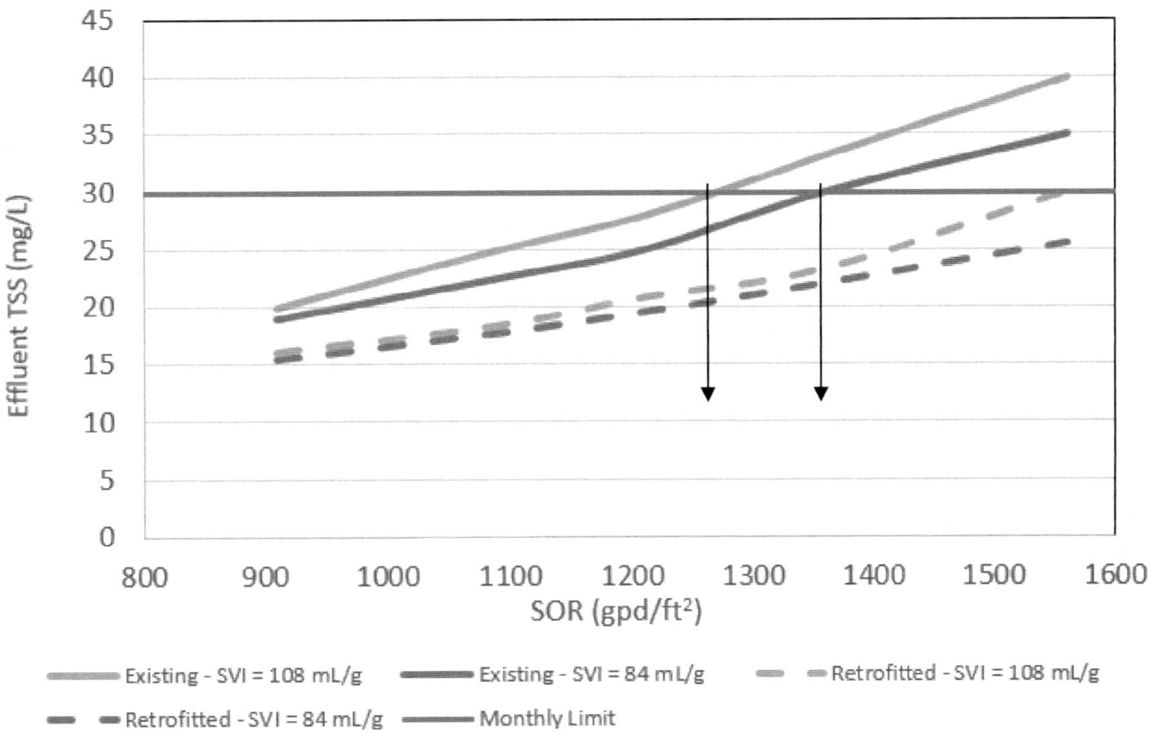

Figure 8. EBMUD MWWTP Increased clarifier capacity

Case Study 3: Clarifier Depth Evaluation

This case study illustrates the application of stress testing and CFD modeling for the evaluation of the effect that tank depth has on clarifier capacity. Existing 11 ft side water depth (SWD) and 15 ft SWD peripheral feed clarifiers were stress tested side by side and evaluated using CFD modeling. Assuming an SVI = 120 mL/g, the clarifier capacities and performance were evaluated for MLSS equal to 3400, 2900 and 2600 mg/L.

This 26.6 mgd Florida facility currently operates six peripheral-feed peripheral-overflow (PFPO) circular secondary clarifiers. There are four 90 ft diameter clarifiers and two 115 ft diameter clarifiers. A new 115 ft diameter secondary clarifier will be added as part of the plant improvements project to treat peak flows up to 53 mgd. The new clarifier will be circular center feed peripheral-overflow. **Table 5** presents the basic dimensions of the existing and proposed secondary clarifiers.

Table 5. Case Study 2 Florida WWTP Secondary Clarifiers – Basic Dimensions.

Parameter	Feature/Dimension (Existing Clarifiers)	Feature/Dimension Existing Clarifiers	Feature/Dimension (New Clarifier)
Number	4	4	1
Type	PFPO	PFPO	Circular Center Feed
Clarifier Diameter (ft)	90	115	115
Side Water Depth (ft)	11	15	16
Bottom Slope (5)	0.5	0.5	0.5
Center Well Diameter	NA	NA	29.0
Center Well Depth	NA	NA	7.25
Effluent Launder	Inboard	Inboard	Inboard
Sludge Withdrawal	Suction	Suction	Suction
Peripheral Baffle	NA	NA	Stamford Baffle

As indicated in the methodology section, the 11 ft and 15 ft SWD clarifiers were evaluated for MLSS equal to 3400, 2900 and 2600 mg/L at an SVI = 120 mL/g. **Figure 9** shows the predicted ESS for the 11 ft SWD secondary clarifiers for different SORs. The results indicate that these tanks could achieve acceptable effluent quality at peak flow conditions and MLSS less than 2,900 mg/L, but the SOR at 3,400 mg/L MLSS needs to be restricted to approximately 820 gpd/ft^2. **Figure 10** shows the predicted ESS for the 15 ft SWD secondary clarifiers. The results indicate that these tanks should be able to achieve acceptable effluent quality even at SOR slightly over 1,000 gpd/ft^2 and MLSS of 3,400 mg/L. The results show that the deeper tanks are able to sustain higher SORs and solids loading rates (SLRs). The deeper tanks have additional capacity to storage sludge during these high loading events. Naturally, the deeper tanks allow higher sludge blanket depths. Higher blanket depths result in more compaction at the bottom and higher RAS TSS concentrations, which is beneficial for preventing thickening failure.

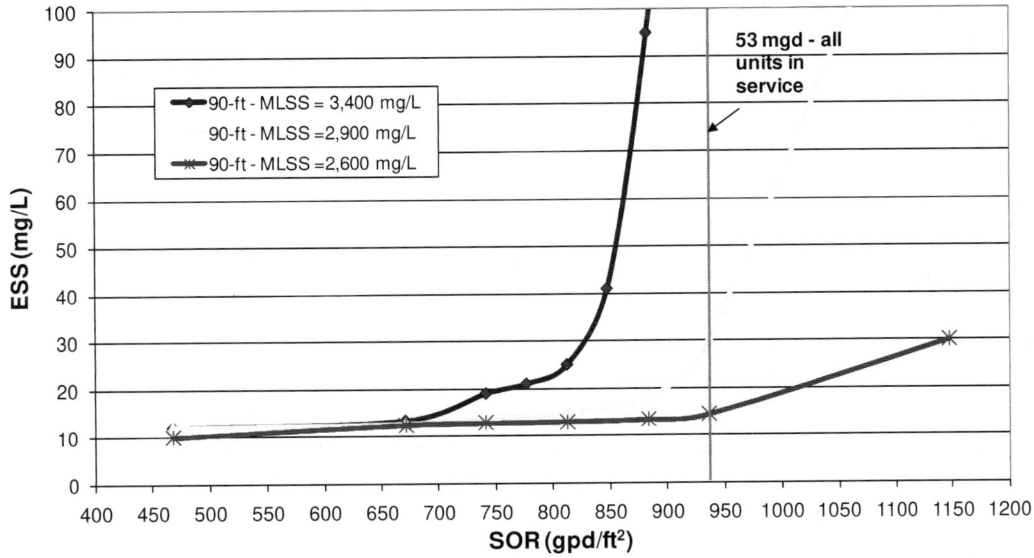

Figure 9. Performance of 11 ft SWD Peripheral Feed Secondary Clarifiers at Different MLSS and Different SORs (RAS = 50%).

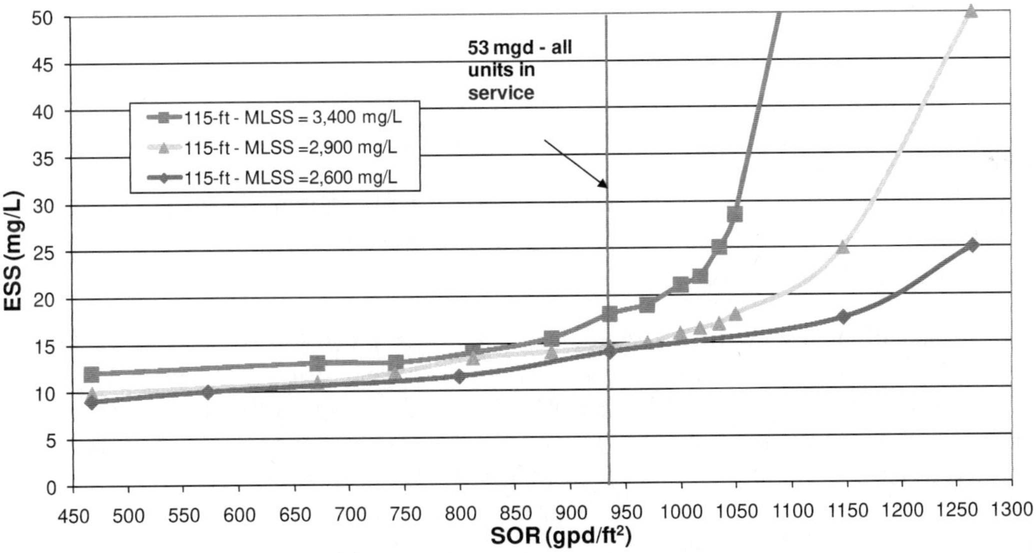

Figure 10. Performance of 15 ft SWD Peripheral Feed Secondary Clarifiers at Different MLSS and Different SORs (RAS = 50%).

Calibrating Glycerin Driven EBPR to Full Scale Results at a 350 MGD Midwestern Plant.

Samuel Ledwell[1*], Mehran Andalib[1]
[1]Environmental Operating Solutions, Inc., Bourne, MA
Email: sledwell@microc.com

ABSTRACT:

This paper compares the experimental data from a full scale EBPR trial at a 350 MGD municipal facility using glycerin (MicroC 2000™) as an electron donor to a calibrated BioWin model using fermentation kinetics of glycerin. The full scale evaluation occurred over a 70 day period in late 2014. Average PO4-P removal was 2.54 mg/L in the test battery and was within 5.2% of the stoichiometric estimates based on an assumed COD:N of 6.1 and COD:P of 15. Fermentation of glycerin yields a higher ratio of propionate (Pr) to acetate (Ac) and as such, three mechanistic parameters in BioWin were altered to reflect a more accurate Pr/Ac ratio. The calibrated dynamic simulations showed greater agreement with the experimental data than the un-calibrated dynamic simulations.

KEYWORDS:

EBPR, phosphorus removal, glycerin, MicroC, fermentation, BioWin

INTRODUCTION:

Recent studies have explored the potential for a dual-benefit from the addition of glycerin (glycerol) to denitrify and promote enhanced biological phosphorus removal (EBPR) [Ledwell, 2013; Cyplik, 2013; Guerrero, 2012]. Furthermore, since it has been proven that fermentation of glycerin results in a different composition of volatile fatty acids compared to biodegradable COD (complex), the kinetics of glycerin-based EBPR and fermentation should be calibrated with revised stoichiometric coefficients and kinetics in a simulator such as BioWin to ensure good modeling practice.

Very few studies have investigated glycerin fermentation end products using mixed cultures. This is an important knowledge gap as wastewater sludges do not behave as pure cultures. The primary objective of this paper is to compare simulated results using BioWin to the data from a full scale EBPR evaluation using a glycerin based carbon source (MicroC 2000™) at a 350 MGD Midwestern plant. First, using default fermentation kinetics, a simulation using glycerin as sCOD is performed. Next, the fermentation kinetics of glycerin as the fermentable substrate were modified, and the model was validated using bench scale fermentation test results from the literature. Dynamic simulations using both the calibrated and un-calibrated models were performed and these results were compared to the experimental data from the full scale evaluation.

METHODOLOGY

Full Scale Evaluation

The facility discussed in this paper is a 354 MGD design average facility with a service area of 300 square miles serving a population of 1.1 million people in the Midwest.

Prior to investigating glycerin addition at the full scale, the plant completed a number of preliminary tests beginning in 2012. First, in 2013, it was determined that there was indeed a carbon deficiency at this facility (average primary effluent BOD:TP of 15). Second, two of the five batteries (Batteries A&B), were converted to an anoxic-anaerobic-oxic configuration in addition to sludge return and mixing channel modifications to try and stimulate EBPR activity without external carbon (Figure 1). Third, SBR bench scale evaluations mimicking the full scale operations determined that with a glycerin dose of 30 mg/L as rbCOD, and a COD:P ratio of 15, sustainable EBPR was established.

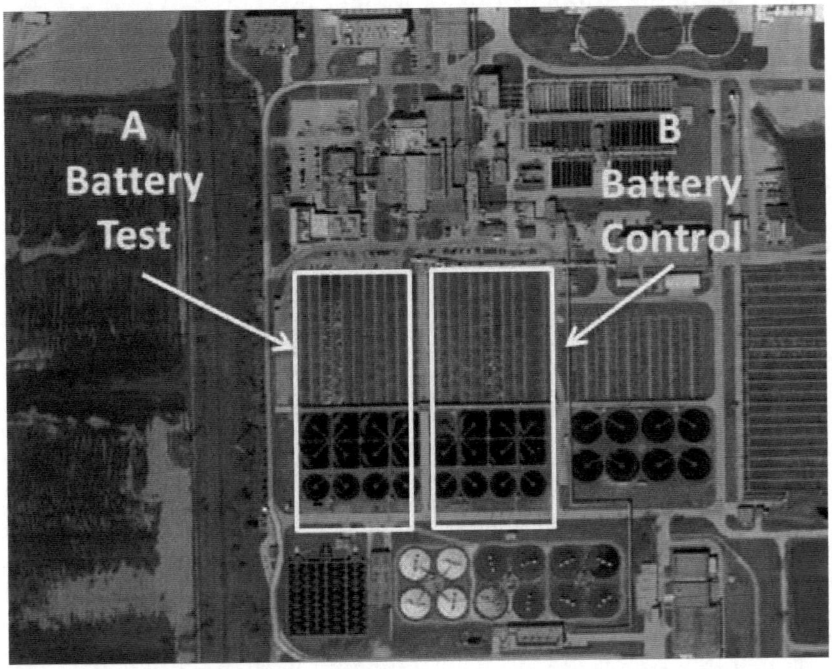

Figure 1: Satellite View of Battery A (test Battery) and Battery B (control Battery)

A 70 day full scale evaluation was completed in October to December of 2014. Two batteries were monitored during the evaluation, Battery A was dosed with MicroC 2000™ and Battery B was used as a control. The primary objective was to maintain an effluent PO4-P of 0.75 mg/L in Battery A. Table 1 describes the conditions of the evaluation.

The system was converted to an anoxic-anaerobic-oxic process by turning off the air to the front end of each of the 11 trains to each battery. Nitrate is returned to the front end of the batteries in the RAS and mixed with primary effluent. If the Nitrate is reduced via carbon in the primary effluent or via an external carbon source, conditions can become anaerobic partway down each of the trains, stimulating phosphorus release if carbon is available to drive the EBPR process. Note that baffles and mixers were not installed and it was difficult to evenly distribute the airflow to each of the 11 trains. Thus, this is an imperfect plug flow process but was suitable for the evaluation.

Table 1: Conditions of the Evaluation

Parameter	Condition(s) During Full Scale Evaluation
Primary Effluent Flow	Fixed at 45 MGD to Batteries A&B
RAS Flow	Fixed at 45 MGD to Batteries A&B
Avg Daily HRT	Fixed at 6.9 hrs (includes settling)
SRT	Fixed at 10 Days
MicroC Dosing Locations	RAS Channel and/or Influent Splitter Box of Battery A
MicroC Dosing Scheme	Manual Feed Rate
Facility Data Collection	• 24 hr Composite Samples • Inline monitoring for DO, NH4, NOX-N • Weekly Nutrient Profiles • Intracellular PHB and PolyPhosphate quantification

Simulation

BioWin Version 4.0 (Envirosim Associates Ltd, Hamilton, Ontario, Canada) was used to perform the simulations in this paper.

RESULTS AND DISCUSSION

Full Scale Evaluation

Over the 70 day evaluation period, phosphorus removal averaged 2.54 mg/L at an average MicroC™ feed rate of 2991 gpd. This reduction was within 5.2% of the estimated stoichiometric requirements assuming a COD:P of 15, a COD:N of 6.1 and COD:DO of 2.1 for MicroC™ 2000. EOSi estimated a daily COD requirement of 24,545 lbs COD/day to remove 3 mg/L of DO, 4 mg/L of NOX-N in the RAS and 2.25 mg/L of PO4-P. Actual removal was 3 mg/L of DO, 4.5 mg/L of NOX-N and 2.54 mg/L of PO4-P with 25,932 lbs of COD or 2991 gpd of MicroC 2000™.

Influent PO4-P to the BNR process (also referred to in this study as Primary Effluent PO4-P) averaged 3.93 mg/L during the evaluation but varied greatly, from as low as 1 mg/L to as high as 9 mg/L (Figure 2). The variability in influent PO4-P to the process was unexpectedly high and rendered some of the original study objectives impossible to achieve in the absence of real time nutrient monitoring or a massive over dose of MicroC™ to consistently achieve the target of 0.75 mg/L PO4-P.

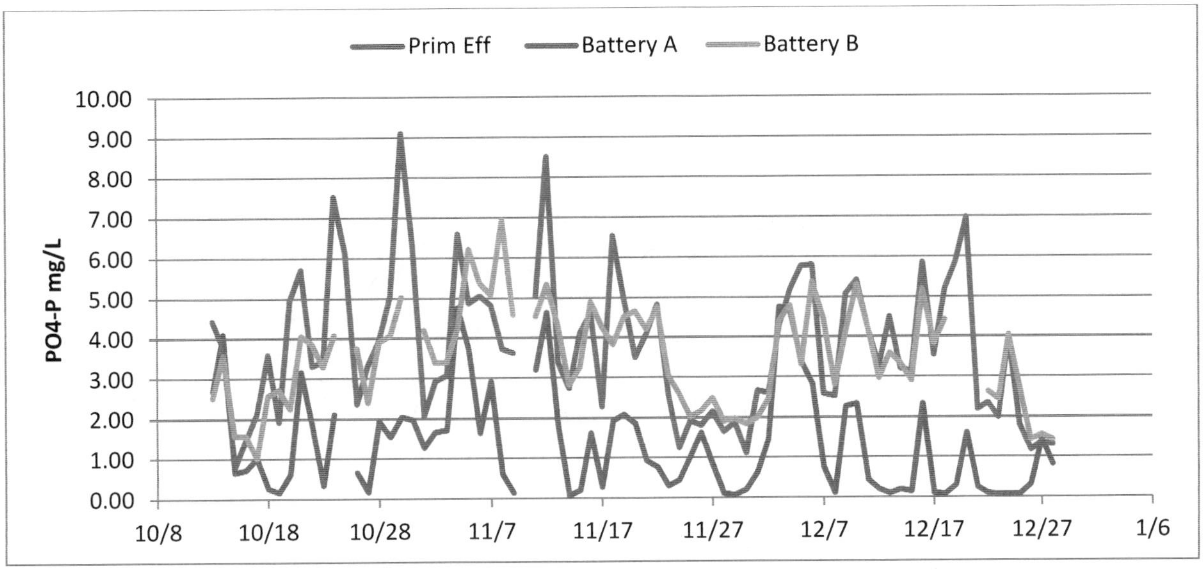

Figure 2: Orthophosphate Concentrations of Influent (Prim Eff), Test Battery (A) and Control Battery (B)

Figure 2 shows the greatest phosphorus removal between the test (Battery A) and control (Battery B) batteries at the end of the evaluation when MicroC™ feed rates were at the highest point of 3428 gpd from 12/6 to 12/23. Also of interest is the deterioration in performance over the holiday weekend from 11/29 to 12/3 during which time the power to the MicroC™ feed pumps was inadvertently turned off. Note that on average there was little to no phosphorus removal in the control battery where no external carbon was added. The manual MicroC™ feed rate was sufficient to remove approximately 2.5 mg/L. In general, Figure 2 shows that on days when influent PO4-P peaks above 4 mg/L the fixed feed rate is not sufficient to maintain PO4-P effluent at or below the target level.

Table 2 summarizes the performance during the period of the highest MicroC™ feed rate and that of the entire study. The desired effluent concentration of 0.75 mg/L PO4-P was not reliably achieved due to the variable influent TP concentration and the manual MicroC™ feed strategy. During the period of highest feed (3428 gpd) to Battery A, 3.30 mg/L of PO4-P was removed (81% removal) maintaining an effluent PO4-P of 0.79 mg/L, close to the initial study objective. During the same period, effluent PO4-P values were less than 0.5 mg/L on 15 of the 21 days and were as low as 0.07 mg/L at times. At a minimum, to consistently meet the desired effluent TP of 0.75 mg/L, the variability in influent Phosphorus would need to be considered.

Table 2: Summary of Phosphorus Removal in Battery A

	Highest Feed Rate 12/6-12/23	Entire Study 10/13-12/23
MicroC Avg Feed	3428	2991
PREF PO4-P	4.09	3.93
Battery A EFF PO4-P	0.79	1.39
P Removed	3.30	2.54
% P Removed	81%	65%
# of Days < 1	13	36
# of Days < 0.75	12	32
<1 Goal Achieved	72%	51%
<0.75 Goal Achieved	67%	46%
BOD:PO4	20	22
BOD:PO4 w/ MicroC	40	40

Increasing the BOD:PO4-P ratio from 20 to 40 was expected to achieve fairly consistent effluent phosphorus performance. Note in Table 2 that the BOD:P with MicroC™ was the same for both periods however the phosphorus removal was much greater at the end of the study during the highest feed rate. Figure 3 shows that regardless of the BOD:P ratio in the control battery (B) the effluent PO4-P is never less than 1 mg/L. In the test battery (A), the effluent PO4-P is less than 1 mg/L 87% of the time when BOD:PO4-P is > 50. Also note that there are many instances when the BOD:P is > 40 in the test battery and effluent phosphorus values deteriorate. This implies that conditions in the process have changed, such as the concentration of electron acceptors in the presence of glycerin, resulting in less effective phosphorus removal. Figure 4 is a graphical representation of the glycerin requirements in this particular system where MicroC 2000™ is dosed into an aerated RAS channel with NOX-N. Only when the DO and NOX-N is reduced will glycerin be utilized to support EBPR.

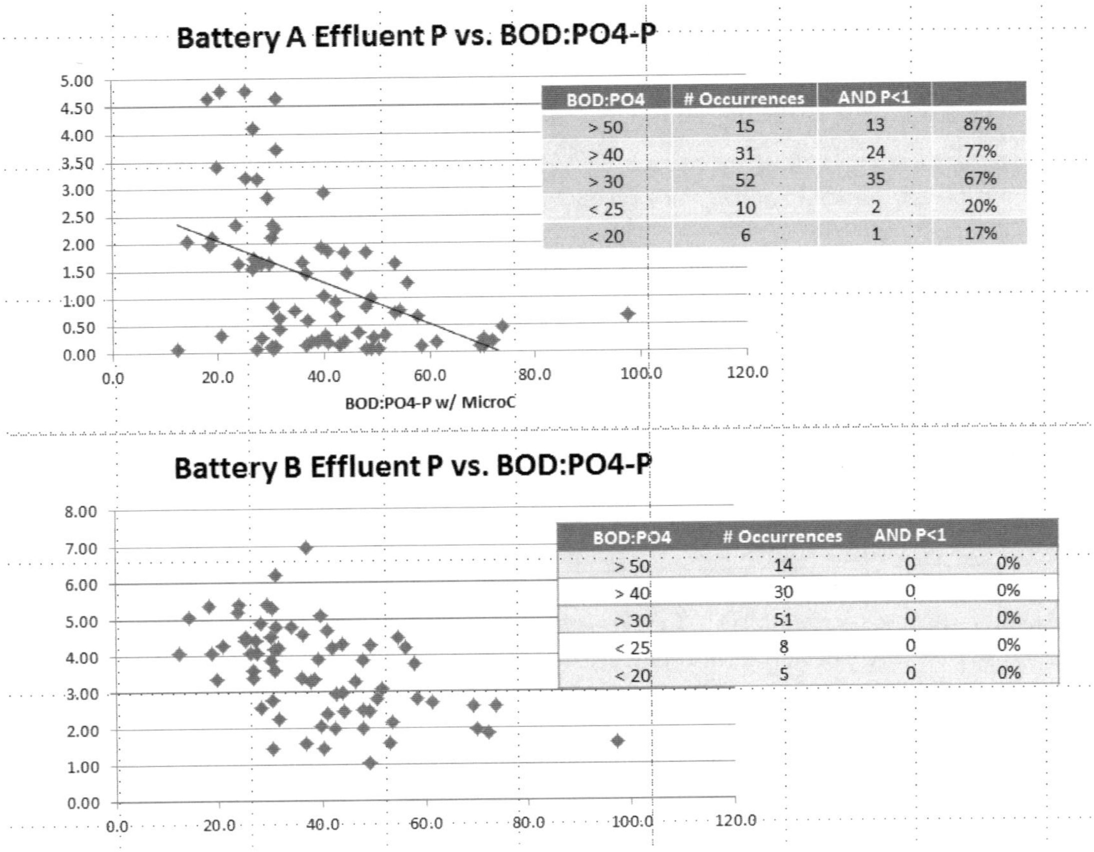

Figure 3: BOD:PO4-P Ratio in Battery A (Test) and Battery B (Control)

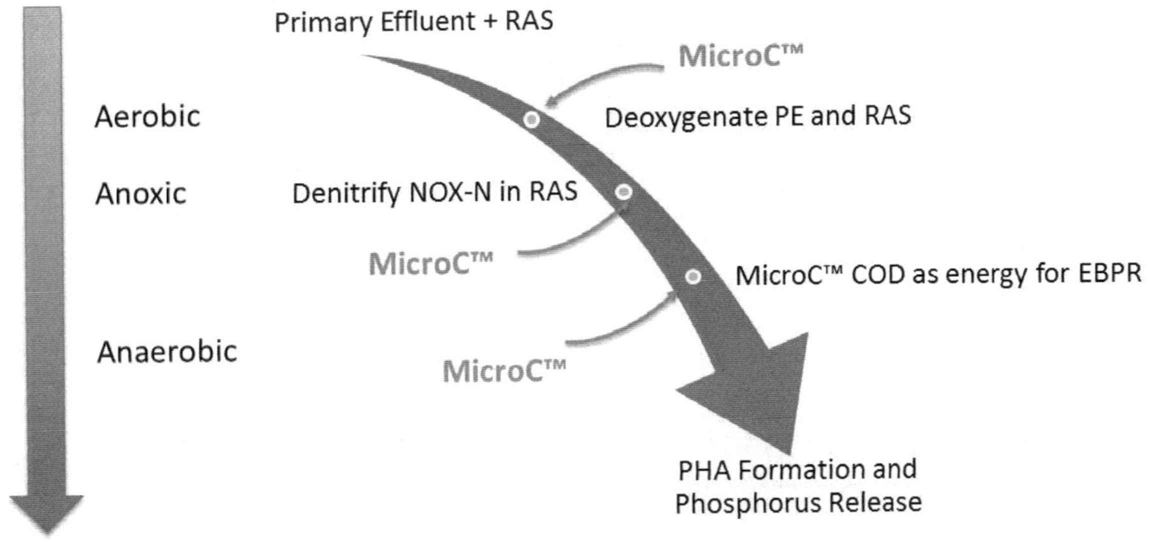

Figure 4: Hierarchy of Glycerin (MicroC™) Utilization in this BNR Process

The following example demonstrates the impact of other electron acceptors on phosphorus removal capacity at this facility. To remove an incremental 1 mg/L of NOX-N in the RAS it would require an additional 2289 lbs of COD as glycerin, impacting the COD available for EBPR and reducing the overall theoretical phosphorus removal by 0.41 mg/L at a COD:P of 15 and constant MicroC™ feed. On average, NOX-N removal in the test battery compared to the control battery was 4.5 mg/L but at times was as high as 8 mg/L. This variability in the presence of electron acceptor at the injection point impacts the COD availability for EBPR.

Modeling of a Glycerin-Driven EBPR Process

The next sections will review the available literature discussing various pathways in which glycerin could be converted to VFA's and the relative abundance of acetate and propionate. Using this research, three parameters in BioWin were changed to reflect the expected end products of glycerin fermentation. This model will be calibrated to a study using a mixed culture with glycerin addition. Finally, using the calibrated model, a comparison between full scale data and a simulation in BioWin will be performed.

Based on a few glycerin-driven EBPR studies in the literature [Guerrero, 2012; Yuan, 2010], it has been hypothesized that glycerin was not directly used by PAOs; rather glycerin was first anaerobically degraded to VFA products, through fermentation, that could be used by PAOs. Propionate is known as the major fermentation product of glycerin metabolism with respect to other compounds such as acetate, butyrate, or propanol [Barbirato, 1997; Himmi, 2000; Yuan, 2010]. Under non-limiting conditions, glycerin was mainly converted to propionate, depending on the microbial culture used, i.e. 0.6-0.8 mol/mol yield for pure cultures [Barbirato, 1997; Himmi, 2000] and a lower yield for mixed cultures [Guerrero, 2012]. Guerrero et al. (2012) studied glycerol as a sole source for EBPR, testing direct replacement of the usual carbon source for glycerol and a two-step consortium development with glycerol anaerobic degraders and PAO. The results obtained in this study with a mixed culture show a yield of around 0.5 mol of propionate per mol of consumed glycerol.

Acetate is internalized by PAOs in the anaerobic zone as mostly poly-hydroxybutyrate (PHB), whereas propionate is internalized as poly-hydroxyvalerate (PHV) [Lemos, 1998; Randall and Liu]. Therefore, correct mechanistic correlations of the fermentation process should be taken into account to appropriately simulate glycerin-driven EBPR processes. The dominant pathway for ordinary heterotrophic growth through fermentation of readily biodegradable (complex) substrate to acetate, propionate, carbon dioxide and hydrogen is the dissolved hydrogen concentration. Three mechanistic parameters in BioWin (the K_S for H_2 and two propionate yields at different H_2 concentrations) were used to adjust and calibrate the glycerin-based fermentation process in the Guerrero study to yield 0.5 mol of propionate per mole of consumed glycerin (Pr/Ac of 1.15-1.25) that would result in a theoretical P/C ratio of 0.21 mol P/ mol $C_{Glycerin}$.

Table 3: Changes to BioWin Parameters

Parameter	Default	Calibrated
H2 low/high Saturation Constant [mgCOD/L]	1	2.5
Propionate Yield (fermentation/low H2)	0	0.3
Propionate Yield (fermentation/high H2)	0.7	0.45

Figure 5 shows the calibration of the glycerin based fermentation process using the altered mechanistic parameters in BioWin and the conditions of the bench scale reactor from the Guerrero study. The proportions of acetate to propionate are roughly equal throughout the SBR cycle with a slightly higher production of propionate, similar to Guerrero's observations.

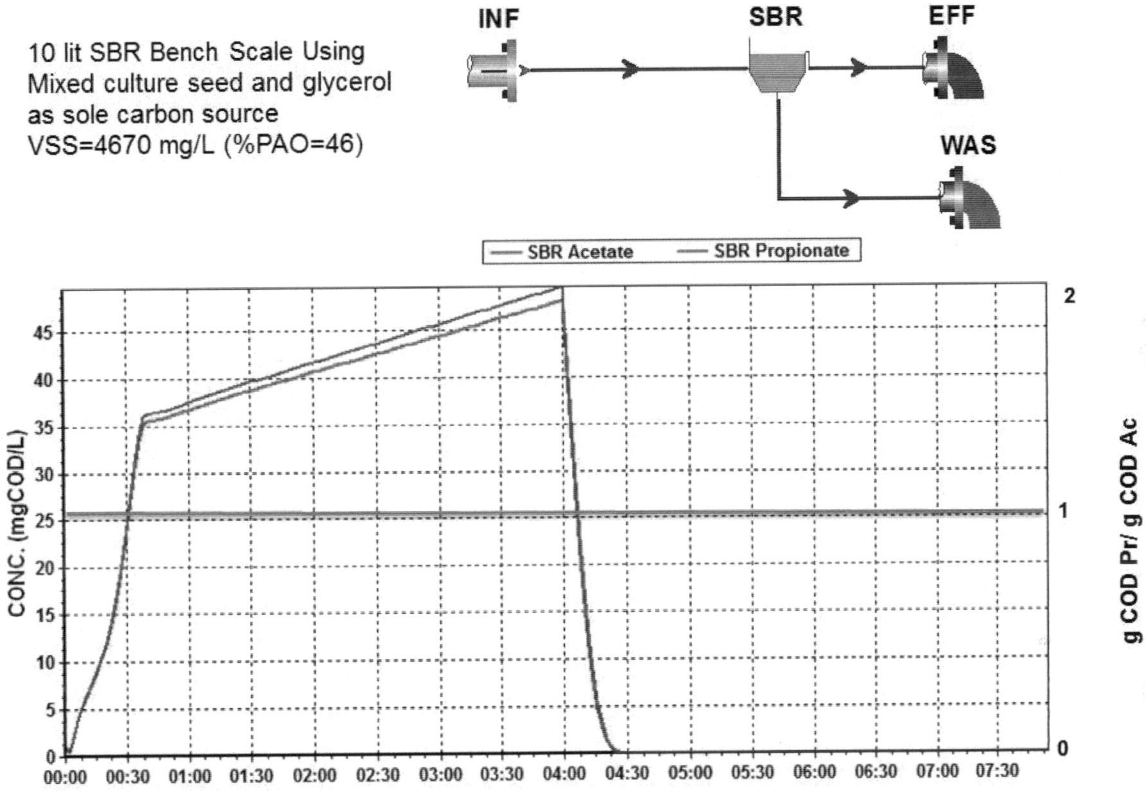

Figure 5: Biowin Calibration to Guerrero's Study Showing Roughly Equal Generation of Acetate and Propionate via Glycerin Fermentation

Next, the model calibrated to Guerrero's SBR using a mixed culture dosed with glycerin was applied to the full scale study. Figure 6 shows the process flow diagram in BioWin for both the control and test batteries. MicroC™ is dosed to battery A in the RAS channel and to the influent splitter box. The sizes of the anoxic, anaerobic and oxic zones were estimated based on profiling along the batteries for nutrients and dissolved oxygen.

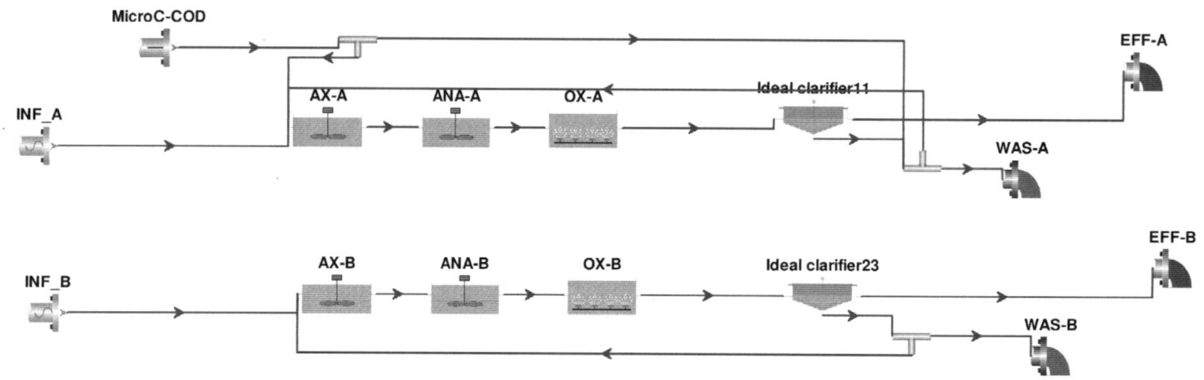

Figure 6: Process Schematic in BioWin for the Full Scale Facility

A 60 day dynamic simulation of the control battery using data from the full scale evaluation and default BioWin parameters was performed. Glycerin was not dosed to the control battery and therefore BioWin allocates the production of propionate and acetate accordingly. Figure 7 shows the simulated anaerobic and oxic soluble PO4-P concentration in addition to the flow weighted PO4-P concentration predicted by BioWin compared to the actual flow weighted PO4-P concentration from the full scale evaluation. Note that there is very good agreement between the predicted and actual values. Also note that as the full scale data showed, there is very little to no phosphorus removal in the control battery.

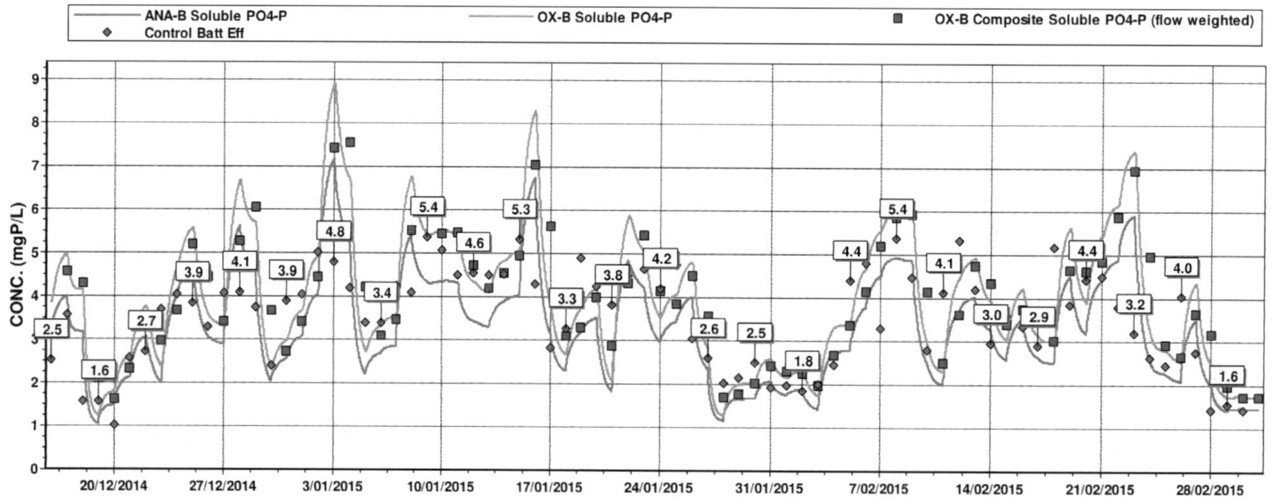

Figure 7: Dynamic Simulation of the Control Battery

Using the default parameters for fermentation of COD as glycerin, Figure 8 shows the anaerobic and oxic soluble PO4-P concentration over a 60 day dynamic simulation. Also shown is the 24 hour flow weighted composite PO4-P value predicted by BioWin in green and the actual 24 hour flow weighted PO4-P from the full scale trial. In general the values predicted by BioWin are lower than the experimental data and overall there is a poor correlation between the two values.

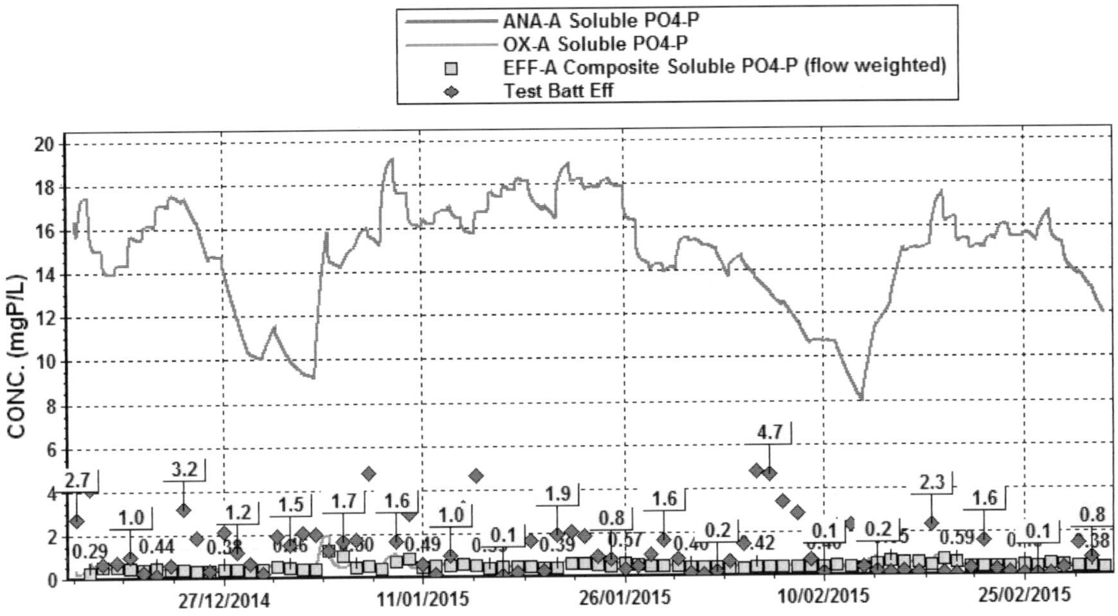

Figure 8: 60 Day Dynamic Simulation Using Default Parameters in Biowin Compared to Full Scale Data

Next, the three mechanistic parameters in BioWin were changed (the K_S for H_2 and two propionate yields at different H_2 concentrations). Figure 9 depicts the simulated Pr/Ac ratio in a 30 day dynamic simulation. The Pr/Ac ratio fluctuated between 0.75-1.27 with an average value of 1.05. This shows the ratio of Pr/Ac via the fermentation of glycerin is similar to the findings by Guerrero and others.

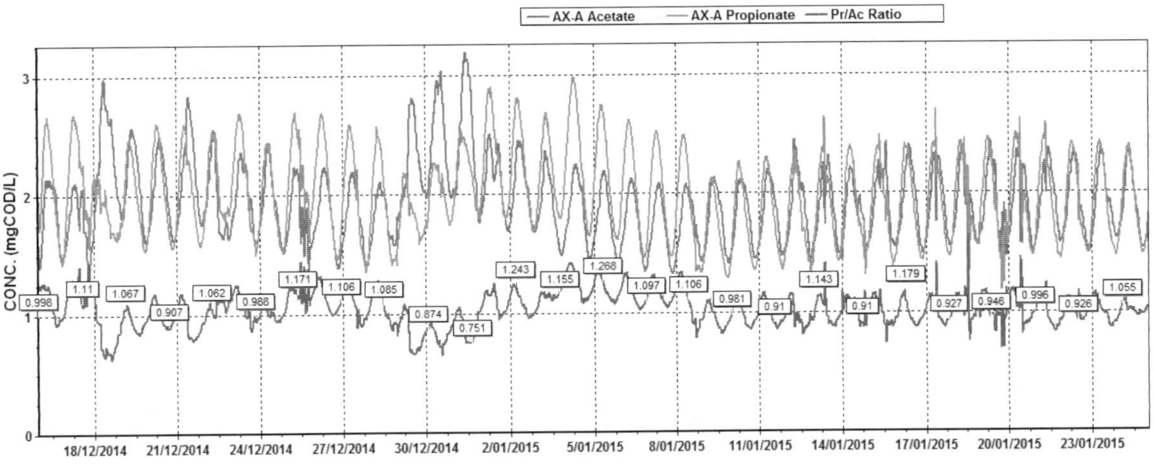

Figure 9: Propionate to Acetate Ratio Using the Calibrated Model and Full Scale Data

Using the calibrated model, Figure 10 depicts a second 60 day dynamic simulation of Battery A. Overall, the predicted 24 hour flow weighted composite PO4-P corresponded better with the experimental data. Even so, there are some excursions that can be explained by a number of changes and events during the evaluation, such as: carbon feed pump failures, availability of electron acceptors at the dosing point, imperfect plug flow and mixing conditions, further model development, etc.

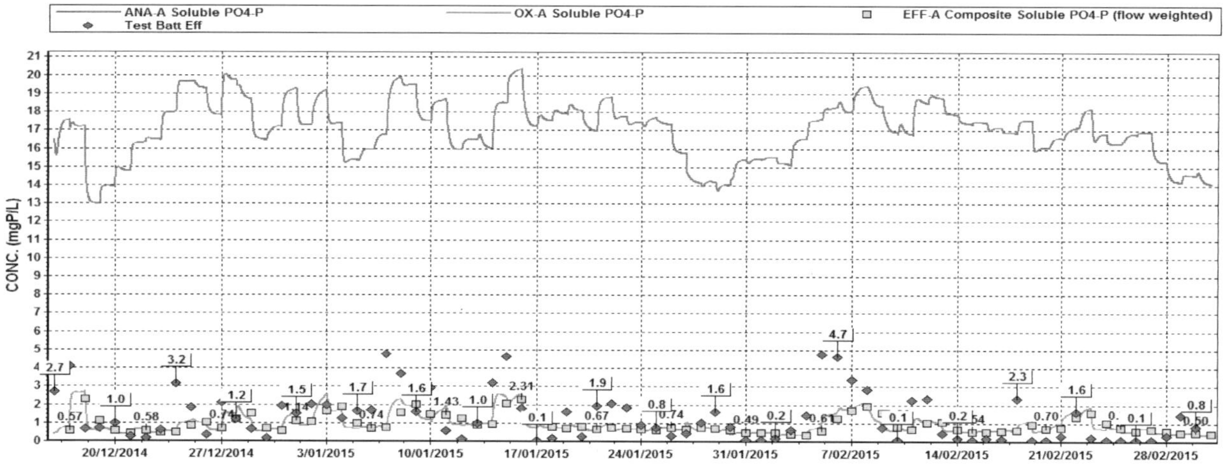

Figure 10: 60 Day Calibrated Dynamic Simulation Compared to Full Scale Data

CONCLUSIONS

The full scale implementation of a glycerin based carbon source (MicroC 2000™) at a large suspended growth facility yielded effective phosphorus removal of 2.54 mg/L compared to the control battery. The removal of dissolved oxygen, nitrate and phosphorus compared to the COD consumed was within 5.2% of the stoichiometric estimates using a COD:N of 6.1 and COD:P of 15 for glycerin. At the highest feed rate towards the end of the evaluation, effluent PO4-P averaged 0.79 mg/L, close to the client's objective of 0.75 mg/L. During the same period, effluent PO4-P values were less than 0.5 mg/L on 15 of the 21 days and were as low as 0.07 mg/L at times. This suggests that with nutrient monitoring, optimized glycerin dosing points and an optimized feed strategy the facility could meet their objectives on a consistent basis.

A review of the literature suggests that fermentation of glycerin yields a higher proportion of propionate to acetate. The parameters in BioWin that control for the production of acetate and propionate were changed and a model was successfully calibrated to results from Guerrero. Dynamic simulations were performed using default and calibrated parameters in BioWin and these results were compared to the experimental results from the full scale evaluation. The simulated results and experimental results from the control battery matched quite well. Overall the calibrated dynamic simulation of the test battery was more accurate than the default dynamic simulation.

Future work includes a comprehensive summary of glycerin fermentation to be presented at WEFTEC 2015. Using data from this and other studies, further refinements to the model calibration can be made to better predict the fate of glycerin in specific environments and evaluate the effectiveness of glycerin for EBPR.

REFERENCES

Barbirato, F., Chedaille, D., Bories, A., 1997. Propionic acid fermentation from glycerol: comparison with conventional substrates. Applied Microbiology and Biotechnology 47 (4), 441-446

Cyplik, P., Juzwa, W., Marecik, R., Powierska-Czarny, J., Piotrowska-Cyplik, A., Czarny, J., Drozdzynska, A., Chrzanowski, L., 2013. Denitrification of industrial wastewater: Influence of glycerol addition on metabolic activity and community shifts in a microbial consortium. Chemosphere 93, 2823-2831

Himmi, E.H., Bories, A., Boussaid, A., Hassani, L., 2000. Propionic acid fermentation of glycerol and glucose by Propionibacterium acidipropionici and Propionibacterium freudenreichii ssp. shermanii. Applied Microbiology and Biotechnology 53 (4), 435-440.

Guerrero, J., Taya, C., Guisasola, A., Baeza, J.A., (2012) Glycerol as a sole carbon source for enhanced biological phosphorus removal. Water Research, 46, 2983-2991.

Ledwell, S., Togna, P., Andalib, M. Giovannone, A., Rooney, K. (2013) The Fate of Glycerin in BNR, A Closer Look at Nitrite Accumulation and Glycerin Specialists. Proceedings of the WEF/IWA Nutrient Removal and Recovery Conference, Vancouver, British Columbia.

Lemos, P.C., Viana, C., Salgueiro, E.N., Ramos, A.M., Crespo, J.P.S.G., Reis, M.A.M., 1998. Effect of carbon source on the formation of polyhydroxyalkanoates (PHA) by a phosphate-accumulating mixed culture. Enzyme Microbiol. Technol. 22, 662–671.

Randall, A. A.; Liu, Y.-H. (2002) Polyhydroxyalkanoates Form Potentially a Key Aspect of Aerobic Phosphorus Uptake in Enhanced Biological Phosphorus Removal. Water Res., 36, 3473–3478.

Yuan, Q., Sparling, R., Lagasse, P., Lee, Y.M., Taniguchi, D., Oleszkiewicz, J.A., 2010. Enhancing biological phosphorus removal with glycerol. Water Science and Technology 61 (7), 1837-1843.

Managing Nutrient Removal at a Four-Sludge BNR Facility

James Ervin[1], Simret Yigzaw[1], Kelley Mayne[2]
[1]City of San Jose,
[2]San Jose-Santa Clara Regional Wastewater Facility
* Email: james.ervin@sanjoseca.gov

ABSTRACT

The San Jose-Santa Clara Regional Wastewater Facility employs Biological Nutrient Removal (BNR) as its secondary return activated sludge process. BNR treatment was attained through adaptation of a large existing facility.

The Facility was originally constructed in 1956 for primary treatment. Secondary treatment, with mirror image A-side and B-side, was added in 1964. Nitrification aeration basins and clarifiers were added in 1979. In the 1990s, Secondary and Nitrification areas were modified to perform parallel BNR treatment. Primary effluent flows are step-fed to first and third quad tanks in basins in both areas. Second and fourth quads are aerobic. Dissolved oxygen is continuously monitored at the end of the second and fourth quads to regulate aeration.

Secondary/BNR treatment is accomplished via four autonomous treatment areas. A-sides and B-sides of two areas are operated to discharge similar concentrations of BOD, TSS, and ammonia. The physical layout and operating experience and parameters are reviewed.

KEYWORDS: Activated sludge, Biological Nutrient Removal.

INTRODUCTION

The Facility and Receiving Water

The San Jose-Santa Clara Regional Wastewater Facility provides tertiary treatment of domestic, commercial, and industrial wastewater for a total service area population of approximately 1.4 million. The Facility serves multiple cities and wastewater districts: the cities of San José, Santa Clara, and Milpitas; Santa Clara County Sanitation Districts No. 2 and No. 3; the West Valley Sanitation District, including Campbell, Los Gatos, Monte Sereno, and Saratoga; and the Cupertino, Burbank, and Sunol Sanitary Districts. Each associated satellite collection system is owned, operated, and maintained independently from the discharger, and collects wastewater from its respective service area (Figure 1).

The wastewater treatment process includes screening and grit removal, primary sedimentation, secondary treatment by activated sludge process, secondary clarification, filtration, disinfection, and dechlorination. The Facility has an average dry weather flow design capacity of 167 million gallons per day (MGD) and a 271 MGD peak hourly flow capacity for full tertiary treatment.

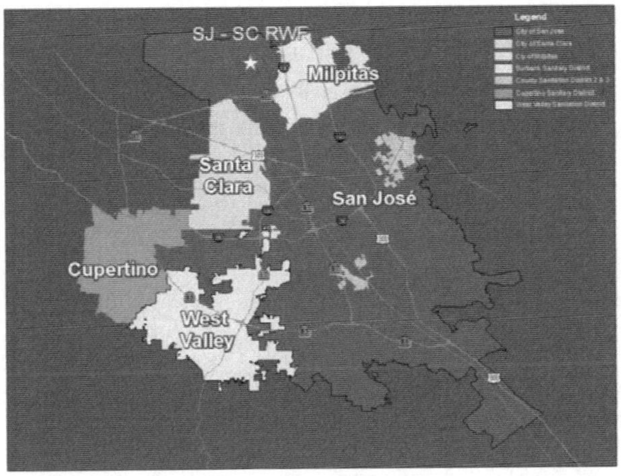

Figure 1. The tributary agencies served by the Facility.

Primary treatment (settling or skimming) removes a significant percentage of both organic and inorganic solids from wastewater. Secondary treatment employing activated sludge removes fine suspended solids, dispersed solids, and dissolved organics through volatilization, biodegradation, and incorporation into sludge. Tertiary (advanced) treatment includes nitrification (the biological oxidation of ammonium with oxygen, then into ammonium, then into nitrite followed by the oxidation of these nitrites into nitrates) and filtration (the removal of minute solids to further improve the effluent quality before it is discharged to the receiving environment). Biological Nutrient Removal (BNR) removes total nitrogen (TN) and total phosphorus (TP) from wastewater, through the use of microorganisms under aerobic and anaerobic conditions in the treatment process. Nitrification and denitrification are the specific biological processes for removing nitrogen. During nitrification, ammonium is oxidized to nitrite, after which nitrite is then oxidized to nitrate under aerobic conditions. Denitrification involves biological reduction of nitrate to nitric oxide, nitrous oxide, and nitrogen gas, under anaerobic conditions. Biological phosphorus removal relies on phosphorus uptake by aerobic heterotrophs capable of storing orthophosphate in excess of their biological growth requirements (Metcalf & Eddy, 2003).

The Facility discharges to the Lower South San Francisco Bay (LSB). The LSB area extends from the Dumbarton Bridge south to the Southern Pacific Railroad Bridge across Coyote Creek. Treated wastes are discharged to Artesian Slough (also called Mallard Slough, 37° 26'23.38" Latitude and 121° 57' 29.18" Longitude), from where it flows by way of Coyote Creek, to the main body of South San Francisco Bay (Consoer, Townsend and Associates, 1968, p. 25). The discharge point is situated approximately two miles from the Creek, 6.5 miles from South Bay proper (Larry Walker Associates, 1987, 1-4).

TREATMENT PLANT UPGRADES AND POLLUTANT REDUCTION

The San Jose-Santa Clara Regional Wastewater Facility (formerly known as the San Jose/Santa Clara Water Pollution Control Plant) was constructed or modified in four successive phases starting in 1956. The Facility operated as strictly as a primary treatment plant from 1957 to January 1964. From that point until February 1979 single-stage return activated sludge was employed to provide secondary treatment. The Facility initiated new nitrification aeration basins and a filtration facility in February 1979. From that point until June 1997 the Facility provided advanced secondary wastewater treatment that employed nitrification and filtration. Finally, starting in June 1997, flows to both the secondary and nitrification aeration basin areas were modified so that both areas operated in parallel as four-stage BNR reactors that provided both nitrogen and phosphorus removal.

Biochemical Oxygen Demand (BOD) removal

Facility effluent BOD, TSS, and nutrient loading decreased with Facility treatment process upgrades. Based on monthly averages, effluent BOD load dropped from 47,138 kg/d (during the period from January 1957 to January 1964) to 10,318 kg/d (during the period from February 1964 to January 1979) to 1,667 kg/d (during the period from February 1979 to June 1997) to 1,439 (during the period from July 1997 to December 2013), corresponding to the wastewater improvement periods of primary treatment, secondary treatment, tertiary/advanced wastewater treatment, and BNR) (Figure 2). The percentage of BOD removal from one treatment era to the next was 31%, 91%, and 99%.

Total Suspended Solids (TSS) removal

Effluent TSS load dropped from 22,555 kg/d (during the period from January 1957 to January 1964) to 14,414 kg/d (during the period from February 1964 to January 1979) to 886 kg/d (from February 1979 to June 1997) to 736 (from July 1997 to December 2013), corresponding to the wastewater improvement periods of primary treatment, secondary treatment, tertiary/advanced wastewater treatment, and BNR (Figure 2). The percentage of TSS removal from one treatment era to the next was 63%, 88%, and 99%.

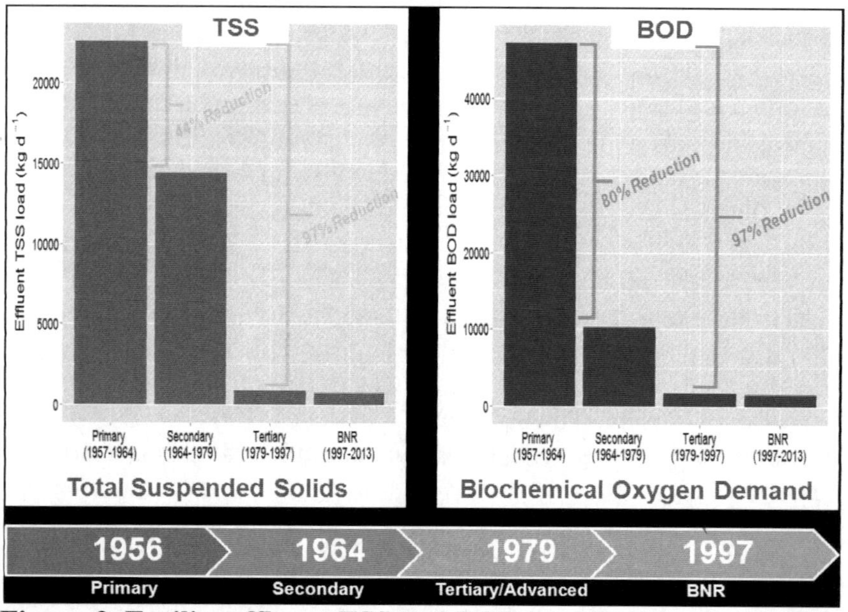

Figure 2. Facility effluent TSS and BOD load changes between treatment eras: primary (1957-1964), secondary (1964-1979), tertiary (1979-1997), and BNR (1997-2013). (From Yigzaw, 2014)

Nitrogen and phosphorus removal

The major component of facility nitrogen load to the LSB is nitrate (83.2%). Only 5.5% of the nitrogen load from the Facility is in the form of ammonium (RWF 2014). There appears to be small upward trend in terms of an increase of ammonium in the past 10 years. The total nitrogen load coming into the Facility is comprised of 57.5% ammonium, 41% organic nitrogen, and 2% nitrate. The Facility removes almost all the ammonium (99%) from influent. However, it also adds100 kg/d of ammonium for Chloramination disinfection (ammonium is added to prolong chlorine effectiveness) to the Facility final effluent.

Effluent ammonium load dropped from 5,386 kg/d (from February 1964 to January 1979) to 539 kg/d (from February 1979 to June 1997) to 275 (from July 1997 to December 2013), corresponding to wastewater improvement periods of secondary treatment, tertiary/advanced wastewater treatment, and BNR (Figure 3). (The primary treatment era is not covered because no data exist for this period.) The percentage of ammonium removal from one treatment era to the next was 26% and 94%.

Effluent nitrate load first jumped from 1,194 kg/d (from February 1964 to January 1979) to 7,389 kg/d (from February 1979 to June 1997), and then declined to 3,886 (from July 1997 to December 2013), corresponding to the wastewater improvement periods of secondary treatment, tertiary/advanced wastewater treatment, and BNR (Figure 3). (There is no data for the primary treatment era.) The percentage of nitrate removal from nitrification to BNR was 47%.

Effluent phosphate load dropped from 2,261 kg/d (from February 1964 to January 1979) to 1,912 kg/d (from February 1979 to June 1997) to 417 (from July 1997 to December 2013), corresponding to the wastewater improvement periods of secondary treatment, tertiary/advanced wastewater treatment, and BNR (Figure 3). (There is no data for the primary treatment era.) The percentage of phosphate removal from one treatment era to the next was 56% and 89%. It should be kept in mind that much of the phosphorus reduction can also be attributed to the reduction in removal of phosphates from detergents, as mandated by California law. Nonetheless levels of phosphorus further declined with BNR, a reduction of 78%, from the previous treatment era.

Facility upgrades resulted in clear overall decreased loads of discharged BOD, TSS, NH4+, and PO4 over time, across treatment eras (primary, secondary, tertiary, and BNR). The same overall trend was true for nitrate (NO3-). Ammonium loadings drastically decreased 89% when treatment upgraded to tertiary (nitrification). Nitrification did not remove total nitrogen, but rather converted most of the ammonium to nitrate. While ammonium loads dropped following nitrification, nitrate loads increased approximately 520%. Upgrade to BNR in 1997 introduced denitrification to the treatment process with a subsequent nitrate reduction of 48%. (Yigzaw, 2014)

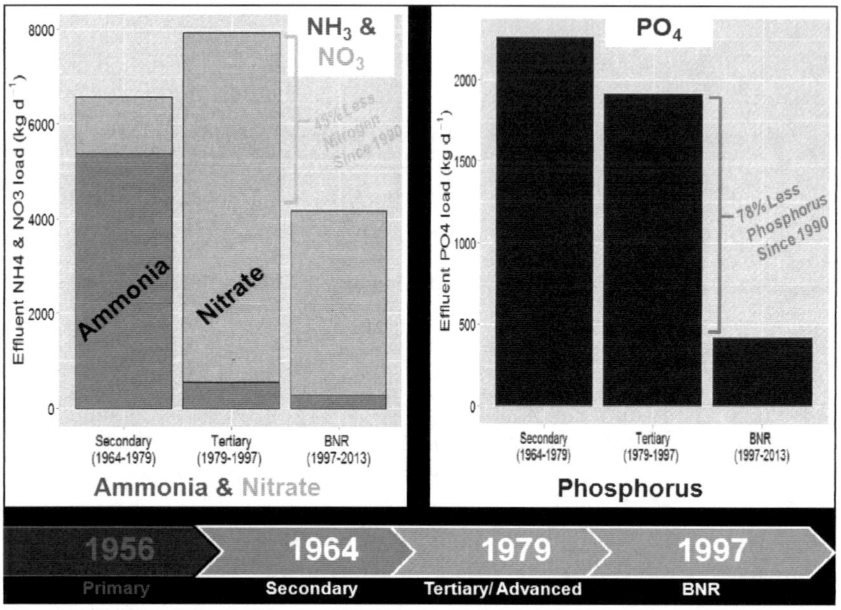

Figure 3. Facility effluent NH4+ and NO3- loads and PO4 load changes between three treatment eras: secondary (1964-1979), tertiary (1979-1997), and BNR (1997-2013). (From Yigzaw, 2014)

Effect of Seasons

As is the case with influent load, effluent load (containing BOD, TSS, NH4+, NO3-, and PO4) shows seasonal variability. During the dry season, with BNR treatment operating efficiently, PO4 removal from influent to effluent can exceed 90%. During the dry season, phosphorus in effluent was very low. In 2012 and 2013, for example, the wet season drop in phosphorus removal corresponded with rains, cooler temperatures, and changes to process operations in response to seasonal changes. Like phosphorus, both nitrification and denitrification are affected by temperature and therefore subject to seasonal effects (San Jose-Santa Clara RWF, 2014a).

BNR AERATION AREAS

Following primary clarification, wastewater flow is split and directed to both BNR aeration areas: BNR 1 and BNR 2. These areas continue to be commonly referred to by their former names and former functions as “Secondary” and “Nitrification” areas.

After construction of the nitrification aeration basins in 1979 until the 1997 adoption of the BNR process, the facility operated as a two-stage activated sludge process. The two aeration areas operated in series. The BNR 1 “Secondary” area was aerated vigorously and maintained 1.5 to 2 day solids retention time (SRT) to provide carbonaceous BOD removal. Secondary effluent then passed to the BNR 2 “Nitrification” aeration basins that afforded up to 15 day SRT to accommodate a nitrification process (oxidation of ammonia to nitrate).

From 1993 through 1996, the facility (at that time known as the San Jose/Santa Clara Water Pollution Control Plant (SJ/SC WPCP)) undertook a series of studies to evaluate optimization of the then existing two-stage activated sludge process (City of San Jose 1998). Ironically, the goal at that time was not to enhance removal of nutrients, but rather to determine empirically if a BNR process would provide a reliable level of copper removal. In 1994 and 1995, aeration basins were pilot tested in four and eight tank configurations. The study determined that reconfiguring all existing aeration basins to a step-feed four-tank (or "quad") configuration employing an anaerobic, aerobic, anoxic, and aerobic aeration scheme afforded a number of benefits in addition to slightly reducing overall effluent copper concentrations. Conversion to the BNR aeration process added greater protection against bulking and foaming, afforded greater hydraulic flow through the units, and recovered some amount of oxygen and alkalinity. It was recognized at the time that the BNR process also reduced effluent nitrogen and phosphorus loads, but since there were no regulatory concerns about those constituents, this was not considered a significant factor for adopting BNR in the later 1990s. (City of San Jose 1998)

Thus, the two BNR aeration areas, BNR 1 and BNR 2, are today operated as parallel, theoretically identical, processes. Each of the 24 aeration basins (16 in BNR 1 and 8 in BNR 2) are configured as four tanks referred to as "quads" or "passes." Primary effluent and Return Activated Sludge (RAS) is introduced into the first tank with a low rate of aeration to maintain mixing but afford as little free oxygen as possible. This creates an anaerobic selector zone that discourages fast growing filamentous bacteria in favor of slower growing nitrifying bacteria and Phosphorus Accumulating Organisms (PAOs). Wastewater progresses through the next tank which is vigorously aerated to stimulate both carbonaceous BOD removal and nitrification. Primary effluent is step-fed to the third tank along with a low rate of aeration to create an anoxic zone that facilitates denitrification. The final fourth tank is again vigorously aerated to remove additional carbonaceous BOD and ammonia. Mixed liquor channels convey wastewater from the final aeration tank to the secondary clarifiers. Mixed liquor channels are also aerated to maintain mixing and this affords additional aeration treatment, albeit not entirely intended. Any small amount of oxygen and nitrate remaining in the RAS is substantially removed by microorganisms during the clarification process before the RAS is circulated back to the first tank in the BNR basins.

Because they were built at different times and for different purposes, the size and configuration of BNR aeration tanks differs between the two areas. BNR 1 was constructed as 16 basins in a four-tank configuration optimized to remove an enormous quantity of BOD using 2-day or less SRT. The BNR 2 area was originally operated as 16, two-tank, basins with a 15-day SRT. As a result of the conversion to parallel BNR processes, BNR 2 was reconfigured into 8 four-tank basins to mirror the process and configuration of BNR 1. Both areas are now generally operated with an SRT of 8 to 10 days. The two BNR areas attain similar levels of wastewater treatment performance even though the tank volumes, aeration efficiency, and flow throughputs vary between the units. The total aerator volume in BNR 1 area is 44 million gallons compared to 24 million gallons in BNR 2. As a result, BNR 1 generally treats about 60 percent of primary effluent with BNR 2 treating the remaining 40% (Figure 4). In practice, the flow split will vary widely depending on availability of aerators, incoming wastewater loads, and time of year. Additionally, both BNR areas were each constructed as two mirror image sides for reliability and redundancy: an "A-side" and a "B-side." RAS is not exchanged between the BNR areas or sides, which means that the four aeration basin segments operate with independent microbial populations, or "four sludges."

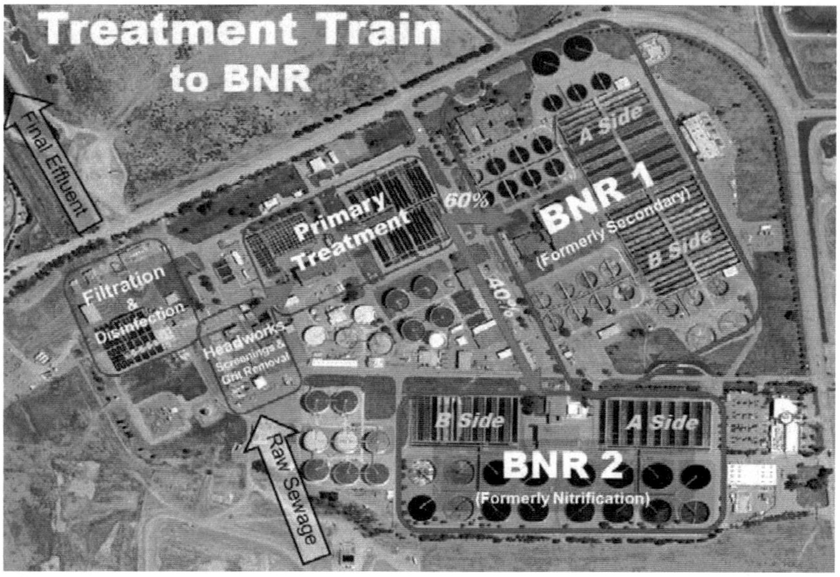

Figure 4. An overhead view of the San Jose-Santa Clara Regional Wastewater Facility highlighting the two secondary/BNR areas: BNR 1 and BNR 2.

Controllable Parameters

The goal of BNR treatment is to provide optimal environment for microbes that consume nitrogen and phosphorus nutrients. When BNR is employed as part of an activated sludge process, the controllable parameters are surprisingly limited.

- A theoretical SRT directly regulates the rate of sludge wasting and hence the age of the microbial population and the balance of food to microbial mass.
- Flows are apportioned and redirected to apportion incoming loads to available tankage.
- Oxygen concentrations are maintained and varied to accommodate the biological consumption of BOD and nutrients.
- A fourth factor, temperature, has a significant effect on BNR effectiveness, but is essentially uncontrollable at this facility; ranging from around 15 degrees centigrade in the cool months to 25 degrees in the summer.

Primary control over the BNR process is exercised through continuous monitoring and adjustment of solids wasting based on SRT (or mean cell residence time (MCRT)) calculation. The facility on-line Distributed Control System automatically monitors parameters for each of the four BNR sludge units (BNR 1A, BNR 1B, BNR 2A, and BNR2B). Parameters monitored are: number of tanks in service, Mixed Liquor Suspended Solids (MLSS) concentration, RAS suspended solids concentration, total flow to the unit in gallons, and Waste Activated Sludge (WAS) flow rate in gallons per minute to attain the desired SRT expressed in days. A final needed factor for the SRT calculation, effluent TSS, is not measured. Instead, it is simply assumed for purposes of the SRT and wasting calculation that a target of 8 mg/l is desired in secondary effluent. It was discovered in practice that reliable monitoring of secondary effluent TSS was not possible at a needed frequency for the four BNR units. Therefore, the reasonable figure of 8 mg/l was inserted as a constant in SRT calculation. All other figures are tracked, averaged over a defined period of time from 2 to 12 hours to screen out random data spikes, and continuously calculated to match established set points. Facility operators monitor equipment operation at all times, make adjustments as needed, and take manual control if or when the automated system is not able to adjust to demands.

Flow equalization is employed to balance flows to the BNR units throughout the 24 hour cycle. Primary effluent flows are partially directed to a 16 million gallon equalization basin during the high flow periods: from 10:00 AM to 2:00 AM the following day. As flows decrease, from 2:00 AM to 10:00 AM, the equalization basin is drained of "settled sewage" that now flows back to the BNR units with continuously incoming primary effluent. The equalization basin provides a good deal of stability for the downstream BNR units. However, owing to the closer proximity of the BNR 1 area and the pre-existing alignment of plumbing, distribution and mixing of settled sewage with primary effluent is less uniform in the BNR 1 aeration basins.

Primary effluent, in combination with settled sewage from the equalization basin, is delivered to the first and third tanks in the BNR 1 and BNR 2 basins. This is a step-feed process whereby roughly 60 percent of incoming primary effluent/settled sewage enters the first quad tank and the remaining 40 percent is fed to the third quad. Minimal aeration is provided to the first and third tanks, enough to provide mixing, but not enough to sustain significant carbonaceous BOD removal. The anaerobic conditions in the first tank stimulate PAO-type microbes to release phosphates to power minimal metabolism while growth of most other forms of bacteria is seriously inhibited. The second tank in each aeration basin is vigorously aerated to a set dissolved oxygen concentration of 2.5 mg/l. The oxygenated environment supports both carbonaceous BOD consumption and nitrification. During this phase, the PAOs also accumulate and store polyphosphates from the wastewater. In the third tank, the combination of low rate of aeration (dissolved oxygen concentration target is 0.5 mg/l), and incoming primary effluent/settled sewage, stimulates rapid depletion of remaining dissolved oxygen. Carbonaceous BOD removal and nitrification continue only to the extent that dissolved oxygen concentrations support those processes. Microbes that can exploit nitrate (NO_3) as an electron acceptor are now favored, and the majority of the denitrification (conversion of nitrate to nitrogen gas) occurs in the third quad tank. The fourth tank is again vigorously aerated to facilitate additional BOD removal and nitrification. Mixed liquor from the aeration basin fourth tank then passes to the secondary clarifier. Any remaining dissolved oxygen or nitrate is consumed from the sludge blanket in the clarifier before RAS returns to the aeration basin first tank. (Figure 5)

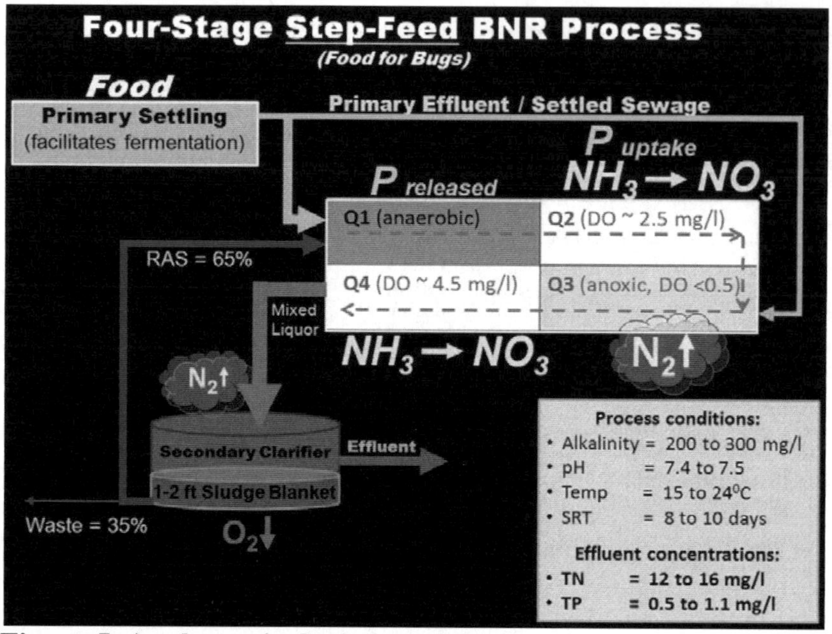

Figure 5. A schematic depiction of the four-stage, step-feed, BNR process. The arrangement of tanks in this diagram matches the BNR 1 area. Adapted from San Jose-Santa Clara RWF 2013 Annual Self Monitoring Report, p.22.

The facility air distribution system could be considered antiquated. Two parallel stainless steel headers in each of the 96 BNR 1 and BNR 2 tanks are connected to Sanitaire coarse bubble diffusers or EDI FlexAir fine bubble diffusers. The number of diffusers in each tank varies from 268 in the shorter BNR 2 tanks to 324 in BNR 1 tanks 2 and 4, and 460 diffusers in BNR 1 tanks 1 and 3. One control valve regulates air flow to both headers in each tank. Meters located near the end of the second and fourth tanks monitor dissolved oxygen concentrations against the established set points of 2.5 and 4.5 mg/l for tanks 2 and 4, respectively.

Until the early 2000s, Sanitaire stainless steel coarse bubble diffusers were exclusively used to disperse air into BNR tanks. In 2002, the facility began pilot testing EDI FlexAir T-Series fine bubble diffusers consisting of 2 and 4 foot PVC tubes of 3-inch diameter encased in perforated polypropylene sleeves. The fine bubble diffusers were adopted to more efficiently distribute air in the aerobic tanks (tanks 2 and 4 in both BNR 1 and BNR 2 areas). The PVC tube design was incorporated because these diffusers could be retrofitted into existing coarse bubble diffuser fittings with no other modifications. The existing system creates a "dual spiral roll" of turbulence in the tanks to mix and distribute air. A disadvantage of this retrofit design is that polypropylene sleeves have a short operational life of between 3 to 5 years, must be protected from extensive UV exposure at all times, and they are not as efficient as newer membrane disc designs.

Pulse aeration, blowing air intermittently to accommodate mixing while reducing air flow, has been employed in the anaerobic and anoxic first and third quad tanks since 2008. This was adopted as an energy saving practice. Theoretically, the anaerobic and anoxic tanks would be better served with mechanical mixers to prevent solids settling without wasting air. Pulse aeration affords a compromise that avoids completely replacing existing aeration basin infrastructure. The durable stainless steel coarse bubble diffusers are not adversely affected by the varying air flows.

Monitoring Performance

Monitoring of BNR performance is carried out by both continuous monitors and hand sampling or calculation.

- As mentioned above, dissolved oxygen concentration is continuously monitored in the second and fourth quad tanks of each aeration basin. The continuous monitoring equipment is verified with a hand-held meter once per day. DO concentrations are used to regulate air flow to the basin.
- Mixed Liquor Suspended Solids (MLSS) is also monitored continuously in the mixed liquor channel of each of the four BNR units: BNR 1A and B, BNR 2 A and B. A hand-held MLSS meter is used to check continuous results twice per day. Both MLSS and volatile solids are tested in the facility laboratory twice per week as another check on the continuous results.
- Depth of secondary clarifier sludge blanket is measured daily using a clarifier sludge judge.
- Filament counts and microbial composition evaluations are performed by microscope twice per week to determine sludge age and assess the condition of the microbial population in each of the four BNR units.
- Sludge Volume Index (SVI), using the 30-minute settling test, is measured and calculated three times per day.
- Daily 24 hour composite samples of influent and effluent of the four BNR units are measured for chloride, ammonia, nitrite, nitrate, sulfate, TSS, and turbidity.

In addition to the above readings, operators continuously monitor flows and compare them to valve readings. This is to check the validity of the automated monitoring systems. It is not uncommon, almost inevitable, that at some point a mechanical or electrical failure will occur. Frequently, the difference between inconvenience or disaster is the amount of time that passes between the failure and its detection by staff. (Lemma, 2013)

For 24-hour operations, the Facility maintains a continuous monitoring system that tracks flows, water levels, and operations of all valves, pumps, blowers, and other equipment. This function is performed by the Facility's Distributed Control System (DCS) which is in the process of being upgraded to an 800xA ABB hardware and software platform. Data displayed by this system is watched at all times by a control room operator, as well as by BNR area operators on duty. (San Jose-Santa Clara RWF, 2014a)

Control Options

Just as the list of BNR controllable parameters is limited, so too are the options that can be exercised to manipulate the BNR process. Ammonia breakthrough is the greatest concern. Ammonia is the only nutrient parameter regulated by the Facility's NPDES permit. Also, incomplete nitrification in the BNR areas results in elevated concentrations of nitrite (NO2+) which disrupts the downstream chlorine disinfection process. For these reasons, Facility operators respond quickly to indications of ammonia or nitrite breakthrough.

The initial response when ammonia or nitrite concentrations start to elevate is to increase aeration. In practice, this means that aeration is increased in the third, anaerobic quad tanks of the affected BNR unit. Typically, the third tank will be taken out of pulse aeration mode and aeration will be run continuously to enhance oxidation of ammonia and nitrite.

If simply increasing aeration does not appear to solve the problem, the next change to operations that is typically considered is to reduce the amount of primary effluent/settled sewage feed flowing to the BNR unit. If feed cannot be simply shunted over to other units, Facility operators may decide to cut off step-feed (primary effluent feed to the third anoxic tank) and run the affected basins in "plug flow" mode. In plug flow, all primary effluent is fed to the first quad tank, and usually, tanks 2 through 4 are continuously aerated at increased air volume.

After aeration, adjustment of sludge wasting is the next control option. Regular twice per week microscopic examination of mixed liquor should detect gross changes to the microbial population, but ammonia breakthrough or other signs of upset may trigger reevaluation. Ideally, RAS and SRT are controlled to maintain sludge at an ideal log growth phase, neither too young, nor too old. Solids wasting can be increased or decreased to adjust sludge age. And, finally additional BNR basins can be put into service, or taken out, if the source of the problem is overall facility loads.

Tracking Nitrogen Trends

An example of long-term performance of the facility's four BNR units is seen in the charted ten-day running average measurements of nitrite, ammonia, and nitrate concentrations (Figure 6). "Good" BNR performance, with respect to nitrogen species typically results in no detectable concentrations of nitrite or ammonia. As the chart shows, the best, consistent performance is usually seen in the later, warmest months of the year. Ammonia and nitrite spikes are more likely to occur and persist in the cooler month when lower temperatures prevent microbes from regenerating as rapidly. It is also evident that the four BNR units occasionally perform differently. Ammonia and nitrite spikes never occur in all four units simultaneously at the same magnitude, which is a very good thing.

Viewing data charted over the past 30 months, it is apparent that the "Nitrification B-side," otherwise known as BNR 2B has been more prone to nitrite and ammonia breakthrough (bright green trend line in figure 6). This has been a long running issue attributed to differences in hydraulic flow, but more recently exacerbated by the fact that BNR 2B is the last of the four units on the rolling 5-year fine bubble diffuser replacement cycle. Hence, when breakthroughs have occurred in BNR 2B, they tend to be a little more intense and take a little longer to resolve. For example, the initial ammonia breakthrough shown in late January 2013 occurred as a result of a mechanical valve failure that was quickly discovered. A second bigger spike occurred in May as BNR 2B tanks were brought back in service following valve replacement and the new BNR 2B sludge required aging before efficient nitrification could be resumed. A similar event occurred in January 2015. It was discovered that the data system was reading incorrect effluent valve positions in one basin of BNR 2B. This led to over wasting of BNR 2B sludge and partial nitrification leading to a nitrite spike. Subsequent ammonia and nitrite spikes reflect draining the affected tanks and manipulation of flows to restore the proper sludge age.

Some caution must be exercised when tracking nitrogen species discharged from a BNR process. The output depends on stability and sludge age. Under good conditions, the microbial population responds to changes over the course of one or two SRT cycles. (e.g. SRT of 8 to 10 days at this Facility). Thus, responding to short duration ammonia or nitrite spikes is not advisable. The ten-day running average of nitrogen data is used so that the results reflect the output of the entire sludge age cohort under the general conditions. It is expected that final resolution of nitrite or ammonia breakthrough in any BNR unit will usually take 10 to 20 days, or longer during the cool season.

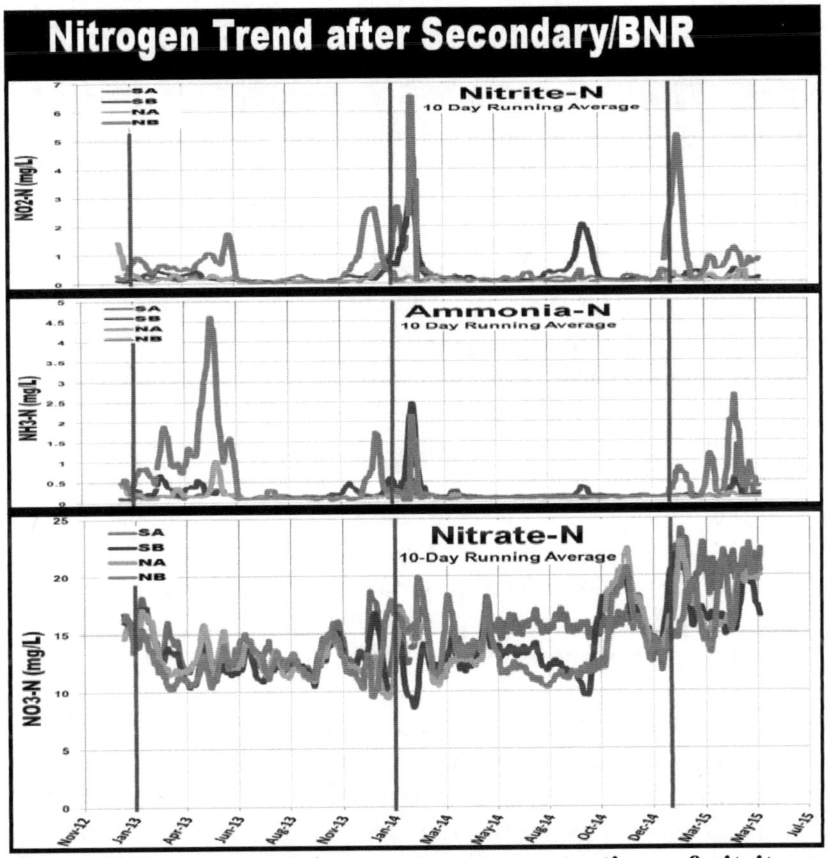

Figure 6. Ten-day running average concentrations of nitrite, ammonia, and nitrate for each of the Facility's BNR units: Secondary A-side (aka: BNR 1A), Secondary B-side (aka: BNR 1B), Nitrification A-side (aka: BNR 2A), and Nitrification B-side (aka: BNR 2B)

The ten-day nitrogen trend data is a central feature of a weekly operations area supervisors' meeting. The scheduled meeting is the main venue for reviewing microbial trends and discussing response options. Frequently, a response to problems or significant maintenance requires shifting wastewater loads from one BNR unit to another. A weekly assessment of the status of microbial population is essential to the smooth running of the facility, particularly during the cool season.

Future

As service population increases, so will influent flow and load. The San Jose-Santa Clara RWF has been embarked in a master planning process since 2007. The "Plant Master Plan (PMP)," as it is called, is used to plan and coordinate major infrastructure repair and upgrade over the next 3 decades. (San Jose-Santa Clara RWF, 2013b).

According to the PMP, the current step-feed mode of operation will have insufficient capacity around 2026, assuming projected increases in load and flow. At that point the facility may have to transition to Nitrifying Activated Sludge (NAS) mode. If future regulations require more stringent nitrogen limits, then the current BNR units may have to be upgraded to a modified Ludsack-Ettinger (MLE) or other type of BNR process requiring internal recycle of mixed liquor and changes to the aeration scheme. A recently concluded PMP project involved evaluation of current and future year 2040 aeration demands for the BNR units. A process simulator was used to examine three different aeration configurations under current (year 2014) and future (year 2040) influent conditions. The results of this evaluation will provide a first step in determining optimal upgrades that may be required. (R. Appleton, personal communication)

SUMMARY AND CONCLUSIONS

The San Jose-Santa Clara Regional Wastewater Facility operates a very effective BNR treatment process, albeit with no chemical addition, and using 1964 and 1979 infrastructure that was not designed to perform as a BNR plant. The Facility has undergone a series of three significant upgrades since original construction as a primary treatment plant: first to secondary, then to add nitrification and filtration, and ultimately to convert to four-stage, step-feed BNR. Each upgrade or modification has resulted in significant reduction in conventional pollutant and nutrient loads.

Implementation of a 4-stage BNR aeration process involves adaptation of existing infrastructure to balance flows and loads that favor nutrient removal microorganisms. At times, this can be frustrating: tiny upsets can seem to linger for weeks and months. Other times it seems uncanny that the microbes seem to know what to do. Manipulation of four independent sludges is coordinated on at least a weekly basis by senior operators who monitor the process.

The current BNR system is more than adequate to handle current loads and requirements, however, long term planners understand that the sewer service area is not static. Population is expected to grow. Our existing BNR facility has limited growth potential and even more limited lifespan, having been built 35 to 55 years ago. Thus our planning includes process simulation and evaluation of alternative aeration schemes to accommodate future loads.

ACKNOWLEDGEMENTS

The authors wish to acknowledge the contributions of Dr. Alex Ekster who, as principal process engineer, developed and managed the mid-1990s implementation of the current 4-stage, step-feed, BNR treatment process. We also thank our facility's current process engineers, Dr. Issayas Lemma and Dr. Rong Liu, performed studies and compile on-going data analysis that support the bulk of our findings.

REFERENCES

City of San Jose. (1998). In-Plant Copper Reduction and Treatment Process Optimization Program at the San Jose/Santa Clara Water Pollution Control Plant. Unpublished Manuscript.

Consoer & Tower and Associates. (1968). A Comprehensive Study of the Waste Treatment Requirements for the Cities of San Jose and Santa Clara and Tributary Agencies: Phase 1. Assimilative Capacity of South San Francisco Bay. San Jose, CA.

Larry Walker Association and Kinetic Laboratories Incorporated. (1987). South Bay Dischargers Authority Water Quality Monitoring Program. (Final Technical Report December 1981-November 1986).

Lemma, Issayas. (2013). Facility Training. San Jose-Santa Clara Regional Wastewater Facility. Unpublished Manuscript.

Metcalf & Eddy. (2003). In Tchobanoglous G., Burton, F.L. and Stensel H. D. and Stensel H. D. (Eds.), Wastewater Engineering: Treatment and Reuse. (4th ed.). New York: McGraw Hill.

San Jose-Santa Clara Regional Wastewater Facility (RWF). (2013a). 2013 Annual Self Monitoring Report. http://www.sanjoseca.gov/Archive.aspx?AMID=161&Type=&ADID=

San Jose-Santa Clara Regional Wastewater Facility (RWF). (2013b). Plant Master Plan, November 2013. http://www.sanjoseca.gov/plantmasterplan

San Jose-Santa Clara Regional Wastewater Facility (RWF). (2014a). 2014 Annual Self Monitoring Report. http://www.sanjoseca.gov/Archive.aspx?AMID=161&Type=&ADID=

San Jose-Santa Clara Regional Wastewater Facility (RWF). (2014b). Nutrient 13267 Monitoring Final Report.
San Jose-Santa Clara Water Pollution Control Plant (WPCP). Brief Historical Background. Unpublished Manuscript.

San Jose/Santa Clara Water Pollution Control Plant (WPCP). (1997). Santa Clara's Relationship with San Jose Concerning Ownership and Operation of the San Jose/Santa Clara Water Pollution Control Plant. Unpublished Manuscript.

Yigzaw, Simret Kesete. (2014) Wastewater and Water Quality Changes in Lower South San Francisco Bay, 1957-2013. Environmental Studies., San Jose State University.

Optimizing Operations To Reduce Sidestream Recycle of Phosphorus

Recycling of Phosphorus at the Trinity River Authority of Texas Denton Creek Regional Wastewater System Plant

By John Bennett, Trinity River Authority of Texas; Mike Roser, Trinity River Authority of Texas; Clifford Woods, Trinity River Authority of Texas; Jeff Sober, Garver

The Trinity River Authority of Texas (TRA) operates the Denton Creek Regional Wastewater System (DCRWS) in Roanoke, Texas. DCRWS is planning for a pending total phosphorus effluent discharge limit of 0.5 mg/L. The DCRWS is an advanced activated sludge facility with, sequencing batch reactors, biological nutrient removal (BNR), final clarifiers, cloth filters, UV treatment, aerobic digester/sludge holding tanks, and centrifuges. The plant staff has optimized the operation of the BNR system and are confident in the plant's ability to remove phosphorus from the liquids stream to meet the pending permit limit. Once removed from the liquids stream, phosphorus (in the form of ortho-phosphate) is transferred to the solids stream through the waste activated sludge (WAS). The WAS, as well as Primary Sludge, is transferred to the Sludge Holding Tank No. 3, which is intended to be an aerobic digester. Standard operating procedures call for the Sludge Holding Tank No. 3 to be decanted once per day and settled sludge transferred to the Sludge Holding Tank No. 2. From Sludge Holding Tank No. 2, sludge is pulled and transferred to the dewatering facility where centrifuges produce a cake to be landfilled. This is shown in Figure 1.

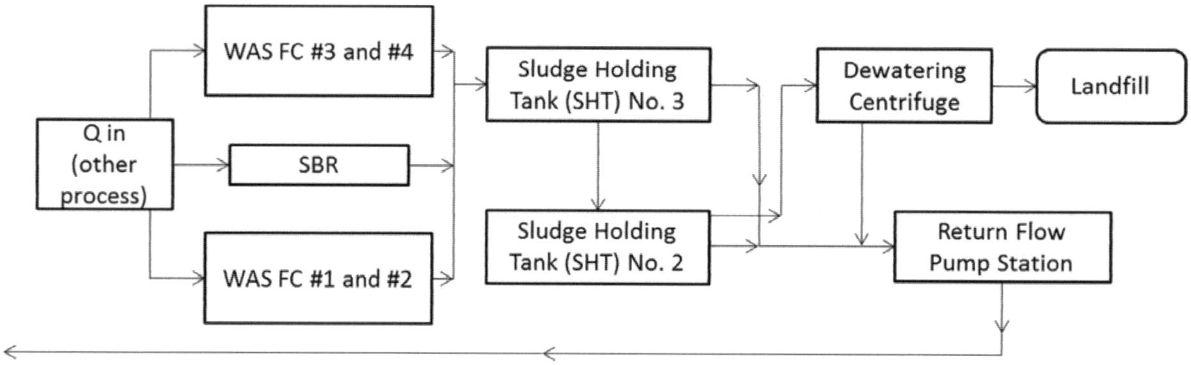

Figure 1 Solids Flow Schematic

The nature of phosphorus accumulating organisms (PAOs) is such that the biomass releases all stored ortho-phosphate when exposed to anaerobic conditions. This reaction makes BNR successful, but can also cause a phenomenon known as phosphorus sink within the solids handling and liquids stream facilities. If there are anaerobic conditions during solids processing, WAS has the potential to release soluble phosphorus. When dewatered, the soluble phosphorous is then recycled back to the head of the treatment plant. This loop causes a sink of phosphorus to build up as it cycles, increasing the load of phosphorus on the liquids stream. The increase in load can create a burden on the biological phosphorus treatment and result in total phosphorus discharge permit excursions.

The plant staff has seen this phenomenon occur at the facility and have determined that the phosphorus recycle in the sidestream is dependent on the operations of the solids processing equipment, specifically the sludge holding tanks. The staff desired to determine exactly what operational changes affected phosphorus recycle, and to what degree they affected the phosphorous limit.

Field Investigation and Historical Data

Analysis

After completing a field investigation, a historical data analysis was conducted to understand the plants design, treatment goals, and current process performance. Data from February 2011 to October 2013 was compiled for all monitored plant parameters.

DCRWS has stringent seasonal permit requirements. The permitted effluent discharge limits are summarized Table 1.

Table 1 Permit Limits

Parameter	Existing Permit Outfall 001		Future Permit Outfall 002	
	Daily Average mg/L	Daily Average lbs/day	Daily Average mg/L	Daily Average lbs/day
CBOD				
Jun to Nov	5	292	5	480
Dec to May	7	409	7	672
TSS	15	876	15	1,439
Ammonia				
Jun to Nov	1.4	82	1.9	182
Dec to May	3	175	3	288
Total Phosphorus	Not in Permit	Not in Permit	0.5	48
E*Coli*	126	N/A	126	N/A
DO	6 (min)	N/A	6 (min)	N/A
pH	6 (min)	N/A	6 (min)	N/A
FLOW	7 MGD AADF	12,917 GPM 2HR PF	11.5 MGD AADF	19,444 GPM 2HR PF

The plant is currently under a permit to discharge to Outfall 001 at a rate of 7 MGD Annual Average Daily Flow (AADF). When the plant expands to a capacity of 11.5 MGD, it will operate with two outfalls and be required to remove phosphorus to a limit of 0.50 mg/L.

The facility was designed based on an influent flow with the constituents and parameters shown in Table 2.

Table 2 Influent Design Criteria

Influent Design Criteria	
Design Flow	11.5 MGD
Maximum Daily Flow	28.0 MGD
Peak 2HR Flow	40.5 MGD
BOD	225 mg/L
TSS	310 mg/L
NH3	40 mg/L
TOTAL P	6.7 mg/L

Significant or important trends were plotted and are presented in the following section. An understanding of the historical plant performance allows the evaluation team to understand how the facility responds under different operational conditions.

Of all the measured parameters, the Ammonia, Total Phosphorus, and pH trends were identified as atypical. The Ammonia trend spiked between December 2012 and May 2013. While the effluent Ammonia concentration was never at risk of exceeding the permit limit, it was an atypical trend for such a consistent facility. Figure 2 illustrates the Ammonia trend, including the aforementioned spike. The Total Phosphorus also had a unique trend in that it decreased in concentration over time and then stabilized. Figure 3 illustrates the Total Phosphorus trend.

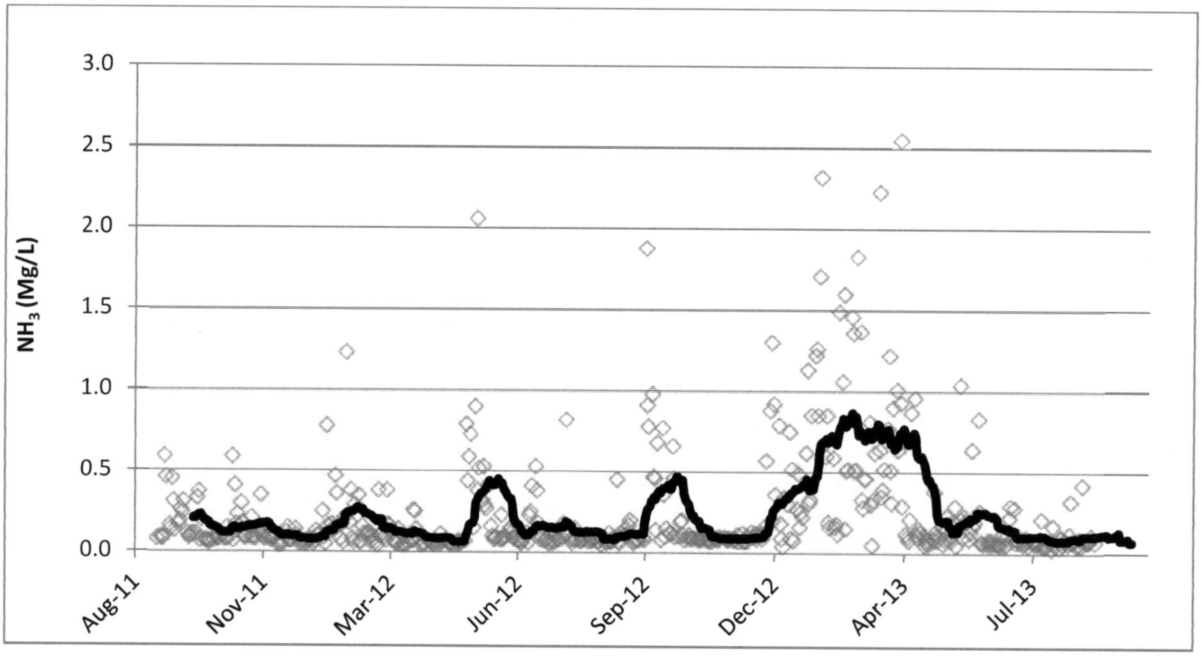

Figure 2 Effluent Ammonia Trend

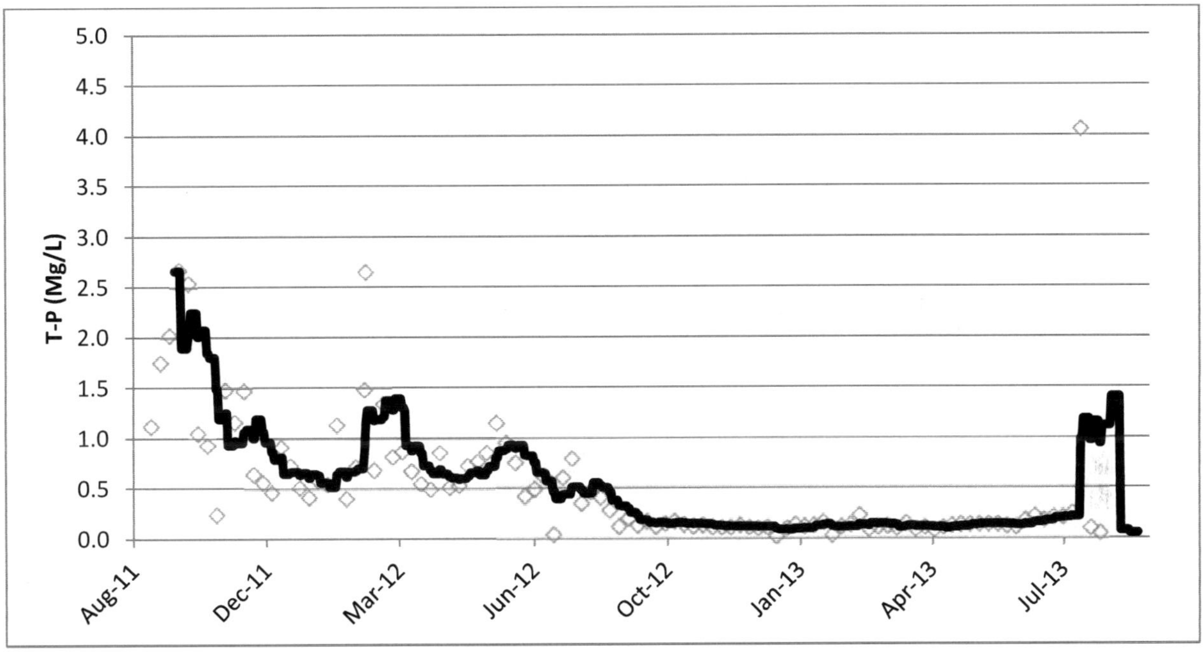

Figure 3 Effluent Total Phosphorus Trend

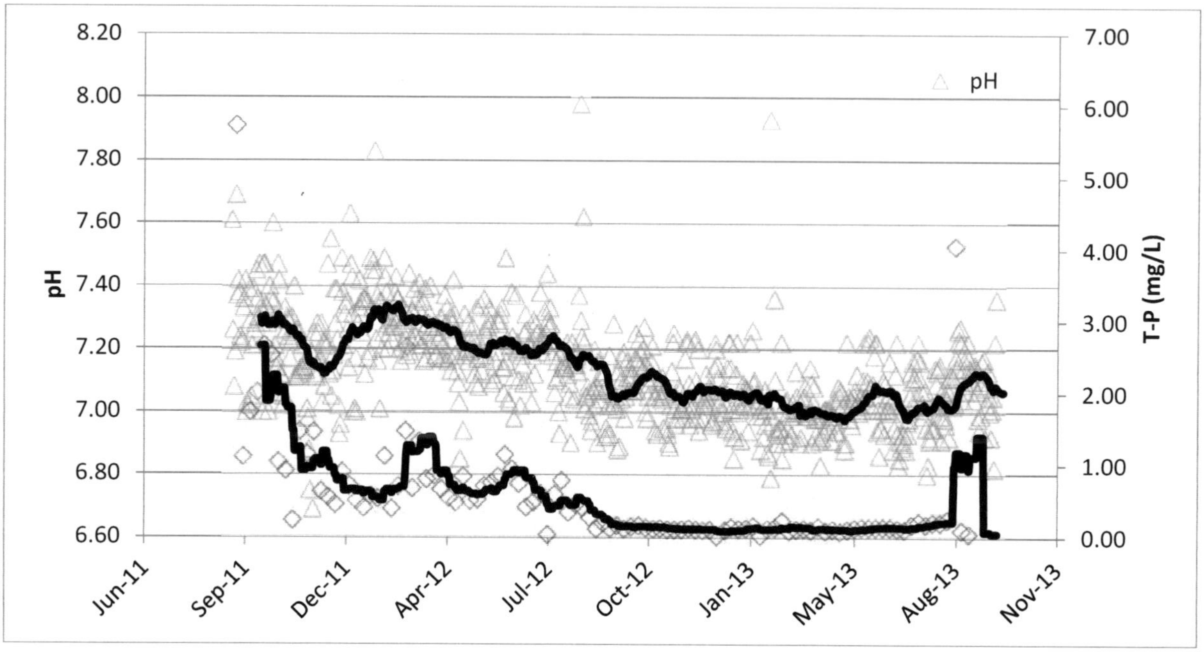

Figure 4 Effluent pH and Effluent Total Phosphorus

The graph of pH and T-P produces a correlation between both pH and T-P but there are too many unknowns to make a claim. Over 2011 to 2013 there is a general decrease in both pH and T-P with T-P significantly decreasing between September and December 2011.

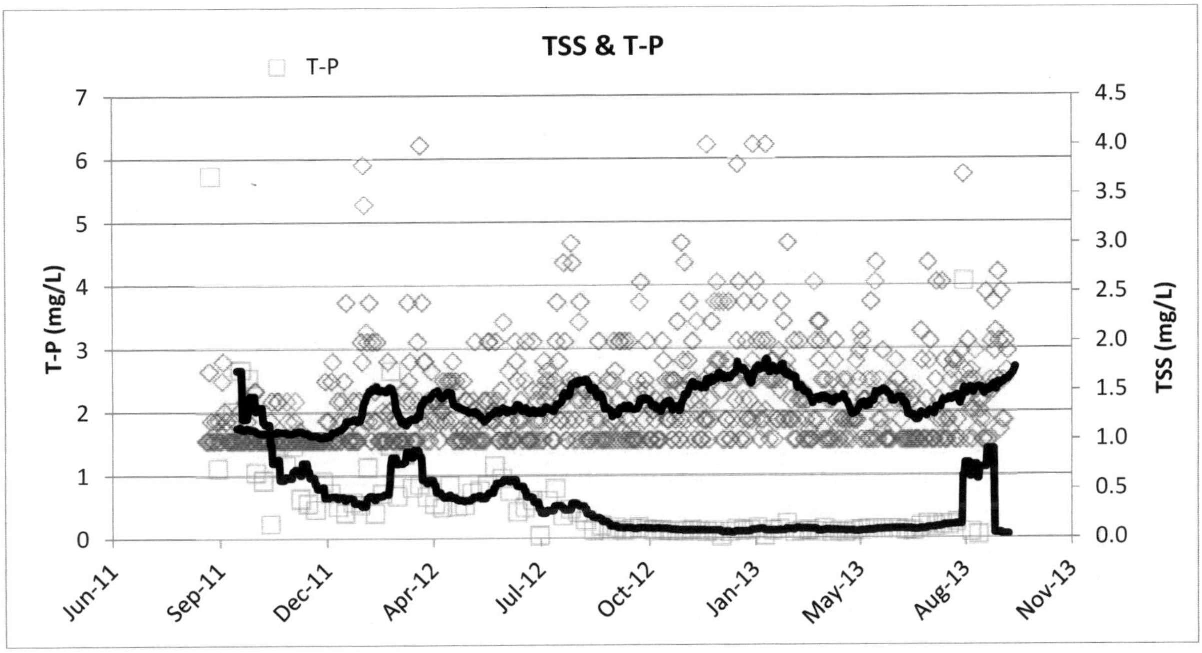

Figure 5 Total Suspended Solids and Total Phosphorus

Graphing Total Phosphorus and Total Suspended Solids can help identify how much particulate phosphorus plays a role in the total phosphorus count. During the period in which the phosphorus removal system was not optimized, it can be noted that the TSS and T-P had similar trends. There is a small correlation that an increase in TSS creates an increase in T-P. When the biological system is optimized, this trend is no longer recognizable.

Sampling Protocol

The project team, led by plant operations staff and assisted by Garver, developed a sampling protocol and test plan to stress test the solids facilities and determine exactly when and how much phosphorus is recycled under specific conditions. Four testing scenarios were developed, focusing primarily on the known sources of the phosphorus release:

- Normal Daily Operations - This sampling scenario will be the study baseline. Sampling will occur at normal average daily flow conditions, under current normal operating procedures of the solids processing train.

- Normal Daily Operations with High Magnesium Hydroxide Feed - This sampling scenario will be based on the historical data evaluation, and will include a stress test of magnesium hydroxide feed to determine the solids train's response.

- Extended Decant Operation - This sampling scenario will include sampling all downstream processes from the digester following an extended (length to be defined in minutes) decant operation.

- Combination High Magnesium Hydroxide/Extended Decant Operation - This sampling scenario will include a combination of Scenarios 2 and 3, and will be used to show a more accurate solids train response to both scenarios occurring at the same time.

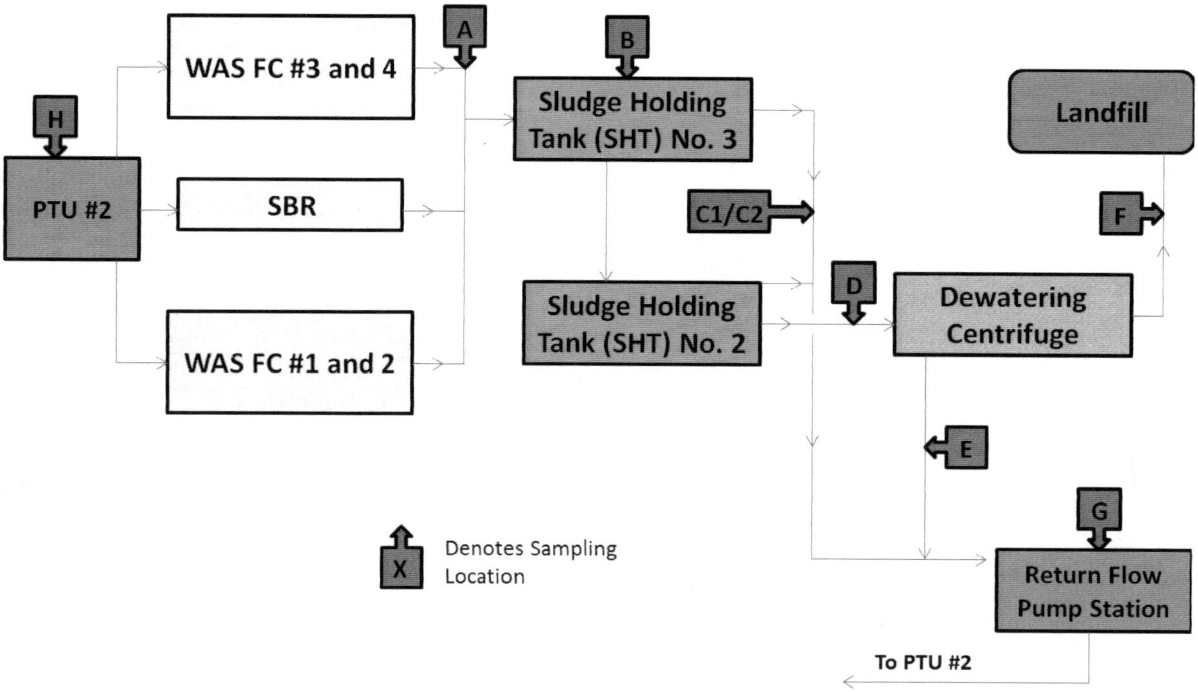

Figure 4 Sampling Locations

This Page Intentionally Left Blank.

Each sample location was fully characterized with the following analyses:

1. Temperature
2. Alkalinity
3. Chemical Oxygen Demand (COD)
4. Soluble Chemical Oxygen Demand (sCOD)
5. pH
6. Total Solids
7. Total Kjeldahl Nitrogen (TKN)
8. Soluble Phosphorus
9. Total Phosphorus
10. Total Volatile Solids (TVS)
11. Dissolved Oxygen (DO)
12. Ammonia
13. Sludge Volume Index (SVI)

An outside laboratory was utilized to analyze the soluble phosphorus samples. An error in sample analysis rendered the soluble phosphorus samples from Scenario 1 unusable. After the end of Scenario 4, with flows back to average daily conditions, new soluble phosphorus samples were collected to obtain baseline data. The rest of the analyses were not run due to budget constraints. The soluble phosphorus data collected will provide a valuable snap shot of the average normal operating conditions of the facility, but cannot be directly correlated with the other data collected under Scenario 1. These data sets will have to be analyzed independently.

There were a few instances where data from sampling did not correlate. For example, in a few samples, the soluble phosphorus concentrations reported for a specific sample were higher than the reported total phosphorus concentrations. This could be due to poor sample analysis on the third party labs part, or due to a specific snap shot in time. This situation made the overall mass balances difficult. To analyze the data sets in detail, mass balances were performed around individual process units and not the entire solids handling train.

The data from the sampling events was collected and analyzed. Each data set was independently evaluated with any outliers (order of magnitude) analyzed in detail and either excluded from the data or observed for a cause and effect relationship. The stress scenarios (No. 2, 3, and 4) were then compared to the baseline scenario (No. 1) as well as to each other. Data that was significantly different was evaluated and conclusions drawn. The data presented below represents data that was statistically different.

Expected Phenomenon

The DCRWS system removes phosphorus biologically in the aeration basins in anaerobic and anoxic zones. The anaerobic zone provides phosphate accumulating organisms (PAOs) with a competitive advantage over other bacteria. The lack a suitable electron acceptor, such as dissolved oxygen or nitrate, in the anaerobic zones prevents aerobic heterotrophic bacteria from assimilating volatile fatty acids (e.g., acetate, propionate, etc.), the preferred food source of PAOs. As a result, non-PAO organisms are starved in the anaerobic zone.

Smolders et al (1994) developed a detailed model explaining the metabolism of PAOs. PAOs store excess amounts of phosphorous as polyphosphates in their cell mass. The hydrolysis of these polyphosphates results in the formation of ATP and the release of orthophosphate (PO43-), as well as the release of magnesium, potassium and calcium cations. ATP, in turn, provides the energy to assimilate volatile fatty acids in the anaerobic zone and store them intracellularly in the form of internal polymers such as polyhydroxybutyrate (PHB) The concurrent hydrolysis of glycogen produces NADH2, another energy carrier. Subsequently, under anaerobic conditions, the PHB content in the PAOs increase while the polyphosphate content decreases and phosphorous is released as orthophosphate (PO43-). Under aerobic or anoxic conditions, the stored PHB is oxidized and releases energy in the form of NADH. The NADH is used to create ATP and provide the energy source for continued polyphosphate and glycogen uptake. Figure X illustrates the metabolism of PAOs under both anaerobic and aerobic conditions.

Anaerobic conditions are the critical puzzle piece as it is generally accepted that chemically precipitated phosphorus will remain in solid form. (Seyfried et al., 1988). Several studies have documented the challenges of downstream solids handling operations and phosphorus return (Pitman et al., 1991.)

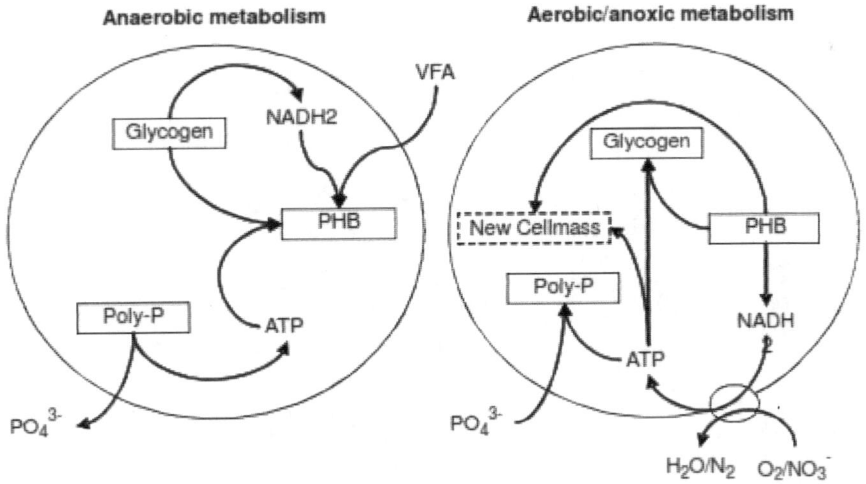

Figure 5 Metabolism of PAO under Anaerobic and Aerobic Conditions

(Source: Smolders, et. Al. 1994)

When these PAOs are wasted, they end up in the solids handling train. Their ability to absorb or release phosphorus is still intact. Therefore, it is critical in the solids train to understand specific site conditions which can act as mechanisms for phosphorus release. For example, an anaerobic zone in a solids handling process, such as an unaerated holding tank, can act as a bio reactor, allowing PAOs to release stored phosphorus as soluble polyphosphate. If the downstream process has a decanting, supernatant, thickening, or dewatering step, this soluble phosphorus is recycled to the head of the plant and re-treated. This phenomenon is expected to occur in at DCRWS during long decant cycles of the solids holding tank.

Locations for Phosphorus Recycle

Only two locations exists at DCRWS for phosphorus recycle. The decant operation from the SHT and the centrate underdrain from the centrifuge. The data analysis will focus on these operations.

Observations

For each sample, the Dissolved Oxygen was measured. The DO concentration never registered as zero in any of the tanks or decant lines, therefore indicating that anaerobic conditions were not realized during our stress tests. DO levels were boosted in two areas, in the solids holding tank where air is supplied, and then again in the centrifuge. The increase in DO concentration across the centrifuge was extremely high. In fact, this increase in DO is most likely responsible for keeping phosphorus returns to a minimum. Figure 17 demonstrates this phenomenon.

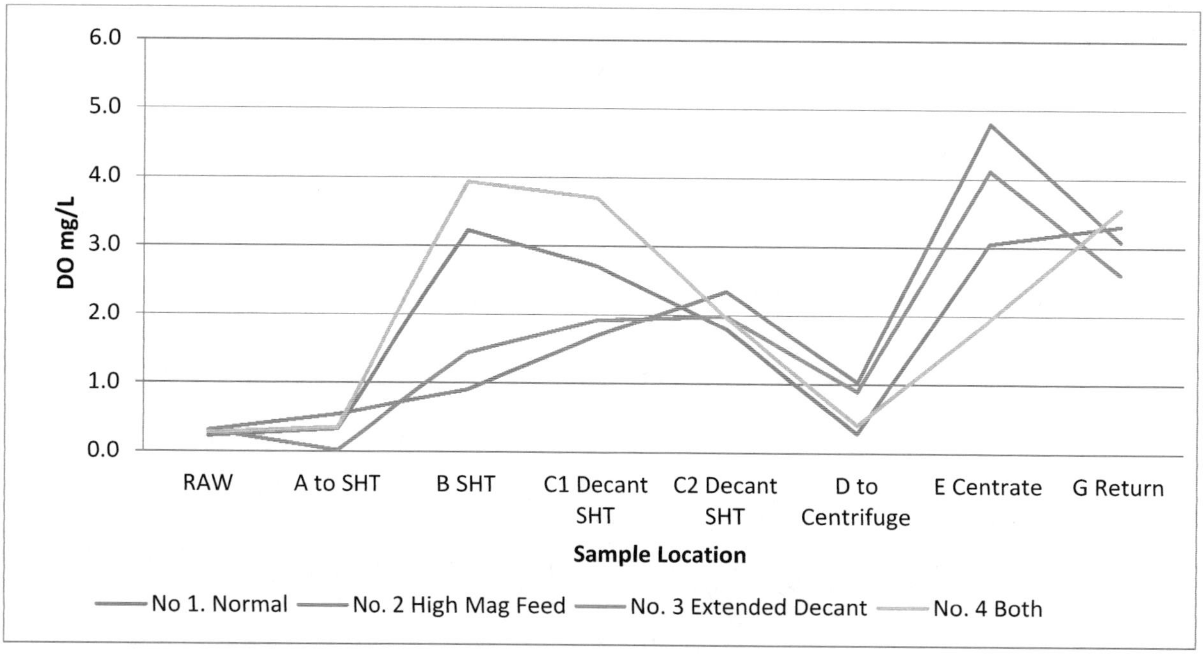

Figure 6 DO Profile Across All Sample Points

The focus of this evaluation is on the phosphorus samples throughout the plant. The profile of the total phosphorus can be seen in Figure 18.

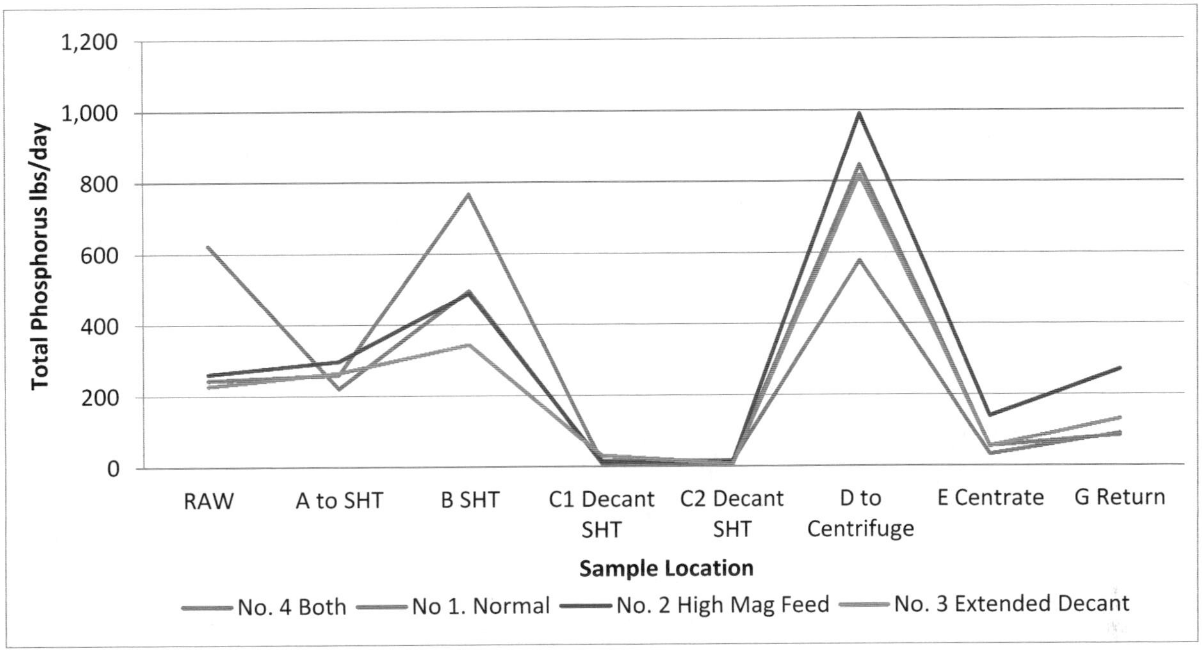

Figure 7 Total Phosphorus Across All Sample Points

The TP is highest in the two scenarios of high magnesium hydroxide feed. This is an interesting find, as magnesium hydroxide can serve as a coagulant for soluble phosphorus removal through precipitation. IN both instances of high magnesium hydroxide feed (Scenario 2 and 4), the phosphorus increases relative to the amount of total phosphorus in the rest of the processes. This is represented in Table 3 which shows the amount of soluble phosphorus gain in each scenario as a fraction of the total phosphorus.

Table 3 Soluble P to Total P Mass Balance

Soluble P to Total P Mass Balance			
Scenario	SHT Inf (TP) (lbs)	SHT Eff (SP) (lbs)	Percent SP to TP
Normal	766.6	195.6	25.52%
High Mag Feed	487.5	284.6	58.38%
Extended Decant	343.2	199.6	58.16%
Both	494.8	247.0	49.92%

Table 3 shows that under baseline conditions, the SP fraction was approximately 25% of the TP. Under Scenarios 2, 3, and 4, High Mag Feed, Extended Decant, and Both, the SP fraction more than doubled, and above or around 50% of the TP. This confirmed the expected phenomenon: operating in an extended decant situation increases the soluble phosphorus percentage and the possibility of a higher phosphorus return to the liquids stream. In both cases of high magnesium hydroxide feed the soluble phosphorus was higher. This is also represented in Figure 19.

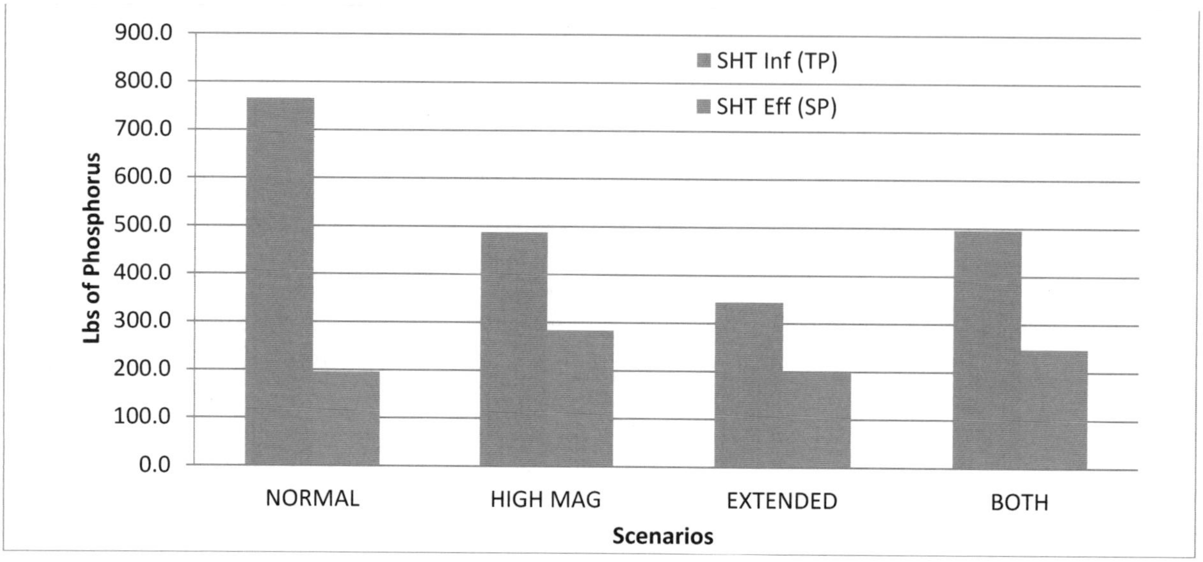

Figure 8 Lbs of SP to TP in Each Scenario

Drilling into the data, another look at the how the scenarios are affecting the decant of phosphorus back into the liquids stream is shown in Table 4.

Table 4 Decant Phosphorus Percentage

Mass Balance of Decant Phosphorus		
SHT Influent (lbs of TP)	SHT Decant (lbs of TP)	Percent of P in Decant
172.9	4.6	2.68%
83.3	7.6	9.09%
92.1	4.2	4.53%
124.3	10.8	8.66%

The table identifies that the return of phosphorus was almost 4 times higher during high magnesium hydroxide feed than normal conditions. Extended decant also had a higher return of phosphorus, almost twice as much. This relationship of a higher soluble phosphorus percentage during magnesium hydroxide addition throughout the process units is shown in Figure 20.

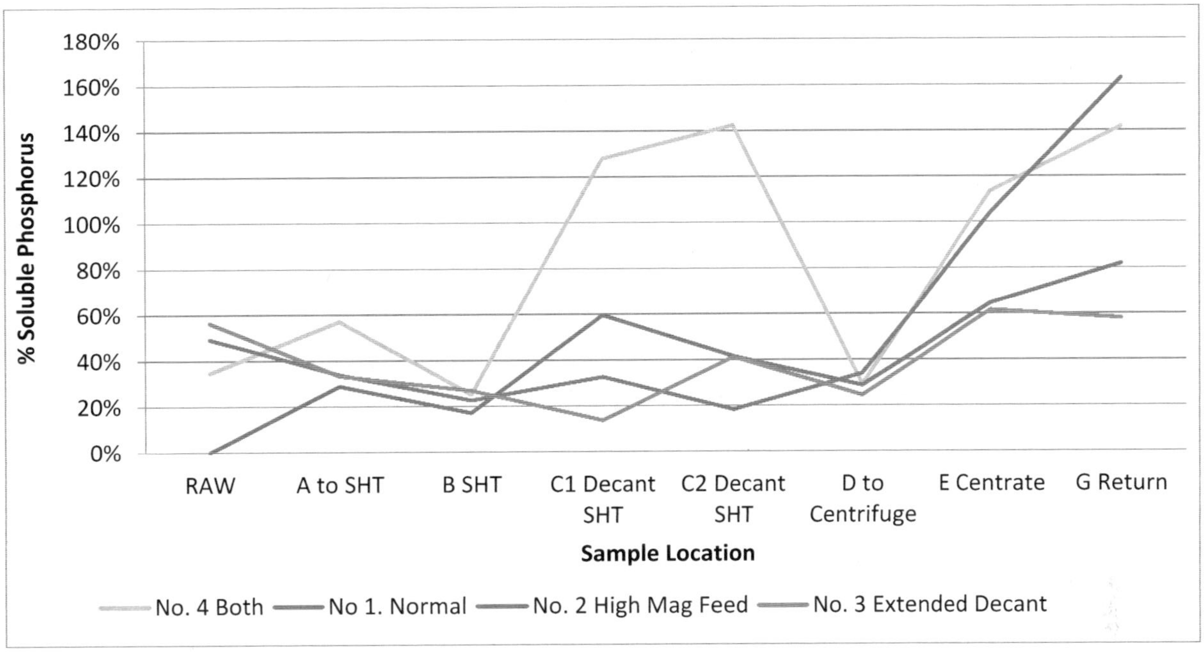

Figure 9 % SP of TP in Each Process

The graph identifies that the % soluble phosphorus increases significantly within the solids holding tank during high magnesium feed operations. The higher return of phosphorus during the addition of magnesium hydroxide addition is unique and not initially expected. The magnesium hydroxide is fed into the Solids Holding Tank for odor control. The simple chemical feed facility is show in Figure 21.

Figure 10 Magnesium Hydroxide Feed

A large change in pH from a low value to higher value due to magnesium hydroxide would have an expected reaction of reducing soluble phosphorus through chemical precipitation. While magnesium hydroxide increased both the pH (Figure 22), it was not enough to affect chemical precipitation reactions.

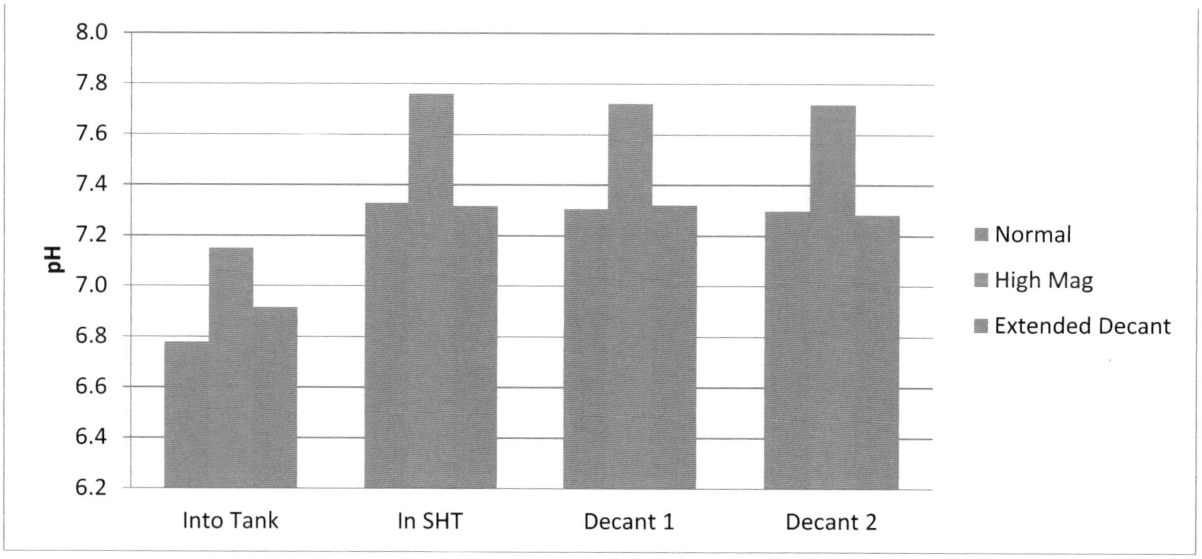

Figure 11 pH Comparison in SHT Tank

One potential explanation for the increase in soluble phosphorus under a higher magnesium hydroxide feed is that magnesium is a nutrient needed for cell growth and expansion. If magnesium is limiting in the Sludge Holding Tank, there could be less volatile solids reduction in the aerobic digester environment. With the addition of magnesium hydroxide providing a limiting nutrient, there could be higher biological activity and a higher volatile solids reduction. With volatile solids reduction in this aerobic digester environment, cells are destroyed and lysed, releasing ortho-phosphate back into the liquid. This condition should be evaluated closer to optimize the magnesium hydroxide feed.

The centrate phosphorus return mirrored that of the decant, in that the scenario of a high magnesium hydroxide feed also provided the opportunity for a higher than normal return of phosphorus. This is shown in Figure 25.

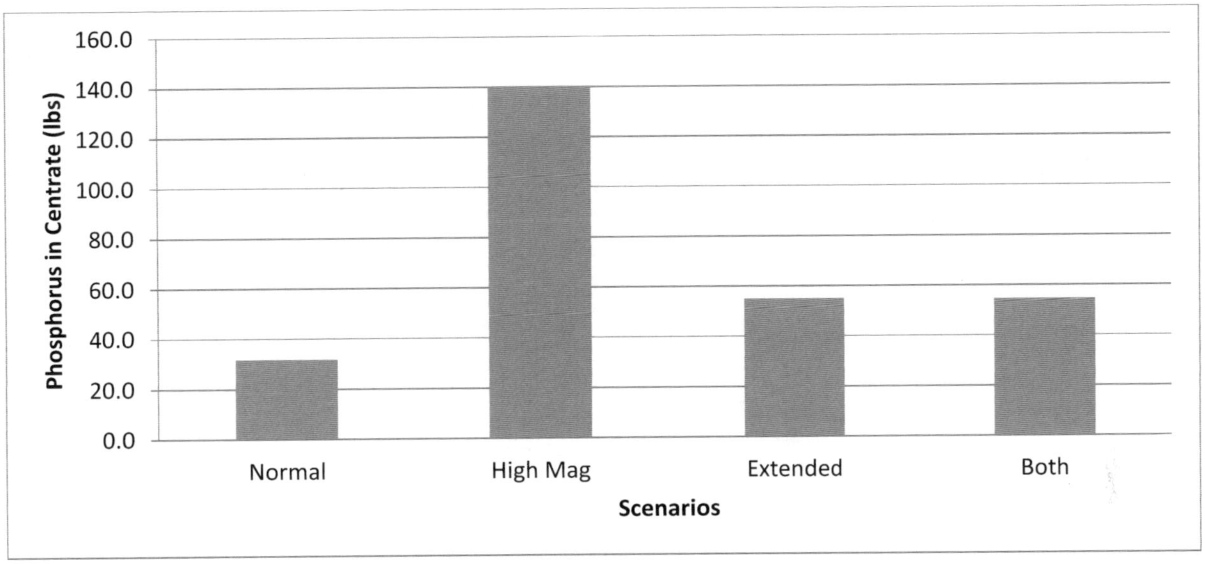

Figure 12 Phosphorus in Centrate

Phosphorus Recycle

The most alarming finding is the amount of phosphorus recycled back to the plant. In all cases, the average lbs of phosphorus recycled back to the facility is 144 lbs per day, of which 141 lbs is ortho-phosphate. This recycle is even in the best case scenarios, under normal operation. The plant was designed for an influent load of 6.7 mg/L, or 642 lbs per day. The recycle stream represents 22.4% of the plants treatment capacity of phosphorus. This is effectively reducing the plants treatment capability to an influent phosphorus concentration of 5.2 mg/L. This diminished capacity is represented graphically in Figure 27.

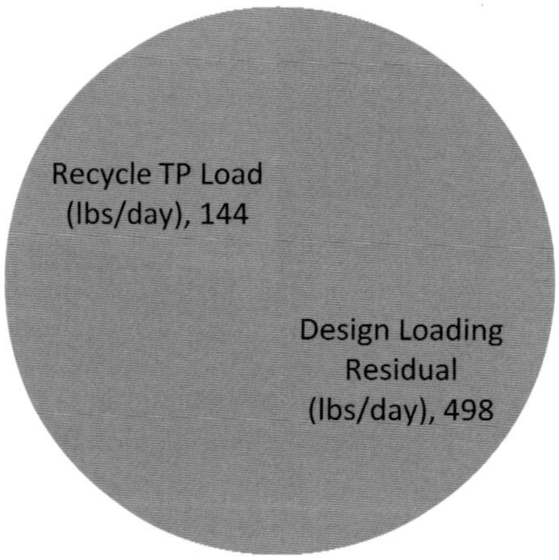

Figure 13 Diminished Phosphorus Capacity

Total Phosphorus in the influent should be kept under a close watch, as the facility has diminished capacity for treatment of phosphorus due to recycle. Phosphorus concentrations above a 5.2 may cause stress on the biological system. There are many options for sidestream treatment available to reduce the phosphorus load. The phosphorus should be monitored and if the influent loading is near or above design capacity, an evaluation for sidestream phosphorus removal should be conducted.

The following conclusions can be made from the sidetream evaluation data analysis:

1. Maintaining dissolved oxygen concentration above 0.1 mg/L reduces the possibility of soluble phosphorus release throughout the solids train.
2. The pH is closely linked with the soluble phosphorus. Watching pH closely can quickly identify a potential soluble phosphorus trend.
3. Extended Decant operation resulted in twice as much phosphorus return than normal operations.
4. Under all conditions, the average phosphorus recycle to the liquids stream from the solids

stream was 144 lbs, or approximately 22% of the plant's liquid treatment capacity.

Conclusions

The following conclusions were made from the historical data review:

1. The plant achieves effluent quality consistently better than required by permit
2. The plant responds as expected to temperature and DO variations.
3. The Staff has optimized the biological nutrient removal system for phosphorus removal.

Recommendations

The following recommendations can be made from the data analysis and the conclusions:

1. Total Phosphorus and Soluble Phosphorus samples should be taken between all process trains (Inf, Raw Plus, PTU Influent, Aeration Basin Influent, Aeration Basin Effluent, Clarifier Effluent, and Filter Effluent) for better process control and optimization of the phosphorus removal system. It is recommended that this be completed once per day for several months to understand the system prior to the new permit.
2. An online in-situ ORP meter in the solids holding tank could provide superior process control to mitigate and have warning prior to any anaerobic conditions and subsequent
3. The SHT should not be allowed to decant for extended periods of time. The amount of time the tank is unaerated should be minimized to mitigate anaerobic conditions and phosphorus release. This could be controlled with an ORP meter or DO meter.
4. Influent Phosphorus concentrations and loading should be monitored. Because the solids stream recycle is 22% of the plants design capacity, an evaluation of potential sidestream removal alternatives should be conducted if treatment performance is affected due to overloading.
5. A solids mass meter on the transfer from SHT No. 2 to the centrifuge would greatly increase process control by allowing more understanding of solids transfer rates and percent solids to centrifuge.

Compliance Strategies to Meet a Zero Net Loading Limit for Phosphorus

Linda K. Sawyer[1*], Rita Miller[2], Mike Prinz[2], Joe Schwall[2], and Dave Smith[3]

[1]Brown and Caldwell
[2]City of Santa Rosa, CA
[3]Merritt Smith Consulting
*Email: LSawyer@BrwnCald.com

ABSTRACT

The City of Santa Rosa implements an integrated approach to achieve a zero net loading of phosphorus to the watershed. No discharge occurs in normal and dry years due to water recycling, primarily via delivery to the Geysers geothermal steam field. Treatment modifications will halve plant effluent phosphorus concentrations. Watershed nutrient offset projects make up the difference. This paper discusses challenges encountered, the probability-based analysis to maintain compliance and the innovative nutrient offset program.

KEYWORDS: Nutrient removal, offset trading, water recycling

INTRODUCTION

The City of Santa Rosa's Laguna Water Reclamation System in Northern California serves about 213,000 people and treats an average flow of 79,500 m^3/d (21 million gallons per day (mgd)). The region has a Mediterranean climate with little to no rainfall during summer months. To mimic natural conditions, Santa Rosa's National Pollutant Discharge Elimination System (NPDES) wastewater permit contains a discharge prohibition which extends from May 15th through September 30th (summer season) and only allows discharges through the remainder of the year, October 1st through May 14th (discharge season).

The plant discharges to the Laguna de Santa Rosa (Laguna), a sensitive ecosystem home to a number of endangered plant and animal species protected under the Endangered Species Act and recently designated as a Wetland of International Importance under the Ramsar Convention, an intergovernmental treaty dedicated to conservation and wise use of wetlands (see Figure 1). The 658 km^2 (254 square miles) watershed of the Laguna de Santa Rosa is the largest tributary to the Russian River. The Laguna is the largest freshwater wetlands complex on the Northern California coast. The Laguna watershed has widespread urban and agricultural development, including cropland, pasture, rangeland, orchards, and vineyards.

The Laguna is determined as impaired for nitrogen and phosphorus in the North Coast Regional Water Quality Control Board (Regional Board)'s 303(d) list. A Total Maximum Daily Load study (TMDL) for nutrients is currently being developed to identify sources of nutrients and determine a water quality action plan to address the impaired condition.

Figure 1. Laguna de Santa Rosa.

In 2006, the City received an interim NPDES permit limitation specifying "zero or no net loading of nitrogen and phosphorus" to the Laguna, with compliance beginning in the 2012 – 2013 discharge season. In 2013, the NPDES Permit was reissued. The renewed permit narrowed the nutrient limitation to a provision specifying no net loading of total phosphorus (TP). Compliance is based on a rolling three-year average comparing the TP mass discharged and TP mass controlled through offset projects. The first three-year averaging period began with the 2014-2015 discharge season and the first compliance determination will occur in 2017 after the 2016-2017 discharge season. The current permit also includes a 10.6 mg/L average monthly limit for total nitrogen (TN) for discharges to the Laguna.

The City has begun implementation of a three-tiered compliance strategy which prioritizes limiting discharges while concurrently reducing TP effluent levels and preemptively implementing approved nutrient offset projects as discussed in the following sections.

REDUCING DISCHARGES

By carefully managing storage capacity and maximizing the beneficial use of recycled water, the City has minimized the volume of water and mass of phosphorus discharged. The City has six storage ponds with a combined volume of 6,250,000 m^3 (1,650 million gallons). Delta Pond is the largest storage pond and the City's main point of discharge occurs from this location. Recycled water produced from the plant is stored during winter months for reuse in summer. The Geysers geothermal steam field 63.9 km (39.7 miles) north of the plant began using recycled water in 2003, consuming an average flow of 47,700 m^3/d (12.6 mgd). The City's water recycling program also includes both agricultural and urban irrigation sites (see Figure 2).

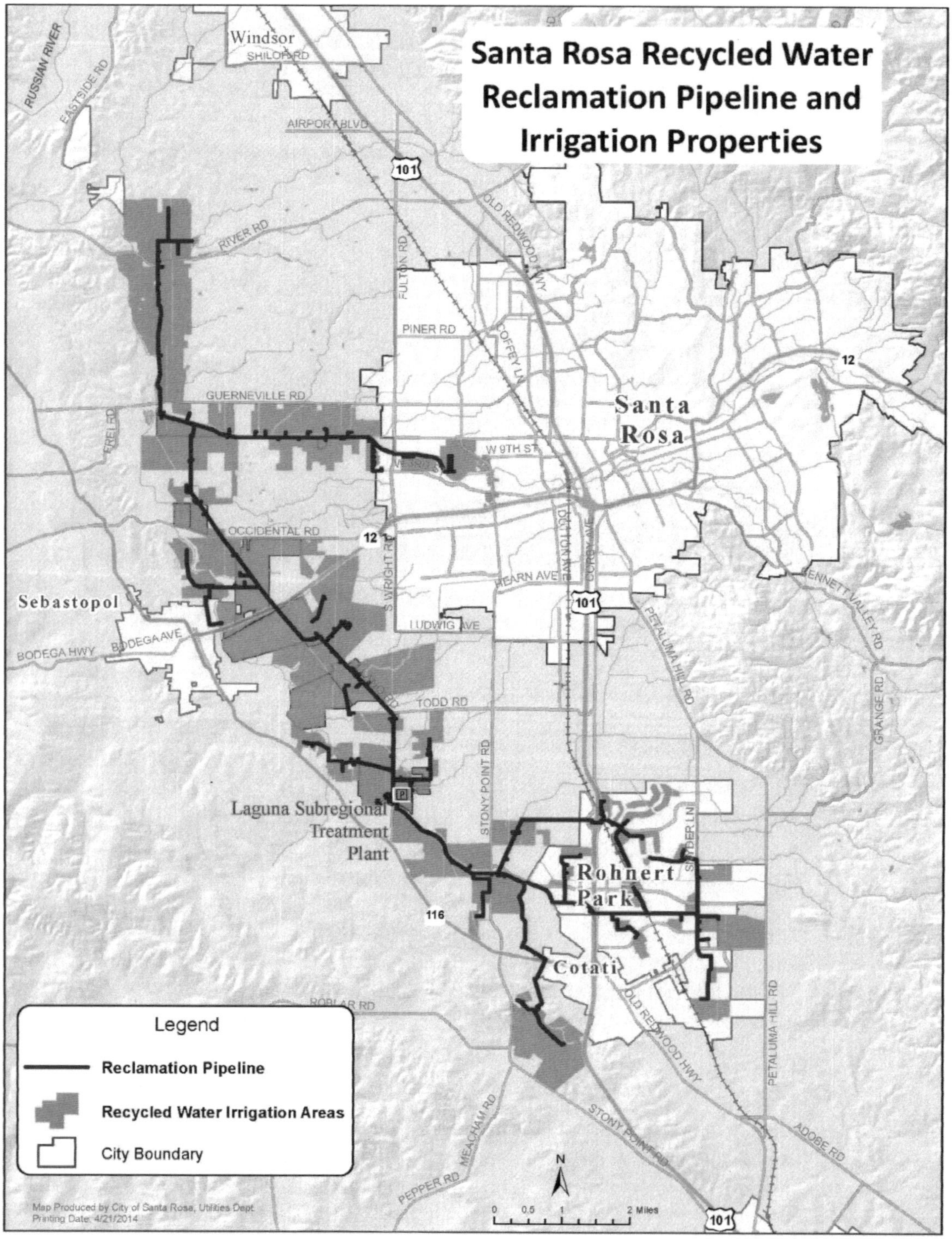

Figure 2. Recycled water irrigation areas.

Figure 3 shows the annual volume of water recycled and discharged. After the Geysers project came online in 2003, the volume of water discharged to the Laguna decreased significantly. No discharges occurred in four discharge seasons (2008-09, 2012-13, 2013-14, 2014-15), and a fifth (2011-12) had only a minor discharge to test the then newly installed discharge control system, which is automated and designed to control the amount of discharge from Delta Pond based on real time water quality data in Delta Pond and receiving waters.

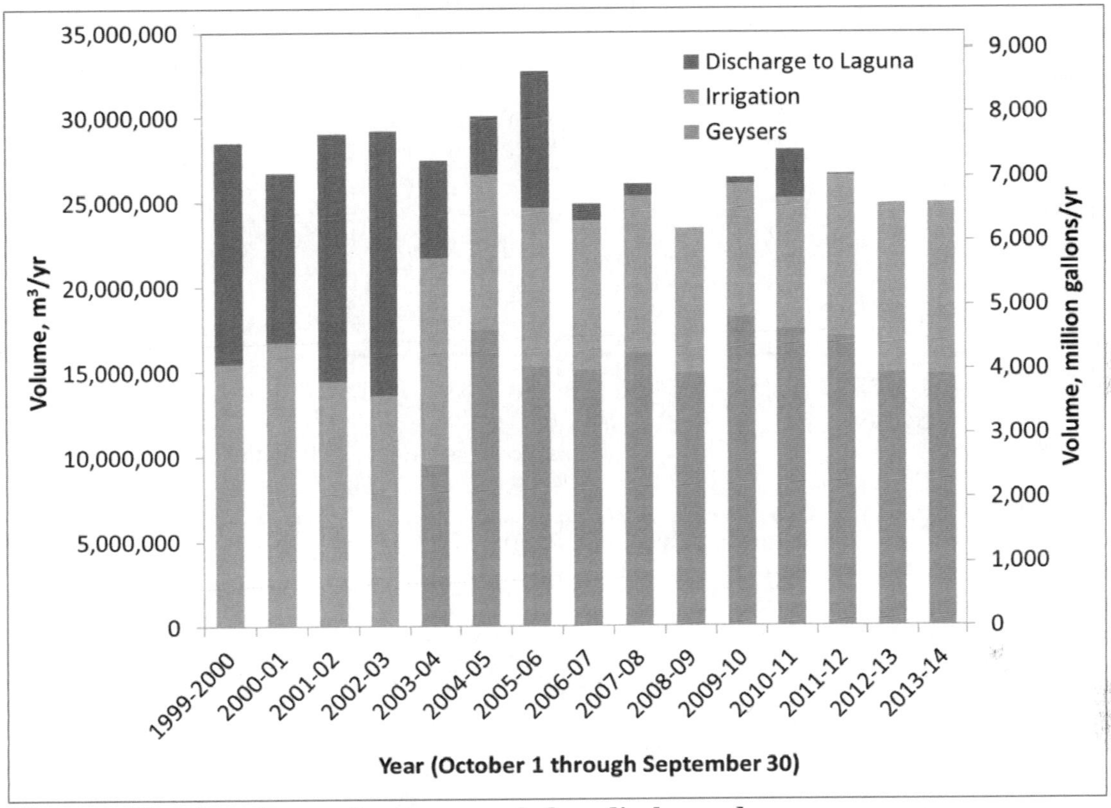

Figure 3. Yearly volume of water recycled or discharged.

Operating the recycled water system to maximize beneficial re-use in the dry season and to avoid discharge in the wet season is challenging due to unpredictable weather, limited storage capacity, and the seasonal nature of irrigation. As illustrated in Figure 4, recycled water demand is high during the dry summer months when irrigation needs are high. Flow to the plant is highest during the winter, and the storage ponds are filled during winter months. Staff continuously monitor the storage pond volume through the winter months, and strategize if and when to discharge. Discharge generally occurs when storage is nearing capacity except in April and May, when full storage is needed to meet re-use demand in the upcoming dry season. As illustrated in Figure 5, the exact timing of pond filling varies from year to year.

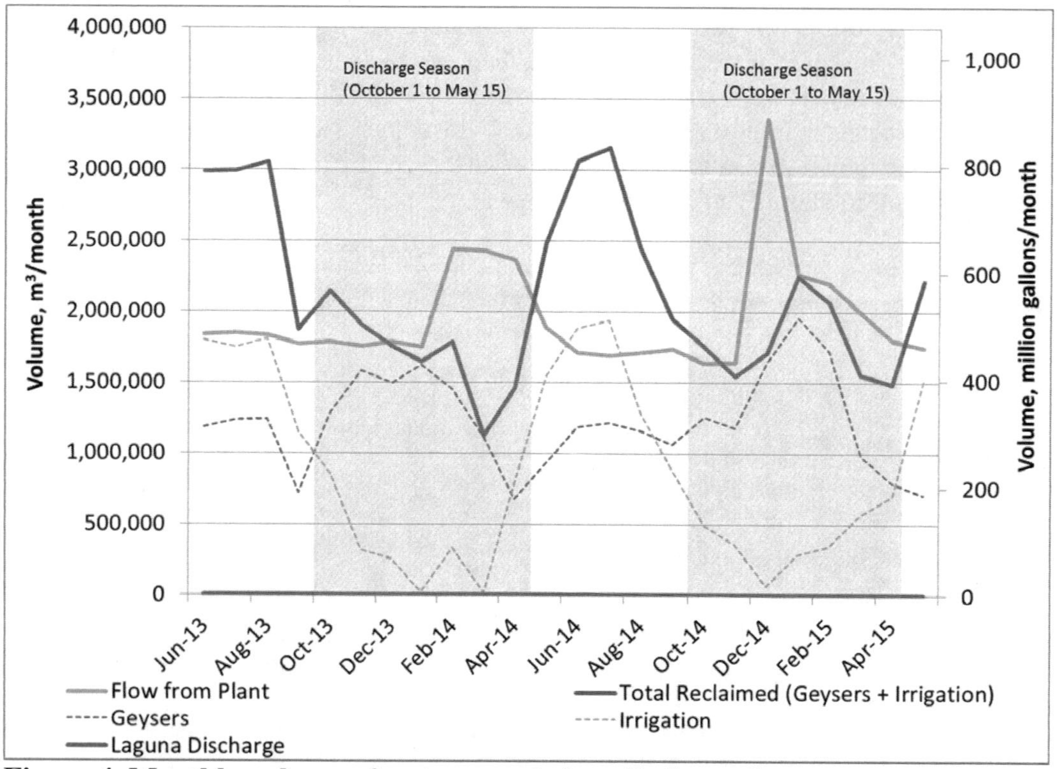

Figure 4. Monthly volume of water recycled or discharged.

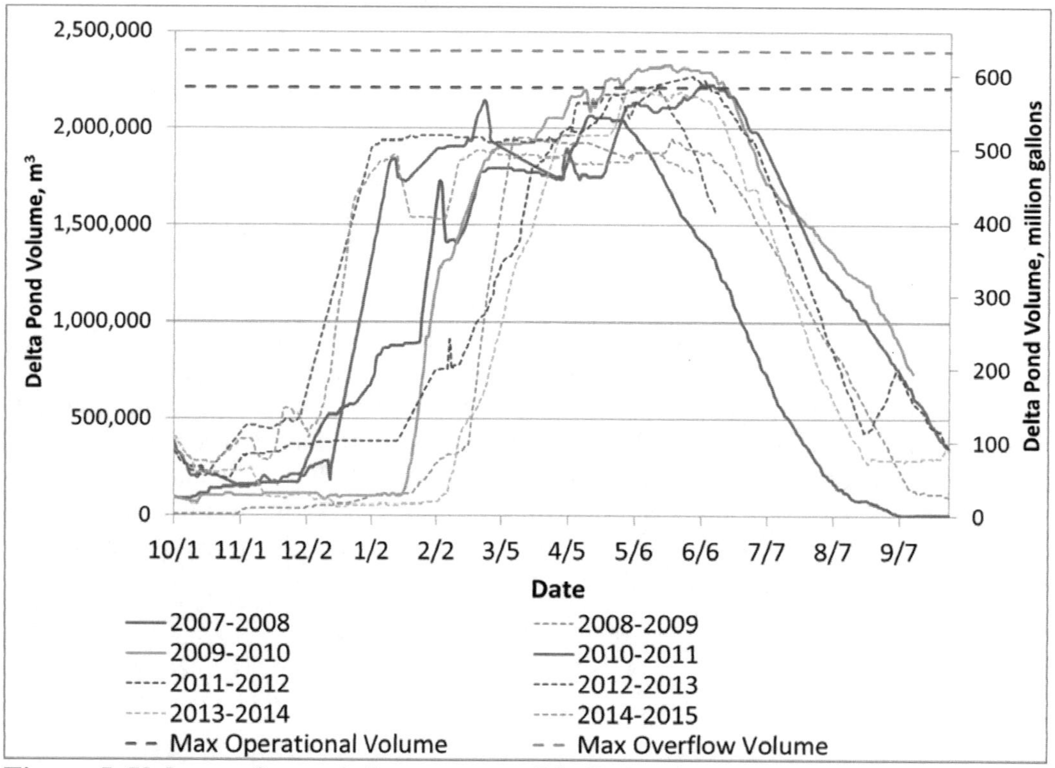

Figure 5. Volume of recycled water stored in Delta Pond.

REDUCING PHOSPHORUS CONCENTRATIONS

Santa Rosa is working to optimize nutrient removal during the discharge season, and thus reduce phosphorus loading when discharge occurs. Santa Rosa's liquid treatment process consists of primary clarifiers, activated sludge (nitrification and denitrification with mixed liquor recycling), tertiary filters, and ultraviolet (UV) disinfection (see Figure 6). Chemically enhanced primary treatment (CEPT) with ferric chloride and polymer addition is used on a limited basis during peak wet weather events to reduce loading to the activated sludge process.

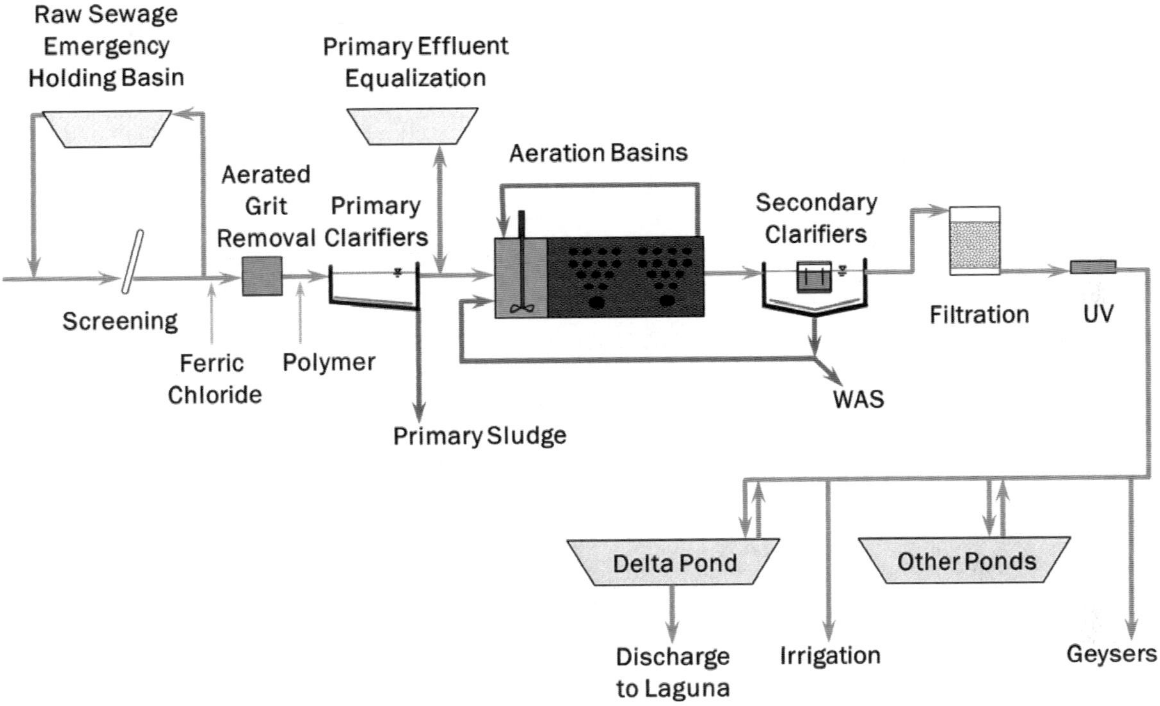

Figure 6. Treatment plant process flow diagram.

Recent plant effluent phosphorus concentrations are shown on Figure 7. Average treatment plant effluent TP concentrations (measured following UV disinfection) are 3.3 mg/L. As noted previously, discharge to the Laguna occurs from Delta Pond. Note that Delta Pond nutrient concentrations are typically only measured during discharge. Average TP discharged from Delta Pond was 2.0 mg/L. Given the limited quantity of data for Delta Pond discharge it is not clear whether phosphorus is removed in Delta Pond.

Recent nitrogen concentrations are shown on Figures 8. Average treatment plant effluent TN concentrations were 15 mg/L. Discharge TN concentrations from Delta Pond are consistently lower than the treatment plant effluent concentrations, indicating that additional denitrification is occurring in the storage ponds. The additional denitrification occurring in the storage ponds is necessary to meet the discharge limitation of 10.6 mg/L for total nitrogen.

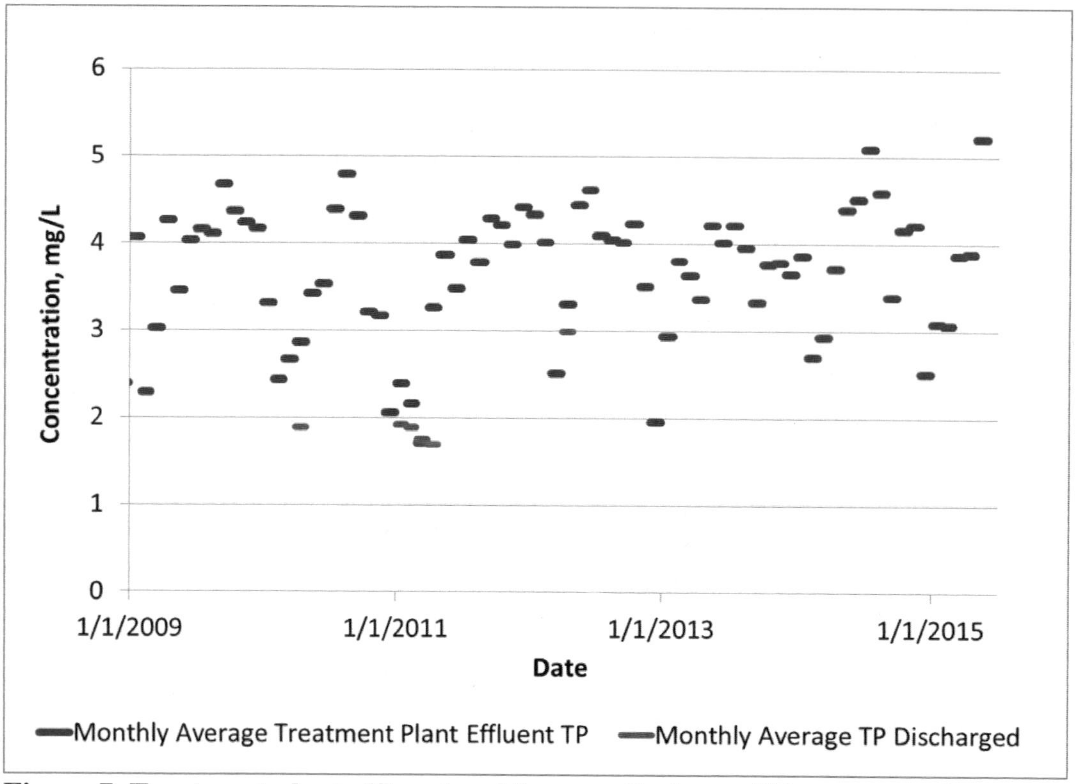

Figure 7. Treatment plant effluent and Laguna discharge TP concentrations.

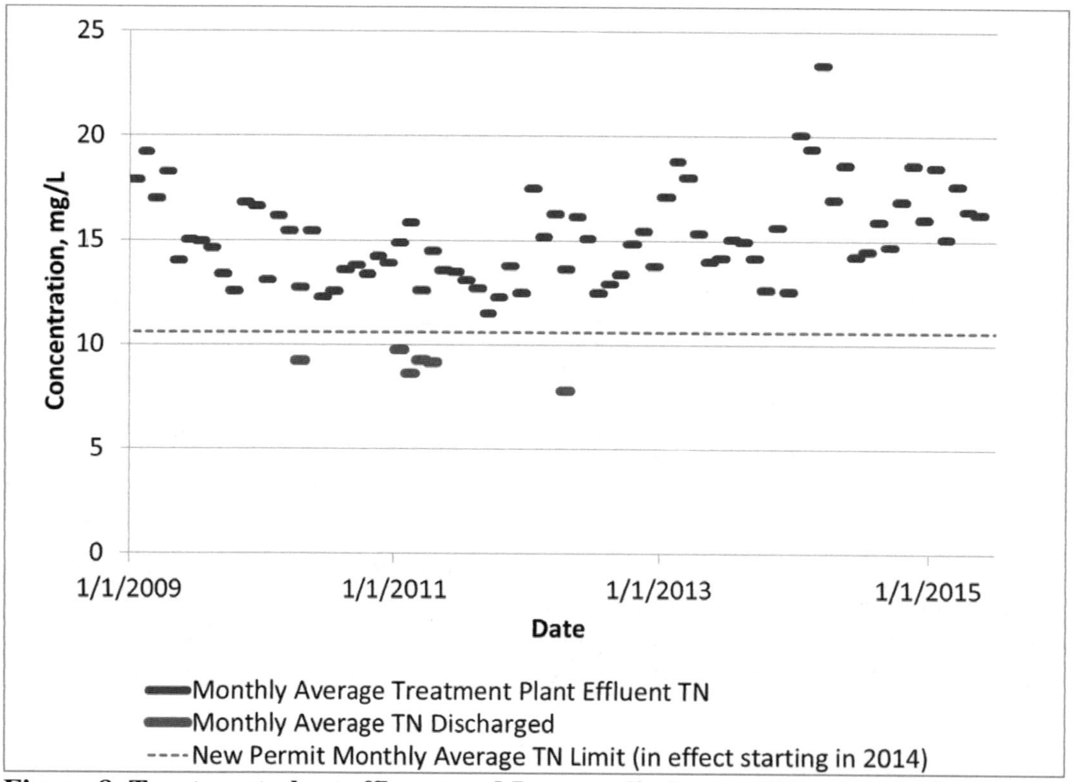

Figure 8. Treatment plant effluent and Laguna discharge TN concentrations.

Plant performance, special sampling and BioWin modeling indicate that the plant's nutrient removal processes are carbon limited, and there is not sufficient readily-biodegradable carbon for both biological phosphorus and nitrogen removal. Nitrogen removal is required to comply with the monthly nitrogen limit for discharges. Two process improvement alternatives were evaluated for the near-term to reduce TP while not increasing TN.

The first alternative was to increase CEPT to remove phosphorus in the primary clarifiers. Increased mixed liquor recycle flow was also included to improve denitrification. Since discharge to the Laguna occurs from Delta Pond, chemical phosphorus removal is needed when the ponds are filling. At the beginning of the rainy season, most of the stored recycled water has been used. Ponds typically fill from about November until staff has determined that discharge to the Laguna will not be needed for the remainder of the rainy season (typically in March or April). It is not feasible to reduce chemical usage by directly discharging plant effluent because plant effluent needs to be routed through storage ponds to achieve the additional denitrification necessary to comply with the TN effluent limit.

The second alternative was to construct a primary sludge fermenter to produce additional carbon on-site. The activated sludge tanks would be modified to add an anaerobic zone for biological phosphorus removal, followed by an anoxic zone with mixed liquor recycle.

The evaluation of alternatives determined that the least expensive improvement in the near-term is to implement CEPT for chemical phosphorus removal in the primary clarifiers. The main cost of this alternative is the increased volume of chemicals. The plant is considering more sustainable options for the longer term.

IMPLEMENTING AN INNOVATIVE NUTRIENT OFFSET PROGRAM

Appropriately managing recycled water storage volumes and achieving maximum TP reductions within the treatment process significantly reduces the amount of TP that could potentially be discharged. However, a third compliance strategy was developed to strengthen the City's ability to comply with its zero net loading limit. This third strategy, the regulatory agency-approved Nutrient Offset Program (NOP) is the first of its kind in California and allows the City to reduce nutrient loads elsewhere in the watershed by an amount at least equal to the amount discharged (and of equivalent bioavailability).

As noted previously, compliance to the zero net loading limit is determined through a rolling three year average mass balance comparing the total amount of TP discharged with the total amount of TP controlled elsewhere in the watershed. Although given three years to retrospectively determine compliance, the time from project planning through completion varies and has proven to be quite lengthy. The three projects approved to date have each taken several years to develop and implement. As a result, the program essentially requires the City to preemptively implement offset projects that reduce phosphorus loads and obtain sufficient offset credits, in case discharge occurs.

Project development requires a number of steps which include:

- Coordinating City contracts for project development and specialized technical assistance

- Identifying potential projects with sufficient credit worthiness
- Determining property owner interest and negotiating project agreements
- Calculating crediting eligibility and value of potential offsets
- Obtaining project go-ahead from the City Board of Public Utilities and obtaining formal credit approval from the North Coast Regional Board
- Implementing the project and verifying nutrient offset credit values
- Annual compliance reporting

Since a sufficient number of offset credits must be accumulated in advance of anticipated annual discharges, the challenge is to estimate how many credits are needed. The City has developed a probability-based approach to determine expected discharge volumes over a three-year period. Combined with the average plant effluent TP concentration, this approach predicts how many pounds of TP are anticipated to be discharged over a three-year period, and thus determines how many offset credits should be obtained. The current range of offset credits needed for the three year period ending in 2017, after the 2016-17 discharge season, averages 1,600 kg/year (3,500 pounds/year).

Each year, the City is required to submit a mass balance compliance report to compare the accumulated number of earned TP offset credits with the cumulative amount of TP discharged over the past three year period. Each approved offset project produces a particular number of credits each year for a particular time period (up to 30 years). Credits earned in a calendar year are applied to the following discharge season. Since compliance is determined based on a rolling three-year credit total versus TP mass discharged, credits may be unused and are no longer useful once the year in which they are earned is no longer in the three-year compliance window – so credits are really more like insurance for which the City pays for a certain level of assurance to meet the zero net loading limit, whether or not actually needed.

The expiration of credits and the costly, time-consuming demands of project development intensify the need for careful evaluation for the number of credits needed because, ultimately, the cost of compliance is borne by local ratepayers and the City is committed to minimizing rate increases. Despite the uncertainty related to actual compliance needs and ongoing programmatic costs related to credit expiration, the NOP provides an essential pathway for City compliance with the zero net loading limit.

Since the inception of the NOP, a myriad range of projects have been considered. Although now primarily focused on agricultural based Best Management Practices (BMPs), the NOP initially evaluated a much broader range of project possibilities including municipal activities like more frequent street cleaning and residential leaf pick-up services. However, initial credit estimating quickly distilled agricultural based BMPs as the most cost-effective and credit-worthy options to pursue.

Thus far, the City has successfully designed and implemented three regulatory approved nutrient offset projects. Considering the complexity of project development, each is considered a notable achievement. Projects have generated credits for dairy manure and agricultural land management improvements; roadway stabilization and culvert replacements at a nature preserve; and manure pond dredging at a former dairy with land application at local vineyards.

How are Offset Credits Calculated?
Regional Board staff assess the eligibility of management practices for offset credits using the following criteria:

- Are the proposed activities required by a Regional Board permit, waiver, or other permit and thus ineligible for credit offsets?
- Will the project improve habitat and ecosystem conditions and provide long term benefit to water quality?
- Are the calculations for nutrient reductions and offset credits supported by scientific literature or other acceptable methodology?

Once a project scope and land owner agreements are developed, detailed credit calculations are developed for each project and then reviewed and formally approved by the Regional Board. Calculations describe expected reductions of TP and propose a reasonable life for credits. Expected TP reductions are dependent on factors associated with the source being controlled (manure production rate, sediment loading, timing) and the BMP type. Credit life first considers whether/when the management practice will be required by any other regulatory provisions, since only practices that are not required by any regulatory mechanism are eligible. Since implementation of local dairy regulations began shortly after adoption of the NOP, only certain dairy best management practices (above and beyond) remain eligible. Credit life also considers the anticipated facility life and the estimated duration of the BMP's benefit to water quality.

Methodologies for calculating nutrient reductions and offset credits are based on a variety of nutrient offset tools including the Pennsylvania Credit Calculator (PCC) (Pennsylvania Department of Environmental Protection, 2008). The PCC offset calculation tool is well established in the Midwest and accepted there by the pertinent regulatory agencies.

Generating credit calculations is complex and exceedingly stringent. In addition to using proven and acceptable methodologies, the NOP specifies that credit calculations utilize appropriate Margins of Safety:

"... To account for any uncertainties in granting reduction credits, all projects proposals must include an appropriate Margin of Safety (MOS), which can be described numerically, or by spatial and temporal aspects of a given proposal. ..."

More specific descriptions of the three projects approved to date and their respective crediting approaches follow.

Beretta Dairy Best Management Practices Nutrient Offset Project
The Beretta Dairy Best Management Practices Project was the first project approved for credits. The project proposed to install three agricultural BMPs for manure and pasture land management in order to reduce potential nonpoint source nutrient discharges into Roseland Creek, a tributary of the Laguna de Santa Rosa that flows through the dairy.

The first BMP proposed installing a large manure stacking pad to allow an alternative way for the dairy to manage manure. The existing land area used for manure stacking would be fenced and revegetated to better manage winter runoff from a portion of the dairy. The dairy had been stacking manure on this land in the summer and then scraping the area clean of manure before winter rains. Providing a new manure stacking pad, locating it within the production area of the dairy and creating a new vegetated area between the former stacking area and adjoining watercourse would buffer the creek from any potential runoff impacts. The manure stacking pad received a credit life of 20 years, reflecting the expected life of a concrete manure stacking pad and long-term maintenance of the buffer as agreed to by the dairy owner.

The second BMP added a vegetated buffer and culvert cattle crossing to protect a drainage swale passing through one of the dairy's heavily used pastures (see Figure 9). The heavy use area was used as a milk cow loafing and staging pen where cows would linger for several hours before being released into grazing pastures. Prior to the project the dairy protected water quality in the adjacent swale by scraping the heavy use area prior to the wet season and restricting cow access from the area until the end of the rainy season. A short term credit life of 2.5 years was granted for the fence and crossing because of forthcoming dairy water quality rules which would require implementation of this or a similar BMP. A life of 20 years was provided for the vegetated buffer as it is above and beyond dairy rule requirements and will be maintained.

The third BMP included relocating a heavy use loafing pen/pasture area to a remote field, interior to the dairy, and converting a heavy use area to a rotational grazing pasture. The heavy use area served as a staging area for milk cows before they were released into nearby rotational grazing paddocks. The water quality benefits of this BMP considered two significant factors. The first was moving the loafing pen to a remote location and reducing the edge-of-field delivery ratio. This ratio could be reduced because greater upland attenuation of nutrients occurs as the distance between the heavy use area and surface waters increases. The second factor involved changing land use conditions within the dairy from denuded soils within the heavy use area during the majority of the year to the use of rotational paddocks, which remain in perpetual vegetation and act as permanent buffering that minimizes the potential for sediment laden runoff. A credit life of 20 years was provided for the third BMP due to an operation and maintenance agreement with the dairy owner to continue this voluntary implementation action.

Figure 9. Beretta Dairy BMP No. 2 – Heavy use area upgrades: Before (left) and after (right)

Pepperwood Preserve Sediment Reduction Best Management Practices Nutrient Offset Project

The Pepperwood Preserve Sediment Reduction Project implemented a series of erosion control treatments on stream crossings and rural roads within the Pepperwood Preserve, a 12.6 km^2 (3,117-acre) nature preserve. These BMPs will serve to prevent long-term sediment delivery into Mark West Creek, a salmonid spawning tributary of the Laguna de Santa Rosa.

The project at Pepperwood repaired and/or stabilized 36 eroding stream crossings with culvert improvements, rock slope protection and armoring; and repaired over three miles of eroding road surface and ditches. Erosion control treatments included decommissioning roads, out-sloping road cross sections, re-rocking road surfaces, and ditch relief culverts. These actions were deemed eligible for credits as they were identified as voluntary initiatives with no current regulatory requirements to address nutrient loading from these sites.

Creditable erosion sites at Pepperwood included road and ditch systems hydrologically connected to stream channels that discharge to Mark West Creek. These systems either threatened to erode, or were actually eroding as the result of poor stream crossing designs, the presence of springs, and/or inadequate drainage facilities.

The following photos (Figure 10) illustrate the type of sites being addressed.

Figure 10. Pepperwood Preserve: Examples of an eroding bank of a headwater stream (left) and poorly designed stream culvert systems (right).

Roads and drainage facilities within the project were either upgraded to accommodate 100-year peak storm flows, or decommissioned. Both options addressed eroding surfaces and prevented or minimized future sediment contributions to adjoining stream channels. Sediment loads not only added phosphorus (and nitrogen) to receiving streams, but also diminished salmonid habitat and increased stream temperature. The project's ancillary biological and physical benefits to restore and protect critical salmonid habitat were particularly appealing to regulatory staff.

The actively eroding stream crossings and other future projected erosion sites were granted a four-year credit life. The eight stream crossing corrections cited for imminent failure represented future erosion potential. The associated sediment loading predictions were equally divided over four years to determine credit values. Thirty-year credits were granted for actively eroding road surfaces and associated ditches.

Estimated potential annual soil loss estimates were based on site surveys at each location. Total soil loss estimates based on these site assessments were the basis for credit calculations as follows:

- Actively eroding stream crossings (26) = 660 m^3/year (864 cubic yards per year)
- Future potential sediment loading sites (8) = 22 m^3/year (29 cubic yards per year)
- Actively eroding road surfaces and ditches [3.7 km (2.3 miles)] = 1,719 m^3/year (2,249 cubic yards per year)

Site-specific sampling data collected from stream crossings at Pepperwood found that soil phosphorus content averaged 0.2 g P per kg soil (0.4 pounds/ton) for stream crossings and other sites and 0.1 g P per kg soil (0.2 pounds/ton) for road surfaces and ditches. These calculations yielded TP load reductions in pounds per year for each site. As required in the NOP framework, nutrient bioavailability was considered and based on a comparison of phosphorus bioavailability from wastewater versus soil erosion components.

Nunes-Ocean View Dairy Best Management Practices Nutrient Offset Project

The Nunes-Ocean View Dairy Project involved emptying two manure ponds (see Figure 11), on-site dewatering of the removed manure, and on-/off-site land-application and incorporation of the dried manure on local vineyards at agronomic rates. This project reduced potential nutrient loading from the former dairy site into Windsor Creek, a tributary to the Laguna de Santa Rosa.

The Nunes-Ocean View Dairy project is located at a former dairy site and was developed to address concerns that the two full manure ponds presented imminent potential to overtop and contribute nutrients to Windsor Creek during precipitation events. Credits were derived from the project by determining both the quantity of manure stored in each pond and the average nutrient concentrations within each.

Figure 11. Nunes-Ocean View: Dredging western manure pond.

Volume estimates of the manure stored in the ponds were based on measurements of pond side slopes and depth after the manure was removed. Nutrient concentrations within the ponds were based on manure sampling both before and during project implementation. After the manure was removed from the ponds, it was stacked for drying.

Close to 41,000 kg (90,000 pounds) of TP were removed, based on pond volumes and manure quality. Based on the landowner's reported application rates and areas, over 39,000 kg (85,000 pounds) of P were land applied. Calculated credits were based on the amount of TP estimated in the ponds and the amount beneficially land applied. To ensure credit calculations were conservative, three Margin of Safety discount factors were applied to the 39,000 kg (85,000 pounds) estimate to calculated credits as follows:

- 30 percent to reflect the expected amount of P in the manure that was not expected to be discharged to waterways if the manure ponds were not emptied in a controlled manner;
- 39.9 percent edge of field factor to reflect potential loss of manure P from vineyards after spreading; and
- 93.5 percent bioavailiability factor to reflect the fact that 6.5 percent of the P in the manure is not considered bioavailable.

Further Development of the Nutrient Credit Program

By limiting its discharges, maximizing TP reductions in the treatment process and successfully implementing its NOP, the City is currently well poised to meet its compliance requirements if anticipated discharges occur within the next four years (see Figure 12).

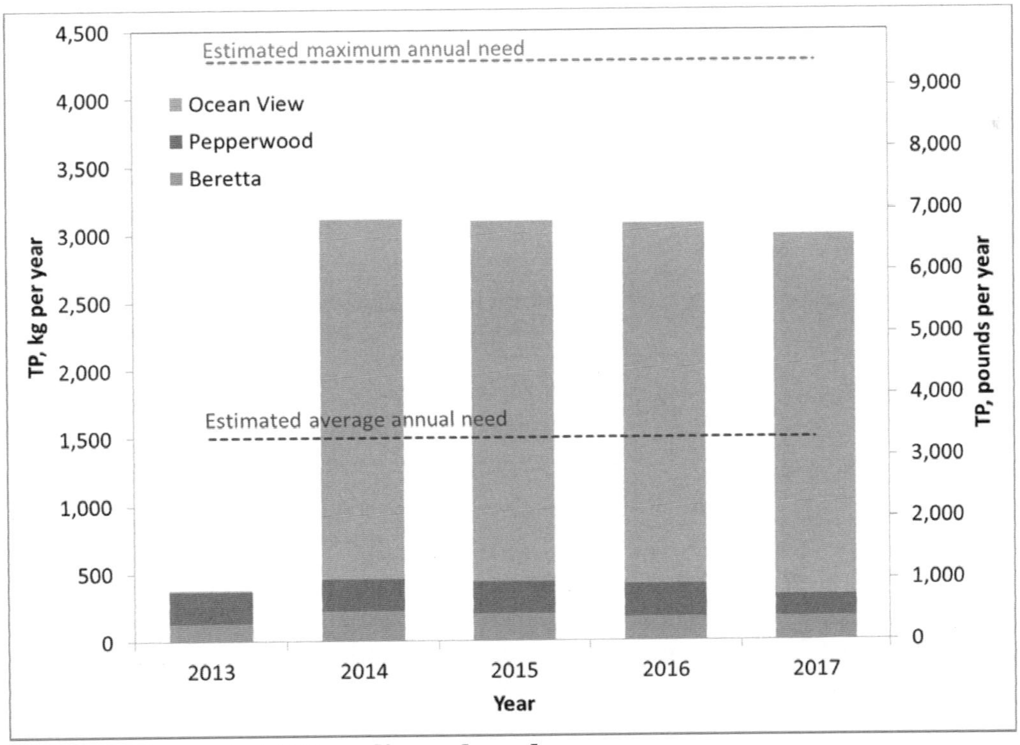

Figure 12. Expected TP credits and needs.

However, considering the length of time to develop projects and coordinate landowner agreements and realizing that many local farmers are hesitant to enter into complex regulatory agreements that may highlight water quality impacts associated with their operations, the City is working to expand the scope of currently creditable activities. The City is now aiming to develop crediting protocols to determine scientifically based nutrient reductions associated with stream restoration BMPs such as increased stream channel complexity, increased conveyance through flood control channel dredging, stream bank repair, restoration of riparian corridors and the like. With this type of nutrient crediting methodology, the potential list of creditable projects will expand significantly. And, the City will be able to successfully integrate its nutrient offset needs with its creek restoration, flood protection and environmental stewardship goals.

CONCLUSIONS

The framework developed and approved for the Nutrient Offset Program has allowed the City to avoid installation of capital intensive upgrades to the treatment plant during the interim period prior to TMDL approval, but the City has faced many challenges. More than a million dollars has been allocated already, with more costs in the future. The credits purchased by the City expire, so in the case of consecutive dry years, the City may have purchased a large number of credits needlessly.

Despite the challenges, it is noteworthy that the Nutrient Offset Program has allowed the City to build collaborative relationships with North Coast Regional Board staff. This collaboration has successfully grounded the concept of nutrient offset crediting within the region and work is currently being conducted to develop a more extensive water quality trading program.

Ultimately, the City's successful development of its three pronged approach in meeting its zero net phosphorus loading limitation not only promises compliance reliability but also minimizes long term cost impacts to ratepayers and provides sustaining environmental benefits.

AKNOWLEDGEMENTS

The authors wish to thank those involved in the development and implementation of the compliance strategies to meet the zero net loading limit for phosphorus, including:

- Robert & Doug Beretta (Beretta Dairy Project)
- GHD, Inc.
- Gold Ridge Resource Conservation District
- Kieser & Associates, Inc.
- Krisilsa Pacific Farms, LLC (Nunes-Ocean View Dairy Project)
- North Coast Regional Water Quality Control Board
- Pacific Watershed Associates Inc.
- Pepperwood Foundation (Pepperwood Preserve Project)
- Sonoma County Resource Conservation District
- The Freshwater Trust

REFERENCES

North Coast Regional Water Quality Control Board. Notice of Approval, Nutrient Reduction Activities for the Beretta Dairy Nutrient Offset Project; September 20, 2012.

North Coast Regional Water Quality Control Board. *Notice of Approval, Nutrient Reduction Activities for the Nunes-Ocean View Dairy Nutrient Offset Project*; January 24, 2013; Amendment dated August 19, 2014; Second Amendment dated April 29, 2015.

North Coast Regional Water Quality Control Board. Notice of Approval, Nutrient Reduction Activities for the Pepperwood Preserve Nutrient Offset Project; October 18, 2012.

North Coast Regional Water Quality Control Board. *Waste Discharge Requirements and Master Reclamation Permit*; Order R1-2013-0001, NPDES No. CA0022764; November 21, 2013.

Pennsylvania Department of Environmental Protection. *Draft Phosphorus and Sediment Credit Calculation Form*; Effective January 30, 2008.

Assessing the Feasibility of Phosphorus Trading Programs

Alex Johnson[1*]
[1]The Freshwater Trust, Portland, Oregon.
*Email: alex@thefreshwatertrust.org

ABSTRACT

Under Clean Water Act provisions, many regulated entities such as municipal wastewater treatment facilities will face strict new limits in upcoming NPDES permits. As populations grow and discharges increase, municipalities seek to address increased loading and lower nutrient limits with a cost-effective solution. To achieve compliance, municipalities can often look beyond traditional infrastructure upgrades at the WWTF and instead explore a nutrient trading program through watershed-scale collaboration with other point and nonpoint sources.

Detailed analyses to assess the feasibility of nutrient trading employ several methods of quantification and provide entities with calculated estimates of pollutant reduction or environmental "uplift" and available credits for potential trading programs in their watersheds. Analyses also consider the political and social aspects of trading programs, including the regulatory environment and nonpoint source outreach. Using these numbers in comparison with costs for infrastructure upgrades allows regulated entities to make informed decisions about reducing impacts on water quality.

KEYWORDS: nutrient reduction, nutrient trading, nutrient offset program, water quality trading

INTRODUCTION

Water quality trading is becoming an increasingly important compliance option for utilities subject to new regulations on phosphorus and nitrogen. These water quality parameters are influenced heavily by agricultural and urban runoff, setting the stage for holistic watershed solutions that were not previously possible.

The Freshwater Trust (Portland, Oregon) is a nonprofit conservation and river restoration organization that works nationally building the science and measurement systems needed to protect and restore freshwater ecosystems. The Freshwater Trust practices "quantified conservation," a new approach to conservation that measures and values the environmental benefits of restoration actions. These quantified actions can be used by regulated entities to meet strict load limits set by regulatory agencies to protect water quality and sensitive habits (e.g., wastewater and stormwater NPDES permittees, power-generation and other facilities); by developers to mitigate environmental impacts and improve aquatic habitat; or by conservation funders to more efficiently and effectively achieve environmental goals.

The Trust is experienced in water quality trading analysis, policy development, and trading program design, implementation, and management. In 2013, the Trust received the prestigious U.S. Water Prize in recognition of its Rogue River temperature credit program, which uses a transparent and transferable water quality trading model. This program, funded by the City of

Medford's wastewater treatment plant, is one of the few active trading programs permitted for National Pollutant Discharge Elimination System (NPDES) compliance in the U.S.

The Freshwater Trust provides two examples of nutrient trading projects:

1) In 2014 The City of Meridian, Idaho, and The Freshwater Trust completed a detailed water quality trading feasibility study for phosphorus reduction. The City will likely consider water quality trading in conjunction with a technology upgrade to meet NPDES permit requirement. The City is considering implementing a pilot project to learn more about the transaction system for trading and the costs of nonpoint source phosphorus credits.

2) In 2014 The Freshwater Trust implemented a five-year nutrient reduction pilot program for the energy utility PacifiCorp in the Klamath River basin, Oregon. After evaluating key sites where the exclusion of livestock from the stream would result in lower nutrient loads, The Trust and local partners installed livestock exclusion fencing on one site. Phosphorus credits were generated and verified then sold to PacifiCorp. Information gained from the pilot project will support future nutrient reduction projects.

METHODS

Specific steps and milestones are needed to develop a viable nutrient reduction program using the watershed approach. These include:

- **Policy Analysis:** Assess key state regulations and guidelines to support water quality trading, establish baselines for obligations, and identify potential barriers to program implementation.
- **Potential Future Pollutant Loading Evaluation:** Evaluate current and/or future wasteload allocations, discharge monitoring report data, and technology options and estimate potential scenarios for future pollutant loading from wastewater discharge to determine the necessary size of a water quality trading program to generate nutrient credits.
- **Supply and Uplift Analysis:** Identify geographic drivers for nutrient loading and analyze best management practices for nonpoint sources that could provide greatest water quality benefit (uplift) to a regulated entity. Calculate total "credits" available through nonpoint source reductions in nutrient loading in the defined trading area within the basin or subbasin to meet the needs outlined in the potential future pollutant loading evaluation.
- **Program Cost Estimates:** Determine number of nonpoint reductions needed to maintain NPDES compliance and estimate cost of implementing and maintaining a trading program in the service area for the life of the nutrient credit.
- **Trading Program Design:** Recommend components for a trading program, including site prioritization for implementation, considerations for implementation milestones and resources for permitting to use in discussion with regulators and permit writers.

These steps are part of the 11 elements that can guide and assist agencies and stakeholders in making key decisions when designing and launching water quality trading guidance, frameworks, and plans consistent with the Clean Water Act, total maximum daily loads (TMDLs), and other relevant regulations. The elements have been proposed by the National

Network on Water Quality Trading (2015), of which The Freshwater Trust is an active participant, and were also outlined in a joint discussion led by Willamette Partnership and The Freshwater Trust (2014) in collaboration with the states of Washington, Oregon and Idaho, and with technical advice from EPA Region 10.

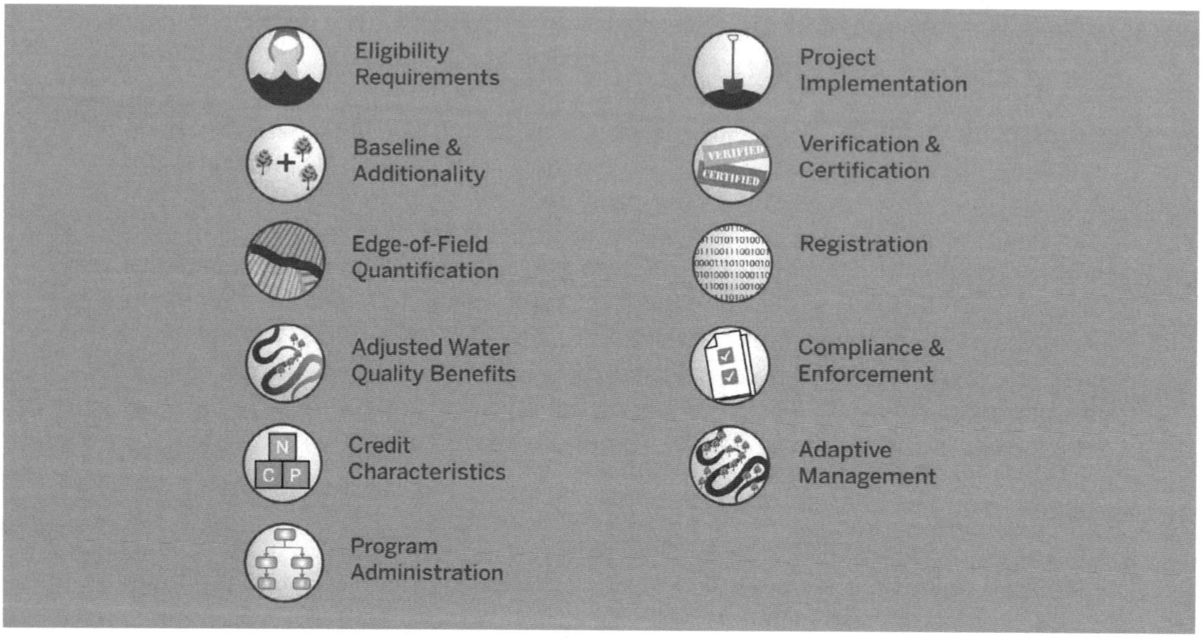

Figure 1. 11 Elements of Water Quality Trading. These elements can guide and assist agencies and stakeholders in making key decisions when designing and launching water quality trading guidance, frameworks, and plans consistent with the Clean Water Act, TMDLs, and other relevant regulations. (*Source: The Freshwater Trust, adapted from Regional Recommendations for the Pacific Northwest on Water Quality Trading.*)

NEW TECHNOLOGY APPLICATIONS

The Freshwater Trust uses geospatial data and analyses in its StreamBank® BasinScout™ Methodology for program analysis and implementation. BasinScout is a set of diagnostic tools developed by The Freshwater Trust that is used to estimate environmental uplift and prioritize restoration sites. Using outputs from BasinScout, restoration efforts and management practices can be targeted on high-disturbance sites that will yield the greatest environmental results.

Within a trading service area, typically a watershed, The Trust first assembles tax lot-scale data including satellite data on land cover, crop type, soils data, local topography, and other key variables. Current farming practices are estimated based on data, aerial photos, and conversations with local experts. The appropriate field scale model is then used to estimate loading from the field under current farm practices and then with locally appropriate BMPs installed. The difference between the two scenarios is an estimate of the water quality benefit or nutrient credit available from each tax lot. This detailed information allows for strategic program implementation to target and recruit the highest pollutant loading tax lots, facilitating a trading program that is both economically and ecologically efficient.

CASE STUDY: CITY OF MERIDIAN, IDAHO

Goal: Reduce phosphorus inputs and improve watershed health in the Lower Boise River Subbasin.

Action: Create a point source to nonpoint source water quality trading program for phosphorus credits.

The wastewater treatment facility (WWTF) for the City of Meridian, Idaho, will face new limits its upcoming NPDES permit. The city is seeking to address increased loading from projected population growth and lower nutrient limits with a cost-effective solution.

With the new permit, the WWTF will be responsible for reducing the amount of phosphorus in the treated wastewater it discharges into a tributary of the Lower Boise River. The city has already invested in upgrades to the treatment facility. However, sediment and nutrients enter the tributary primarily from nonpoint sources (IDEQ 2001). To achieve compliance, Meridian is interested in assessing how a water quality trading program through watershed-scale collaboration with other sources could fit into its overall facility compliance plan.

In 2014 The Freshwater Trust completed a water quality trading program feasibility study for Meridian. Guidance from the Idaho Department of Environmental Quality (2010) allows water quality trading as a compliance solution. Other municipal sewer districts and permitted treatment facilities in the Lower Boise River Subbasin are similarly evaluating water quality trading solutions. To meet the TMDL targets, the ultimate goal is to reduce the nutrient load to the Lower Boise River-Snake River system, thereby reducing the downstream algal growth that impacts water quality and recreation (IDEQ 1998).

Calculating Current Load

The Trust used past discharge monitoring report (DMR) data from the city to estimate the current total phosphorus loads from the WWTF's discharge. The total phosphorus load is calculated as:

> Effluent flow (mgd)*Effluent concentration (mg/L)*CF = Effluent phosphorus load (lb/day)
> *where CF is a constant; a conversion factor of 8.34 was used to convert mgd*mg/L to lb/day*

To calculate the daily total phosphorus load, The Trust used facility's design flow of 14.1 mgd (20-year flow based on facility planning documents (CH2MHill and HDR)). The Trust also factored in three potential concentrations that could be achieved through biological nutrient removal technologies as described in the facility plans. As the effluent total phosphorus concentration is reduced through technology, the size of a water quality trading program decreases. But because technology can get exponentially more expensive at the margins, trading can be an important option for cost-effective compliance once basic technologies are installed.

Implementing a new Regulatory Paradigm to support Effective Nutrient Management in California

Thomas Grovhoug[1*], Adam Link[2], Mitch Mysliwiec[1]

[1]Larry Walker Associates, Davis, California
[2]California Association of Sanitation Agencies, Sacramento, California
*email: tomg@lwa.com

ABSTRACT

Ongoing activities are underway in California to develop and implement a new, more effective regulatory approach for nutrients. The new regulatory paradigm is designed to complement and encourage the development of nutrient management plans in surface water bodies throughout California, including: San Francisco Bay, the Sacramento-San Joaquin Delta, inland surface waters, and other coastal estuaries. The approach is also designed to avoid common pitfalls and inefficiencies that occur through implementation of a "business as usual" regulatory approach.

Important elements of the new paradigm include: establishment of an overarching solution-oriented framework that clearly identifies and addresses goals, objectives, and common interests of key stakeholders; development of scientifically defensible information and necessary modeling tools in a joint fact-finding approach; and synthesis of information to develop a common understanding of management outcomes, in terms of ambient water quality and beneficial use improvements. The approach is especially applicable to nutrient regulation given the complexity of factors that determine the eutrophic effects of nutrients on different water bodies.

KEY WORDS: nutrients, management, collaboration, regulatory paradigm, outcomes

BACKGROUND

Since the adoption of the Clean Water Act in 1972, USEPA, States and the regulated community have implemented numerous approaches in an attempt to meet the goals of the Act. One long standing regulatory paradigm has been the practice of establishing water quality "standards" (the California nomenclature is "water quality objectives") as an early step in the regulatory process. (The California terminology will be used hereinafter since this paper focuses on the California regulatory regime.) Water quality objectives have been used as a primary tool to drive action through NPDES permits, effluent limitations, 303(d) listings, and TMDLs. This focus on water quality objectives has placed a primary focus on point sources, since point sources are regulated through NPDES permits and a strong connection exists between water quality objectives and the water quality-based effluent limits that are established in permits.

In years of implementing the current regulatory paradigm in California, the ability to attain objectives has not been a focus of the objectives-setting approach. Typically, objectives have been adopted without any meaningful assessment of the ability to achieve them in ambient waters. Also, objectives have been adopted with the presumption of resultant benefit, in the absence of any direct assessment of expected benefits. The common approach has been to rely on use attainability analyses and phased TMDLs as means of refining objectives to take attainability into account and to produce beneficial outcomes.

While the current approach has yielded positive results in terms of load reductions, upgraded wastewater treatment plants and a limited number of cooperative trading programs, the current approach has also led to a number of problems: unattainable effluent limits, uncertainty regarding the benefit of major expenditures by point sources to meet effluent limits, administrative appeals and litigation over NPDES permits and TMDLs, and others.

In particular, with regard to regulation of nutrients, the traditional approach to setting objectives and implementing those objectives through NPDES permits and TMDLs has proven unsatisfactory and inefficient. California regulators and stakeholders have recognized the need for a different approach to nutrient regulation for more than 15 years.

The difficulty with regulation of nutrients has occurred as a result of the complexity of eutrophication and other nutrient-related impacts. Depending on the water body in question, those impacts may include depressed dissolved oxygen levels, harmful algae and cyanobacteria proliferation, nuisance algae blooms, modified food webs, and other effects. Ambient nutrient concentrations are but one of numerous factors that affect these biological conditions. The ability to affect changes in ambient nutrient concentrations through load management and the ability to produce beneficial changes in biological conditions through such changes are major questions that should be addressed through the regulatory process.

PURPOSE

The purpose for this paper is to articulate the common themes and approaches being utilized in three significant, nutrient policy development efforts in California.

Ongoing activities underway in California to develop and implement a new, more effective regulatory approach for nutrients are described. This new regulatory paradigm is designed to complement and encourage the development of sound nutrient management plans in surface water bodies throughout California, including: San Francisco Bay, the Sacramento-San Joaquin Delta, inland surface waters, and other coastal estuaries. The approach is also designed to avoid common pitfalls and inefficiencies that occur through implementation of a "business as usual" regulatory approach, i.e. triggering regulatory action through early adoption of water quality objectives.

Important elements of the new paradigm include: establishment of an overarching solution-oriented process and framework that clearly addresses goals and common interests of key stakeholders; development of scientifically defensible information and necessary modeling tools in a joint fact-finding approach; synthesis of information to develop a common understanding of management outcomes, in terms of ambient water quality and beneficial use improvements; and the development of a sustainable and effective management plan with attainable goals and objectives.

CALIFORNIA CASE STUDIES

The paper will describe the new regulatory paradigm in the context of three ongoing nutrient policy development case studies in California: the Sacramento-San Joaquin Delta Nutrient Management Strategy and Science Plan, the Nutrient Management Strategy in San Francisco Bay, and the Nutrient Policy for Inland Surface Waters of California.

Sacramento-San Joaquin Delta Nutrient Management Strategy

This effort in the Sacramento-San Joaquin Delta estuary is being led by the Central Valley Regional Water Quality Control Board, in coordination with a number of stakeholders, including publicly owned treatment works (POTWs), agriculture, stormwater agencies, non-governmental organizations, water supply interests, USEPA, San Francisco Bay Regional Water Board, and other state and federal agencies. While the current focus of the effort over the next several years is on the development of a Delta Nutrient Science and Research Plan, the Stakeholder and Technical Advisory Group (STAG) for the process has developed a charter to capture goals, objectives and guiding principles for an overarching nutrient management strategy effort.

The charter outlines the overall process for development of a Delta Nutrient Management Strategy and Plan. It specifically addresses the question of whether nutrient load management can resolve specific problems in the Delta. The process addresses the scientific research and planning to enable sound decision making regarding future regulatory policy determinations. It calls for a transparent process, robust stakeholder involvement and independent scientific peer review of science work products.

The goal is to develop a framework that will lead to effective nutrient management decisions and actions that achieve reasonable protection of beneficial uses in surface waters of the Delta. The framework is to include the clear definition of problems to be addressed, potential drivers of those problems, and evaluation of management options. Important problems to be addressed in the Delta include cyanobacteria, macrophyte blooms, food web impacts, and drinking water concerns associated with algal blooms and taste and odor issues. Ultimately, the stated intention is to develop attainable management goals and implementation plan based on an understanding of management capabilities and resultant benefits.

Over the next year, Science Work Groups comprised of stakeholder technical staff and outside experts will assist the Regional Water Board and the STAG in the development of several white papers to describe the current state of knowledge regarding the role of nutrients and other factors in the proliferation of cyanobacteria and invasive macrophytes in the Delta, and the development of nuisance algal blooms in the Delta and downstream water supply conveyance and storage facilities. Data gaps and proposed research will be identified.

Concurrently, a work group will assist in the selection of water quality and ecosystem modeling tools to be used as tools in the nutrient management planning process.

The current plan is to develop and execute a science research plan by 2018, at which time decisions will be made regarding the need for nutrient water quality objectives for the Delta and further steps in the nutrient management plan for the Delta.

San Francisco Bay – Nutrient Management Strategy

The nutrient management effort in San Francisco Bay is being spearheaded by the San Francisco Bay Regional Water Quality Control Board. It is being performed in coordination with Bay Area Clean Water Agencies, USEPA, and numerous other stakeholders.

Development and execution of a science plan is being led by the San Francisco Estuary Institute (SFEI), with assistance from the Southern California Coastal Water Research Project (SCCWRP) and other technical experts. Key elements of the science plan include development of a conceptual model of nutrient sources, loadings and effects; development of water quality and ecosystem models; an assessment framework to examine nutrient-related problems; data collection to support modeling and to address key scientific questions, and special studies in specific sub-embayments of San Francisco Bay.

A Nutrient Watershed Permit (NPDES No. CA0038873) was adopted in April 2014. The permit is applicable to 34 NPDES permittees discharging to San Francisco Bay and its tributaries. An early requirement of the watershed permit is the development of a Bay-wide study to determine nutrient load reduction capabilities that are achievable through either process optimization or major facility upgrades at Bay area POTWs. It is planned that the information from this study will be used to determine the potential capability to modify ambient nutrient levels in the Bay through nutrient load management.

The Nutrient Watershed permit also requires permittees to contribute to the funding for the science plan to support the long range nutrient strategic plan for the Bay. The Bay Area Clean Water Agencies (BACWA), which represents Bay area POTWs, supported the watershed permit and its requirements as part of a collaborative approach to nutrient management with the Regional Water Board and USEPA.

A Steering Committee has been formed, comprised of BACWA, the Regional Water Board and other agencies contributing funding to the overall nutrient management effort in the Bay. The Steering Committee has adopted a charter to establish goals and guiding principles. The Steering

Committee also has adoptive a governance structure and funding plan to support the management strategy effort over the next ten plus years.

The charter, adopted in June 2014, articulates the vision for the Bay process as a

> "locally supported, multi-interest, long term science strategy and associated implementation program to provide information to support nutrient management decisions in the Bay."

The long range plan is to implement the science plan through a collaborative effort and, ultimately, to develop a sustainable nutrient management plan for the Bay. It is currently anticipated that it will take 10 to 15 years (2 to 3 watershed permit cycles) to complete a suite of studies to support development and implementation of a long range nutrient management strategy for the Bay.

Nutrient Policy Development for Wadeable Streams in California

The third ongoing nutrient regulatory policy effort in California is being led by the staff of the State Water Resources Control Board (SWRCB). The process includes a Stakeholder Advisory Group (SAG), which is comprised of a diverse group of stakeholders representing municipalities, agriculture, water supply interests, and non-governmental organizations. Although an official charter has not yet been established to guide the stakeholder process and policy development activities, the SWRCB has invited stakeholder input throughout the process and has provided commitments to the stakeholders to establish goals and guiding principles for the multi-year process. Written documents to capture goals and principles are under development by the SWRCB. Goals of the process are to provide scientific information and stakeholder input to ensure that proposed water quality objectives are reasonably attainable and will produce desired benefits in their application to different surface water categories in California.

The SWRCB process includes an Independent Science Panel, comprised of four national nutrient experts. The role of the Independent Science Panel is to review the technical work by the Science Team. For this project, SCCWRP has been retained by SWRCB to develop and execute a science work plan. Work produced under that plan is being reviewed by the SAG and the Independent Science Panel.

Unlike the previous two case examples, this effort is not focused on a single defined water body. Instead, the SWRCB effort is applicable to numerous inland surface waters throughout California. As such, several different options are being considered for the wadeable streams nutrient policy that consider either watershed-specific or water body category-specific approaches to setting objectives that are reasonably attainable and provide reasonable protection of beneficial uses. Stakeholders, led by the California Association of Sanitation Agencies, have put forward a watershed specific approach. The Independent Science Panel has suggested application of tiered aquatic life uses and objectives to arrive at reasonable regulatory goals.

SWRCB and stakeholders have agreed that early adoption of numeric nutrient water quality objectives based on currently available science, while an option, would not produce desired outcomes in terms of beneficial use protection.

The stated plan is to complete work on the science plan, continue engagement with the SAG and the Independent Science Panel, and develop a draft nutrient policy for wadeable streams in 2016 with a primary focus on effective nutrient management. Depending on the outcomes from that process, a nutrient policy and objectives may be adopted in 2017.

COLLABORATIVE ADAPTIVE MANAGEMENT

Although not originally structured as such, the approach to nutrient regulation that is occurring in California appears to be a good example of Collaborative Adaptive Management (CAM), which, as described by Lynn Scarlett (2013) is a convergence of two decision processes: collaborative decision making and adaptive management. CAM is described by Scarlett as a process that provides a tighter intersection of science and decision making, greater public engagement in knowledge building and decision making, and improved decision outcomes. Each of these features describes the nutrient regulatory approaches being implemented in California.

Key steps in the CAM process that are consistent with the nutrient regulatory paradigm described in this paper include:

- Define problems to be addressed with Participants
- Establish Clear Goals for the process
- Ensure early engagement of
    - Scientists
    - Stakeholders
    - Decision makers
- Provide a process for Joint Learning
- Develop information on Effectiveness of Actions
- Provide an ongoing venue for communication and adaptive management in response to change

Given the complexities and uncertainties associated with nutrient management in aquatic systems, CAM provides an appropriate, and well recognized, context for the regulatory approach being employed in California.

SUMMARY

The key in these three, quasi-independent, ongoing regulatory policy development efforts, is that stakeholders have been engaged to work collaboratively with regulatory agencies to develop a long term management plan for nutrients. In each effort, attention is being paid to attainable outcomes and the management actions that would need to be taken to achieve commonly defined goals. The efforts are collaborative, with mutually defined goals and objectives and a vision for the process. Each process places a premium on joint fact finding and the development of appropriate tools to allow informed decisions regarding management options. Water quality objectives development plays a role in each process, but is seen as a tool in nutrient management, not an end in itself. Objectives are also recognized to be a potential impediment to efficient and effective nutrient management planning and implementation, if adopted prematurely.

An additional aspect of each of the approach being employed in each of the case studies is the freedom to consider innovative, and in many instances, necessary watershed-scale approaches in developing cost-effective solutions to meet common goals and objectives. The approach is especially applicable to nutrient regulation given the complexity of factors that determine the eutrophic effects of nutrients on different water bodies. A shift from sole reliance on nutrient load management to the full consideration of other factors affecting eutrophication and resulting ecosystem impacts is a clear strength of the new regulatory paradigm being implemented in California.

ACKNOWLEDGEMENTS

The authors gratefully acknowledge the many regulatory agency staff and other stakeholders who are making positive contributions to develop sustainable nutrient management plans and policies in California.

REFERENCES

Central Valley Regional Water Quality Control Board (2015) Charter - Process to Develop a Delta Nutrient Management Strategy. http://www.waterboards.ca.gov/centralvalley/water_issues/delta_water_quality/public_involvement_stag_meetings

San Francisco Regional Water Quality Control Board (2014) Waste Discharge Requirements for Nutrients from Municipal Wastewater Dischargers, Order No. R2-2014-0014, NPDES No. CA0038873

Bay Area Clean Water Agencies (2014) Charter – The San Francisco Bay Nutrient Management Strategy: Purpose, Organization and Governance of the Nutrient Management Strategy. http://bacwa.org/nutrients/

State Water Resources Control Board (2015) Development of Nutrient Objectives for California, Stakeholder Advisory Group, http://www.waterboards.ca.gov/water_issues/programs/nutrient_objectives/stakeholder_advisory

Scarlett, L. (2013) Collaborative Adaptive Management: Challenges and Opportunities. *Ecology and Society*., 18(3):26.

Six Municipalities, One Watershed: A Collaborative Approach to Remove Phosphorus in the Assabet River Watershed, Massachusetts

Luke Cole, ORISE Fellow at US EPA, Lisa Larimer & Corey Buffo, US EPA

In an effort to meet a phosphorus TMDL issued in 2004, four wastewater treatment facilities (WWTFs) in Hudson, Marlborough, Maynard, and Westborough—and the six Massachusetts communities they serve—partnered to develop and implement phosphorus removal strategies. Their goal was to improve water quality in the Assabet River by reducing phosphorus loads on a watershed scale. Each WWTF selected a different phosphorus removal technology, which was implemented with other facility improvements. After the upgrades, all WWTFs have been meeting effluent limits of 0.1 milligrams per liter ($mg\ L^{-1}$) on a 60-day rolling average, representing over 90% reduction in total phosphorus loads to the river during the months from April through October. The facilities are also meeting effluent limits of 1 $mg\ L^{-1}$ the rest of the year. This case study describes the joint efforts of the WWTFs and key elements of each WWTF's improvements, and provides a summary of lessons learned for use by other watersheds and communities seeking to work together to improve water quality. It also provides a summary of the significant water quality improvements measured in the Assabet River from 1995 through 2013.

Seeing Red, but 'in the Black' – Diverse Case Studies Demonstrate Substantial Cost Savings by Converting to Deammonification

T. A. Constantine*, D. Houweling*, P. Sanjines**, J. Sandino***

* CH2M HILL Canada, 425 Consumers Road, Toronto, Ontario, Canada (E-mail: Tim.Constantine@ch2m.com), Ph: 416-499-9000, ** CH2M HILL, Alexandria, VA, *** CH2M HILL, Kansas City, MO

Abstract

The drivers for implementation of deammonification technologies vary between facilities depending on factors such as effluent requirements, mainstream process performance and configuration, and local electricity prices. At the Ejby Mølle WWTP, relatively high electricity prices combined with even higher electricity prices for sale of "green" electricity provide significant energy savings from implementation of sidestream and mainstream deammonification. In addition, lower effluent TN is expected as an outcome of sidestream deammonification which will translate into operational cost savings due to the effluent TN tax. The combination of these factors provide an excellent direct payback for both sidestream and mainstream deammonification at this facility. At both the Skyway WWTP and AlexRenew WRRF, sidestream deammonification is the lowest cost alternative to achieving the process benefits of removing ammonia in the sidestream load. At the Skyway WWTP, the benefits of treating the sidestream load relate primarily to energy savings and alkalinity recovery. At the AlexRenew WRRF, sidestream treatment is required to help meet very stringent effluent TN limits. In addition, seeding of anammox organisms from sidestream treatment at the AlexRenew WRRF is an integral part of a mainstream deammonification strategy that is being implemented in 2015.

1.0 Background

Anammox (anaerobic ammonium oxidation) describes a novel biological process for biologically converting ammonia directly to nitrogen gas (≈90%) and nitrate (≈10%) using nitrite as the electron acceptor. Although only discovered in the 1990s, anammox has been rapidly adopted by the wastewater industry for use in sidestream reactors to treat dewatering return streams that contain both high temperatures and high concentrations of ammonia. Coupled with partial nitritation (the oxidation of half the ammonia to nitrite), anammox provides for ammonia removal in an overall process called "deammonification" that uses only 38 percent of the oxygen compared to conventional nitrification. In addition, deammonification provides the benefit of no alkalinity requirements and very little nitrate production. The oxygen and carbon savings of the deammonification process are illustrated in Figure 1.

To date, deammonification in sidestream applications has been successful because dewatering return flows offer conditions that mitigate the primary challenges of the process: (a) the slow growth rate of the Anammox organisms and (b) the requirement for partial nitritation i.e. retaining ammonia oxidizing bacteria (AOBs) while suppressing or washing out nitrite oxidizing bacteria (NOBs). Several named sidestream deammonification technologies have been commercialized and are in operation throughout the world including: DEMON™, Anita™Mox, DeAmmon®, Cleargreen™ and ANAMMOX®.

While deammonification in the sidestream is by now an established process, deammonification in the mainstream is at the beginning stages of development and implementation. Currently, investigations at pilot and full-scale suggest that suppressing or washing out NOBs (while retaining AOBs) is the greatest challenge to mainstream deammonification. Growth and retention of anammox organisms has been shown to be achievable provided that a minimum temperature is maintained (in the range of 10 to 20 degrees Celsius) and there is not too much competition for nitrite with heterotrophic organisms. This latter requirement implies that a mainstream deammonification process may need to be implemented as a tertiary stage or, if implemented in secondary treatment, downstream of enhanced primary treatment.

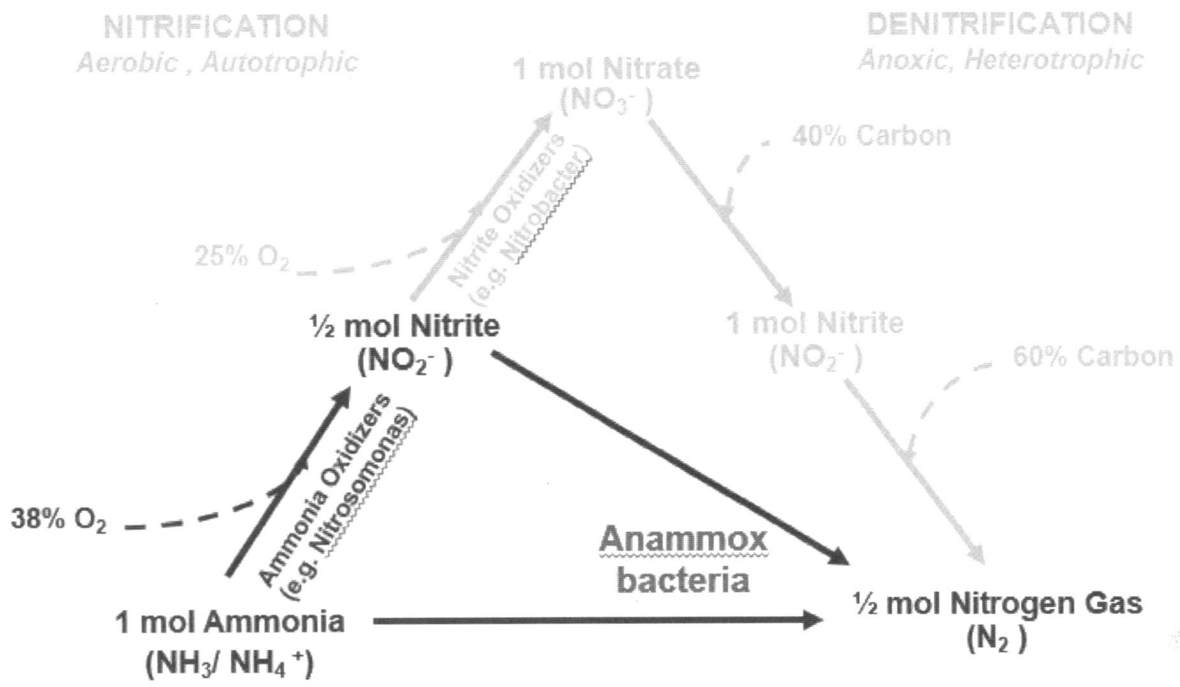

Figure 1. Pathways of the deammonification process as compared to conventional nitrification-denitrification.

2.0 Identifying Potential Savings

As presented in Figure 1, the oxygen requirements to remove ammonia through deammonification are approximately 60% lower than for conventional nitrification. Whether this translates directly into energy savings for aeration, however, will depend upon the efficiency of the aeration system (kWh required to transfer a kg of O¬2) in the sidestream process as compared to the mainstream, as well the potential to denitrify in the secondary process. As an example, a facility that operates high efficiency turbo blowers for the mainstream process but lower efficiency positive displacement blowers for the sidestream will not see a 1:1 aeration energy savings from implementation of deammonification in the sidestream. Also, facilities that are able to fully denitrify all nitrate in the mainstream process without the addition of an external carbon source would not be expected to see any decrease in oxygen or aeration energy requirements from implementation of deammonification in the sidestream. Nevertheless, for many facilities that have a significant sidestream ammonia load and a requirement to meet an effluent ammonia limit, implementation of sidestream deammonification would be expected to translate into overall energy savings.

By eliminating carbon from the pathway to convert ammonia to nitrogen gas, deammonification can also provide savings for facilities that rely on dosing of external carbon supplies to meet an effluent TN requirement. For facilities that do not need to dose an external carbon source but do have an effluent TN requirement, deammonification provides an opportunity to capture more carbon in primary treatment so that it can be directed to anaerobic digesters and converted to biogas.

Alkalinity recovery is also a significant savings that is achieved from deammonification. There is typically only sufficient alkalinity to nitrify half of the ammonia in sidestream dewatering loads and thus full treatment of a sidestream load must rely on either (a) alkalinity addition (b) alkalinity recovery through denitrification using an external carbon source or (c) deammonification. In addition, for facilities that are alkalinity limited in the mainstream process, deammonification provides an important opportunity to recover alkalinity.

3.0 Case Study #1: Ejby Molle WWTP

The Ejby Mølle wastewater treatment plant (EMWWTP) is the largest treatment facility in Odense, Denmark, and has a treatment capacity of 385,000 population equivalents (PE) which is equivalent to approximately 60 ML/d (16 mgd). VandCenter Syd (VCS) Denmark, is the provider of water and wastewater services for Odense, and operates the EMWWTP. The load to the plant is mainly of domestic origin, and it discharges to a small river with very stringent effluent requirements. The plant is currently producing on an annual average basis, effluent concentrations of 1.9 mg/L of BOD5, 4 mg/L of total nitrogen as N, and 0.1 mg/L of total phosphorus as P. Effluent TN and TP are subject to an effluent which creates a strong incentive to optimize biological nutrient removal (BNR) at this facility

Redefining Being Green: Extended Piloting of an Algae-Based Technology for Low-Level Nutrient Removal and Recovery

Alexandra Doody[1*], Jane Madden[1], Karla Sangrey[2], Mark Johnson[2], Rick Johnson[3], Will Holm [3]

[1] CDM Smith, Boston, Massachusetts.
[2] Upper Blackstone Water Pollution Abatement District, Millbury, Massachusetts.
[3] Clearas Water Recovery, Missoula, Montana.
*Email: DoodyAT@cdmsmith.com

ABSTRACT

The Upper Blackstone Water Pollution Abatement District piloted the innovative algae-based advanced biological nutrient recovery (ABNR™) technology, which uses algae to remove nutrients, in an effort to find a sustainable solution to meet new total nitrogen and total phosphorus permit limits. Extended piloting indicates that the ABNR technology shows promise for nutrient recovery (in the waste algae) and for the removal of phosphorus to very low levels (less than 0.03 mg/L) without metal salt chemical addition, while also producing oxygen and sequestering carbon dioxide, a greenhouse gas. The pilot set-up, results, and conclusions are presented.

KEYWORDS: Nutrient recovery, algae, low-level phosphorus removal, greenhouse gases.

INTRODUCTION

The Upper Blackstone Water Pollution Abatement District (the District) located in Millbury, Massachusetts owns and operates an enhanced biological nutrient removal (EBNR) water resource recovery facility designed for an average daily flow of 170,000 m^3/day (45 mgd). The District's facility receives wastewater from a combined sewer system that has a mix of residential, commercial, and industrial dischargers. The District operates the EBNR facility in an anaerobic/anoxic/oxic (A^2/O) mode to remove nitrogen and phosphorus without metal salt chemical addition or filtration and it achieves effluent with less than 6 mg/L total nitrogen (TN) and less than 0.5 mg/L total phosphorus (TP). However, tertiary treatment will be required to consistently meet new seasonal permit limits of 5 mg/L TN and 0.1 mg/L TP.

The District has screened conventional technologies available for both nitrogen and phosphorus removal. All of the conventional phosphorus technologies require the addition of metal salts to achieve the 0.1 mg/L TP limit, which produce residuals that could pose problems with the District's multiple hearth incinerators. In addition, most of these technologies would operate at the edge of their capabilities to meet the stringent TP limit. The District became interested in the innovative Advanced Biological Nutrient Recovery (ABNR™) technology patented by

CLEARAS WATER RECOVERY® (Clearas, Missoula, Montana), which uses algae to consume nutrients, in an effort to find a sustainable solution to meet the new permit limits.

Figure 1 presents a process flow schematic of the ABNR process, which includes three main steps: (1) mixing, (2) nutrient uptake/recovery, and (3) separation. The Clearas ABNR technology operates in a manner similar to a suspended growth activated sludge process in that it consists of a suspended growth reactor that promotes biological growth and nutrient uptake, followed by a liquid/solid separation step from which the solids are returned to the bioreactor. A key difference is that the ABNR process is seeded with microalgae (including members of the genus *Scenedesmus*) that have a high growth rate. The algae grow rapidly when four main factors are present: light, dissolved reactive nitrogen (ammonia, nitrate, or nitrite), dissolved phosphorus (orthophosphate), and dissolved inorganic carbon (carbon dioxide).

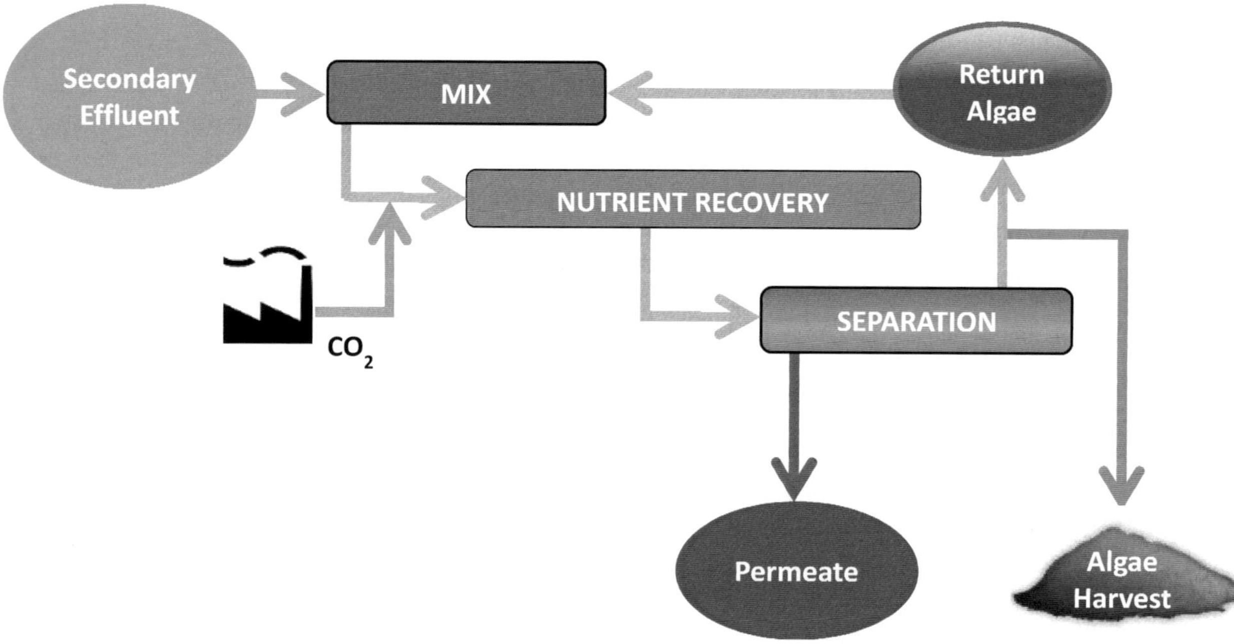

Figure 1. Schematic of the ABNR™ Process

Several pilot studies of the ABNR technology had already taken place in EPA Region 8 that demonstrated effluent TP concentrations with less than 0.03 mg/L. These pilots included one 18-month long pilot at an industrial pulp and paper mill and multiple pilot studies over a 13-month period at a municipal water resource recovery facility with primarily residential dischargers. A bench-scale nutrient recovery test of the District's secondary effluent performed by Clearas showed it had characteristics that were suitable for achieving low-level effluent phosphorus concentrations. Since the ABNR technology appeared to show potential for the District to meet its new low-level TP effluent limit without metal salt addition, and the process also has the ability to remove dissolved nitrogen, the District elected to undertake piloting of the ABNR technology in summer 2014.

PILOT METHODOLOGY

During the pilot, the District pumped 2.5 m^3/hour (11 gpm) of secondary effluent through a 50 μm pre-filter to remove large suspended solids and then to a 378 liter (100 gallon) mix tank, where the secondary effluent mixed with the 2.5 m^3/hour (11 gpm) recycled algae stream. From the mix tank, the combined flows were pumped through a photobioreactor (PBR) consisting of six fences of clear 50 mm (2 inch) diameter piping located in a greenhouse, as illustrated in Figure 2. This arrangement allowed the PBR to be irradiated with natural sunlight during the day and with light-emitting diode (LED) lights at night, thus promoting nutrient uptake by the algae throughout the day regardless of ambient light conditions. Carbon dioxide gas was injected prior to the PBR to provide a carbon source for the algae and to control pH.

Figure 2. The mix tank (left) is where the District's secondary effluent (pilot influent) mixed with the return algae, similar to how raw wastewater mixes with return activated sludge in a conventional activated sludge process. The photobioreactor (right) included six fences of piping and LED lights located in a greenhouse.

Downstream of the PBR, the algae were separated from the pilot effluent using a flat sheet ultrafiltration membrane system with a nominal pore size of 0.04 μm. Figure 3 shows a photograph inside the membrane separation unit before the system was placed into operation. The membrane unit had an outside-in flow configuration, with the influent water from the PBR spraying into the top of the tank. The thickened biomass (called "return algae") flowed by gravity from the membrane separation unit back to the mix tank. In order to maintain a target density of algae within the PBR, a small slipstream of the return algae was "harvested", similar to how waste activated sludge is removed from a suspended growth activated sludge system to maintain a mixed liquor concentration. Figure 3 shows a photograph of the harvested biomass after thickening. Once thickened and dried, the harvested algae from the pilot were analyzed for a variety of characteristics, including nitrogen content, phosphorus content, potassium content, crude protein, carbohydrate, fat, and ash content. These characteristics were assessed so that potential beneficial end uses for the harvested biomass product could be identified.

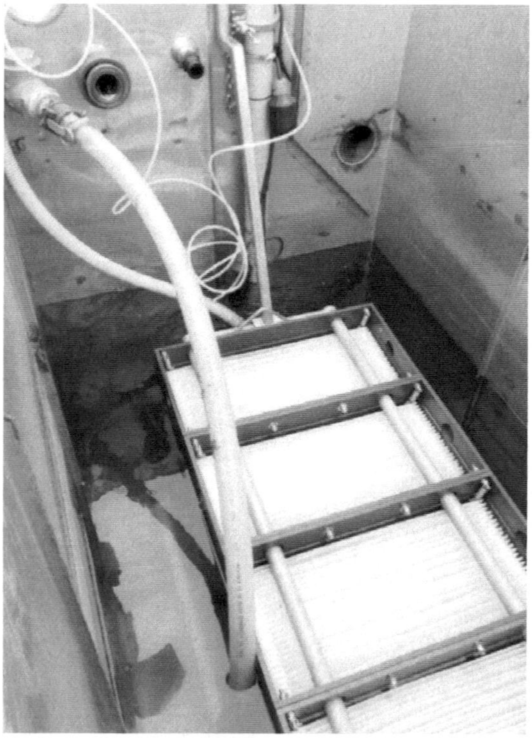

Figure 3. The separation technology used for this pilot was a flat sheet ultrafiltration membrane unit (left). The harvested biomass (pictured on the right after thickening with a clay-based polymer) has the potential for a variety of beneficial end uses.

The main pilot objectives and success criteria are summarized in Table 1. Demonstration of effluent phosphorus consistently below 0.05 mg/L TP was the primary objective, while additional reduction of nitrogen species was a secondary focus. The hydraulic retention time (HRT) for the PBR and membrane tank was less than 1.5 hours total. These design criteria were determined based on an average daily secondary effluent concentration of 0.45 mg/L TP, although the system was designed with the ability to handle short-term spikes up to 1.0 mg/L TP.

Table 1. Summary of Pilot Objectives and Success Criteria

Objective	Metric	Target
Demonstrate the ability to reduce TP from the secondary effluent to < 0.05 mg/L in a continuous flow operation	TP	< 0.05 mg/L
Demonstrate reduction in secondary constituents, particularly nitrogen species	BOD_5, NO_2-N, NO_3-N, TKN, NH_4^+-N	50% reduction
Confirm acceptable performance of the membrane separation technology	Transmembrane pressure (TMP) and membrane flux rate	26 L/m^2-hr (15 gpd/ft^2)
Confirm biomass quality produced	Protein, carbohydrate, fat, ash, and nutrient content analysis	Report

In order to assess the ability of the pilot system to meet the performance objectives shown in Table 1, time-paced composite samples were collected of pilot influent (i.e. the District's secondary effluent) and pilot effluent (permeate) for daily analysis of TP, orthophosphate (ortho-P), ammonium as N (NH_4^+-N), total Kjeldahl nitrogen (TKN), nitrite as N (NO_2^--N), and nitrate as N (NO_3^--N). Weekly analyses were also conducted for 5-day biochemical oxygen demand (BOD_5), chemical oxygen demand (COD), total suspended solids (TSS), and total dissolved solids (TDS). All of these samples were sent to an independent laboratory for analysis at the outset of the pilot. This level of data collection occurred from mid-June through August 2014. After that period of time, ongoing monitoring of TP was conducted using a handheld Hach DR900 colorimeter, with split samples analyzed by the District's lab or an alternative third-party lab at least twice weekly.

Some additional key variables monitored on a daily basis included:

- *Culture density in the photobioreactor,* measured using a TSS test. The density measurement is analogous to measuring mixed liquor suspended solids from grab samples in an activated sludge bioreactor tank. The PBR culture density varied significantly during the pilot depending on the nutrient loads into the system.

- *CO_2 consumption*, since there was a possibility that the District's incinerator flue gas could be a CO_2 source for a permanent installation, thus reducing the facility's on-site greenhouse gas (GHG) emissions.

- *Dissolved oxygen* (DO) in the photobioreactor and in the permeate, since the algae generate oxygen during photosynthesis. The permeate DO levels regularly exceeded 10 mg/L.

The pilot encountered several challenges throughout the summer, which resulted in the decision to extend piloting beyond the originally anticipated duration, as discussed below. The pilot operation to date can be broken into four distinct phases, which are summarized in Table 2.

Table 2. ABNR Pilot Phases

Phase	Description	Duration
1	Startup and Summer Operation with Secondary Effluent	June through mid-August 2014
2	Troubleshooting Performance Challenges	August and September 2014
3	Winter Operation with Outfall Water and Separation Unit Optimization	December 2014 through March 2015
4	Spring/Summer Operation with Outfall Water	Underway

RESULTS AND DISCUSSION

Before the official pilot testing began, control tests were conducted in which the District's secondary effluent bypassed around the PBR directly to the membrane separation tank. The purpose of these tests was to determine a baseline for nutrient removal by the membranes without algae present in the system. Figures 4 and 5 show the measured phosphorus and nitrogen species, respectively, from grab samples collected in three different locations during the control tests: (1) secondary effluent, (2) the top of the membrane separation tank, and (3) permeate.

The results shown in Figure 4 demonstrate that although the membranes were capable of removing the particulate fraction of the TP (a reduction of 0.05-0.06 mg/L), the membranes alone could not achieve TP permit compliance because no reduction in the soluble orthophosphate fraction occurred across the membranes. The increase in TP in the separation tank compared to the secondary effluent shows how the separation tank concentrates the solids, so that the return algae have a higher solids content than the photobioreactor. Similarly, Figure 5 demonstrates that minimal reduction in the nitrogen species occurred across the membranes when no algae was present with only the particulate fraction of the TKN (0.08 mg/L in Run 1 and 0.24 mg/L in Run 2) being removed. The soluble nitrogen species (ammonia, nitrate, and nitrite), which represent the majority of the nitrogen remaining in the District's secondary effluent, were not removed across the membranes.

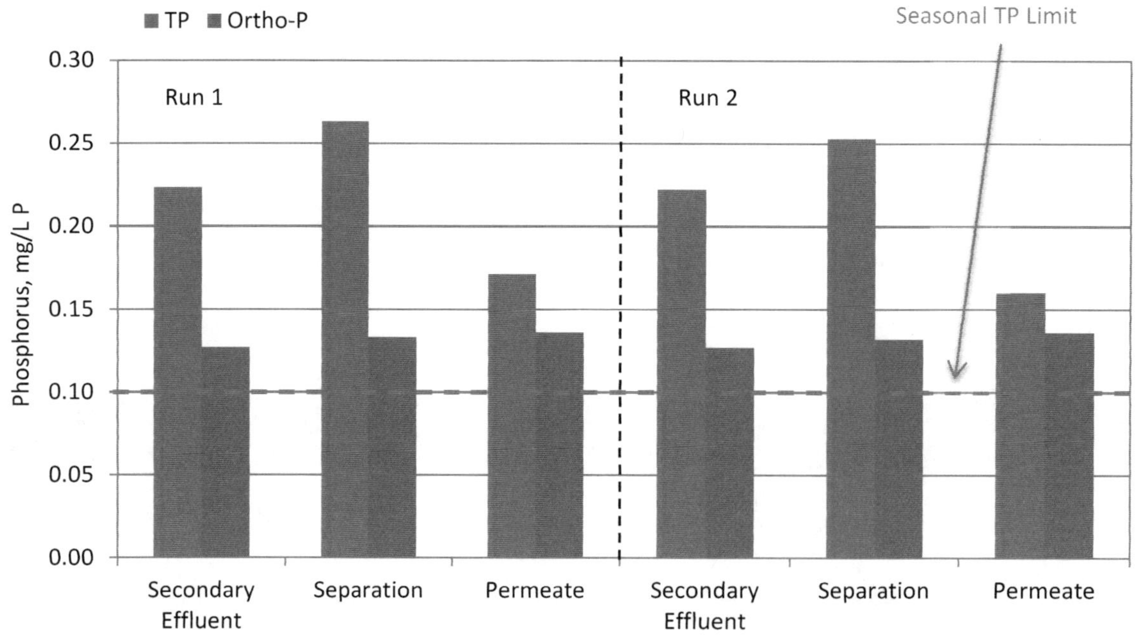

Figure 4. Control test results for phosphorus species before seeding the system with algae

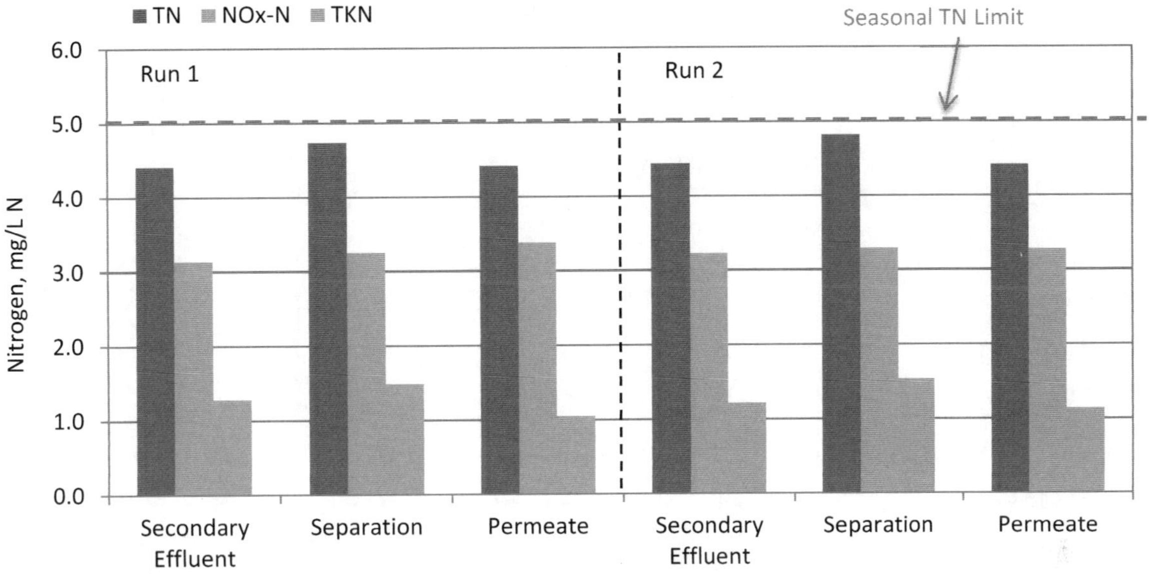

Figure 5. Control test results for nitrogen species before seeding the system with algae

Phase 1 – Startup and Summer Operation with Secondary Effluent

After the control tests were complete, the system was seeded with algae. The PBR operated in a recirculation mode with a synthetic nutrient feed (F/2 Algae Food) to promote algae growth to a suitable density for treatment of secondary effluent. Once the measured density in the PBR exceeded 250 mg/L data collection commenced in mid-June 2014, marking the start of Phase 1 of the pilot.

During several periods of operation in the summer, the ABNR pilot performed well with permeate below the pilot target of 0.05 mg/L TP and well below the TP permit limit of 0.1 mg/L. Figures 6 and 7 summarize the secondary effluent and permeate results for TP and TN, respectively, during those periods of healthy, steady-state operation. (Note that the TP concentrations are presented on a logarithmic scale to help distinguish the very low concentrations evaluated for this study. In one particular period, the permeate samples had reached the lower limit of detection of 0.005 mg/L TP for the method used by the third-party lab. In each case, the stable operation lasted for 5 to 8 days.

On other occasions, the pilot did not perform as expected due to numerous challenges. Figure 8 shows the full timeline of TP results for the secondary effluent and permeate composite samples collected from June through mid-September 2014, while Figure 9 shows a similar plot for TN concentrations, except that TN measurement was suspended for a portion of August and most of September.

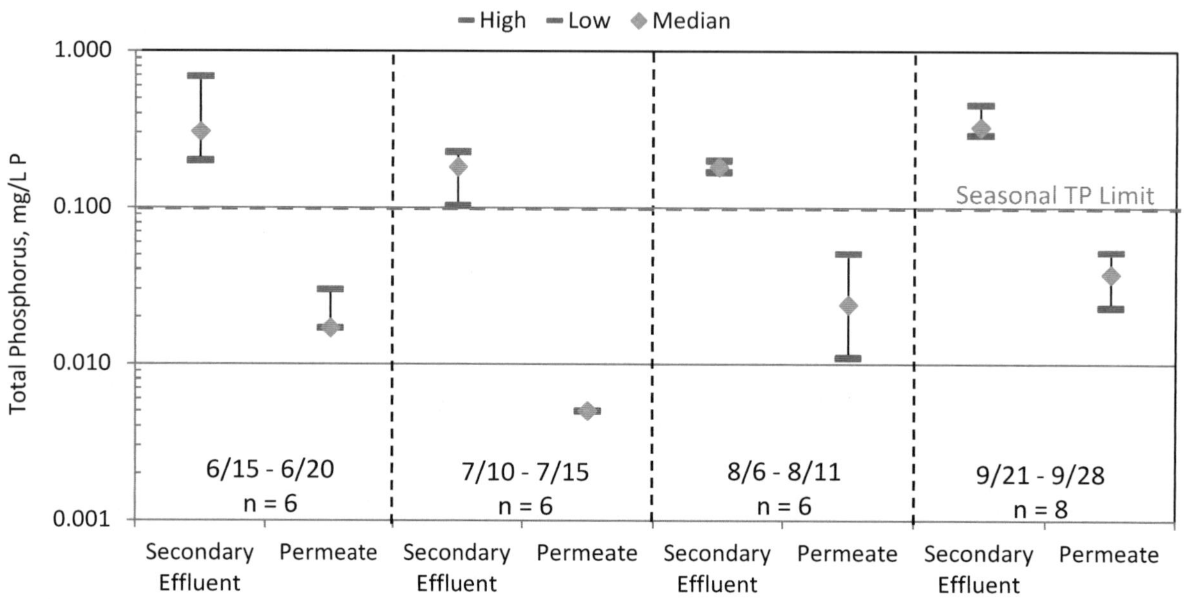

Figure 6. Pilot TP results when system was operating with healthy culture under steady state operating conditions

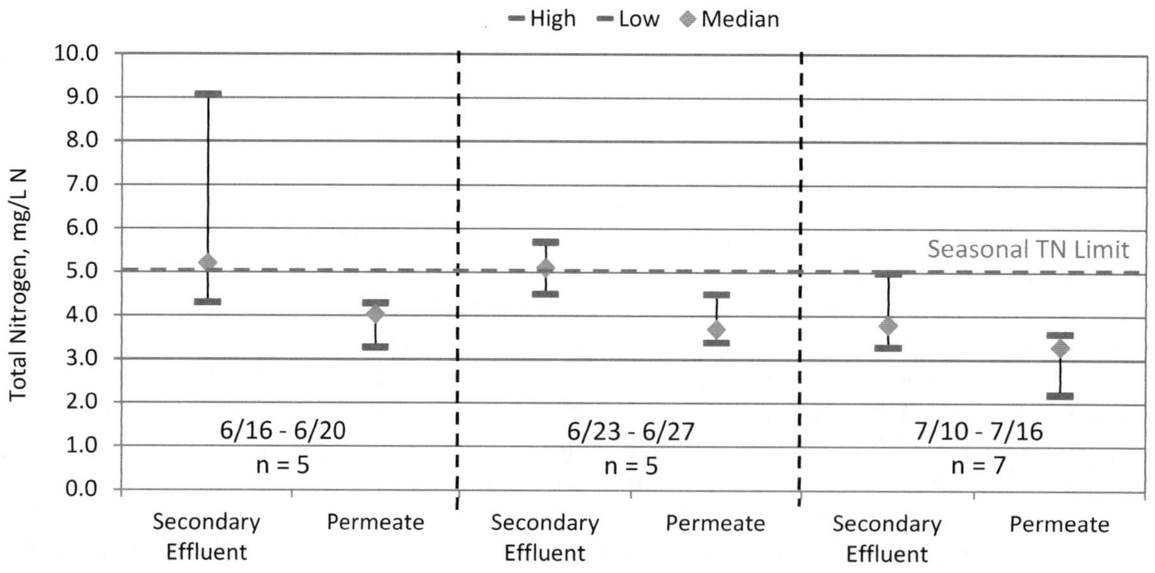

Figure 7. Pilot TN results when system was operating with healthy culture under steady state operating conditions

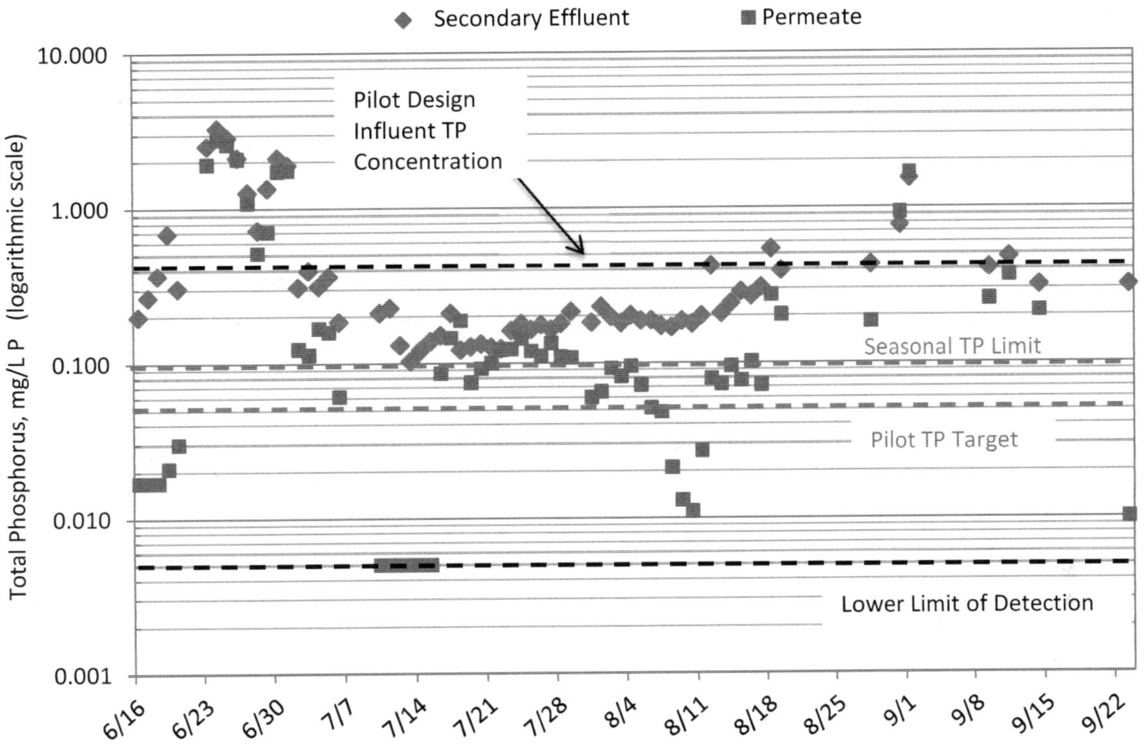

Figure 8. TP Concentration Results for Phases 1 and 2, June through September 2014

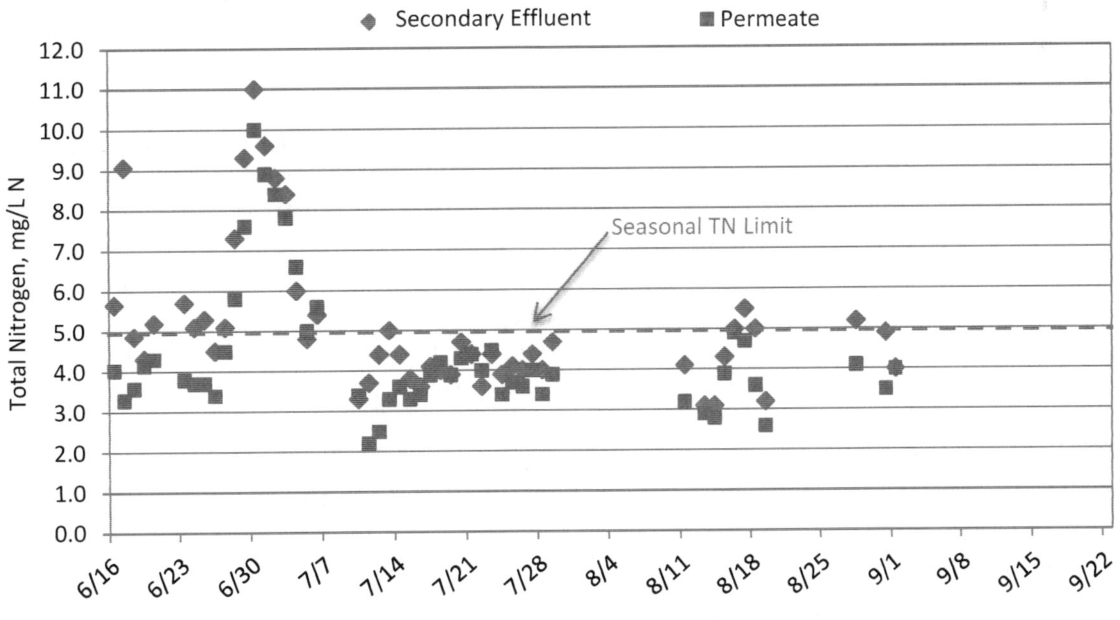

Figure 9. TN Concentration Results for Phases 1 and 2, June to September 2014

Key challenges encountered during operation of the pilot in summer 2014 included:

- Secondary effluent nutrient loads that were outside of the pilot system's design criteria;
- Filamentous algae growth;
- Lack of bio-available nitrogen; and
- Agglomerated biomass with poor nutrient uptake.

The following section provides a brief description of these challenges and some of the resulting investigations that were conducted to mitigate them (constituting Phase 2 of the pilot).

Phase 2 – Troubleshooting Performance Challenges

Nutrient loads in the secondary effluent that were outside of the pilot system's design criteria. Two separate upsets to the District's enhanced biological phosphorus removal system occurred during the summer (one in late June lasting 13 days and another in late August/early September lasting 24 days) that resulted in secondary effluent concentrations that exceeded 0.45 mg/L TP. In addition, the upset in June also resulted in elevated ammonia concentrations. The pilot system was not designed to be able to treat concentrations exceeding 0.45 mg/L TP for such long durations, and as such elevated TP concentrations passed through the system into the permeate. At larger commercial-scale demonstration facilities Clearas has installed a stock culture tank that maintains a standby reserve of available algae. In the event of sudden influent TP spikes (generally up to 1.0 mg/L TP), the standby reserve of algae can be injected into the system to boost the PBR density and associated nutrient uptake capacity. However, this pilot-scale installation did not initially have a stock culture fence that would have allowed for this to be tested.

Filamentous algae growth. The presence of filamentous algae species was a consistent problem throughout the summer. On many occasions long stringy algae clogged the pre-filter, the influent pump, and a rotameter for flow indication. In addition to being a nuisance for the mechanical equipment, the filamentous algae also are potentially problematic for the treatment process performance because: (1) the growth rate is three times slower than the desired *Scenedesmus* organisms, and thus reduce the capacity of the system to uptake nutrients to very low levels at a given PBR density and (2) filamentous algae are not stable under shearing forces and thus are susceptible to releasing phosphorus, as well as bio-polymers that could clog the membranes.

Strategies implemented to limit filamentous algae growth included filtration of the return algae, pH control, and modified harvesting practices. Filtration of the return algae was ultimately not successful in removing the undesired species, whereas control of pH with CO_2 addition combined with improved and more consistent harvesting practices helped to limit filamentous algae growth while maintaining a healthy cell count of *Scenedesmus* algae.

Lack of bio-available nitrogen. Starting in mid-July, the District's A^2/O process began achieving complete nitrification, with secondary effluent ammonia concentrations ranging from the lower limit of detection of 0.075 mg/L up to 0.2 mg/L NH_4^+-N. Most of the nitrogen remaining in the secondary effluent was present in the form of nitrate (ranging from 2 to 5 mg/L NO_3-N) and organic nitrogen (typically around 1 mg/L organic-N) which can be present in both soluble and particulate forms. It was during this time that permeate TP levels unexpectedly started to increase to 0.1 to 0.2 mg/L TP. It was hypothesized that insufficient nitrogen availability was the culprit for the poor TP uptake performance, so a supplemental ammonia dosing study was conducted to test this hypothesis.

Figure 10 displays the result of the supplemental ammonia dosing study. The permeate data shown with red squares corresponds to dates when only secondary effluent was fed to the ABNR system, while the yellow squares represent dates when supplemental ammonia was also fed to the ABNR. When supplemental ammonia was added to the secondary effluent upstream of the mix tank the permeate TP levels decreased to 0.02 – 0.05 mg/L TP. The results from this experiment suggested that:

- The other non-ammonia reactive nitrogen species that remained in the District's secondary effluent in the summer (nitrate-N and nitrite-N) were not bioavailable to the algae; and
- Addition of supplemental ammonia (such as from ammonia-rich sidestreams generated within the District's facility) may be necessary in the late summer and early fall months (when the District's A^2/O process achieves complete nitrification) in order to achieve the very low levels of TP in the permeate.

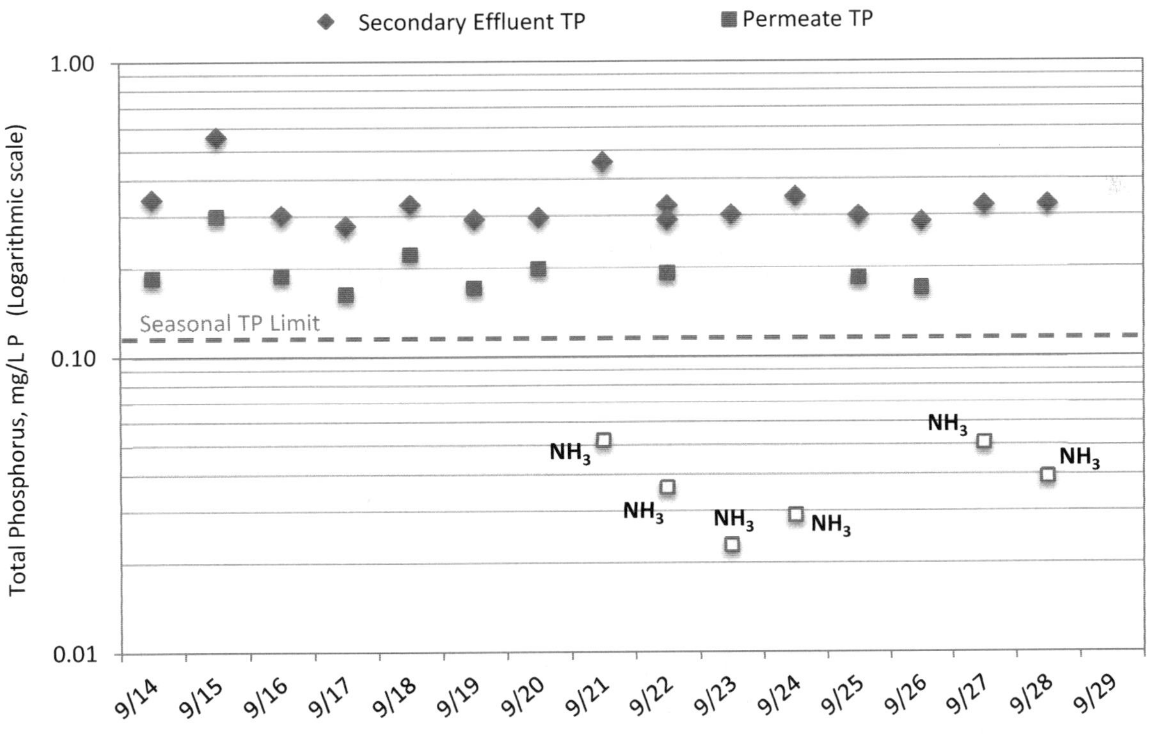

Figure 10. Impact of ammonia deficiency on TP removal to low levels

Agglomerated biomass with poor nutrient uptake. After the supplemental ammonia dosing was confirmed to help with TP removal to low levels, pilot testing resumed in mid-September with supplemental ammonia dosing. However, in October a shift in the culture population appeared to take place that was indicative of an unhealthy culture. Figure 11 displays microscope images of several types of *Scenedesmus* organisms with additional material agglomerated to it. The appearance of these agglomerated biomass particles coincided with poor nutrient uptake.

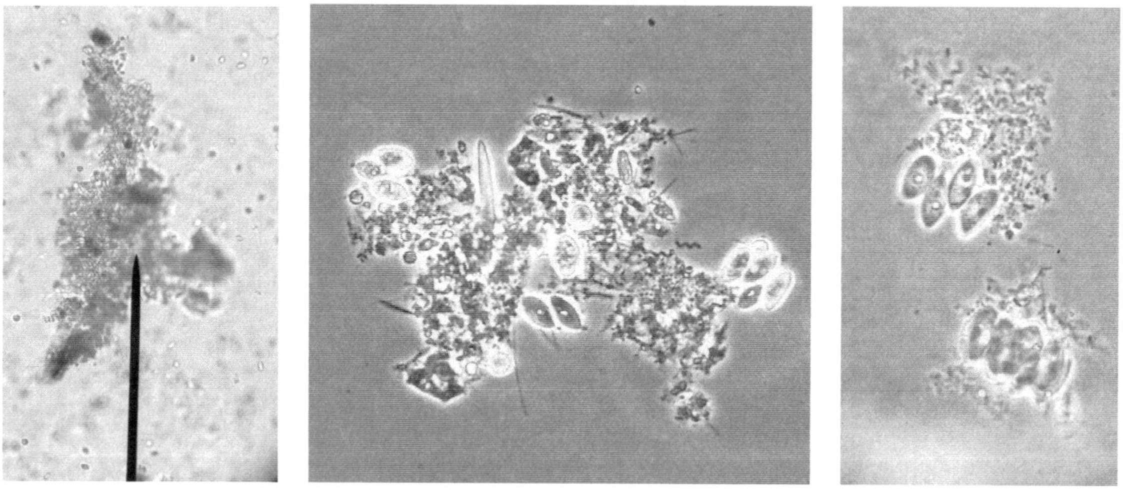

Figure 11. Microscope images of agglomerated biomass that exhibited poor nutrient uptake

One hypothesis regarding this occurrence was that the ABNR system had been operating with a low harvesting rate ("waste rate") and thus with a high solids retention time (SRT) in the system (6 days on average). It was speculated that higher SRTs might have left the algae subject to predation by higher species, resulting in cell death. For this reason, it was hypothesized that the pilot might achieve more consistent nutrient uptake performance if disinfected and dechlorinated water was fed to the pilot rather than secondary effluent.

Phase 3 – Winter Operation with Outfall Water and Separation Unit Optimization

Phase 3 started in December 2014 by re-seeding the reactor, allowing the algae to reach the target density, and going into production using disinfected and dechlorinated water from the District's outfall channel. The main objective of this phase was to determine whether the source of incoming water to the pilot had an influence on the reliability of TP removal and recovery. However, after testing had started, several concerns were raised surrounding the membrane separation unit, several of which merited further investigation. This section discusses the results of outfall water testing as well as some of the investigations conducted around the membrane separation unit installed at the District.

Conditions during this phase of testing that differed from the previous two phases included:

- The District's EBPR system during the winter months was more consistent than the previous phase with effluent ortho-P averaging less than 0.1 mg/L and effluent TP averaging less than 0.2 mg/L. In order to test the pilot for levels of phosphorus closer to the design concentration of 0.45 mg/L TP and to ensure adequate reactive nitrogen for the algae, a small amount of F/2 algae food was fed to the system. Note that the supplemental ammonia was not strictly necessary since the cold temperatures during the winter resulted in some residual ammonia in the effluent (albeit well within the District's permit requirements for the winter months). On average, about 80 percent of the phosphorus in the feed water to the pilot was from the District's final effluent and the remaining 20 percent was from the F/2 supplement.

- Daily TP measurements on the pilot feed water sample and pilot permeate effluent sample were taken daily using a handheld Hach DR900 colorimeter and analyzed in triplicate. Split samples were also sent to the District's laboratory and an independent laboratory, both of which had a much lower limit of detection for TP compared to the handheld colorimetric analytical method. As a result, the feed water TP measurements from the DR900 and the independent laboratory agreed well with each other, whereas the permeate TP measurements from the independent laboratory usually were much lower than the results from the DR900 measurements.

- Since the focus was on maintaining very low levels of effluent TP during this phase, limited nitrogen measurements were collected during this period.

- Chemically-enhanced backwashes (CEBs) of the membrane system were performed only when the system was taken out of production and drained. In prior phases, CEBs were performed in situ with the algae culture remaining inside the tank while sodium hypochlorite was pulsed into the interior pockets of the membranes to remove organic fouling. Out-of-tank CEBs were performed during this phase to rule out the potential for hypochlorite to come into contact with the algae.

- During the first out-of-tank CEB, settled biomass in the corners of the membrane tank was discovered. It was apparent that the diffused air system for the membrane tank was not maintaining the algae fully in suspension, leaving the settled biomass susceptible to dying and releasing phosphorus. The separation unit was modified to eliminate this occurrence and all data collected in Phase 3 reflects performance with this modification.

Figure 12 displays a summary of the phosphorus data and PBR density data that were collected during Phase 3 testing with outfall source water in January 2015. The phosphorus content of the feed water to the pilot ranged from 0.2 to 0.5 mg/L TP, while the TP in the permeate remained well below the target of 0.05 mg/L TP, except on January 24 when the permeate TP rose to 0.066 mg/L. This rise in permeate TP appears to have occurred because the feed water TP suddenly rose at the same time that a lower density of algae than designed was being maintained. The temporary decrease in algae density occurred immediately after performing an out-of-tank CEB.

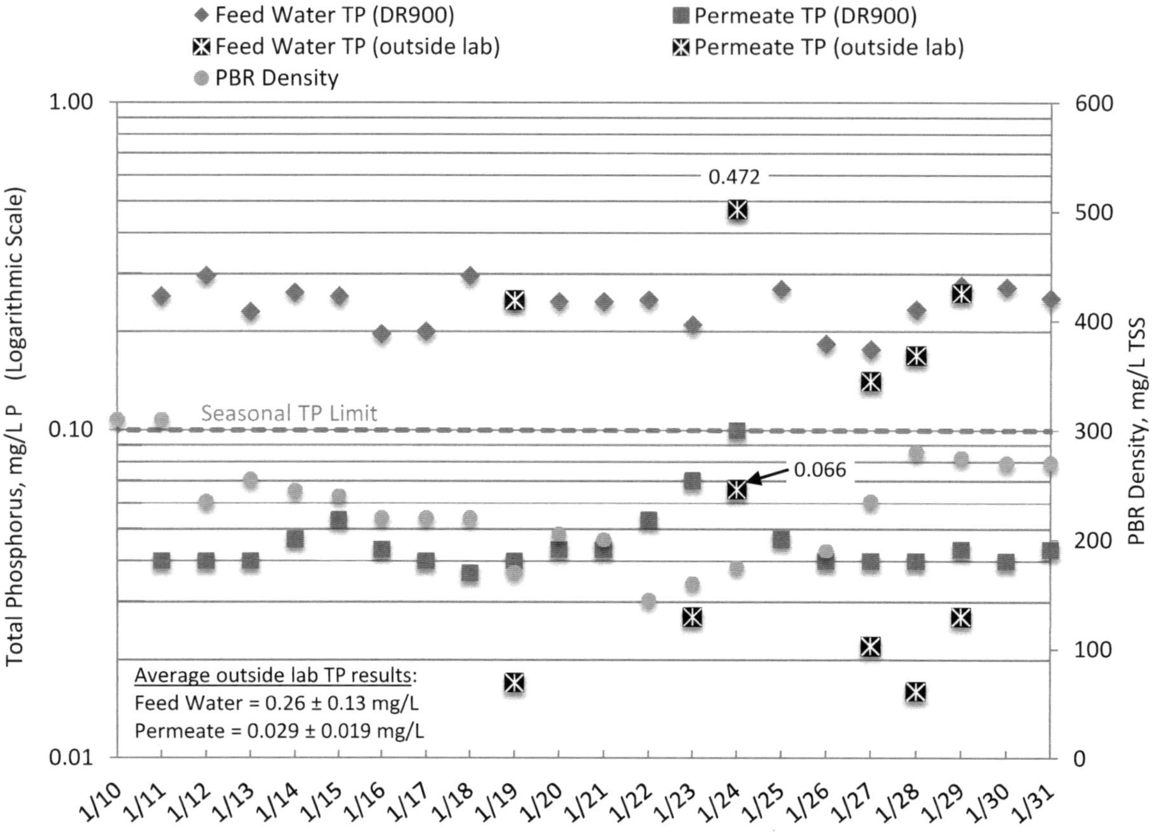

Figure 12. TP results for Phase 3 testing with outfall water using out-of-tank backwashes

Although the results from January 2015 showed good performance with respect to TP removal and recovery, conducting out-of-tank CEBs was less desirable because draining the membrane tank is a time-intensive process and this CEB protocol made it more difficult to maintain a constant algae density in the system. As a result, February and March 2015 were spent investigating the effects of different cleaning protocols on system performance. At first, half-strength bleach was used for in-situ CEBs. Figure 13 presents the phosphorus performance when operating the system with half-strength bleach. TP removal performance remained excellent with all permeate samples analyzed by an outside lab measuring less than 0.03 mg/L TP. However, this alternative cleaning protocol resulted in only partial transmembrane pressure recovery, and thus CEBs were required about once every day. In March, CEBs using full-strength bleach were resumed and this mode of operation was tested for two additional weeks. Figure 14 presents the phosphorus performance when operating with full-strength in situ CEBs. Average permeate TP remained excellent, with an average of 0.012 mg/L TP, and CEBs were required far less frequently.

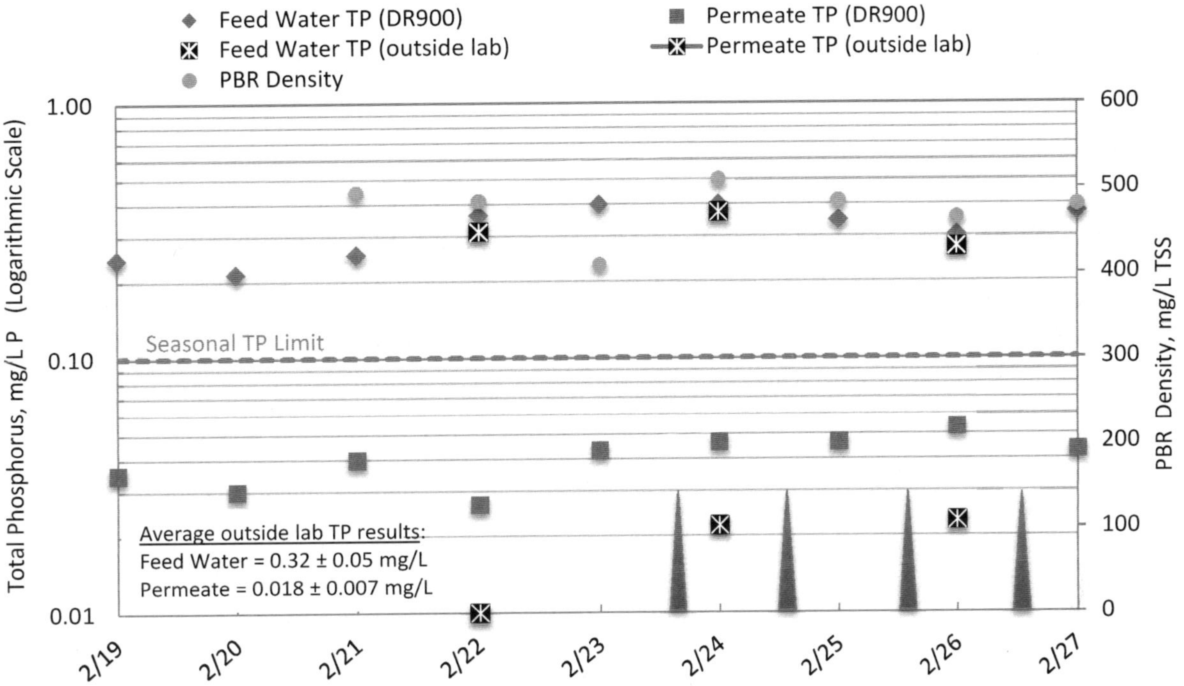

Figure 13. TP results for Phase 3 testing with outfall water using in situ, half-strength backwashes (blue triangles indicate when backwashes were performed)

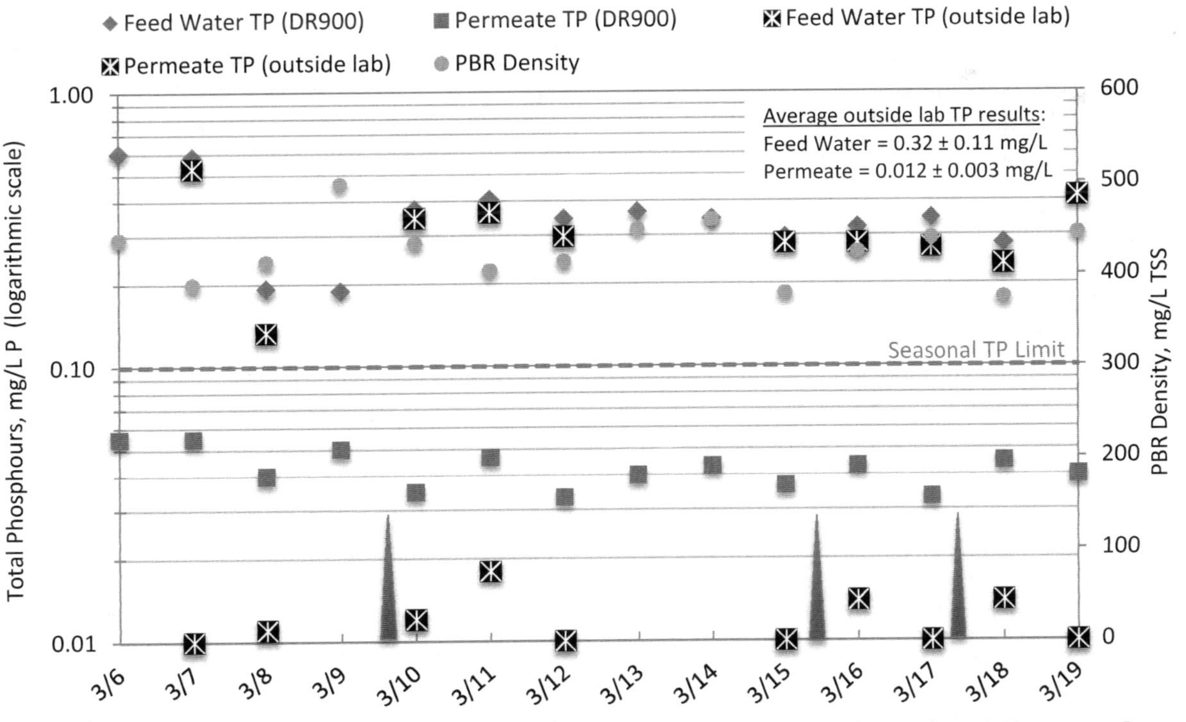

Figure 14. TP results for Phase 3 testing with outfall water using in situ, full-strength backwashes (blue triangles indicate when backwashes were performed)

Overall, Phase 3 testing with outfall water showed that the ABNR process was able to achieve effluent phosphorus to very low levels (0.01-0.03 mg/L). In addition, investigations on the separation process during this phase indicated that performance was negatively impacted by improper mixing inside the membrane tank and this problem was eliminated with some modifications to the pilot unit.

Phase 4 – Spring and Summer Operation Using Outfall Water

Phase 4 consists of continued testing of the ABNR pilot system with outfall water during the District's permit season months in order to demonstrate process reliability for extended periods of time. This phase began in May 2015 and is currently under evaluation. Data collection for this phase includes not only phosphorus, but also key nitrogen species (ammonia, TKN, combined nitrate/nitrite), BOD_5, COD, TSS, and total organic carbon (TOC). TOC is being measured to assess the potential for using the permeate for reclaimed water applications. Preliminary third-party lab data from this phase are presented in Table 3. Recall that the phosphorus and ammonia concentrations in the pilot feed water are higher than the ammonia in the District's effluent due to the influence of F/2 addition. Permeate TP performance has remained very good, with average effluent lower than 0.03 mg/L. Fewer samples have been analyzed for the other constituents, but the preliminary results demonstrate elimination of any oxygen demand remaining in the District's effluent as well as a reduction of 1.5 mg/L TN and 1.3 mg/L TOC.

Table 3. Preliminary Phase 4 Data Summary

Constituent (mg/L)	Sample Count	Average and Standard Deviations	
		Feed Water	Permeate
Chemical Oxygen Demand	4	23.5 ± 4.1	Non-detect
5-day Biochemical Oxygen Demand	4	3.8 ± 0.9	Non-detect
Total Dissolved Solids	4	475 ± 72	473 ± 63
Total Suspended Solids	4	Non-detect	Non-detect
Total Kjeldahl Nitrogen	4	2.5 ± 1.9	1.8 ± 0.6
Ammonia-N	4	1.1 ± 0.9	0.8 ± 0.5
Nitrite-N	4	0.30 ± 0.16	0.10 ± 0.09
Nitrate-N	4	5.4 ± 0.5	4.8 ± 0.4
Total Nitrogen	4	8.2 ± 1.2	6.7 ± 0.3
Total Phosphorus	16	0.279 ± 0.054	0.026 ± 0.031
Total Organic Carbon	4	6.2 ± 0.7	4.9 ± 1.6

CONCLUSIONS

Following is a list of the major conclusions obtained from pilot testing of the algae-based ABNR process at the District, with a particular focus on the pilot objectives listed in Table 1.

- The ABNR pilot system demonstrated the ability to reduce TP to less than 0.03 mg/L when operated with outfall water and with proper mixing inside the separation process. Testing of outfall water was conducted using a supplemental ammonia feed and additional phosphorus feed to bring the feed phosphorus up closer to the design concentration of 0.45 mg/L. Ongoing testing in summer 2015 will focus on demonstrating process reliability without any supplemental nutrient feeds.

- The objective of 50 percent reduction in secondary constituents was achieved for some constituents (including BOD_5, TSS, etc.), but reduction in nitrogen species could not be fully evaluated because supplemental ammonia was added in excess. Preliminary data indicate that ammonia is the primary nitrogen constituent consumed by the algae, whereas statistically significant reductions in nitrate and nitrite do not appear to occur in the ABNR process. More testing is being conducted in Phase 4 to assess nitrogen uptake and to confirm this finding.

- The separation technology performed as expected with respect to transmembrane pressure and membrane flux rate, but the pilot unit needed to be modified in order to maintain the algae in suspension. Optimization of the separation technology is ongoing.

- The biomass produced when the pilot was operating with secondary effluent has been assessed by potential off-takers and it has been accepted for use as a feedstock for the bioplastics industry. Biomass produced from when the pilot was treating outfall water is still under evaluation to determine potential beneficial end uses and market potential.

Extended piloting of the innovative algae-based ABNR process at the Upper Blackstone Water Pollution Abatement District has demonstrated the ability to achieve effluent with very low levels of phosphorus without the addition of any metal salts. The ABNR process also offers the following potential benefits:

1. Recovery of nutrients in a biomass that has a variety of beneficial end uses, some of which have revenue-generating potential;
2. Sequestration of carbon dioxide;
3. Generation of dissolved oxygen in the permeate; and
4. Production of high quality effluent with potential for a variety of water reuse applications.

Although further optimization and study of the ABNR process is required to confirm process reliability under variable influent conditions, these potential benefits warrant further advancement of this technology.

ACKNOWLEDGEMENTS

This pilot was conducted as a public-private partnership that included the District, CDM Smith, Clearas, and the Lyndon Paul Lorusso Charitable Foundation. The authors would like to thank the Lyndon Paul Lorusso Charitable Foundation for its generous financial support for the pilot, as well as the CDM Smith Research and Development program for the funding it provided to allow for CDM Smith's pro bono participation. We also gratefully acknowledge the assistance of the District's staff who helped to continually improve the pilot installation and run the laboratory analyses reported in this paper: Randy Komssi, Debra LaVergne, and Denise Prouty.

PilotTesting of Struvite Recovery from Centrate of a German Municipal WWTP Through Electrochemical Precipitation (ePhos Technology)

Iosif Mariakakis[1*], Jennifer Bilbao[1], Siegfried Egner[1], Thomas Hirth[1]

[1]Fraunhofer Institute for Interfacial Engineering and Biotechnology, Stuttgart, Germany
[*]Iosif.mariakakis@igb.fraunhofer.de

ABSTRACT
In this study a pilot plant investigation for phosphorus recovery in the form of magnesium ammonium phosphate (MAP or struvite) by the innovative ePhos technology was carried out. Phosphorus was recovered from centrate-water separated from the digested sludge dewatering of a German wastewater treatment plant (WWTP) performing enhanced biological phosphorus removal (EBPR). The pilot plant had been incorporated into the operation of the WWTP and was successfully operated for 3 months in a sidestream. The average phosphorus removal performance obtained was approximately 84.2% and the average outflow concentration reached was 20.8 mg/L PO_4^{3-} P. The average specific phosphorus recovery was determined to be 0.004 mg P/(L min cm^2). The pH was successfully increased by the electrochemical reaction to the required struvite-precipitation range from 7.6 to 9.0-10.0 without any addition of chemicals.

KEYWORDS: Phosphorus recovery, electrochemical precipitation, pilot study

INTRODUCTION

One of the most advantageous approaches to recover P from wastewater treatment plants is the crystallization of Mg^{2+}, NH^{4+} and PO_4^{3-} in form of magnesium ammonium phosphate hexahydrate ($MgNH_4PO_4{\bullet}6H_2O$), commonly called struvite or MAP [1-4]. The advantage of this approach is that the obtained product, struvite, can be applied directly to the soil since it is an excellent slow releasing fertilizer [2, 5-8] which has the potential to become a marketable product. Struvite crystallization has been widely studied in the last 15 years [1, 2] and many crystallization reactors have been developed, including a fluidized bed reactor [3] and a stirred tank reactor [9] that successfully work at full-scale. Most of these processes re-covered struvite from sludge liquors after sludge anaerobic digestion and utilize an inorganic compound as source of Mg^{2+} to reach the minimum molar ratio Mg:P 1:1 for the precipitation[1-4].
Various sources of Mg^{2+}, such as $MgCl_2$ (magnesium chloride), $MgSO_4$ (magnesium sulfate), $Mg(OH)_2$ (Magnesium hydroxide), and MgO (Magnesium oxide) have been used for the precipitation of struvite [2]. The disadvantage of the addition of $MgCl_2$ or $MgSO_4$ as a solution is the dilution of the wastewater that must be treated (by the use of low concentrated solutions typically 32% $v{\cdot}v^{-1}$ [9]) and an increase in the salinity of the water due to extra chlorine (Cl^-) or sulfate (SO_4^{2-}) ions that enter the solution. In order to avoid an increase in salinity of the wastewater, $Mg(OH)_2$ and MgO can be used as Mg-source. $Mg(OH)_2$ has a low solubility in water ($Ksp= 1.8\times10-11$) [10] and MgO reacts very quickly in water to form $Mg(OH)_2$. Therefore, these compounds must be added to the wastewater as suspensions. The drawback

of adding any of those compounds for the struvite precipitation is that for a complete struvite precipitation these compounds must be added in great amounts (Mg:P =1.2 to 2) [11-13]. Moreover, this mixture must be agitated continuously. Thus, the use of $Mg(OH)_2$ and MgO considerably increases the costs and decreases the efficiency of the process.
In addition, for the production of struvite the pH value of the wastewater must be increased to 8.5 to 9.5, since this is the ideal pH for this precipitation [14-15]. Usually, for this purpose either a base, such as NaOH (sodium hydroxide) is added [16-17] or the wastewater is aerated to force the CO_2 (carbon dioxide) to escape the solution[18-20]. The first method increases the salinity of the wastewater. The second method is very energy-intensive and both methods increase the costs of the process.
Consequently there has been considerable interest in identifying a new technology for the formation of struvite, which overcomes the major disadvantages of the state of the art technologies. For this purpose, a novel approach to recover struvite from wastewater has been developed in this study using an electrochemical process with a magnesium electrode working as a sacrificial anode. The main advantages of this process are:

1. No use of chemical compounds for the precipitation process, which prevents the increase of water salinity
2. High efficiency in the Mg^{2+} dosage, since Mg^{2+} is immediately available for the precipitation
3. Through the electrochemical process the ideal pH value for the struvite precipitation is reached. Thus, there is no need to add a base or strip the CO_2. Therefore, considerable costs can be saved.
4. Low energy requirement without high pumping costs
5. Fast start-up, low volume and modular configuration and operation

METHODOLOGY

The ePhos technology

The patented ePhos technology is an electrochemical phosphorus-precipitation process. It takes place in an electrolytic cell consisting of an inert cathode and a sacrificial anode made of magnesium (Mg). By the cathodic reduction water splitting takes place: OH^- ions are formed that raise the pH while hydrogen gas (H_2) is evolved. At the anode, the oxidation of metallic Mg takes place: Magnesium ions dissolve and react with the phosphate and ammonium contained in the water to form struvite (Figure 1).

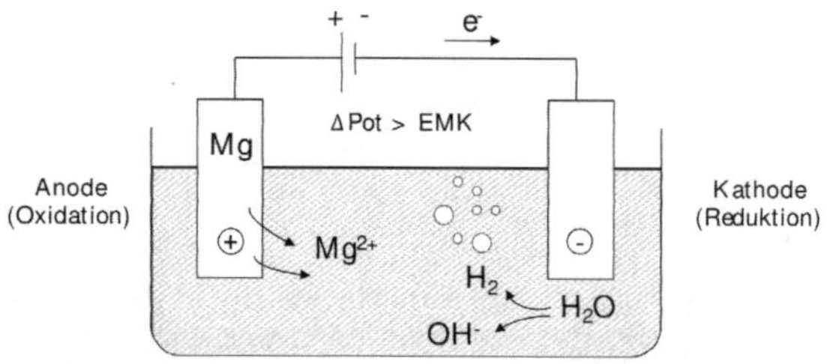

Figure 1: Schematic of an electrolytic cell for metal dissolution with a Mg sacrificial anode

Pilot plant operation

The pilot plant was integrated and operated on a 24 h/d basis at a Wastewater Treatment Plant in northern Germany performing Enhanced Biological Phosphorus removal (EBPR). The pilot scale plant had a hydraulic capacity of 1 m^3/h. It comprised of 4 reactors out of stainless steel that served as the inert cathodes. Magnesium sacrifice electrodes with a total mass of 4 kg were mounted centrally in each of the reactors. Its oxidation was regulated galvanostatically.

The operation was fully automated and each reactor could be independently controlled (hydraulic loading, applied current).

Inflow pH, conductivity, temperature into the plant and outflow pH and conductivity of each reactor separately were continuously monitored. Orthophosphate phosphorus (PO_4^{3-}-P) and magnesium (Mg^{2+}) in inflow (pilot plant) and outflow (each reactor separately) were determined photometrically (Hach-Lange LCK 049, LCK326). Samples from the inflow and the outflow of all reactors were taken daily in regular intervals.

The pilot plant treated the phosphorus-rich centrate-water of the dewatered digested sludge.

The plant was operated in the beginning 8 h/d, in order to optimize the operation. The stability was then tested during long term operation 24 h/d, which was conducted twice, in order to demonstrate the reproducibility.

The Wastewater Treatment Plant (WWTP)

This particular WWTP receives a high phosphorus load during summer time, destabilizing the process, so that a simultaneous chemical P-precipitation is carried out along EBPR, in order to comply with the phosphorus effluent requirements. The WWTP receives an annual load of 9,284 kg/ P/a and the centrate hydraulic load is 39,504 m^3/d. The PO_4^{3-} - P concentration in the centrate-water ranged between 200 mg/L and 300 mg/L during sludge dewatering.

Table 1: Annual centrate production

	Centrate
P-load centrate [kg/a]	9.284
Centrate [m^3/a]	39.504
Peak load centrate [m^3/d][1)]	300

RESULTS AND DISCUSSION

Operation 8 h/d

In Figure 2 the phosphorus elimination efficiency of the 8 h/d operation are shown. In the reactor with the 5 A P-removal efficiency was over the entire test period less than 50%. The current capacity of 5 A is not sufficient for a high degree of elimination. The trials with higher currents have shown significantly better results. The experiment with 20 A had shown a higher efficiency than the trials of 10 A at the beginning. However, it decreased steadily from the 2nd day and resulting to a lower efficiency than that of the reactor with with 10 A. The declining performance may be due to rapid oxidation of the magnesium electrode that has led to the formation of a thick passivation layer, which on its turn has reduced the performance or Mg^{2+} release. Experiments with 10 A exhibited initially P-elimination rates of about 75%. Over time, the latter increased significantly and has stabilized after about 4 d of operation at a value of 95%. It could be concluded that the optimum current was 10 A

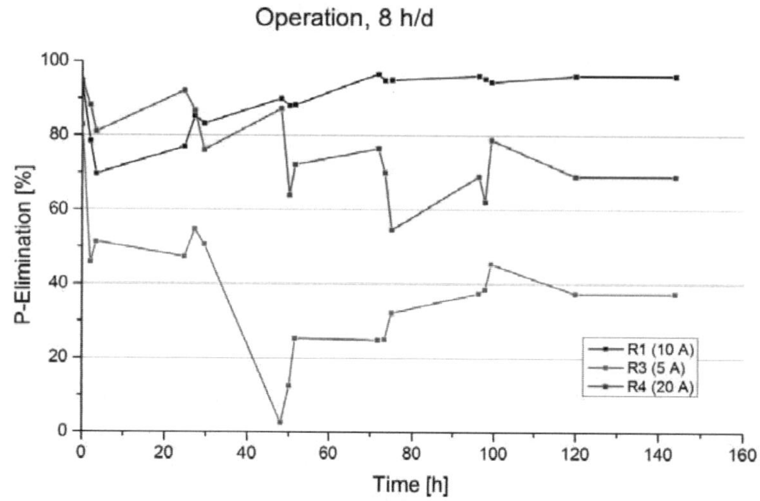

Figure 2: Variation of the P-elimination efficiency for the operation 8 h/d

Operation 24 h/d

Figure 3 shows the operating parameters and results of a typical week of piloting. All reactors were operated under the same conditions, which were determined as the optimal during the 8 h/d operation.

The inflow concentration of phosphorus exhibited sharp fluctuations within an operating day. In daytime during the operation of the sludge dewatering the pilot plant was operated with fresh centrate with PO_4^{3-}-P concentrations between 200 mg/L and 300 mg/L. During night-time, the concentrations exhibited an average concentration of 70 mg PO_4^{3-}-P/L, since the sludge dewatering was not used and wastewater was taken from a different point on the wastewater treatment plant. Throughout the weekend, the phosphorus concentration was similar to the night operation, since in this case also no sludge dewatering took place. The conductivity has a similar course as the phosphorus concentration. With rising PO_4^{3-}-P concentrations both the conductivity and the pH at the inlet increased. At the start of the dewatering, the conductivity was about 150 µS/cm and the pH 7.2. Throughout the day, the conductivity rose partially up to 500 µS/cm and the pH to 7.8. After the maximum values were reached both decreased back to the initial levels. This trend can be explained by the higher ion concentration like PO_4^{3-}-P and NH_4^+ -N present in the centrate.

As can be seen from Figure 5, the varying feed concentrations affected reactor operation. With declining conductivity the voltage increased through the galvanostatic process control automatically, so that the resulting increased resistance could be compensated. The pH during reactor operation also fluctuated significantly. The pH increase at the beginning of the day was higher as that than at the end. This suggests that the reduction in voltage/ of the electric field strength has a negative effect on the release of Mg^{2+} and hydroxy ions. Nevertheless, the pH in the reactors was always in the appropriate range for the precipitation of phosphorus as struvite.

The elimination rate of PO_4^{3-}-P fluctuated between 70% and 95% with an average of 85%. The results of the 8 h/d operation could be confirmed by the continuous operation of 24 h/d. Abrupt efficiency drops as at 29.09.2014 were due to malfunctions such as clogging of the pipping. The concentration of PO_4^{3-}-P at the effluent exhibited a similar pattern to the inlet concentration. With increasing inflow concentration, the effluent concentration of the pilot plant also increases. The increase of the outflow concentration caused by the reduction in the elimination efficiency could not be fully compensated. Nevertheless, even for unfavorable conditions during the night (low conductivity and low PO_4^{3-}-P and NH_4^+ -N content) elimination grades of more than 80% could be achieved, which speaks for the robustness of the method.

In Figure 4 the result of the 24 h/d operation until almost complete consumption of the Mg-electrodes is shown (1. Run). After 40 d of operation the pH-value could not be increased to a level suitable for phosphorus precipitation (pH > 8). In addition, the P-removal efficiency dropped below 70%, and the current and the voltage have reached the limit of the power supply. These events signaled the renewal of the magnesium electrode.

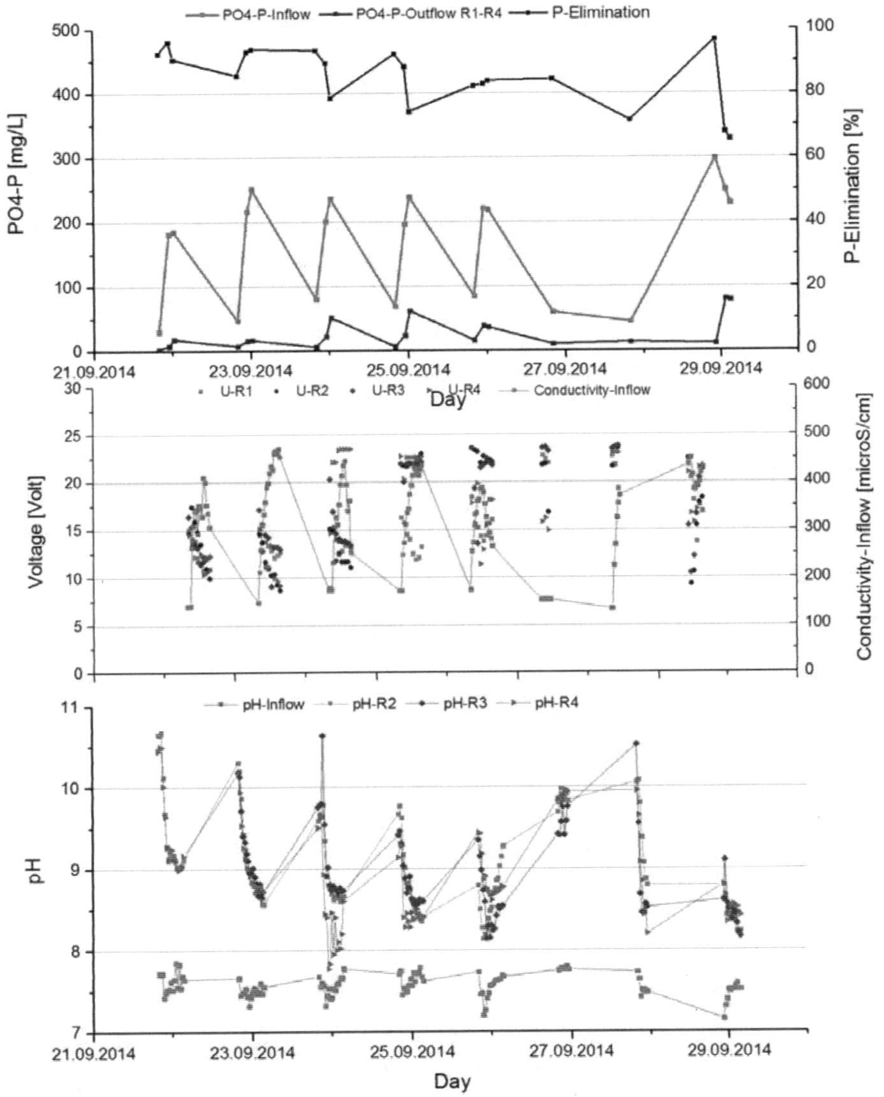

Figure 3: Typical operation parameters of the 24 h/d operation

The P-elimination performance fluctuated during the 1. Run between 70% and 95%. In the period 09.10.2014 and 25.09.2014 average P-elimination was about 90%. The average outflow concentration of PO_4^{3-}-P was approximately 18 mg/L. As of 25.09.2014 a declining

Aerobic Granular Sludge Bench-Scale Testing to Full-Scale Conceptual Design

Roland Jezek[1*], Ed Kobylinski[1], Mark Steichen[1], James L. Barnard[1], Belinda Sturm[2]

[1] Black & Veatch, 8400 Ward Parkway, Kansas City, MO 64114 (*jezekr@bv.com)
[2] Department of Civil, Environmental & Architectural Engineering, The University of Kansas, 1530 W 15th St, Lawrence, KS 66045

ABSTRACT

A bench scale pilot study at the Tomahawk Wastewater Treatment Plant in Overland Park, Kansas was conducted to evaluate the feasibility of an aerobic granular sludge (AGS) process as an alternative technology for the plant upgrade. Growth and cultivation of aerobic granules was successful, counter to concerns that the wastewater at Tomahawk was too weak compared to wastewater used in other AGS applications around the world. In this study, AGS was evaluated using three bench-scale reactors operated for six months with raw influent wastewater supplemented with fermented primary sludge (0, 10, and 20% influent v/v). After granulation was achieved, the performance and characteristics of the maturing granules' were evaluated. The pilot process achieved an optimum effluent quality of 5.6 mg-N/L and 2.8 mg-PO_4-P/L for the reactor supplemented with fermented sludge supernatant. The results from the bench-scale testing will be presented and sludge characteristics relative to land are also discussed.

Keywords: aerobic granular sludge, BNR, nitrogen removal, phosphorous removal

INTRODUCTION

The Tomahawk Creek Wastewater Treatment Plant in Overland Park, Kansas, operated by Johnson County Wastewater, is a trickling filter plant that must upgrade to biological nutrient removal. The site is located in a flood plain and is land-limited. Small footprint technologies are needed to meet discharge goals of 10 mg-N/L and 1.0 mg-P/L or 8 mg-N/L and 1.5 mg-P/L. Currently, the majority of dry weather flow is diverted to a neighboring jurisdiction at significant cost, so there is a need to increase plant capacity and maximize biological treatment on-site. With these site challenges, small footprint technologies are needed, and recent full-scale applications of aerobic granular sludge (AGS) by Nereda, DHV, have made aerobic granular sludge (AGS) a promising option.

Aerobic granulation is a process that is induced by several external ecological conditions. Spontaneous formation of granules has been reported since 1999, mainly in sequencing batch reactor (SBR) systems. The first full-scale application of AGS began operation in the Netherlands in 2012, and there have been many full-scale plants built since on three continents,

The advantages of granular sludge compared to ordinary activated sludge are many.

- The compact granules allow for high biomass concentrations and fast pollutant removal rates in a smaller reactor.
- The high removal rates of granules allow for treatment of raw influent, eliminating the need for primary clarification and processing of primary sludge.
- The dense granules settle very fast, which eliminates final clarifiers and increases treatment capacity by freeing time previously needed for sedimentation in an SBR configuration.
- Simultaneous nitrification and denitrification, and phosphorous removal occur within the granule depths, where environmental conditions are favorable for these specialized bacteria and biological processes.

Because the Tomahawk Creek influent wastewater has low soluble chemical oxygen demand (COD) and volatile fatty acid (VFA) concentrations, a primary sludge fermenter was operated for supplemental VFA. This study is among only a few trials of aerobic granular sludge on low-strength wastewater in the United States. All reported installations of Nereda treat wastewaters with higher soluble COD concentrations (i.e., 880 mg COD/L (Giesen et.al., 2013)).

METHODOLOGY

Three SBRs (3L each) were operated with a 4-8 hour HRT and a 50% volumetric exchange ratio. A schematic of the bench scale reactors is shown in Figure 1 and a photo of the pilot system is shown in Figure 2. The custom-built pilot reactors consisted of a Plexiglas cylinder column supported on a PVC platform and an aluminum frame on casters. Each reactor column was made of interchangeable stackable units to achieve various settling distances. For this study there were three reactors in parallel (labeled: red, white and blue) with two units stacked on top of each other. Each unit has a 480 mm height and 65 mm inner diameter. The units were constructed with a double wall water jacket for temperature control. The temperature was held constant at 20°C. During aeration phases, compressed air was fed through a fine bubble diffuser. Mixing (without aeration) was performed by bubbling nitrogen gas. Gas flow was measured and adjusted manually to approximately 1-3 L/min using a 0.4-5 L/min Cole Parmer rotameter.

The reactors were operated in an SBR mode, meaning that the reactor was batch fed and treated in sequential phases (Fill, Aeration, Settling, and Decant) that repeated every 2-4 hours. Each repeat of the sequential phases is a cycle. Figure 3 shows an example of the sequential phases comprising one cycle. The cycling was controlled through a 4-circuit Chrontrol programmable timer unit, and cycle phases were identical for all three reactors. All the reactors were initially seeded with activated sludge from a nearby wastewater treatment plant with biological nitrogen and phosphorus removal. The only difference between the three reactors was the influent feed during the process performance evaluation component of the study. Throughout the entire study, the only wasting of biomass occurred via washed out effluent total suspended solids (TSS).

The influent to the AGS reactors was raw wastewater collected before primary clarification, and influent containers were mechanically mixed. The low-strength influent wastewater at Tomahawk Creek was supplemented with fermented primary sludge supernatant. Fermentation of primary sludge converted particulate biochemical oxygen demand (BOD) into soluble and readily biodegradable forms. The primary sludge was collected from the primary clarifier underdrain and was allowed to ferment for 2-4 days. The supernatant was then added to the raw influent wastewater in order to increase its strength (by 10-20 percent by volume), and the primary sludge solids were discarded.

The primary sludge fermentation process was initially operated automatically as a timer-controlled process and temperature controlled at 30°C. Complications with the automatic operation, such as clogged pumps and poor mixing, created methanogenesis and consumption of readily biodegradable COD (rbCOD). As an alternative to automatic operation, the primary sludge was collected and fermented in the collection bucket, which was mixed once per day, which represents the more likely full-scale scenario. The automatic operation was conducted from October 2014 through December 2014, after which the passive bucket fermentation method was used. Both automatic and bucket fermentation methods produced approximately 5-10 L/day of fermentate.

Pilot samples were delivered to the Johnson County Wastewater Laboratory for analysis. Influent and effluent samples of the Fermenter were analyzed for soluble COD, TSS/VSS and volatile fatty acids (VFAs). Influent and effluent samples of the reactors were analyzed for TSS/VSS, total and soluble COD and BOD, ammonia-N, nitrite + nitrate-N, orthophosphate, and VFA.

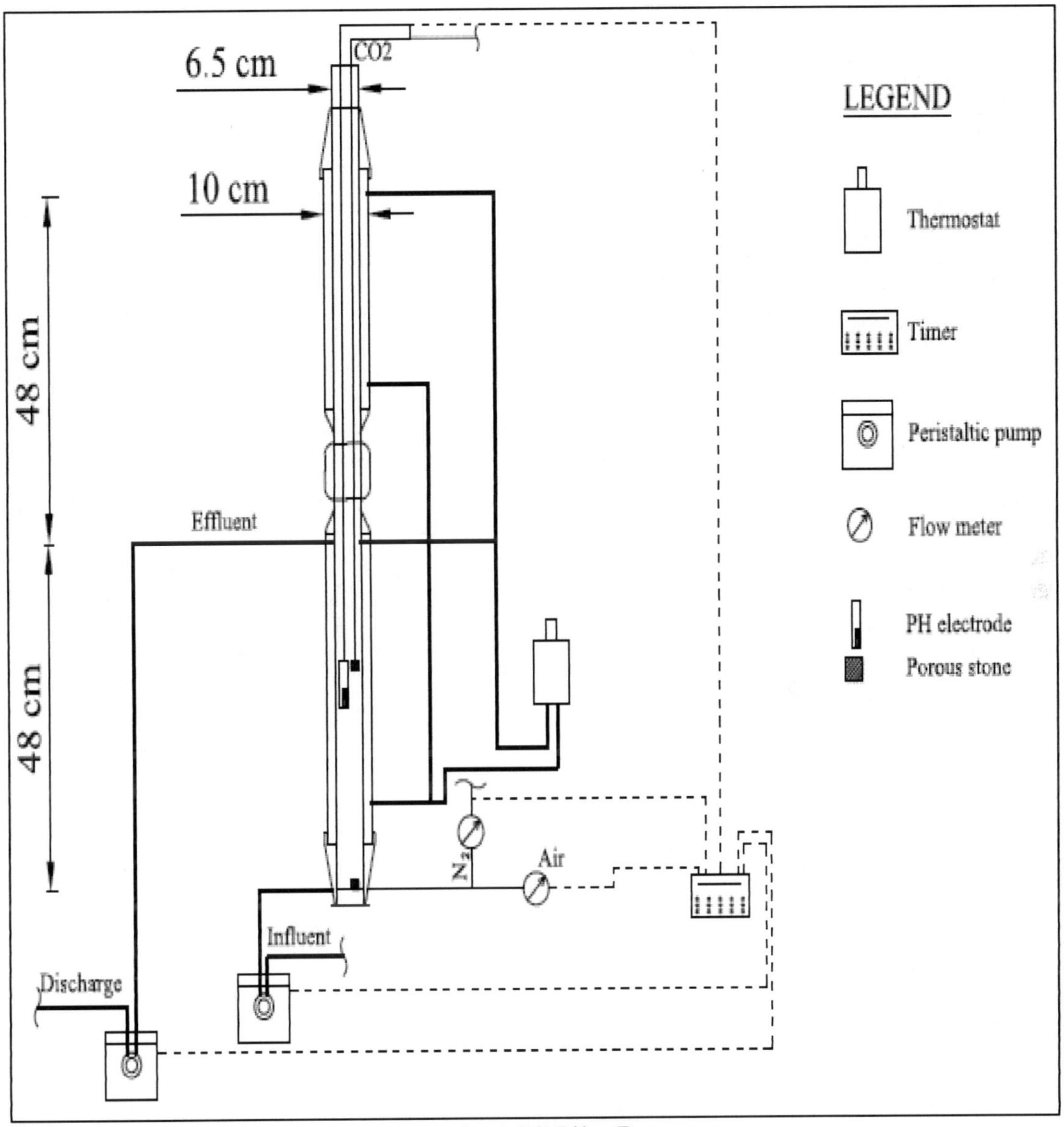

Figure 1 – Schematic of the Bench Scale AGS Pilot Reactor

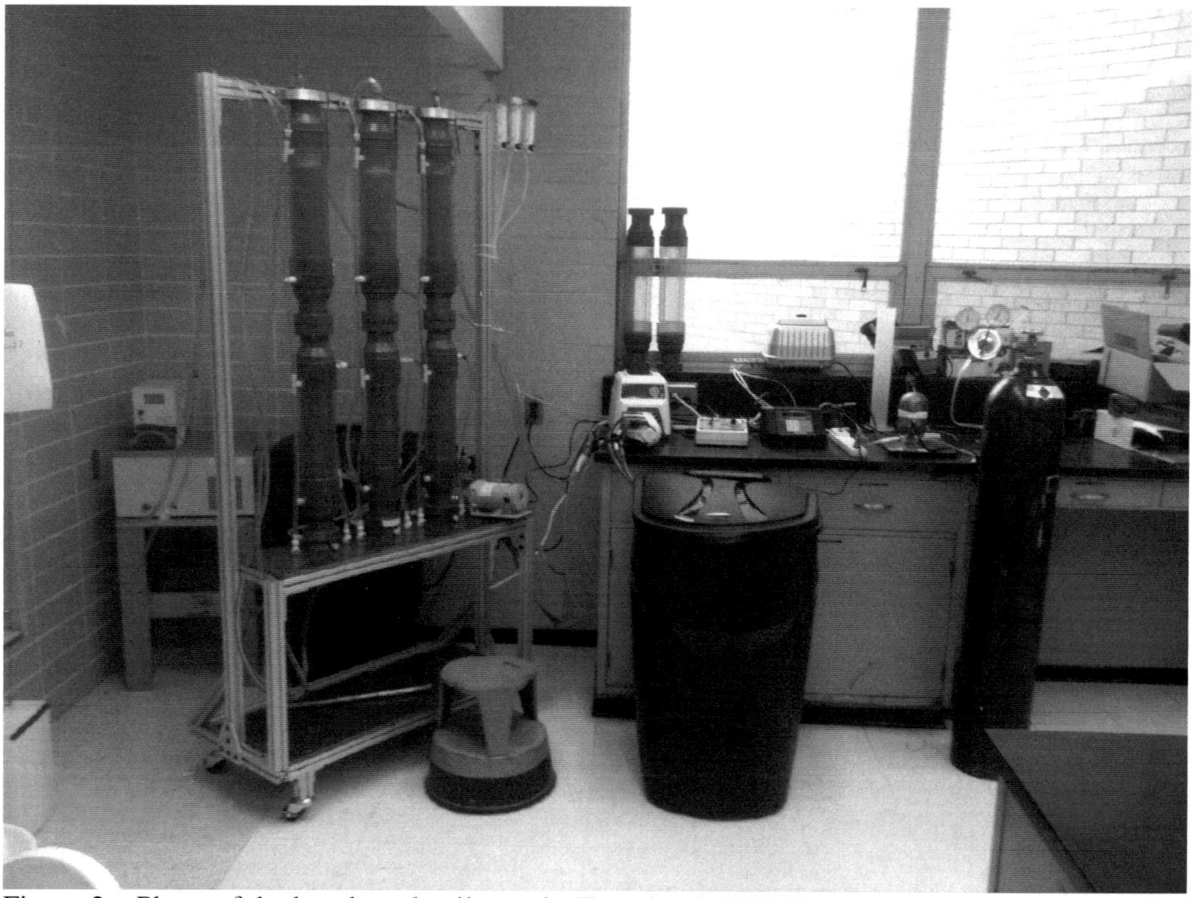

Figure 2 – Photo of the bench scale pilot at the Tomahawk WWTP (the reactors from left to right were labeled – red, white and blue)

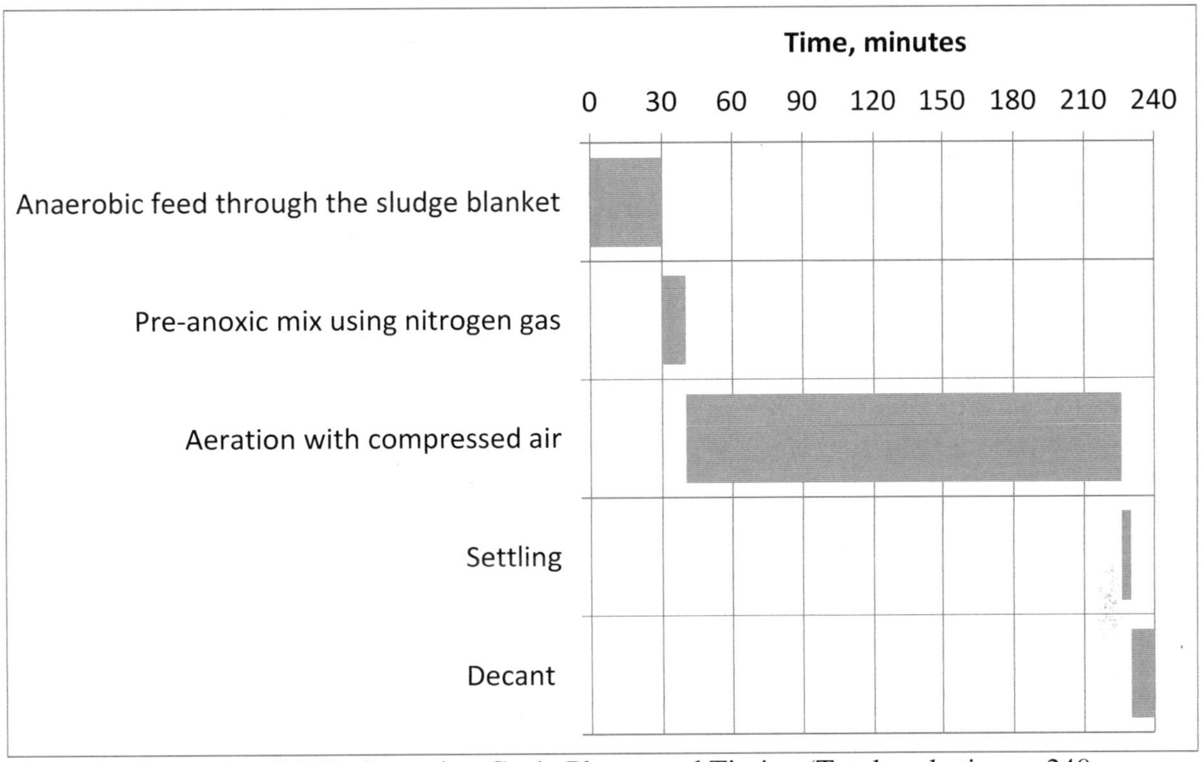

Figure 3 – Example of SBR Operating Cycle Phases and Timing (Total cycle time = 240 minutes (4-hours) consisting of: anaerobic feed – 30 min, pre-anoxic mix – 10 min, aeration – 186 min, settling – 4 min, and decant – 10 min)

RESULTS AND DISCUSSION

The AGS pilot was operated from October 6, 2014 through April 22' 2015. The pilot operation was broadly divided into two phases, 1) granule enrichment (start-up) and 2) process performance evaluation. The first phase of granule enrichment was necessary to develop granules starting from activated sludge, which took approximately 2-months after resolving operational problems and reseeding (4-months into the study). The second phase, which lasted the remaining 2-months, involved an evaluation of process performance and other characteristics of the AGS process in terms of settleability, nitrification, BNR performance, and biosolids characteristics relative to land application. Although the entire pilot duration was a general evaluation of the technology; the second phase examined reactor performance as a function of increasing fractions of rbCOD in the influent. Table 1 shows a chronological list of major pilot changes and activities.

Table 1 – List of Major Pilot Operational Changes and Activities

Date	Day No.	Activity
10/6/2014	1	Started with 2-hour cycle and 15 min settling time
10/14/2014	8	Reduced settling time from 15 min to 5 min
10/24/2014	18	Reduced settling time from 5 min to 2 min
11/14/2014	39	Increased cycle time from 2 hours to 3 hours
11/15/2014	40	Increased settling time from 2 min to 5 min
11/25/2014	51	Added fresh activated sludge
12/2/2014	57	Changed fermenter operation from automatic to bucket method
12/20/2014	75	Reduced settling time from 5 min to 4 min
12/20/2014	75	Increased cycle time from 3 hours to 4-hours
1/19/2015	105	Installed scum control
2/2/2015	119	Granules become predominant form of biomass
3/3/2015	148	Started testing feed composition on reactor performance
3/12/2015	157	Nitrification test on floc and granules
3/20/2015	165	Added pre-anoxic phase to the SBR cycle

Startup

The reactors were initially charged with activated sludge from the Douglas L. Smith Middle Basin (DLSMB) WWTP, which is a BNR plant with biological phosphorous and nitrogen removal. Our initial strategy was to wash out the lighter biomass by gradually reducing the settling time. Typical settling velocities of AGS range from 10-50 m/hr (Haandel et. al, 2011). During the first 20 days, the units were operated with a 15 minute settling time, then 5 minutes, followed by 2 minutes, which corresponds to settling velocities of approx. 2 m/hr, 6 m/hr and 15 m/hr, respectively. A settling velocity of 15 m/hr proved to be too aggressive too soon, and too much biomass was wasted from the system. The biomass was slow to recover, so after 51 days of pilot operation, the reactors were recharged with fresh biomass from the DLSMB WWTP. This was the only time the reactors were recharged with fresh activated sludge biomass, and a 4 minute settling time corresponding to a settling velocity of 7.5 m/h was selected and used for the remainder of the study.

During the first 4-months of operation, all three reactors were treated identically. All reactors received feed from the same influent reservoir; they had the same biomass; and they had the same operating cycles. The cycle times during the study ranged from 2-hours to 4-hours. For every cycle, each reactor was fed 1.5 L of wastewater. The reactors were loaded more heavily during the first 2.5-months using a 2-hour and 3-hour cycle, and then the cycle time was increased to 4-hours for the remainder of the study. During this time, on a weekly basis, the reactors' contents were combined, the SVI was measured, and the biomass was split between the reactors again. When the granules reached the target number and size, the reactors were tested individually to ensure and confirm that they behaved similarly to one another, and the contents of each reactor were kept separate.

One issue with expediting the startup of an AGS process, especially at bench scale with low-strength wastewater, was to ensure that biofilm growth on the reactor walls does not consume substrate intended for granule growth. Since the wall growth was a major issue with this type of reactor, the reactors were manually cleaned weekly. Figure 4 shows a dirty vs. clean reactor and the modified mop used for cleaning the reactors' inner wall of biofilm.

Another issue during the startup was foaming and the accumulation of biomass onto a scum ring that would build up on the reactor wall at the water surface. The scum ring presented a major problem in regard to monitoring the progress of the process start up. The biomass in the reactor tended to accumulate on the scum ring during aeration to such an extent that most of the biomass in the reactor would deposit on the scum ring within a few days. The reduction of suspended biomass in the column due to continuous deposition on the scum ring made representative sampling a challenge. Eventually, several modifications to the reactor operation and procedures were made to control the scum ring, and these modifications worked reasonably well.

Figure 4 – Comparison of dirty column with scum ring vs. a cleaned column (left photo), and modified mop cleaning tool (right photo)

At the 4-month mark, the process performance evaluation phase of the study was initiated, where the three reactors were fed different amounts of fermentate by volume.

- Red Reactor was fed 80 % raw influent and 20 % fermentate
- White Reactor was fed 90% raw influent and 10 % fermentate
- Blue Reactor was fed 100 % raw influent without any fermentate.

Granule Enrichment

One main challenge of the aerobic granular sludge system is a long startup time, especially when treating weaker wastewater such as found at the Tomahawk Creek WWTP. The full-scale Nereda applications have all been designed for much higher influent wastewater strengths. The Nereda facilities are located in parts of the world where more water conservation is practiced, or where the per capita water use is much lower than typical US practices. The net result is an influent wastewater roughly twice as concentrated as typical domestic wastewater in the US.

Since the reactors were started using activated sludge, a period of enrichment was needed to develop sufficient granular biomass to begin a meaningful evaluation. During the granular enrichment phase, all three reactors were treated identically. The reactors were fed a common influent wastewater consisting of primary influent plus a small amount of fermented primary sludge supernatant (0-20% fermentate by volume).

During the granule enrichment phase, efforts were focused on consistent and careful pilot operation in order to maximize the bacterial secretion of extracellular polymeric substances (EPS), while maintaining sufficient suspended biomass within the reactor to achieve treatment objectives, such as full nitrification. Several difficulties were encountered. EPS is produced in response to high growth rates and excessive F/M conditions during start-up; EPS production is also a common response to stress conditions. While EPS aids granule formation, it has the negative effect of increasing biofilm growth on the walls of the reactors. If excess EPS is not carefully controlled, then it can have deleterious effects on the suspended biomass in two ways.

The first is excessive wall growth and accumulation of biomass on the scum ring. Since the reactors are long and slender, they have a large wall surface area to volume ratio. Instead of adhering particles together, the biofilm tended to adhere to the reactor walls. The excess EPS also created a challenge in that the biomass in suspension was lost because it was sticking to a scum ring which formed at the liquid-air interface at the surface of the reactor, quite severely at times. However wall growth and scum would not be a problem with the full scale reactor because of the much lower wall surface area to volume ratio.

The second is that EPS hinders biological coagulation and subsequent sedimentation. EPS is needed for granule formation but excess EPS acted to increase solution viscosity and inhibits TSS settling. The excess EPS was observed to hinder biological coagulation and sedimentation during the startup process. Initially, the reactor was operated with flocculated biomass (activated sludge from DLSMB). Normally, under quiescent conditions, activated sludge will biologically coagulate, and a zone settling regime is created, which squeezes out the interstitial fluid as the sludge blanket compresses under gravitational forces. Under the same quiescent conditions, this time with excess EPS, the biomass did not coagulate normally but instead remained a colloidal suspension that did not settle, even at much longer settling times. The aggressive settling velocity of 15 m/hr therefore created a positive feedback loop which deteriorated the biomass' settling ability. In the positive feedback loop, biomass was washed out/wasted - EPS was created - settling ability was deteriorated - causing more biomass to be washed out/wasted - more EPS was created - and so on, until almost all biomass was lost within a few cycles.

It was found that a settling velocity of 7.5 m/hr was sufficient for selective wasting without hindering biological coagulation and sedimentation. Recent research at the Delft University in the Netherlands has shown that the substance responsible for granulation is a polymer termed alginate-like-exopolysaccharides (ALE) (Roest, 2015). The ALE polymer is known to strongly bind with water and can thicken or gel liquids. The ALE was not directly analyzed; however, the effect on the matrix was observed to impede startup causing the initial biomass to be washed out/wasted before granules were actually formed.

Upon observing this phenomenon, approximately 2-months into the study, the reactors were reseeded with fresh activated sludge biomass and a less aggressive settling velocity of 7.5 m/hr was selected. The reactors were reseeded in order to accelerate granule enrichment and nitrification. Theoretically, sufficient granule enrichment could still have been achieved had the reactors not been reseeded, as evidenced by the early observation of granules in the photo progression in Figure 5, however, recovery of the reactors likely would have exceeded the pilot schedule. The particle size distribution of the granular sludge, performed at the University of Kansas, indicated that the average particle size was approximately 0.51 mm ± 0.76mm. The optimal diameter of the aerobic granule is less than 1.7 mm, considering the absence of the layer of obligate anaerobic bacteria (IWA, 2005). Overall, the depth of biofilm activity zones due to diffusion limitations are a function of bulk concentration, relative surface velocity, bacterial cell density, among many other factors.

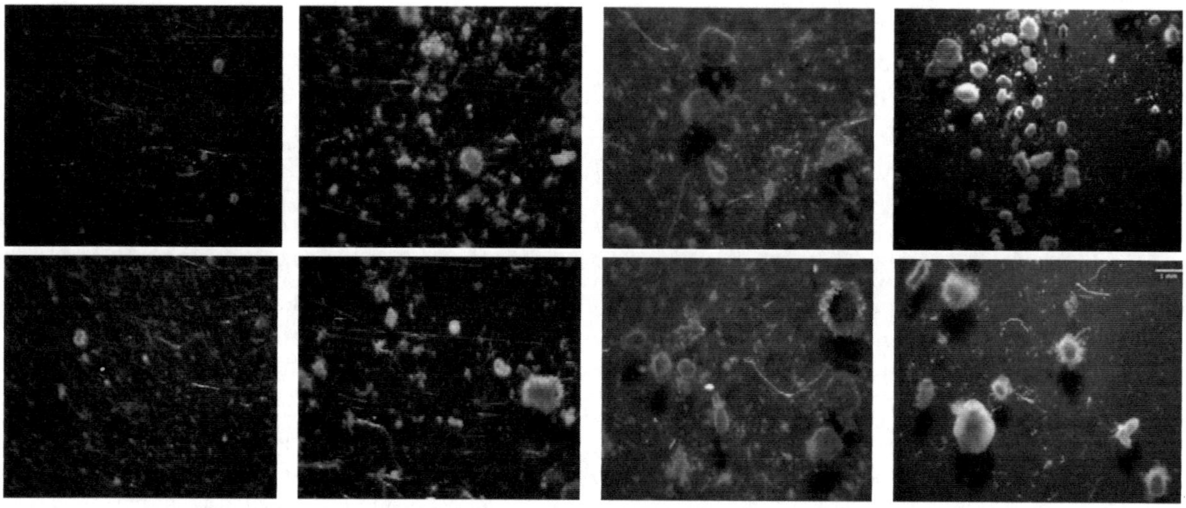

Figure 5 – Photo Progression of granule enrichment. From left to right the total days the reactors were in operation before the photo was taken: 32 days (11/7), 50 days (11/25), 86 days (12/31), and 126 days (2/9).

Settleability

The settleability of the granular sludge was measured weekly using sludge volume index tests (SVI). The SVI_{30} tests were conducted according to Standard Methods 2710 D. The results from the SVI_{30} tests, in Figure 6, show the decreasing trend of SVI_{30}, indicating improved settleability.

Typically, the AGS SVI at 5 minutes should be similar to the SVI at 30 minutes indicating that the settled sludge at 5 minutes is already compressed to the same compression achieved after 30 minutes of settling. Figure 7 shows the variability in the difference between the SVI_5 and SVI_{30}. The variability in the difference between SVI_5

and SVI_{30} was due to fugitive primary sludge solids that were transferred with the fermenter supernatant to the reactor influent containers and repeated failure of the influent reservoir mixer, which caused solids to accumulate on the bottom of the influent reservoir near the feed inlet and were subsequently pumped into the reactors. Influent BOD, either soluble or particulate, tended to increase flocculent biomass which greatly affected the SVI_5.

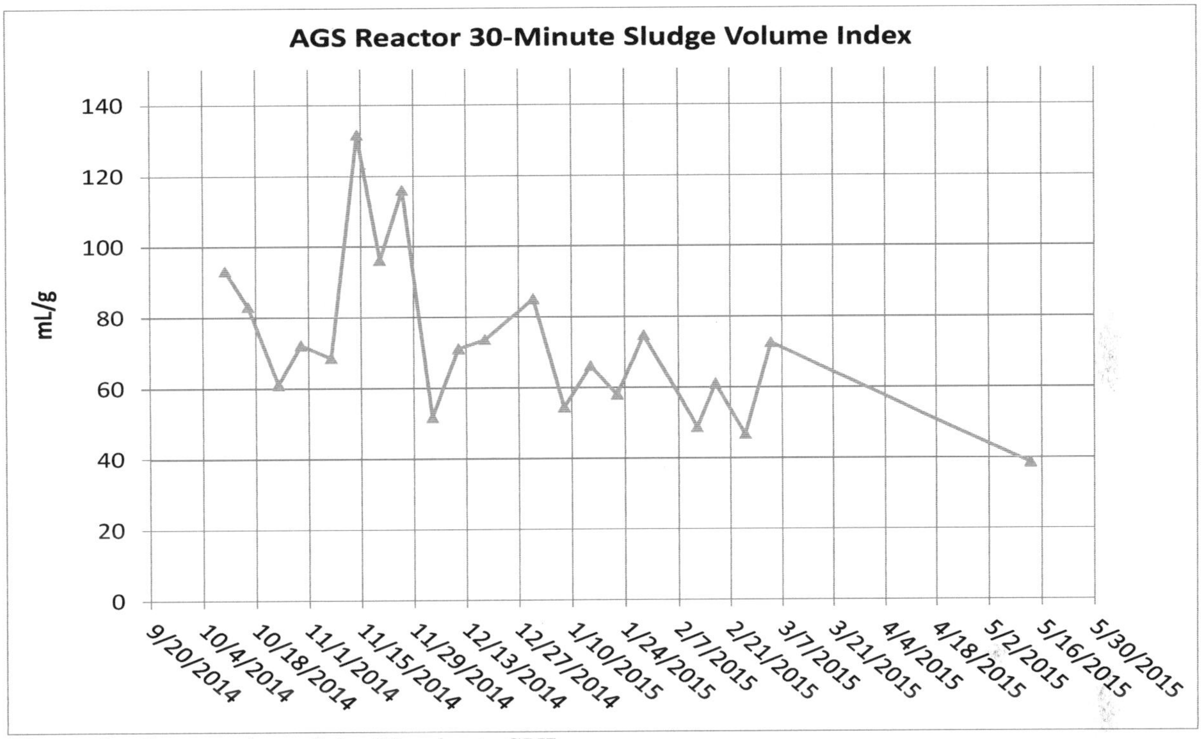

Figure 6 – Progression of the 30-minute SVI

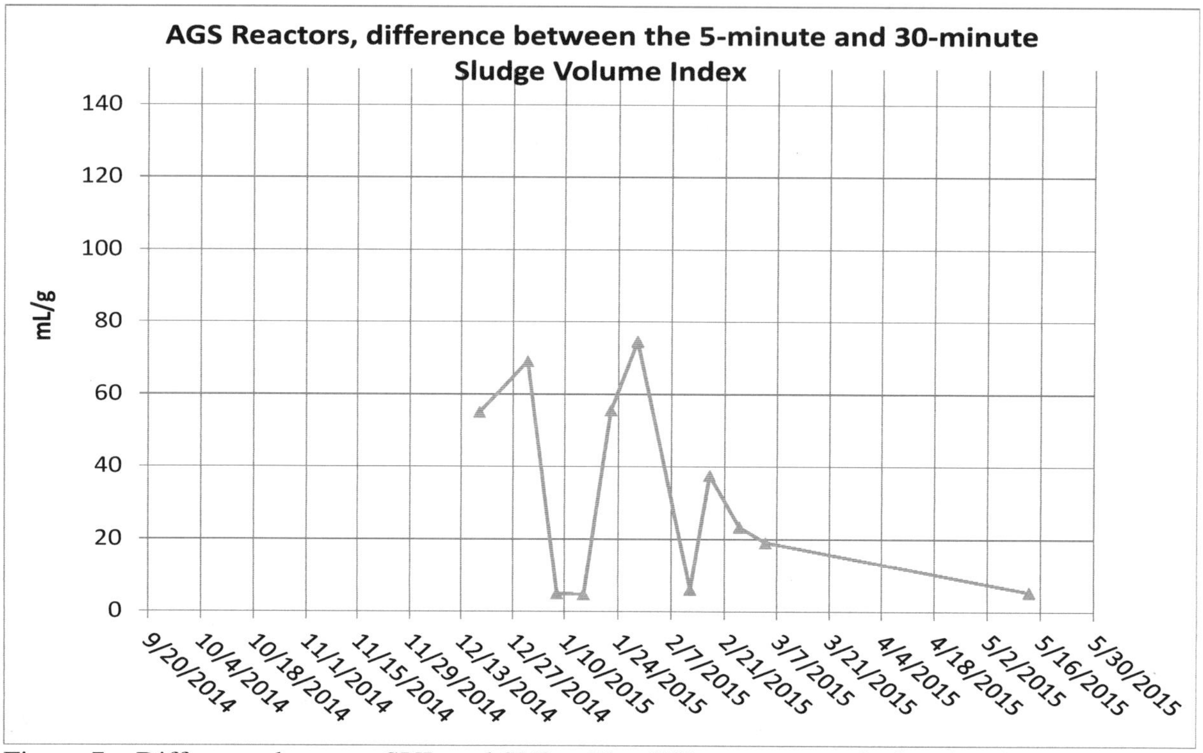

Figure 7 – Difference between SVI_5 and SVI_{30} -The difference was calculated as SVI_{30} subtracted from SVI_5.

Nitrification Rates

Figure 8 shows the experimental results of nitrification rate experiments performed on the two forms of suspended solids (SS) in the reactor, the flocculent (floc) and granule. One experiment was performed with floc SS (thickened washed out solids without granules), and the other was performed with total reactor SS (total reactor contents including floc + granules). The results showed that the specific nitrification rate (normalized to VSS concentration) was 50% greater in the floc SS (0.066 mgN/mgVSS (floc)-d) compared to the floc + granules (0.044 mgN/mgVSS (floc+granules)-d). The difference in normalized rates can be explained by the fact that only the perimeter volume of a granule is expected to be aerobic and house nitrifier populations.

Of the total reactor contents, only a small portion of the mixed liquor is actually washed out as floc during each cycle. The floc concentration was 86 mgVSS/L compared to the total floc + granules concentration of 1770 mgVSS/L. Therefore, within the reactor, the actual nitrification contribution of the floc, which presumably slough off the granule, is approximately 10 percent, and the remaining 90 percent is from the granules themselves. This comparison of nitrification proves that the nitrifiers thrive in the granule structure, and the granules provide a long system SRT, which may not be obvious with mass-balances.

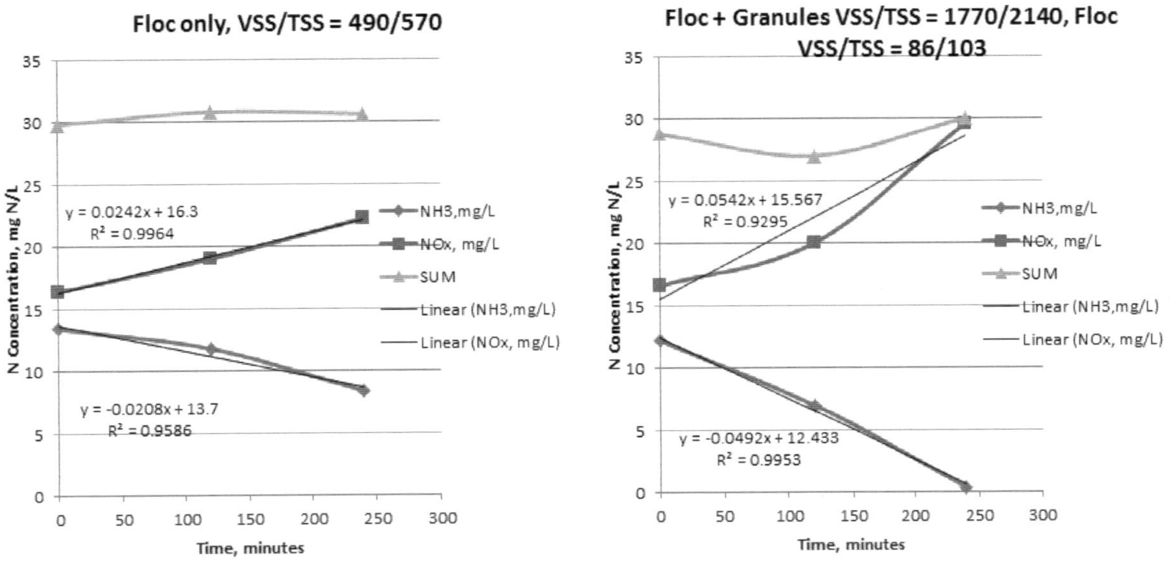

Figure 8 –Nitrification experiment results testing the rate of nitrification separately for floc and granular biomass

Fermenter Performance and Reactor Influent sCOD

The AGS reactor feed was supplemented with sCOD produced by the fermentation of primary sludge. The primary sludge fermenter performance, in terms of soluble COD (sCOD) production, is shown in Figure 9. The influent sCOD was very consistent with an average of approximately 100 mg/L. However the variability in effluent sCOD was a result of variability in the primary sludge solids concentration, which was typically low, in the range of approximately 1000 – 3000 mg/L.

A period of high primary sludge solids concentration to the fermenter the week of 3/17/15, as shown in Figure 9, were carried through to the following week and is seen as a spike in sCOD in the White and Red reactors, shown in Figure 10. All three AGS reactors performed well in terms of sCOD and sBOD removal, with effluent sCOD consistently less than 50 mg/L and average sBOD was always less than 4 mg/L.

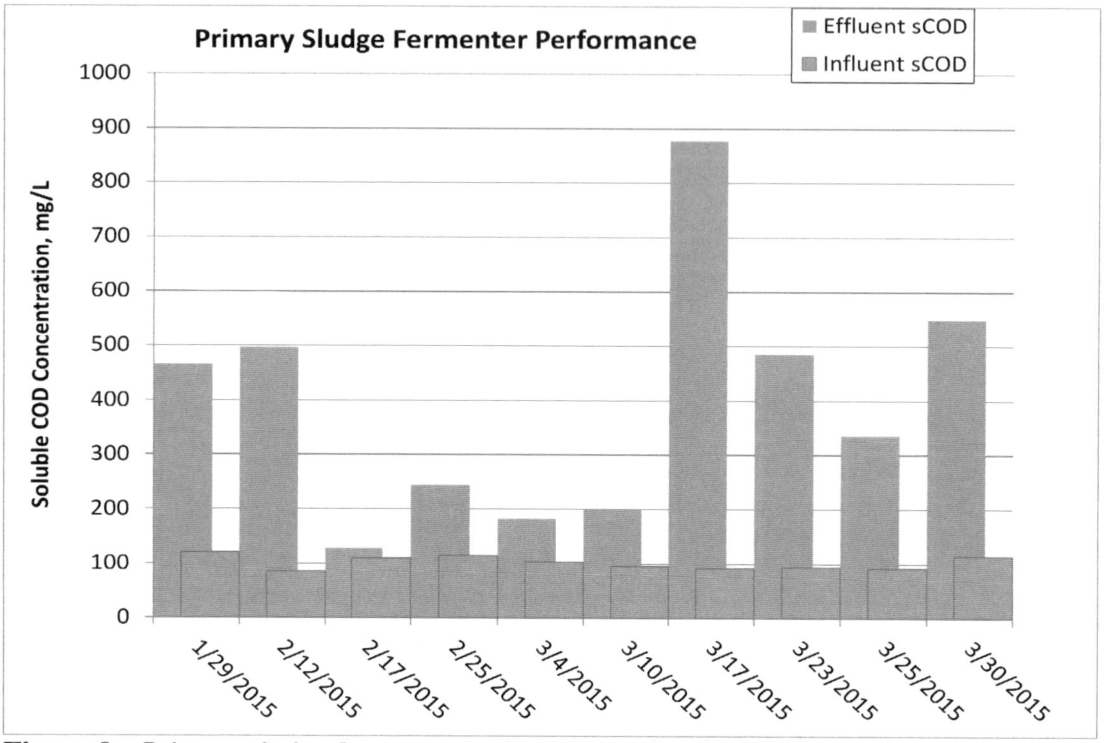

Figure 9 – Primary sludge fermenter performance using the bucket method

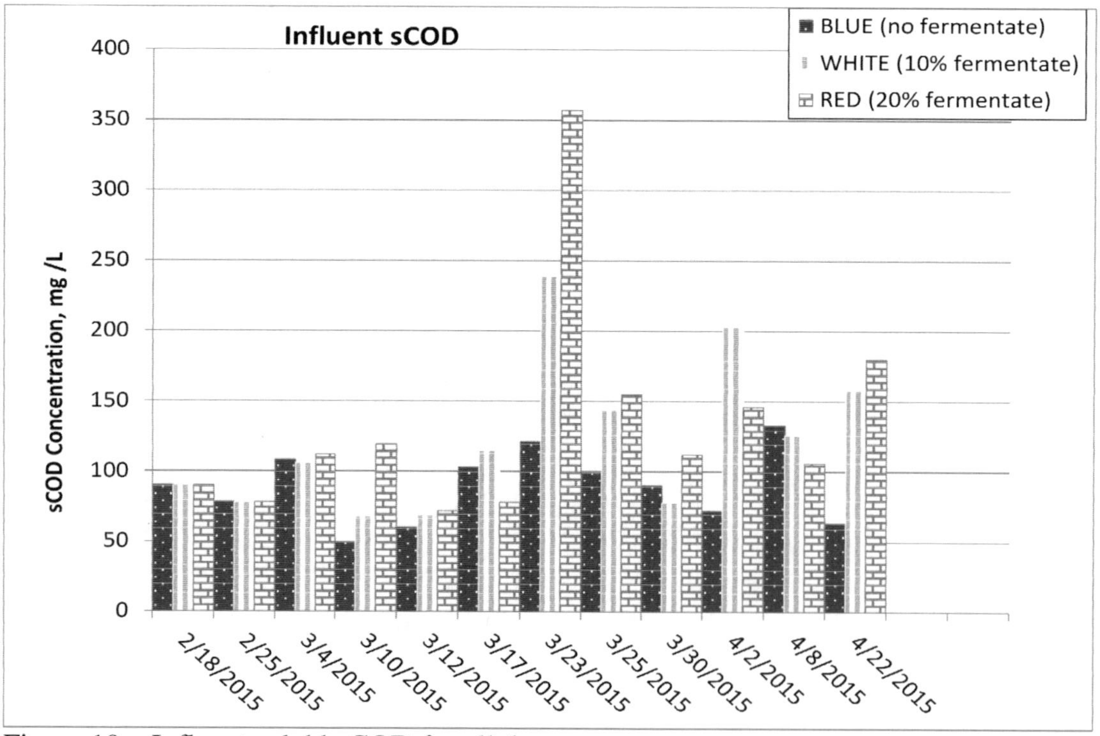

Figure 10 – Influent soluble COD for all three reactors.

BNR Performance

BNR was intentionally delayed until after the initial granule enrichment phase. During the granule enrichment phase, several operational conditions were utilized to speed up the growth of granules that were not conducive to the selection of denitrifiers and phosphorus accumulating organisms. The initial cycle time was 2-hours, which allowed for aggressive biomass growth and wasting. The initial short cycle and resulting heavy loading was selected to compensate for the project schedule and low-strength wastewater. The intention was to heavily load the biomass to achieve an adequate level of granule enrichment as soon as possible in order to move on to the next phase of BNR evaluation.

As mentioned earlier, the aggressive settling velocity and heavy loading in a short cycle time, during startup, created conditions where biomass didn't settle and thus were wasted from the reactor. These conditions resulted in excess EPS that dispersed into the liquid as a gel, inhibiting sedimentation and resulting in biomass loss. For the first 2-months there was insufficient biomass to achieve complete BOD removal or nitrification of all the ammonia. After granule enrichment was well advanced, BOD removal and nitrification were easily achieved with average effluent concentrations of 3 and 0.4 mg/L for BOD and NH_3-N, respectively. The final 2.5-months of BNR performance are shown in Figures 11 for nitrogen and Figure 12 for phosphorous.

After 3/17/15, day 162, a pre-anoxic phase, using nitrogen gas mixing, was added to the cycle. The pre-anoxic phase had a significant impact on nitrogen removal, presumably due to heterotrophic denitrification. The pre-anoxic cycle also had an impact on effluent phosphorous. The organisms that accumulate phosphorous (PAOs) are slow growing and are believed to dwell within anaerobic inner cores of the granules, rather than in the flocculent biomass. Because the granules were small (averaging 0.5 mm in diameter) compared to mature granules reported in the literature, a longer pilot that allowed granules to increase size is expected to improve phosphorus removal. The summary of average BNR performance is shown in Table 2

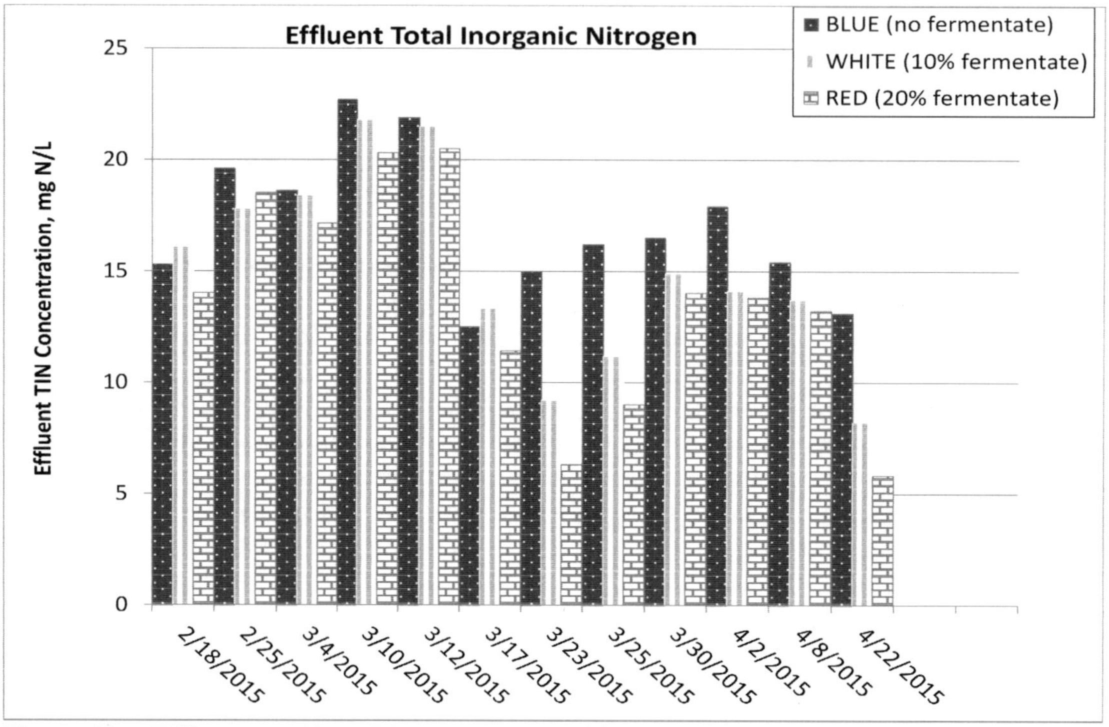

Figure 11 – Effluent quality with respect to total inorganic nitrogen (TIN)

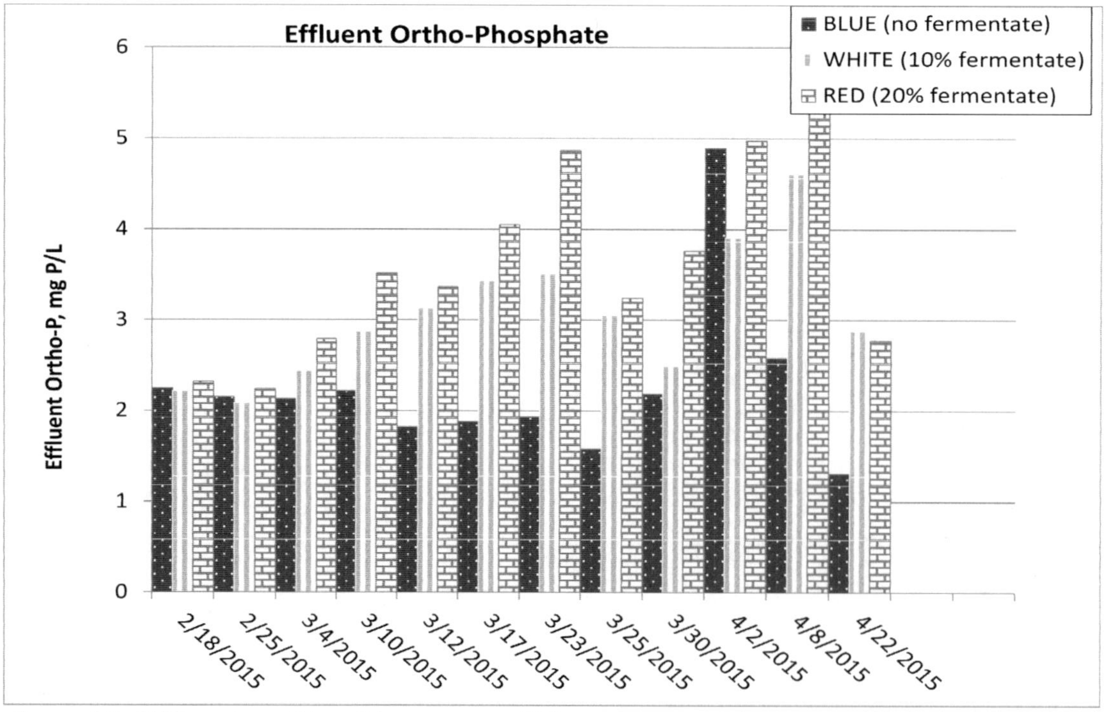

Figure 12- Effluent quality with respect to PO_4

Table 2 – Summary of Average BNR performance after 3/17/15

Pilot Reactor	% N removal	% P removal
RED (20% fermentate)	47 ± 13	25 ± 23
WHITE (10% fermentate)	36 ± 8	26 ± 14
BLUE (0% fermentate)	18 ± 10	16 ± 17

Influent and Effluent Total Suspended Solids

The average influent and effluent total suspended solids of the three reactors are shown in Figure 13. The influent TSS of the White and Red reactors was slightly higher than the Blue reactor due to high TSS concentrations of the fermentate added, and the increase in the amount of fermentate and sCOD addition increased the amount of effluent TSS. Given the settling rate of 7.5 m/hr, the minimum TSS was for the Blue reactor at 77 mg/L, which would require subsequent effluent polishing in a full scale application.

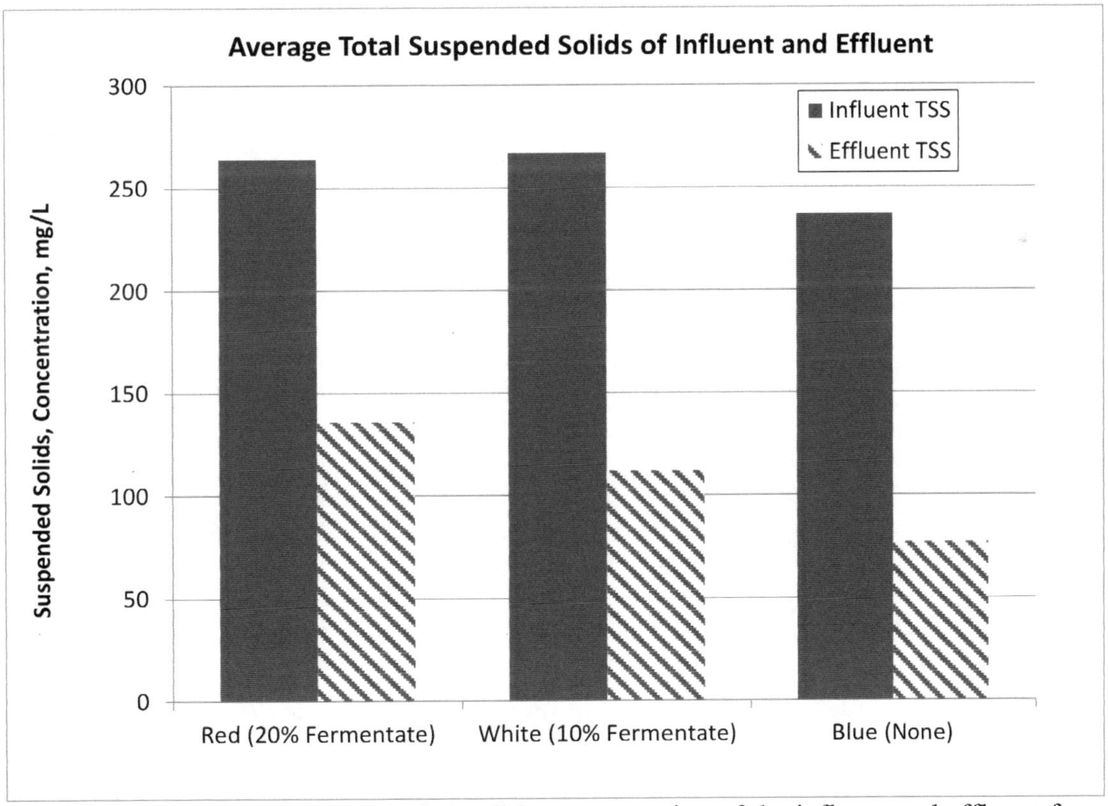

Figure 13 – Average total suspended solids concentration of the influent and effluent for all three reactors.

Sludge Characteristics for Land Application

The sludge characteristics of the AGS reactors were examined in regard to land application. According to USEPA 40 CRF Part 503 regulations for Class B sludge pathogen requirements, the monitoring results of fecal coliform density shall be less than 2 million colony forming units (CFUs) per gram of total solids.

Both types of reactor suspended solids, floc and floc + granules, were tested separately; however, only one sample was collected for this testing due to the low solids production and small scale of the pilot unit. While typical activated sludge aerobic stabilization is defined as 40 days of aeration at 20°C and 60 days at 15°C (to meet regulations by definition) more frequent sampling may prove compliance with the regulation at a shorter aeration time of the AGS sludge.

Figure 14 shows the most probable number (MPN) of CFU in both the floc and floc + granule sludges mixed liquor samples, without any stabilization efforts. The fecal density of the floc + granule sludge in the BLUE Reactor was above the 2 million CFU limit and reached 3 million CFU. Since the test is probabilistic, more routine tests are needed for a better statistical sample. Also, a determination of the fecal die off rate would establish the time and volume needed for stabilization with post aerobic digestion.

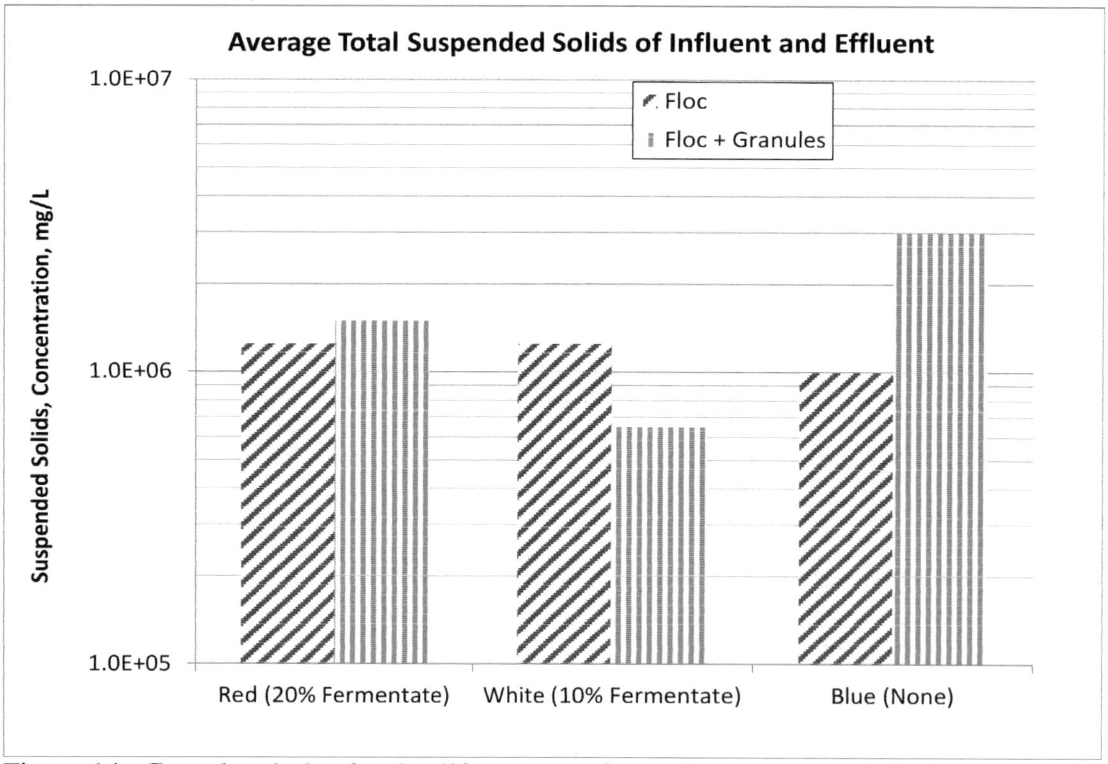

Figure 14– Granular sludge fecal coliform count for each reactor (y-axis in log scale).

Oxygen uptake Rates (OUR)

Oxygen uptake rate experiments were performed on the three reactors during the final week of the study. The experiments were conducted according to Standard Methods 2710 B, and the experimental results are shown in Figure 15. According to the 40 CFR Part 503 regulations for sludge pathogen and vector attraction reduction requirements under the option of aerobic digestion, the specific oxygen uptake rate must be less than 1.5 milligram of oxygen per hour per gram of total biosolids. Table 3 shows the actual specific oxygen uptake rate (SOUR) of the granular sludge ranged from 8-16 mg DO/g TSS-hr.

Based upon these results, it is evident that the granules are still quite active for oxygen uptake. It is recommended that if any other AGS piloting is pursued, further investigation of stabilization of wasted AGS sludge should be performed. The tests performed in this pilot effort were to define the condition of the granular sludge with respect to the 40 CFR Part 503 regulations, as wasted from the process. The testing did not define how quickly the system would endogenate and stabilize the sludge to less than 1.5 mg DO/g TSS-hr. The Blue reactor had the highest SOUR. This may be due to diffusion differences in the sludge particles between reactors, or differences in bacterial density in the granule aerobic zone, which may be a result of higher surface substrate loading of rbCOD.

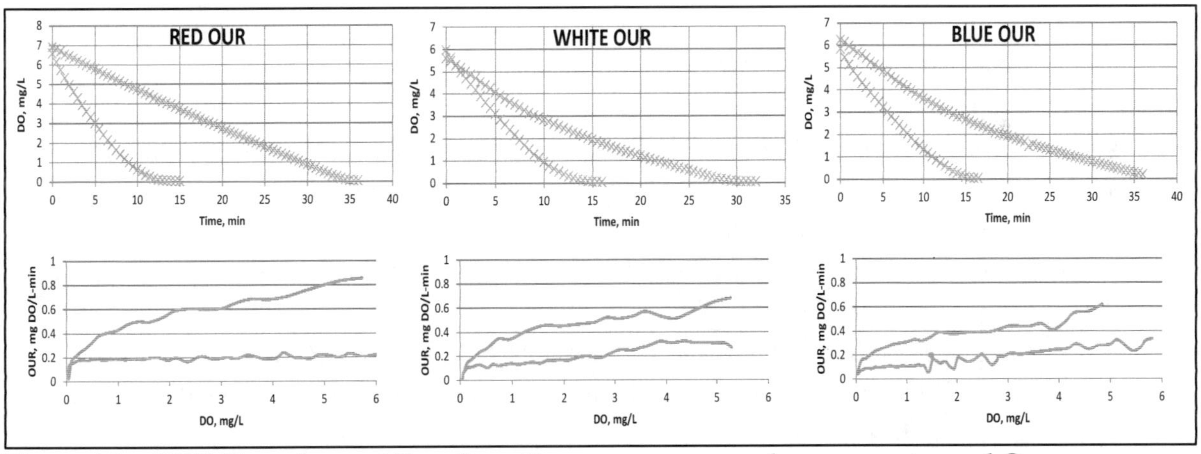

Figure 15 – OUR Experimental Results – Green curves are endogenous rates and Orange curves are spiked with influent sBOD.

Table 3– Table of endogenous Specific Oxygen Uptake Rates (SOURs)

Pilot Reactor	TSS/VSS	OUR	SOUR
	mg/L	mg DO/L-min	mg DO/mg TSS-hr
RED	1720/1460	0.21	8
WHITE	1300/1140	0.30	14
BLUE	1020/860	0.27	16

CONCLUSION

The AGS technology is currently a hot topic of research. New developments are continually being elucidated as to the best method to optimize the AGS process. The main objective of this pilot study was successful in that granulation was achieved with the low strength wastewater found at the Tomahawk WWTP.
Highlights that detail the major points gained from the study are:

- Pilot testing showed that the wastewater characteristics at the Tomahawk WWTP are suitable for treatment using aerobic granular sludge.
- A granule enrichment phase was used to speed up granule formation through aggressive loading and selective wasting. The granule enrichment phase was carefully controlled and monitored to ensure sufficient biomass was retained while maximizing EPS formation. Foaming and scum accumulation was manageable.
- Sufficient size and number of granules were achieved in approximately 2-months. Once the level of granulation was satisfactory, the granules were found to be stable, and the process was robust. The reactors efficiently removed ammonia and BOD with average effluent ammonia and soluble BOD for all three reactors of 0.4 mgN/L and 3 mg/L, respectively.
- With the supplementation of fermentate, the optimum total inorganic nitrogen (TIN) and orthophosphate removal reached 47 percent and 25 percent, respectively.
- Fermented primary sludge supplementation was shown to improve both nitrogen and phosphorous removal. Increased addition of fermentate up to 20% by volume correlated with increased nitrogen and phosphorous removal.
- The pilot reactor effluent TSS at the settling velocity of 7.5 m/hr, would require subsequent effluent polishing
- The pilot reactor effluent and waste would require subsequent stabilization to meet EPA requirements for land application based on testing of fecal coliforms and SOUR.

ACKNOWLEDGEMENTS

This pilot was funded by Johnson County Wastewater (JCW), a utility of Johnson County, Kansas. This pilot is also contributing to the Water Environment Research Foundation Grant U1R14.

REFFERENCES

Giesen, A.; van Loosdrecht, M. C.; de Bruin, B.; van der Roest, H.; Pronk, M. *Full-scale experiences with aerobic granular biomass technology for treatment of urban and industrial wastewater*; Nereda: 2013.

A.C. van Haandel and J.G.M. van der Lubbe. *Handbook of Biological Wastewater Treatment, Design and Optimization of Activated Sludge Systems - Appendix No. 9: Aerobic Granulated Sludge.* IWA Publishing. (2011)

H. van der Roest, M. van Loosdrecht, E. J. Langkamp, C. Uijterlinde. *Recovery and Reuse of Alginate from Granular Nereda Sludge.* Water 21, IWA, April 2015.

S. Bathe, M.K. de Kreuk, B.S. McSwain, and N. Schwarzenbeck. *Aerobic Granular Sludge. Selected proceedings of the 1st IWA-workshop Aerobic Granular Sludge.* IWA Publishing. 2005.

Hybrid Adsorption Biological Treatment Systems (HABiTS) for Improved Nitrogen Removal in Onsite Wastewater Treatment

Sarina J. Ergas[1], Lucie Krayzelova[2], Laura Rodriguez-Gonzalez[1], Karl Payne[1], Maya Trotz[1]

1. Department of Civil & Environmental Engineering, University of South Florida, Tampa
2. Department of Water Technology & Environmental Engineering, University of Chemistry & Technology Prague, Czech Republic

Decentralized wastewater treatment systems (DWTs) treat approximately one third of the wastewater in the United States (USEPA, 2002), with the majority of DWTs located in rural and suburban areas (USEPA, 2008). Advantages of OWTs include simplicity of operation, low cost when compared to centralized treatment, low maintenance requirements and the ability to reuse effluent onsite or recharge groundwater resources. Major challenges of conventional OWTs include: application limitations due to water table elevation and proximity to drinking water supplies and environmentally sensitive areas (FDOH, 2013); variable water usage and long idle times (e.g. during vacations) result in highly variable loading rates, which affect the biological treatment process and limited nitrogen (N) removal, resulting in contamination of ground and surface water.

A combination of ion exchange (IX) and biological treatment has the potential to provide enhanced treatment under the variable N loading conditions observed in OWTs (Hirst et al., 2013; Krayzelova et al., 2014). During periods of high loading, cation (NH_4^+) or anion (NO_3^-) loads in excess of the system biodegradation capacity are adsorbed by the medium. During low loading periods, the ions are desorbed and utilized by the microbial population. IX materials, such as natural zeolites (e.g. chabazite and clinoptilolite), have the ability to adsorb NH_4^+ (Wen et al., 2006). Hirst et al. (2013) showed that addition of clinoptilolite in the nitrification stage of a passive N removal OWTs resulted in 94% removal of TKN. The use of the combined IX/nitrification process coupled with sulfur oxidizing denitrification resulted in average effluent TN < 3 mg/L (Hirst et al., 2013). In a similar manner, scrap tires have a high IX capacity for NO_3^- (Lisi et al., 2004), and can be bioregenerated by denitrification (Krayzelova et al., 2014).

The goal of this research is to improve N removal performance and decrease reactor size requirements in OWTs using Hybrid Adsorption Biological Treatment Systems (HABiTS). Results of bench-scale tests with two different HABiTS are presented: 1) a natural zeolite material, clinoptilolite, which is used as an NH_4^+ IX medium in aerobic nitrifying packed bed reactors (PBRs) followed by 2) upflow saturated PBRs containing scrap tire chips as a NO_3^- IX medium with sulfur pellets and crushed oyster shells to promote sulfur oxidizing denitrification.

Nitrifying HABiTS

A number of materials previously utilized in OWT applications (sand, lava rock, expanded clay, glass, extruded plastic) were tested in batch studies with synthetic septic tank effluent (Fig. 1). The natural zeolite material, clinoptilolite, was found to be the best NH_4^+ exchanger. A moderate NH_4^+ IX capacity was observed for vermiculite; however, the material was not investigated further due to its friability. Batch IX equilibrium and kinetic studies were performed with clinoptilolite using USEPA protocols. The maximum IX capacity (Fig. 2a) was 11 mg NH_4^+-N/g clinoptilolite. A surface diffusion model described kinetics well (Fig 2b). Adsorption of K^+ and Ca^{2+} and desorption of Na^+ were also observed. Packed columns were used to determine the adsorption/desorption "wave" for NH_4^+ in columns containing a mixture of clinoptilolite and expanded clay (Fig. 3). The results showed that for the bench scale tests, a 6 cm clinoptilolite layer will be sufficient to maintain N levels < 10 mg/L for 2 days with a nitrogen loading of 632 mg/day. NH_4^+ desorption is slow, which should allow time for bioregeneration when applied in OWTs. PBR studies with this medium are currently underway and will be presented at the conference.

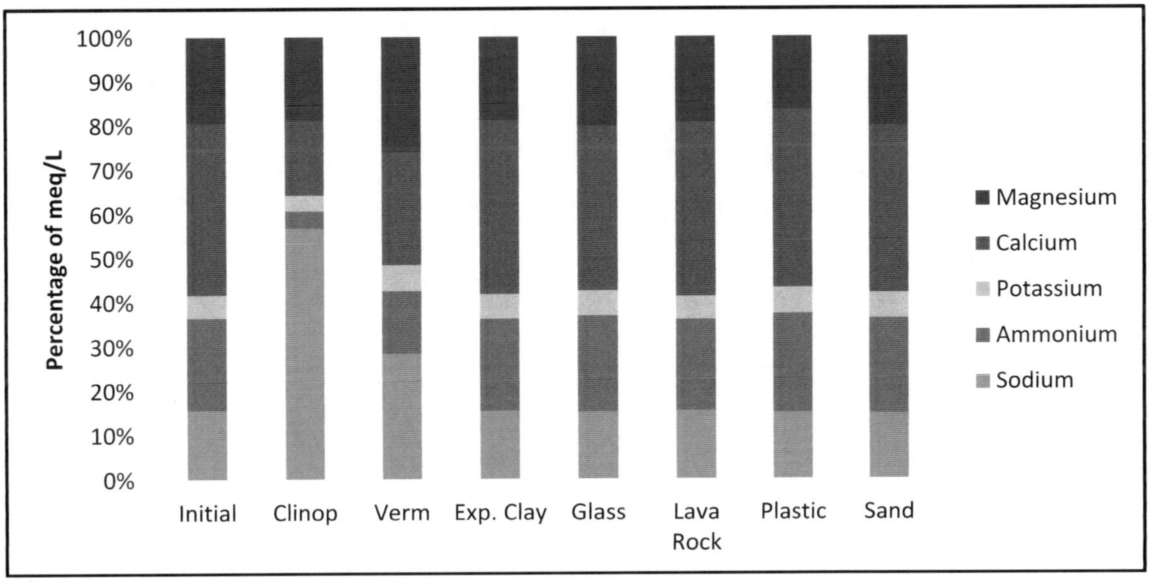

Figure 1: Cation concentrations in the bulk liquid for adsorption studies with varying media materials.

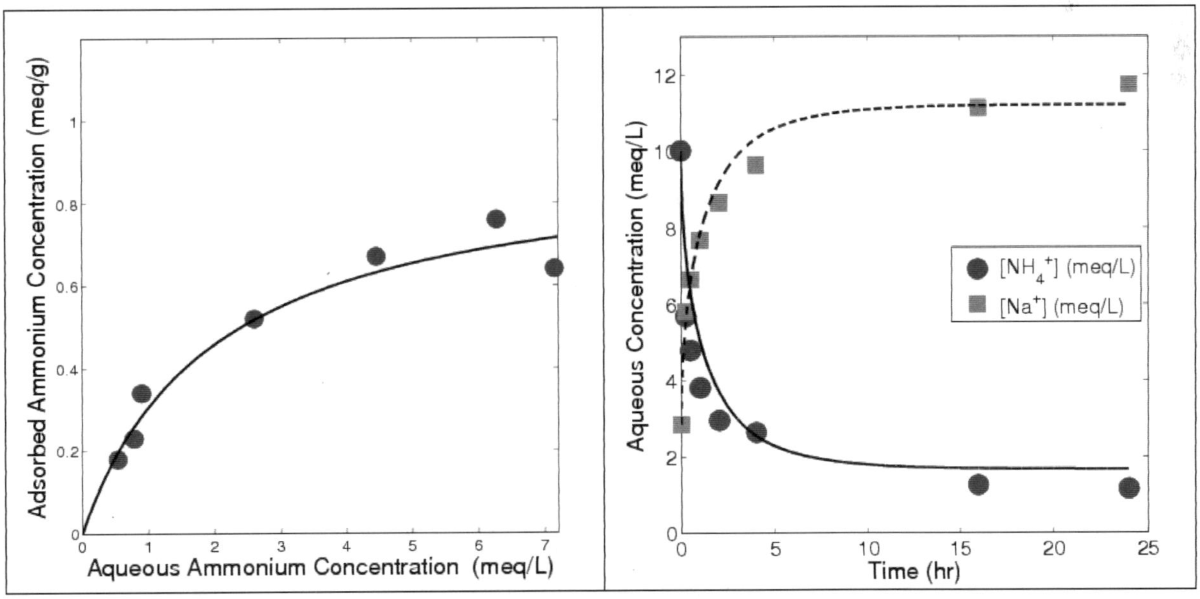

Figure 2. Results of: a) batch adsorption isotherm studies with NH_4^+ and clinoptilolite and b) batch kinetic studies with surface diffusion model.

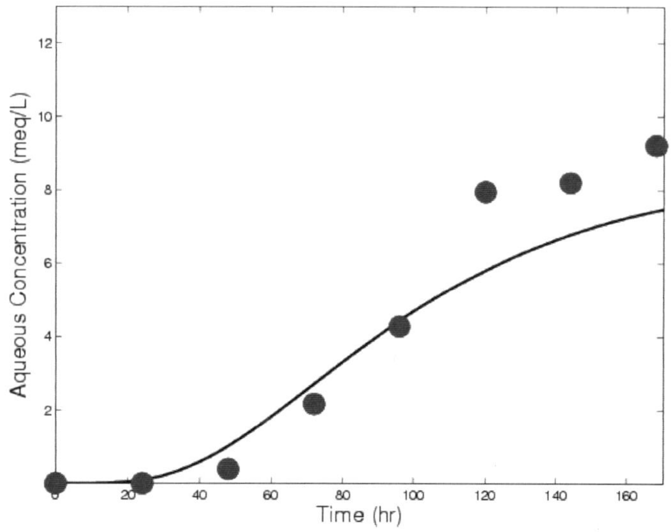

Figure 3. Results of abiotic column study showing NH_4^+ adsorption/desorption wave through a clinoptilolite and expanded clay column fed synthetic septic tank effluent (SSTE).

Denitrifying HABiTS

Batch adsorption equilibrium studies were carried out with synthetic nitrified septic tank effluent and scrap tire chips using EPA protocols. The maximum adsorption capacity of the scrap tire chips (Fig. 4) was 0.66 mg NO_3^--N/g scrap tires. Adsorption of PO_4^{3-} and Mg^{2+} and release of SO_4^{2-} and Ca^{2+} was also observed (data not shown). PBRs were packed with a mixture of scrap tire chips, elemental sulfur pellets and crushed oyster shells. Flow rates and concentrations were set up to simulate loading rates observed from OWTs. Even with the highly variable loading rates applied, average effluent NO_3^--N concentrations were < 5 mg/L and none of the effluent samples exceeded the drinking water standard for NO_3^- (Fig. 5). Average effluent COD concentrations were 0.1 mg/L, indicating that COD leached from the scrap tires was consumed by biological activity. Average effluent SO_4^{2-}concentrations (~60 mg/L) were well below USEPA drinking water guidelines and prior studies of sulfur oxidizing denitrification without tire chip media (Sengupta et al., 2007). Detailed results can be found in Krayzelova et al. (2014).

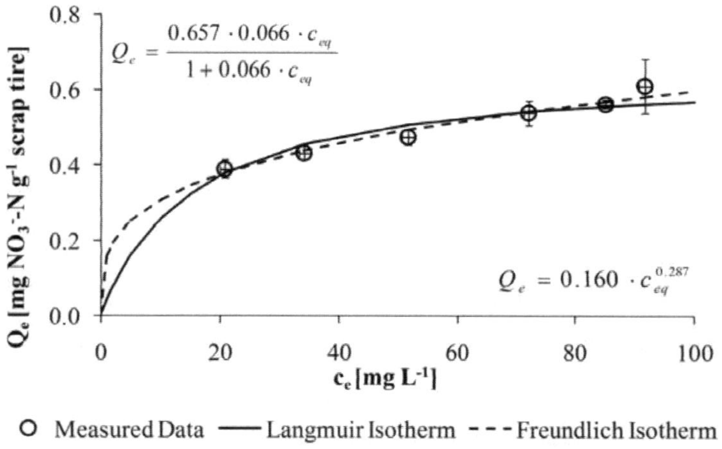

Figure 4. Results of batch adsorption isotherm studies with NO_3^- and scrap tire chips.

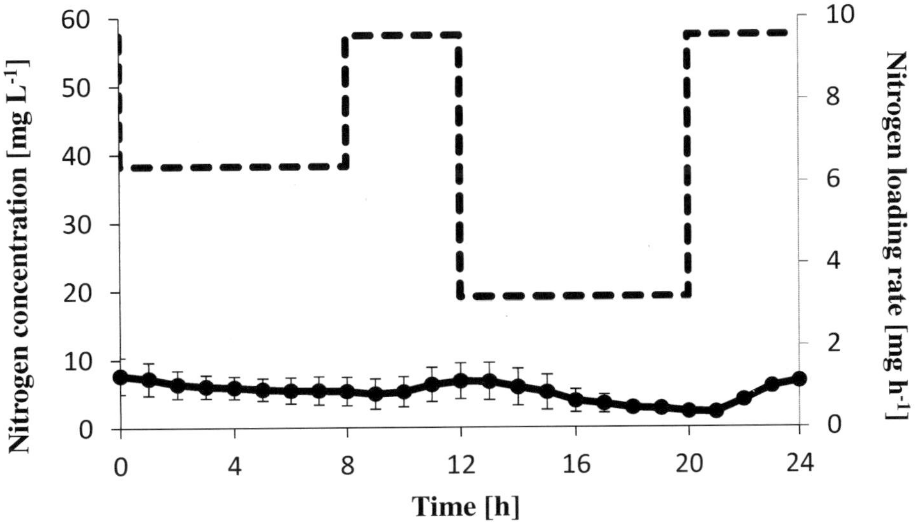

Figure 5. Results of denitrifying PBRs studies with scrap tire chips, elemental sulfur pellets and crushed oyster shells. Low effluent NO_3^--N concentrations were observed despite highly variable loading rates (---) typical of OWTs.

Acknowledgements

This material is based upon work supported by the Environmental Protection Agency Center for Reinventing Aging Infrastructure for Nutrient Management (RD835569). Any opinions, findings, and conclusions or recommendations expressed in this material are those of the authors and do not necessarily reflect the views of the funding agencies.

References

FDOH (2013). Standards for Onsite Sewage Treatment and Disposal Systems, Chapter 64E-6, Florida Administrative Code. Florida Department of Health (FDOH).

Hirst, J.E., Anderson, D.L., Smith, D.P. (2013). The Florida Onsite Sewage Nitrogen Reductions Strategies (FOSNRS) Study: Project Overview and Preliminary Results. Florida Water Resources Conference, Orlando, Florida.

Krayzelova, L., Lynn, T.J., Banihani, Q., Bartacek, J., Jenicek, P., Ergas, S.J. (2014). A Tire-Sulfur Hybrid Adsorption Denitrification (T-SHAD) process for decentralized wastewater treatment. *Water Research* 61, 191-199.

Lisi, R.D., Park, J.K., Stier, J.C. (2004). Mitigating nutrient leaching with a sub-surface drainage layer granulated tires. *Waste Management* 24, 831-839.

Sengupta, S., Ergas, S.J., Lopez-Luna, E. (2007) Investigation of solid-phase buffers for sulfur-oxidizing autotrophic denitrification, *Water Environment Research*, 79(13): 2519-2526.

USEPA (1999). *Decentralized Systems Technology Fact Sheet for Septic Tank and Soil Absorption Systems*. EPA 932-F-99-075. United States Environmental Protection Agency (USEPA). Office of Water, Washington D.C., USA.

USEPA (2002): *Onsite Wastewater Treatment Systems Manual.* EPA 625-R-00-008. United States Environmental Protection Agency (USEPA). Office of Research and Development, Washington D.C., USA.

USEPA (2008). *Septic Systems Fact Sheet.* EPA 832-F-08-057. United States Environmental Protection Agency (USEPA). United States Environmental Protection Agency (USEPA). Office of Wastewater Management Decentralized Wastewater Program, Washington D.C., USA.

Wen, D., Ho, Y.S., Tang, X. (2006). Comparative sorption kinetic studies of ammonium onto zeolite. *J. Hazardous Materials* B133, 252-256.

Industry Update for WASSTRIP® impacts on nutrient flows, digestion and dewatering of EBPR solids

Authors: Ahren Britton*; Steve Wirtel; Ram Prasad; Nate Turner
Ostara Nutrient Recovery Technologies Inc.
*abritton@ostara.com

ABSTRACT

As effluent phosphorus limits are becoming more common, more plants are operating or considering enhanced biological phosphorus removal (EBPR) and phosphorus recovery systems, and their impact on side stream nutrient loads, digestion and dewatering is becoming the focus of additional study. It has been known for some time that biological phosphorus removal coupled with anaerobic digestion can result in significant sidestream nutrient (ammonia and phosphate) loadings as well as proliferation of nuisance struvite formation. More recently it has become apparent that this combination of processes can also alter the chemistry or/or biology in digestion systems in ways that are detrimental to digested sludge dewatering properties. Results from facilities that operate WASSTRIP at full scale, as well as bench and pilot studies carried out at other plants are indicating that WASSTRIP can substantially reverse many of these adverse effects, while also substantially reducing biosolids production.

KEYWORDS: WASSTRIP®, nutrient recovery, digestion, dewatering, Pearl®, sidestream treatment

INTRODUCTION

As the global population grows and becomes more affluent, agricultural practices are becoming more intensive and sophisticated to keep up with the demand for food. This is resulting in an intensification of the phosphate cycle on a global scale - evidenced in both accelerated depletion of ore resources in certain geographies and increased nutrient flows in the biosphere. This in turn is resulting not only in increased crop yields but also in greater environmental leakage of phosphates. These factors have led to both geopolitical concerns about the stability of supply for phosphates (Steén, 1998; Jasinski, 2015; European Commission, 2014) and an increasing concern regarding environmental impacts resulting from overabundance of nutrients in waterways (Millennium Ecosystem Assessment, 2005). This combination of effects drives a need in society to use and re-use phosphorus resources more efficiently in order to minimize dependence on geographically isolated resources, and impacts on our aquatic environments.

Nutrient recovery from various points along the production cycle has been shown to be technically feasible, and a number of technologies have now been put into production at commercial scales. Recovered phosphate fertilizer production capacities in excess of 10,000 tonnes per year are already in operation and additional projects already under development indicate signs of exponential growth in this market (O'Callaghan, 2015).

Struvite recovery represents a unique opportunity to address both supply and waste issues related to the phosphate cycle, because it is both economically recoverable from a variety of waste streams (phosphoric acid production process cooling water, agricultural wastes, bio-energy byproduct streams, livestock manures, food and beverage industry wastewater, and municipal sewage), and is a sparingly soluble fertilizer which has been shown to improve phosphate nutrient use efficiency in agriculture. Struvite is an orthorhombic crystal composed of equal parts of magnesium, ammonium and phosphate in a crystal form including six waters of hydration, with the chemical formula $MgNH_3PO_4$-$6H_2O$. It is sparingly soluble in water (~0.3% water soluble) but completely soluble in mildly acidic solutions (100% ammonium citrate soluble).

Ostara has been developing the Pearl® nutrient recovery process for the past ten years, with a particular focus on producing a high quality fertilizer product and maximizing its value when used in turf, ornamental and specialty agriculture markets. The company now has eight facilities in operation in North America and Europe, with a further seven in various stages of design and construction.

Since 2011, four of Ostara's full scale nutrient recovery facilities have begun operating WASSTRIP at their biological phosphorus removal plants. Four additional WASSTRIP facilities in construction or final design and two facilities that were originally built with the Pearl nutrient recovery system prior to development of WASSTRIP are expanding to accommodate the technology.

The WASSTRIP process complements nutrient recovery for plants that use EBPR by stripping phosphorus, magnesium and potassium from waste activated sludge (WAS) through anaerobic fermentation prior to thickening and digestion. This results in the bypassing of a large fraction of the P, Mg and K found in EBPR WAS around the digester, thus changing the biological and mineral properties of sludge in the anaerobic digester. The WASSTRIP process is based on the retention of WAS under anaerobic conditions to maximize phosphorus release and employs the same principles as the anaerobic zone of EBPR plants, wherein phosphorus accumulating organisms (PAOs) release stored phosphorus in exchange for carbon in the form of volatile fatty acids (VFA). The soluble phosphorus is separated during sludge thickening, then the thickening liquor ("WASSTRIPate") is combined with ammonia rich post-digestion dewatering liquor in the Pearl Nutrient Recovery reactor where P, N and Mg are converted to Crystal Green fertilizer. Figure 1 presents a simplified schematic of the WASSTRIP process integrated into an EBPR facility.

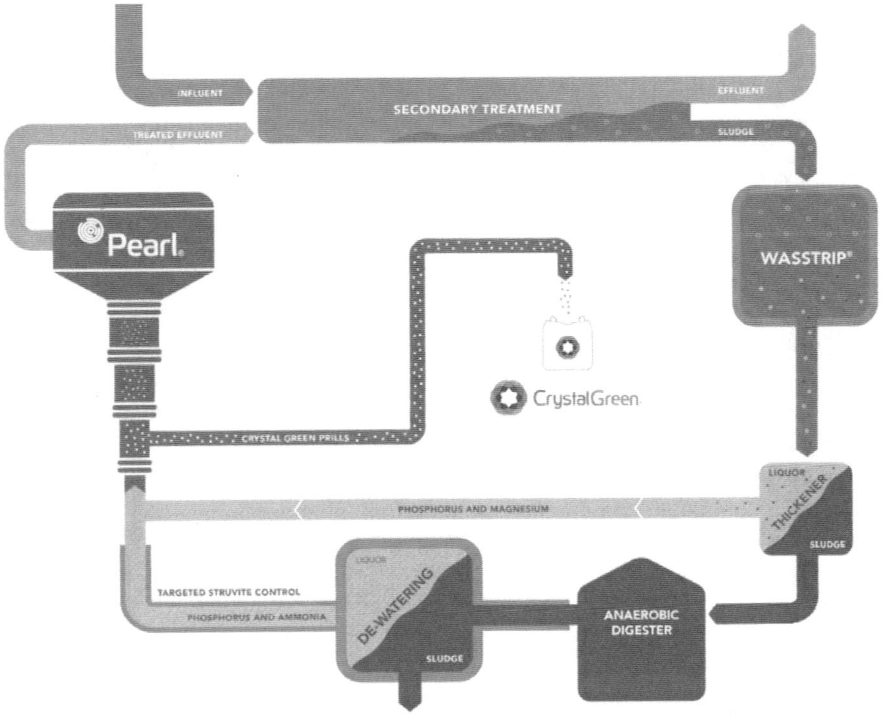

Figure 1: Schematic showing how the Pearl process typically integrates into a municipal wastewater treatment plant.

Since the startup of the first WASSTRIP facility in 2011 at the Durham Advanced WWTP in Tigard OR, the sludge digestion and dewatering properties have been monitored closely, and several improvements have been implemented in the WASSTRIP system resulting in increased stability and better performance of the system. At the same time, additional researchers have been carried out studies which show similar results from treatments comparable to the WASSTRIP process at the Empire WWTP in Minneapolis/St Paul (Alm et al CSWEA 2014), the Atlantic Plant operated by HRSD (Bott et al WEF RBC 2014) and an additional study is underway as part of a process evaluation for the Nansemond WWTP in Suffolk, VA.

With the reversal of EBPR related dewaterability, and nuisance struvite concerns, this technology can help industry professionals understand how optimize nutrient removal and recovery objectives while minimizing costs of operation.

DURHAM WASSTRIP DEVELOPMENT HISTORY

The WASSTRIP technology was developed in 2008 by Clean Water Services in Oregon shortly after the startup of North America's first struvite based nutrient recovery facility at the Durham Advanced WWTP in Tigard, Oregon (Baur et al 2008).

The process was tested at bench and pilot scale from 2008 to 2010, and implemented at full scale starting in 2011 at the Durham plant. The system was installed into a converted WAS storage tank, and the process implementation involved conversion from aeration mixing used to prevent P release, to submerged mixers intended to maintain anaerobic conditions. Initially the WASSTRIP tank was fed with a mixture of WAS from the secondary clarifier underflow with approximately 0.75% solids, and a primary sludge fermentate as a source of VFA to accelerate the release of phosphate and magnesium.

Although this configuration was successful, it typically only resulted in the release of about 30% of the total phosphorus in the WAS. The system resulted in an approximately 80% increase in phosphorus load and four to five times increase in the flow to the Pearl nutrient recovery system, compared to initial operation with post digestion centrate only. Although the Pears system was able to handle the increased flow, it was recognized that increased load was beyond the design capacity of the system, resulting in increased return loads of phosphorus to the headworks of the treatment plant when the treatment capacity was exceeded. As a result the Pearl system was subsequently retrofitted to double its capacity in the first half of 2012.

Even with the expanded capacity, it was found that the return phosphorus loads to the head of the treatment plant from the expanded Pearl system were still substantially higher than before the implementation of the WASSTRIP system, primarily due to the increased flow through the system. The increased flow through the Pearl system, along with the more dilute concentrations was also resulting in increased caustic consumption to maintain target pH in the reactors for phosphorus recovery. Meanwhile the increased phosphate load returned to headworks due to the increased flow, although at similar or lower phosphorus concentrations as before implementation of WASSTRIP, was resulting in increased need to use alum to meet the facility's seasonal 0.1 mg/L TP limit. The increased use of alum in turn resulted in reduced reliance on EBPR processes for P removal, which in turn resulted in less P which could be released in the WASSTRIP system, leading to a vicious cycle in the system.

In mid-2013, the WASSTRIP process was further improved by pre-thickening the WAS to 1.5-2% solids in a high rate gravity thickener, before transferring it to the WASSTRIP tank. This resulted in a 2-3 times reduction in the flow through the WASSTRIP system, and a comparable increase in soluble P concentration leaving the system. Because of the reduced flow, the retention time in the WASSTRIP tank increased from less than 12 hours to approximately 20-30 hours. This eliminated the need to supplement the tank with primary sludge ferementate since the thickened WAS now produced enough VFA to drive phosphate release through endogenous fermentation of the WAS. As a result, more VFA was available for the EBPR process from the primary sludge fermenters, and in fact some supplemental VFA was produced from the WASSTRIP process, further helping to stabilize the EBPR process.

WAS pre-thickening in the gravity thickener also resulted in a substantial reduction in flow to the WAS thickening centrifuges and to the Pearl system. The lower flow and higher phosphate concentrations going to the Pearl system allowed for significant reductions in caustic usage for maintaining target pH for nutrient recovery, while also significantly reducing the phosphate load returned to headworks in the Pearl effluent. This combination of factors appears to have improved the plants ability to optimize its EBPR operation and minimize alum consumption in

2014. Most interestingly these changes have resulted in substantial improvements in biosolids dewaterability and reduced the biosolids production at the facility by much more than would be expected due to reduction in alum usage alone.

Since completion of the WASSTRIP optimization in 2013, biosolids production has been reduced by >20%, both in terms of dry and wet solids measurements. This result correlates with results from a pilot study carried out by Met Council using sludge from the Empire WWTP reported by Alm et al (2014). In this study the reduction in solids production appears to have resulted from an increase in VS destruction in the digester when the WAS is first subjected to an anaerobic holding period, similar to performance enhancements experienced through the use of acid phase digesters.

Impacts of upstream and sidestream processes on digestion and dewatering processes can be difficult to isolate in full scale operations, due to long time scales and cumulative impacts of other process changes, seasonal changes, and wastewater characteristic changes. As a result, we have attempted to pull together results from other studies conducted at bench, pilot or full scale to try and better understand the results observed at Durham.

RESULTS

P release

Phosphorus release, or soluble phosphorus leaving the WASSTRIP tank as a percentage of the total phosphorus entering the WASSTRIP tank is a good indicator of the efficiency of the process, or the potential for diverting phosphorus around the digester and towards phosphorus recovery. The phosphorus release performance of the WASSTRIP system at Durham AWWTP can be seen in Figure 2 below. The figure clearly shows the increase in phosphorus release experienced, particularly since completion of the system upgrades in 2013. Initially when the system was first commissioned, only about 30% phosphorus release was experienced, however in 2014, phosphorus release was kept over 50% for extended periods of time. Seasonal variations can be seen in figure 2, resulting from the seasonal 0.1 mg/L TP permit the Durham AWWTP has. During summer months when the phosphorus limit is in effect, substantial amounts of Alum are dosed to the secondary and tertiary treatment systems at the plant. This alum sludge ends up combined with the WAS in the WASSTRIP tank, therefore limiting the P release during summer months due to partially suppressed EBPR and soluble P scavenging by the alum sludge in the WASSTRIP tank. Ongoing efforts to minimize alum dosages are dependent on bio-P stability, which may be enhanced by increased VFA generation.

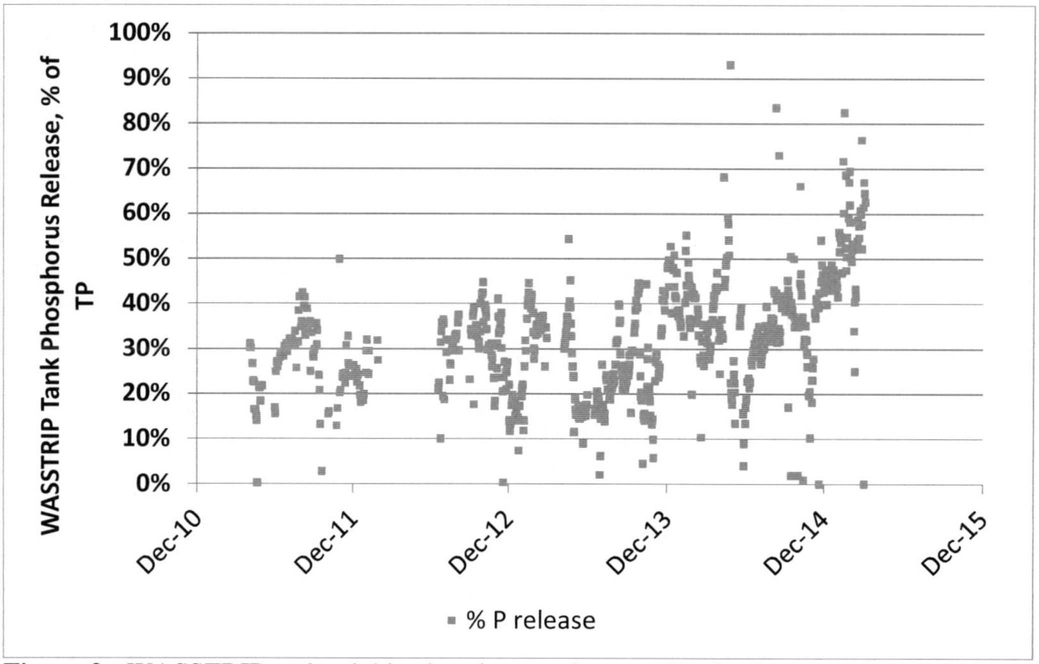

Figure 2: WASSTRIP tank soluble phosphorus release as % of influent total P. (Schauer 2015)

VFA generation

In the early years of operation (2011 and 2012) VFA was supplemented to the WASSTRIP tank from primary sludge fermenters to accelerate P release. This was necessary due to the relatively low retention time in the WASSTRIP tank (<12 hours) but it did not result in optimal P release and actually provided substantial dilution to the WAS, as well as reducing the VFA available for the main EBPR process which may have contributed to instability in the process during those years and increased use of alum to meet phosphorus discharge permit in the summer months. Starting in 2013 once the WAS pre-thickening was started, the VFA supplement was no longer necessary, and as can be seen in Figure 3, the WASSTRIP tank actually started to generate excess VFA, which was recycled to the main EBPR process in the Pearl reactor effluent. This additional VFA source is expected to provide improved EPBR stability in the coming years, allowing further optimization of alum dosages.

The accumulation of VFA in the WASSTRIP tank also indicates that a certain amount of WAS fermentation is occurring. This WAS fermentation may be playing a role similar to acid phase digestion in enhancing digestion performance and hence yielding the reduced biosolids generation that has been observed since 2013.

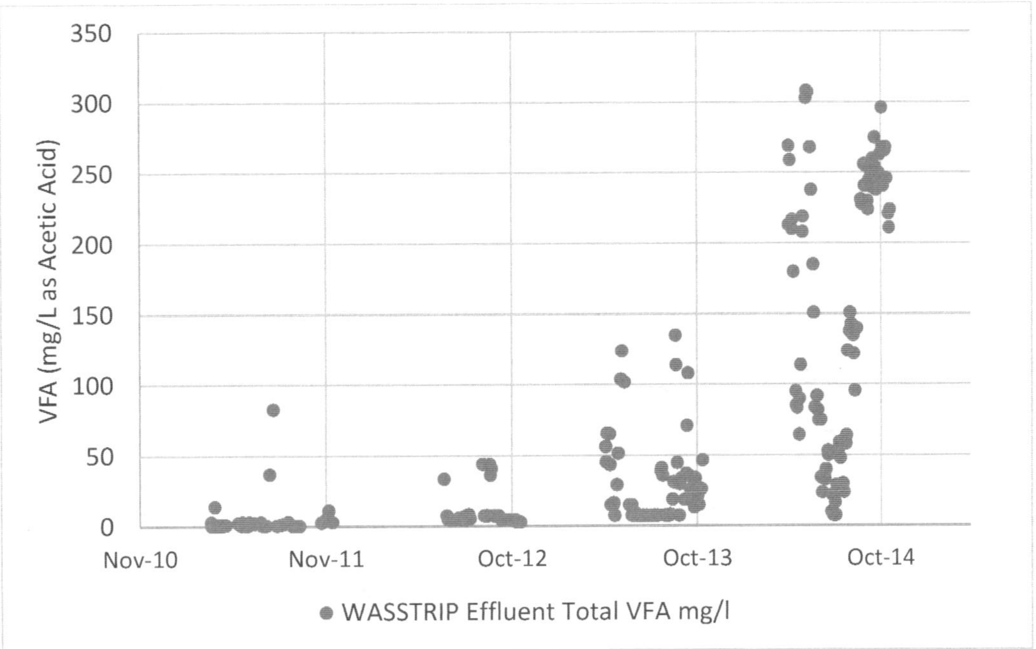

Figure 3: WASSTRIP tank effluent volatile fatty acids concentration

Biosolids production

The Durham AWWTP experienced a substantial decrease in dry solids in its biosolids when the Pearl process was first started up in which was largely thought to be due to reduced usage or alum for P removal, increased reliance on EBPR, and reduced use of lime for alkalinity supplement. This decrease in dry solids was not however accompanied wits a significant reduction in wet solids hauled from the plant, due to decreased dewaterability, likely resulting from the increased reliance on EBPR, as experienced at other facilities.

Figure 4 shows the biosolids production at the Durham AWWTP as both dry and wet solids from 2009 until May of 2015. It can be seen that since completing the WASSTRIP system upgrades in 2013, biosolids production has declined dramatically at the site, with wet solids reduced from 60 tons/day in 2012 to 47 tons per day in 2014, a reduction of more than 20% in cake hauled. 2015 data to data shows that the reduced solids production seem to be maintaining for a second year. The decrease in wet solids production is due to a combination of reduced dry solids and improved dewatering experienced at the site since 2013.

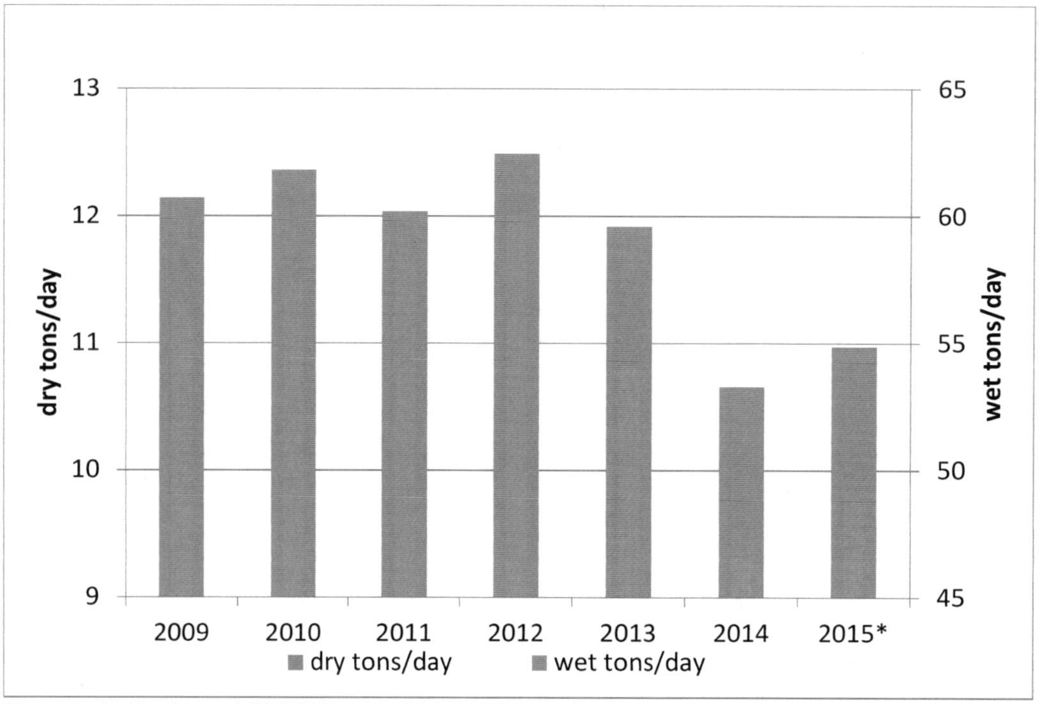

Figure 4: Durham AWWTP annual biosolids production as dry and wet solids. Note 2015 contains data only for Jan-May

Dewaterability

Figure 5 shows the 12 month moving average for biosolids cake solids. As with the biosolids wet solids, there has been a notable increase in biosolids dewaterability as indicated by cake solids since 2013. Cake solids improved from approximately 20.75% solids in 2011 and 2012 to approximately 22.75% solids in 2013 and 2014. It be noted that the polymer consumption in lbs/ton dry solids did not change significantly during this time period, although a number of dewatering optimization trials took place over the years.

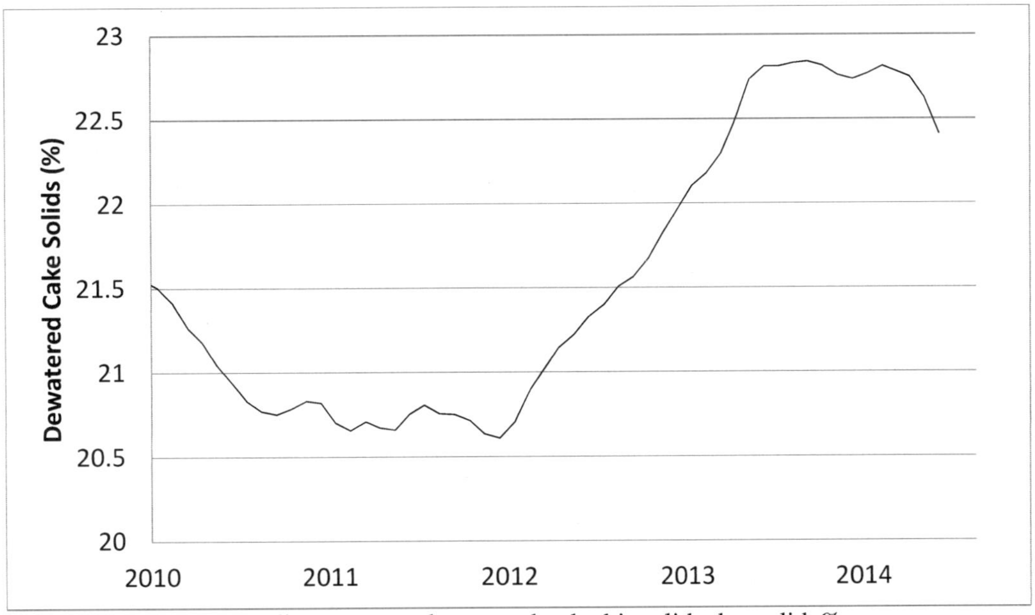

Figure 5: 12 month rolling average dewatered cake biosolids dry solids%.

DISCUSSION

Overall a large number of results appear to be accruing as a result of the optimization efforts undertaken with the WASSTRIP system at the Durham AWWTP. Results show improved dewatering by about 2% solids, a reduction in wet solids hauled from the site by over 20%, while also increasing the P release rate and returning a substantial mass load of VFA to the main EBPR process. This combination of effects appears to be the result of the implementation and optimization of the WASSTRIP system at the site.

Results presented by Schauer (2015) indicate that the improvement in dewaterability correlates to the change in the monovalent to divalent cation ratio in the digester effluent. This agrees well with results observed by various researchers such as Higgins (2014) Alm (2014) and Bott (2014) investigating dewaterability impacts in EBPR systems.

Due to the seasonal nature of the phosphorus limits at the Durham AWWTP, data analysis for this site is always somewhat more challenging than for a site with a more constant permit level and operational objectives.

We look forward to analyzing results from some of our more recently started up facilities in the coming years in order to validate the promising results obtained at the pioneering site for the WASTRIP technology.

CONCLUSIONS

Improvements to the process design of the first WASSTRIP installation at the Durham AWWTP in Tigard, OR have resulted in significant improvements in the performance of the system, in terms of the phosphorus release achieved in the WASSSTRIP tank, VFA generation and return to the EBPR process, reduction in alum usage to meet permit, and most significantly, improved dewatering and reduced biosolids production. These performance indicators are better than had been anticipated, but appear to be supported by results obtained elsewhere when similar WAS anaerobic treatment is employed with or without nutrient recovery.

REFERENCES

Alm, Rebecca; Sealock, Adam; Nollet ,Yabing; Koo, Alex; Sprouse George (2014) Results from Investigations of the Dewaterability of Anaerobically Digested Bio-P Sludge. Central States Water Environment Association 87th annual meeting, St. Paul, Minnesota. Session M.

Baur, R., Prasad, R. and Britton, A. (2008). Reducing ammonia and phosphorus recycle loads by struvite harvesting. Conference Proceedings WEFTEC 2008.

Bott, Charles (2014). Biological P Removal, Struvite Recovery, and Dewatering. Workshop Presentation, WEFTEC 2014.

European Commission, (2014). Report on critical raw materials for the EU. http://ec.europa.eu/enterprise/policies/raw-materials/files/docs/crm-report-on-critical-raw-materials_en.pdf

Higgins, M.J. (2014) Does Bio-P Impact Dewatering after Anaerobic Digestion? Yes, and not in a good way! WEF Biosolids and Residuals Conference, Austin TX

Jasinski, Stephen M. (2015) U S Geological Survey. Mineral commodity summary for phosphate rock. http://minerals.usgs.gov/minerals/pubs/commodity/phosphate_rock/mcs-2015-phosp.pdf

Millennium Ecosystem Assessment, (2005). Ecosystems and Human Well-being: Synthesis. Island Press, Washington, DC, USA.

O'Callaghan, P. (2015). Struvite recovery – reference plant mapping. www.bluetechresearch.com/struvite-recovery-reference-plant-mapping/

Schauer, Peter (2015) Operation of WASSTRIP Process on Nutrient Recovery and Dewatering Characteristics. WEF Biosolids and Residuals Conference 2015, Wasington, DC.

Steén, I. (1998). Phosphorus availability in the 21st century: Management of a non-renewable resource. *Phosphorus & Potassium*, Sept.-Oct. 1998, 25-31.

Implementation of Ammonia-Based Aeration Control (ABAC) at full-scale Wastewater Treatment Plants

Kshitiz Uprety[1,2*], Amanda Kennedy[1,2], William Balzer[2], Rick Baumler[2], Ryan Duke[2], Charles Bott[2]

1 Old Dominion University, Norfolk, VA
2 Hampton Roads Sanitation District, Virginia Beach, VA

*Kuprety@hrsd.com

ABSTRACT

The preliminary results of operating ABAC full-scale have been promising, as the treatment plant has seen a savings on daily supplemental carbon usage and a decrease in energy consumption, while maintaining consistent removal of total nitrogen partially due to simultaneous nitrification-denitrification (SND), as well as some possible suppression of nitrite oxidizing bacteria (NOB) for nitrite shunt.

INTRODUCTION

In the near future, many public utilities will be required to meet stringent effluent total nitrogen (TN) and total phosphorus (TP) limits while simultaneously becoming more efficient and cost effective. This motivation has led to research and development of efficient process control solutions (Rieger et al., 2013) and reliable nutrient sensors, allowing wastewater treatment plants (WWTP) to automate and optimize plant operations. Aeration is known to be the largest consumer of WWTP energy (approximately 60%) (Amand, 2014; Rieger et al., 2013), and therefore considered to be an ideal parameter to optimize. Currently, the most prevalent and conventional aeration strategy is dissolved oxygen (DO) control, where airflow is controlled to maintain a DO setpoint (approximately 2 mg/L). However, an emerging and promising control strategy is ammonia-based aeration control (ABAC), in which partial/incomplete nitrification is encouraged by maintaining lower DO in order to retain a desired ammonia concentration at the end of the aerobic zone.

The main incentive for implementing ABAC at a full-scale facility was to decrease the usage of supplemental carbon for denitrification while simultaneously decreasing aeration energy usage, with the potential for decreased chlorine demand downstream (chloramination) as well as decreased alkalinity demand.

There are many potential impacts, both beneficial and potentially detrimental, of operating ABAC. Therefore, the multiple objectives for this project are:

(i) to decrease energy consumption by optimizing aeration, (ii) to decrease supplemental carbon usage, (iii) to decrease chlorine demand and ammonia addition to the contact tanks, (iv) to decrease alkalinity demand, (v) to verify whether low DO selects for Polyphosphate Accumulating Organisms (PAO) over Glycogen Accumulating Organisms (GAO) and enhances or potentially inhibit (due to low aerobic zone DO) biological phosphorus removal (bioP), (vi) to optimize overall plant operation, (vii) to determine any impact on settleability and effluent quality, (viii) to determine reliability of sensors, (ix) to optimize and tune the feedback Proportional, Integral, and Derivative (PID) controller, and (x) to optimize the supplemental carbon feed controller to be able to support ABAC dynamics.

This project will conduct intensive sampling to evaluate the objectives in Summer 2015.

METHODOLOGY

The Hampton Roads Sanitation District Nansemond Treatment Plant (NTP), a 30 mgd facility operating a 5-stage Bardenpho process with a large industrial waste contribution, approximately 20% of total flow (resulting in low influent COD/TKN), was consuming a considerable amount of supplemental carbon in order to meet a TN treatment objective of 5-8 mg/L (the plant is part of a multi-plant bubble permit, where there is no ammonia limit). The plant operated conventional DO control, prior to ABAC. The main incentive for implementing ABAC at NTP was to decrease the usage of supplemental carbon for denitrification while simultaneously decreasing aeration energy usage, with the potential for decreased chlorine demand downstream (chloramination) as well as decreased alkalinity demand. This plant has typical primary treatment (bar screens, grit vortex separator, and rectangular and circular primary clarifiers). The flow goes to secondary treatment (5-stage Bardenpho) and then to circular secondary clarifiers. Tertiary treatment consists of sodium hypochlorite for disinfection, sodium bisulfite for dechlorination, and then is discharged into a river which is part of the Chesapeake Bay watershed. Primary solids and waste activated solids (WAS) are joined together and thickened by gravity belt thickeners and then sent to two mesophilic anaerobic digesters. The digester solids are dewatered by centrifuge and the cake is hauled offsite for incineration. The centrate is redirected into a sidestream struvite recovery facility, which removes 80% and 30% of the total phosphorus and ammonia from the centrate respectively, before it is returned to the headworks.

The BNR process is a 5-stage Bardenpho with seven parallel tanks/trains (Figure 1). Each tank in the anaerobic/anoxic/aerobic (AAA) has six cells. Cell A is the anaerobic zone. Cells B-D are the first anoxic zone with the nitrate recycle (NRCY) entering in cell B of each tank. Cell E and F are considered aerobic zone 1. The flow from all seven tanks enters into an aeration influent channel, is mixed with fine bubble diffusers, and is split evenly among seven aeration tanks (AT), of which there are 3 small tanks (AT 1-3) and 4 larger ones (AT 4-7). These tanks are split into two zones, zone 2 and zone 3. After the aerobic tanks, flow enters the second anoxic zone where a waste methanol is added for denitrification. All flow is then mixed together in a reaeration channel and sent for secondary clarification.

Currently, ABAC is only operated in AT 4-7, with future considerations for implementation in AT 1-3. NTP uses a feedback PID controller for ABAC to adjust the DO setpoint in all aeration basins and zones. Each aeration tank has three zones of aeration. An ion-selective electrode (ISE) ammonia probe at the aerobic zone effluent of each tank continuously sends a signal of the ammonia measurement to an ammonia PID controller, which calculates a DO setpoint based on the difference between the ammonia probe feedback and the desired ammonia setpoint. This PID calculated DO setpoint is relayed to a DO controller, which manipulates the aeration input to zones two and three of the aeration tanks (Figure 3) in order to meet the calculated DO setpoint, which is allowed to range from 0 to 2 mg DO/L. The average of all the ammonia probes is used as the ammonia feedback signal for zone 1 (Figure 2). The DO setpoint range in zone 1 ranges from 0.5-2 mg DO/L. DO and airflow to each zone are continuously measured in all tanks through the use of DO probes and airflow meters. To prevent prolonged settling of mixed liquor during periods of low aeration rates, if the airflow for any zone in any tank drops below a minimum airflow set point for a certain amount of time (currently set at 15 minutes), the valve for that zone is opened to 30% (purge) for one minute to prevent extended settling of biomass and to ensure continuous sensor accuracy.

There are also a number of safety features in place while in ABAC control. For example, if any one of the ammonia probes at the end of the aerobic zone goes above a "Maximum NH4 Setpoint", then zones 2 and zones 3 in that respective tank will revert back into DO control. This is not an instantaneous time measurement, to account for spikes in measurement due to faulty measurements. The signal has to be above the setpoint for a certain amount of time (generally 30 minutes), in order to ensure that it is indeed a correct signal. This safety is to prevent ABAC from operating when the ammonia probe starts to drift, or loses calibration. This probe can then be identified as "bad quality" on the Distributed Control System (DCS) and the tank can be placed back into ABAC, with the signal now being the average of the other ammonia probes in operation. Another safety is in the ammonia wet chemistry analyzer in the reaeration channel. If this ammonia signal (Max NH4 Safety Setpoint) measures above a certain setpoint for 30 minutes, then all zones in all tanks will revert back into DO mode. The only way to return to DO mode is to wait until this signal is below the setpoint for 1 hour. This safety helps to stabilize the process, especially during wet weather events leading to elevated flow. Although phosphorus control is not part of ABAC, there are phosphate wet chemistry analyzers in the aerobic effluent as well as the reaeration channel. Since this is a biological phosphorus removing plant (bio-P) and there are TP effluent limits, the entire process will revert back into DO mode if either analyzer goes above a certain setpoint). If the plant ever does revert back into DO control for any of the above reasons, there is a program in the DCS that will scan to see if all the above criteria are met. If all the ammonia and phosphorus signals are below their respective setpoints for a certain time setpoint (2 hours), then the program will switch all the zones in all tanks into ABAC.

ABAC is operated at this facility primarily in order to save on supplemental carbon usage. Therefore, a redesign of the existing supplemental carbon feed system was modified. The old method utilized a feedback control, where supplemental carbon was added based off nitrate sensor signals at the end of the second anoxic tank. Supplemental carbon would be added to maintain a nitrate concentration exiting the second anoxic zone. A new program termed "NIC/NOT" (Nitrate In Calculated/ Nitrate Out Trim), utilizes both feedforward and feedback control. The dosage of COD required for denitrification is calculated based on the nitrate entering the second anoxic zone (NIC) and the COD content and expected anoxic yield of the carbon source being used. The NOT feedback will then either trim the dosage up or down based on the disparity between the nitrate concentration setpoint and the measured nitrate concentration in the second stage anoxic zone effluent.

The oxygen supplied in the reaeration channel is also automated. There are two motor operated valves (MOV) which control each half of the reaeration channel. There is a DO probe in the middle of the reaeration channel. The valves will either open or close based on the difference between the DO setpoint and the actual DO signal. This is considered as conventional DO control. The purpose of this control is to allow enough oxygen for biomass suspension, while at the same time, preventing complete ammonia oxidation in the channel to allow for chloramination downstream, which in turn, decreases chlorine demand.

This ammonia addition to the contact tanks was also redesigned to be automated. Previously, supplemental ammonia was constantly fed into the contact tanks to combat any presence of nitrite. Currently, with the new design, when the ammonia signal from the reaeration channel is above an "Ammonia Pump Setpoint" for a certain time setpoint, then the ammonia pumps will slowly ramp down and feed lower amounts of ammonia into the contact tanks. When the ammonia signal from the reaeration channel goes below the setpoint, then the ammonia pump will ramp up the dosage. This program is intended to effectively and efficiently utilize the ammonia in the influent. This setpoint can vary depending on the ammonia setpoint in the aerobic effluent.

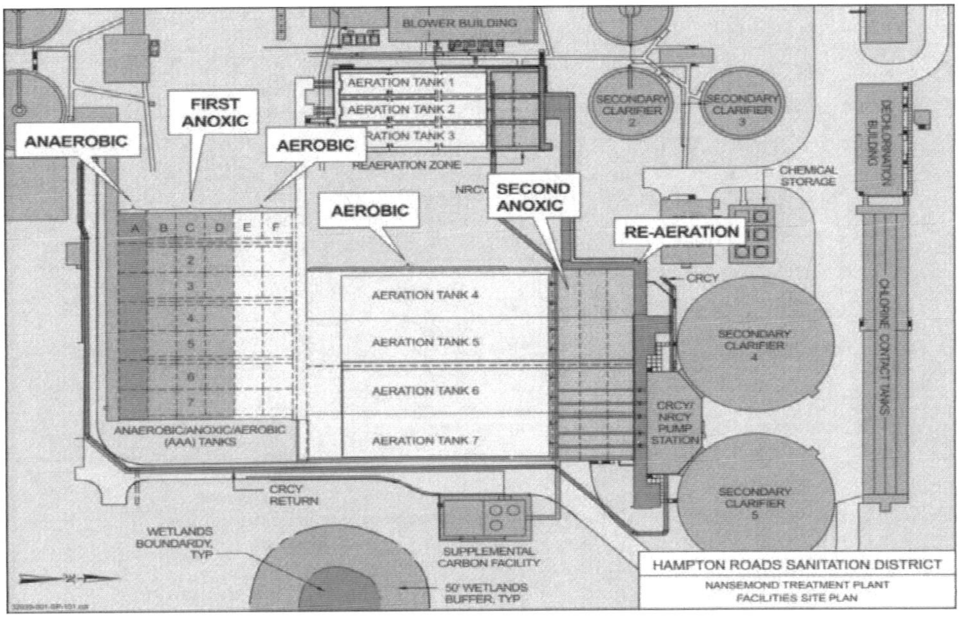

Figure 1: NTP 5 Stage Bardenpho Layout

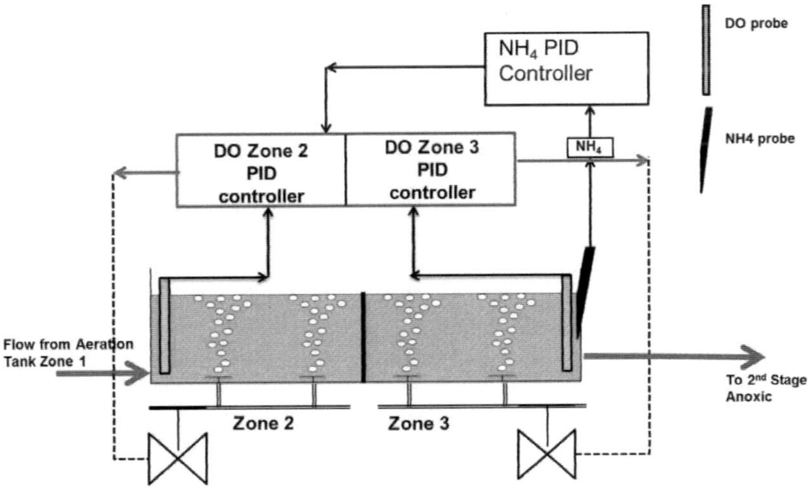

Figure 2: Zone 1 ammonia feedback control

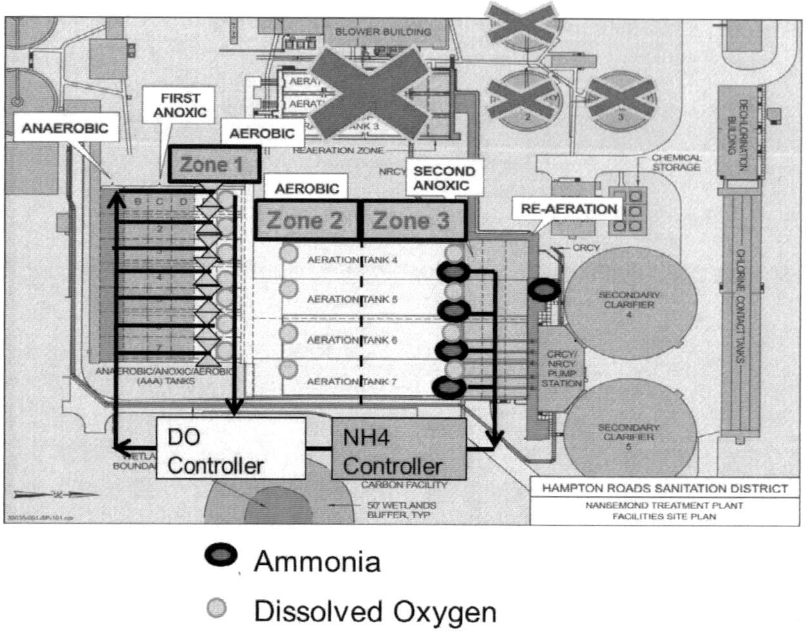

Figure 3: Zones 2 and 3 ammonia feedback control

Many different scenarios were modeled in BioWin and evaluated prior to implementing ABAC at this facility, and the one chosen was based on the criteria of being the most logical, in terms of energy and methanol savings (Figure 4). Scenario 2b was evaluated as being ABAC with ammonia controller calculating the DO setpoint through proportional, integral, derivative (PID) control with a DO setpoint range of 0-2 mg/L. This is chosen and deemed the desired ABAC operation.

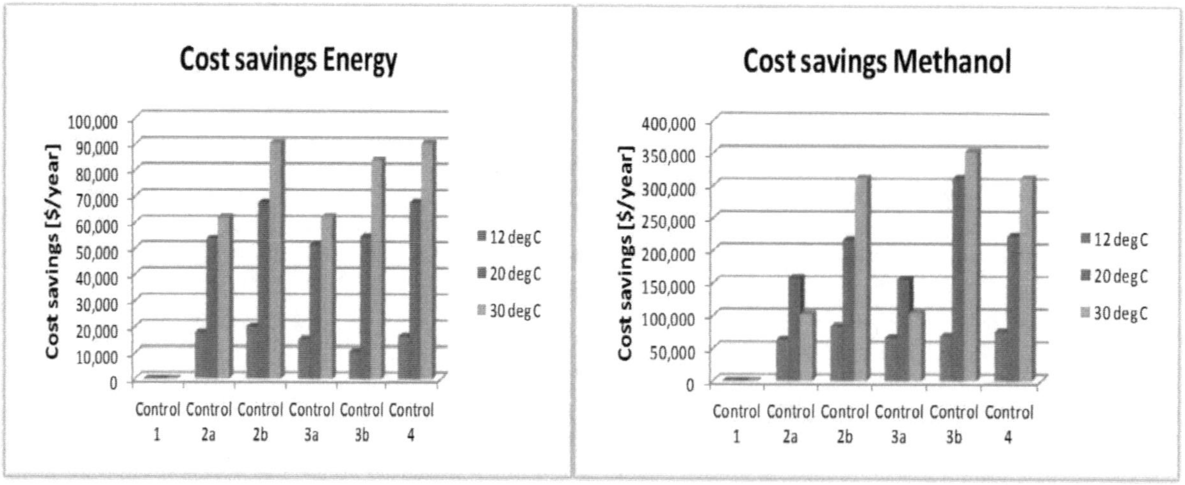

Figure 4: Results from BioWin modelling

The Auto DO setpoint that is calculated via the ammonia PID controller is used as a means of SRT control at this treatment plant. The PID calculated DO setpoint is recorded in 15 minute intervals and the daily average is calculated. The daily averages are tallied into a seven day running average. The plant targets an average Auto DO setpoint of between 0.5 – 0.8 mg DO/L in order to achieve optimal conditions for SND. If the average DO setpoint is above this range, then wasting is increased to allow for operation at lower SRT. When the average Auto DO setpoint is below the target range, then higher SRT is desired and wasting is decreased. Figure 5 below shows the graph of this operation.

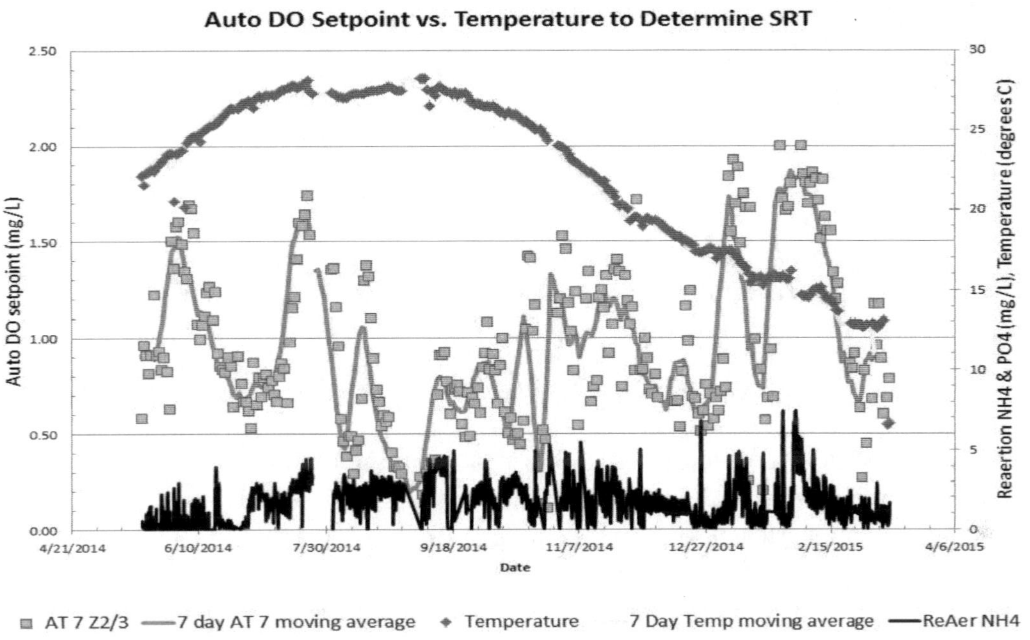

Figure 5: Auto DO Setpoint vs. temperature for SRT Control

RESULTS

The plant implemented ABAC in AT 4-7 beginning in May 2014. NIC/NOT supplemental carbon feeding also was developed and the PID loop was tuned during this time. Upon analysis of preliminary data, the plant has seen an average monthly savings of 47% in terms of the required supplemental carbon usage necessary for denitrification (Figure 6) when comparing ABAC to conventional DO control.

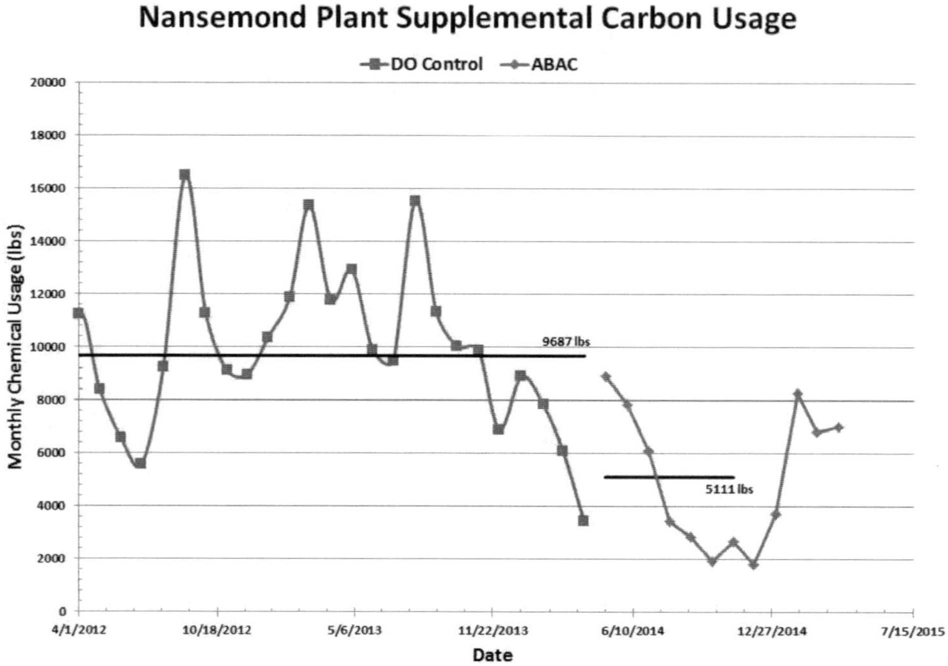

Figure 6: Average Monthly Supplemental Carbon Usage, as lbs COD added

The plant month electricity cost and average kWh used per month comparing ABAC and DO control is shown in Figure 7 below. Plant electricity costs decreased 8% during ABAC. Average Kilowatt-hours per month also decreased by about 4% during ABAC.

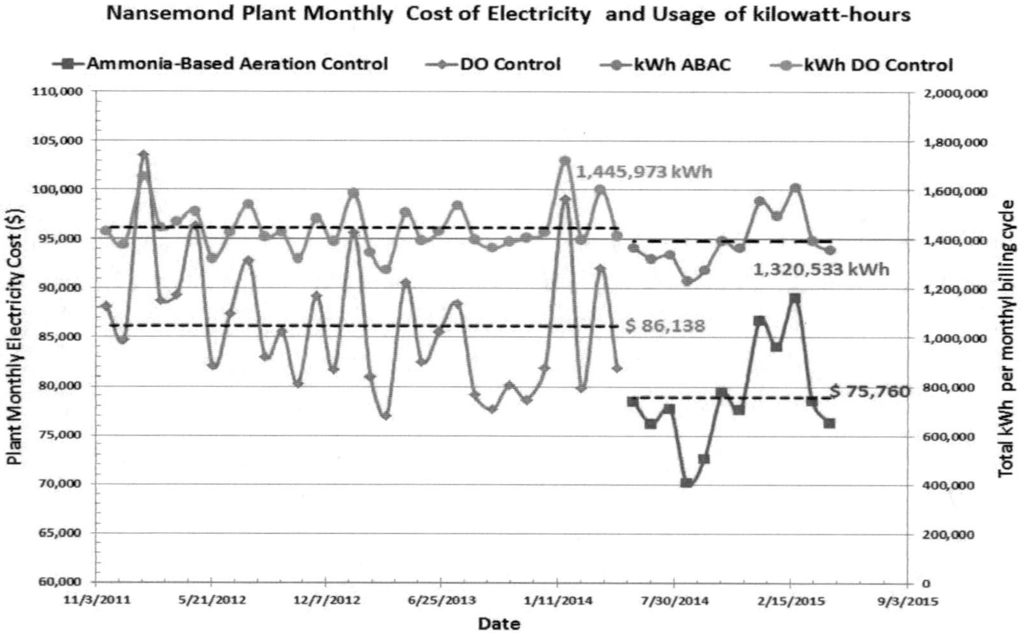

Figure 7: Average Monthly Electricity Cost and average kWh used per month

Figure 8 evaluates the final effluent (FNE) characteristics of the plant comparing ABAC to DO control. FNE total kjeldahl nitrogen (TKN) is increased during ABAC, which is expected, and also contributes to slightly higher TN in the effluent. Effluent TP concentrations are slightly lower during ABAC than during DO control and well below permit limits during both modes of operation. During ABAC, the average monthly effluent TN, TKN, NOx, and TP is 7.7 mg/L, 4.2 mg/L, 3.5 mg/L, and 1.2 mg/L, respectively. During DO control, the average monthly TN, TKN, Nox, and TP is 6.0 mg/L, 2.9 mg/L, 3.1 mg/L, 1.5 mg/L, respectively.

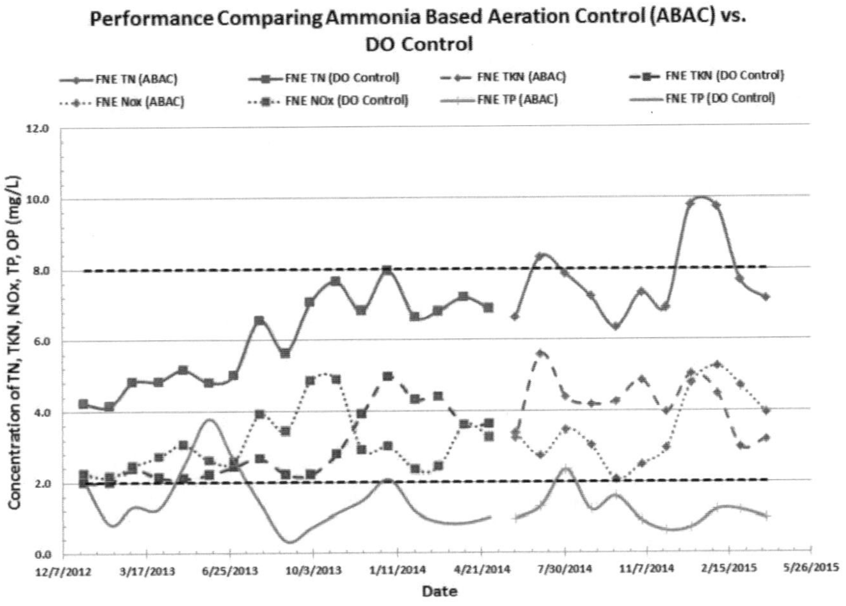

Figure 8: ABAC vs. DO control Final Effluent Characteristics

DISCUSSION

During DO control, the supplemental carbon usage per month pounds (lbs), of methanol as COD, was higher than when compared to the average monthly usage during ABAC. This decrease in usage is due to lower concentrations of nitrate present in the second anoxic influent, mostly due to SND which occurs by making the aerobic zone operate at low DO conditions. This is done by managing the waste rate to obtain an ideal Auto DO setpoint range.

The average monthly TN concentration in the final effluent was higher during operation of ABAC, which can be attributed to the increase of TKN in the effluent. This is probably due to the lowered amount of ammonia nitrified, leading to some residual in the effluent. The automated supplemental ammonia feed system was still in the process of being developed and being tuned correctly. The average NOx presence in the effluent was also slightly higher, but with a great reduction in supplemental carbon usage.

The monthly electricity bill and kilowatt-hours consumed monthly both decreased as a result of ABAC. This was due to operation of low DO demanding a lowered horsepower output from the blower. The reduction of both electricity usage and kilowatt-hours can be further optimized, but blower turndown capability still needs to be evaluated.

TP removal stayed consistent during times of both DO control and ABAC. Further testing will be conducted in to order determine the presence of denitrifying phosphate accumulating organisms (DPAO) and whether polyphosphate accumulating organisms (PAOs) outcompete glycogen accumulating organisms (GAOs) in low DO environments.

CONCLUSION

Preliminary results of ABAC testing resulted in the lowered demand of supplemental carbon usage for denitrification, and a lowered consumption of energy without a significant change in effluent quality. Testing will continue and fine tuning of feedback controllers will be evaluated and optimized. Further work planned work includes further tuning and optimizing the system, a careful overview of potentially negative impacts and related data, determining reliability of sensors, and evaluating the process for overall plant optimization.

References

Åmand, L. 2014. Ammonium Feedback Control in Wastewater Treatment Plants. *Uppsala Dissertations from the Faculty of Science and Technology* 104. 256 pp. Uppsala: Acta Universitatis Upsaliensis. ISBN 978-91-554-8900-7.

Rieger L., Takacs, I., Siegrist. H. 2013. Improving Nutrient Removal While Reducing Energy Use at Three Swiss WWTPs Using Advanced Control. Water Research (84), Number 2 (170-188)

Seeking the Optimum Approach to Generate VFA/RBCOD to Ensure Reliable EBPR at the Robert W. Hite Treatment Facility

Kimberly Cowan[1*], Kurt Carson[1], Cynthia Lynch[1], Jim McQuarrie[1]

[1] Metro Wastewater Reclamation District, Denver, Colorado
* Corresponding author: KCowan@mwrd.dst.co.us

ABSTRACT

Reliable and effective enhanced biological phosphorus removal (EBPR) depends upon a consistent and ample supply of readily biodegradable chemical oxygen demand (RBCOD) and volatile fatty acids (VFA) to the anaerobic zone of the basin. Due to the carbon deficiency of many facilities, the pursuit of augmenting influent VFA/RBCOD concentrations through various means has become a crucial step in EBPR implementation. Laboratory bench scale testing and process modeling campaigns were used to investigate five feasible methods for increasing VFA/RBCOD, thereby improving EBPR performance. Direct acetic acid dosing, addition of acid phase digester effluent, addition and/or fermentation of a locally generated high-COD recycle stream, gravity thickener underflow elutriation (with and without settled mixed liquor addition), and primary sludge fermentation were evaluated for maximum VFA/RBCOD production and weighed against financial and operational concerns each option would present.

KEYWORDS: EBPR, VFA, readily fermentable substrate, volatile fatty acids, RBCOD, phosphorus.

The Metro Wastewater Reclamation District (Metro District) in Denver, Colorado is underway with full-scale conversion of its 220 million gallon per day (833 MLD) Robert W. Hite Treatment Facility (RWHTF) to enhanced biological phosphorus removal (EBPR). Fundamental to consistent and reliable EBPR performance is ample supply of volatile fatty acids (VFA) and readily biodegradable chemical oxygen demand (RBCOD) in the anaerobic zone of the process (*Carlsson et al.,1996; Oehmen et al., 2007*). For reliable high-performing EBPR, the Metro District has targeted an additional 2 to 3 tons/day (1,800 to 2,700 kg/day) of short-chain volatile fatty acids (SCVFA). For this paper, SCVFA is considered to be the sum of acetic, propionic, butyric, valeric and caproic acids, expressed as COD and converted to acetic equivalency using the definition of SCVFA and COD-equivalents presented by *Rossle and Pretorius (2001).* Five possible options have been evaluated to help address the SCVFA deficiency:

Option #1 – Dose commercial-grade acetic acid.
Option #2 – Recirculate a portion of Acid Phase Digester (APD) effluent back through the gravity thickener (GVT) for resettling and release of SCVFA in the GVT overflow effluent.

Option #3 – Use a locally generated recycle stream of comingled expired food grade beverages that is either (a) directly added to anaerobic zone, or (b) pre-fermented before adding to anaerobic zone.

Option #4 – Recirculate GVT underflow or stir the GVT sludge blanket to liberate more SCVFA/RBCOD in the GVT effluent, with or without the addition of settled mixed liquor to increase fermentation in the unit.

Option #5 – Route GVT underflow sludge through a small side fermentation tank, then return the tank effluent back to the GVTs for solids separation and elutriation.

The purchase of acetic acid (Option #1) is convenient in terms of its availability, ease of operation, and its consistent quality. However, given the scale of the RWHTF, cost is a major drawback with this approach. To satisfy the target SCVFA requirement, the estimated annual cost for acetic acid is in the range $1.0M (+/- $0.3M). Furthermore, dosing pure acetic acid may not result in favorable phosphorus accumulating organism (PAO) community dynamics for reliable EBPR. More specifically, a mixture of propionic acid alongside the acetic acid has been proposed to be important for stable PAO performance in mixed culture systems and helpful in generating out-selection pressure on undesirable glycogen accumulating organisms (GAOs) (*Lopez-Vazquez et al. 2009*).

The second option (Option #2) of recirculating a portion of APD effluent back through the GVTs to increase SCVFA in its effluent was evaluated by conducting special sampling of the stream. The APD receives thickened primary sludge from the GVTs and thickened waste activated sludge from dissolved air flotation thickeners. The APD operates at a 1 to 2 day solids retention time at 35° Celsius (95°F). Characterization of the APD effluent found that SCVFA averaged 10,564 mg-COD/L, with 25% in the form of acetic acid and 38% presenting as propionic acid. However, the APD effluent also contained 442 mg-P/L of soluble reactive phosphorus and 569 mg-N/L of total ammonia nitrogen. The recycled phosphorus load carried in the APD effluent would need 3,094 mg-COD/L SCVFA based on a SCVFA:P requirement of 7:1, resulting in a net gain of 7,470 mg-COD/L. This would require 90,000 to 130,000 gallons per day (0.33 to 0.49 MLD) of APD recirculated through the GVT to achieve the target SCVFA requirement assuming a 75% elutriation efficiency of releasing the SCVFA into the GVT effluent. The return ammonia load would be 609 lb/day (276 kg/day) and the return solids load would be 16 dry tons/day (14,500 kg/day), which is a 19% increase of the current solids loading to the GVTs.

This paper focuses on the evaluation of the remaining three options conducted through a series of analyses and bench-scale experiments. The goal is to determine whether a better possible solution than Options #1 and #2 exists for the RWHTF.

FACILITY BACKGROUND

In general, the RWHTF is divided into equally-sized treatment complexes referred to as North and South. The north secondary (NSEC) process consists of modified Ludzack-Ettinger (MLE) configured aeration basins with sidestream anaerobic return activated sludge (RAS) reactors to instigate EBPR as described by *Cavanaugh et al (2012)* and *Carson (2012)*. These anaerobic RAS side reactors receive the overflow effluent from the GVTs, which furnishes

SCVFA/RBCOD for the EBPR process in the north complex (See Figure 1). Thin sludge underflow is pumped from both the north and south primary clarifier complexes to the central GVTs for thickening to about 5% total solids (TS). The south secondary process consists of A2O configured aeration basins where the majority of the SCVFA/RBCOD enters the anaerobic zone with the primary effluent. Provisions allow some of the GVT effluent to be directed to the SSEC anaerobic zones as needed to increase EBPR performance reliability.

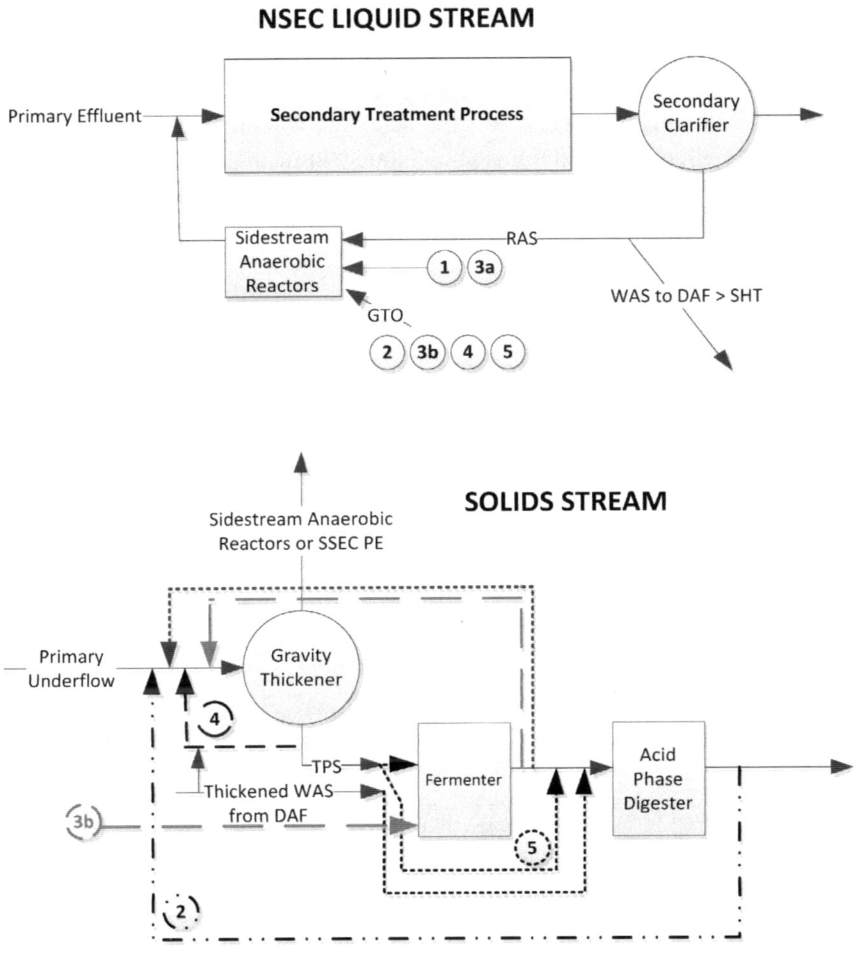

Figure 1 – Schematic of the NSEC (above) and of the relevant solids stream processes (below).

MATERIALS AND METHODS

Testing Options #3 - #5 required the development of bench-scale methodology for the fermentation of GVT sample streams both alone and with augmentation via external carbon sources. Initial method development consisted of approximating anaerobic fermentation tanks on a small scale in the laboratory, while maintaining consistent mixing and temperature throughout multi-day retention times. Details regarding the bench-scale fermentation experimental methods are described in the bench scale testing for Option #3. Deviations in methodology between individual bench tests are outlined in the section for each individual option.

Option #3 – *Use a locally generated recycle stream of comingled expired food grade beverages that is either (a) directly added to anaerobic zone, or (b) pre-fermented before adding to anaerobic zone.*

Option #3 evaluated a locally generated recycle stream for its potential to supply SCVFA/RBCOD. The stream is generated at a local recycling company where pallets of expired packaged beverage products are crushed for container recycling. The contents of the packaged goods consist primarily of beer and soft drinks with some bottled water and mouthwash. The investigation of this option began with a detailed analysis of the stream's constituents which are presented in detail in the Results and Discussion section of this paper. Analytical methods for mentioned analyses throughout this paper were performed via methodology found in Table 7 in the References section. The question of whether any of the carbon and/or ethanol in the beer waste could be further broken down in fermentation was posed, and the following experiment was performed for this task. For comparative purposes, a 2mL spike of 95% ethanol was tested against the locally generated recycle stream. The estimated initial concentration of ethanol in this sample was 1,722 mg/L or 3,600 mg-COD/L in COD equivalents (2.09 g-COD/g-ethanol).

Bench-scale Fermentation Tests

The bench testing approach described below was similar to that used for Options #4 and #5 which are discussed later. Set up of this experiment followed the same protocol as for Options #4 and #5 except that filtered secondary clarifier effluent and distilled water were used to bring all of the samples to the 900 mL test volume. The resulting sample dilutions for the bench test experiment are summarized in Table 1 below.

Table 1 – Dilution and makeup of fermentation test samples for the locally generated recycle stream

Constituent	Control	30% Recycle Stream	60% Recycle Stream	90% Recycle Stream	2.5% Ethanol
GVT Underflow, mL	80	80	80	80	80
Local Recycle Stream, mL	0	34	120	720	0
95% Ethanol, mL	0	0	0	0	2
Secondary Effluent, mL	10	10	10	10	10
Distilled Water, mL	810	776	690	90	808
Total, mL	900	900	900	900	900

This reactor makeup was based upon the desired ratios of sludge to recycle stream to be tested, while keeping the sludge volume constant throughout the reactors. Aliquots for total COD, TS, VS, and VFA (for SCVFA) were taken at each target retention time (24, 48 & 72 hours), while pH and temperature were monitored immediately via probe. Each sample reactor was run in duplicate, and all protocols (mentioned below) were followed for bottle preparation, nitrogen purging, and periodic mixing. The original sample vessels were prepared in duplicate sets and then thoroughly mixed via manual shaking where sample constituent volumes were measured into individual Teflon-lined lidded 1 liter amber glass jars. The headspace of each bottle was then purged with N_2 gas by attaching a glass burette to the N_2 tap in the lab, then submerging the burette to the bottom of the sample, allowing the nitrogen to bubble through the sample for 30 seconds. After bubbling the sample, the N_2 gas was then used to purge the headspace of the bottle by lifting the burette to the top of the sample liquid level, a clean gloved hand closed mostly over the top of the bottle, and the nitrogen gas purged out of a small opening of the gloved hand for 5 full seconds. The bottle was then quickly capped after nitrogen purging. Throughout the sampling, measuring, and gas purging processes, all equipment that came into contact with the samples consisted of acetone-rinsed glass to avoid organics contamination (in consideration of later VFA [SCVFA] analysis).

All prepared reactor bottles were then set up in a large shallow container of water with a depth reaching about 1/3 of the bottle. No temperature control was available due to the lack of a sufficiently sized water bath at the time of this experiment. The water bath was set up as a means to help normalize temperature between all vessels and buffer against any large temperature swings in the lab. Approximately every two hours (during work hours, 07:00 - 17:00), all reactor bottles were inverted 180 degrees three times for mixing, and had their cap slightly loosened (not removed) to relieve pressure from any gas production. The temperature of the water bath was also recorded at the time of each mixing/pressure-release period.

At the end of each prescribed retention time set-point (24, 48 or 72 hours after experiment start time), the contents of an entire bottle for each sample (in duplicate) were used for the creation of analysis saves (mentioned previously). The practice of using separate bottles for each sample time set-point period eliminated the risk of oxygen contamination during the extraction of sample volume for analysis aliquots. Temperature and pH data were taken immediately upon opening each reactor bottle by pouring off a small aliquot and analyzing via Thermo Orion 4 Star meter with a Ross Ultra pH/ATC Triode probe.

Option #4 - *Recirculate GVT underflow or stir the GVT sludge blanket to liberate more SCVFA/RBCOD in the GVT effluent, with or without the addition of settled mixed liquor to increase fermentation in the unit.*

Investigations into the technical merit of adding settled mixed liquor to the GVT to increase SCVFA/RBCOD were evaluated. Process modeling software (BioWin 4.1, EnviroSim Associates, LTD.) was used to predict sensitivity of SCVFA production as a function of increasing contributions of settled mixed liquor to theoretically inoculate the GVT with fermentation bacteria. The modeling exercise examined four volume ratio scenarios using typical constituent characteristics as summarized in Table 2. Results of process modeling suggested that not until a volume ratio of about 1:100 did adding settled mixed liquor to the GVT influent significantly increase SCVFA. At this 1:100 ratio, SCVFA increased by 33 percent compared to the control. These results are summarized in Table 2. Full-scale approaches under Option #4 would require elutriation or some means of liberating SCVFA and RBCOD from the GVT blanket into the effluent of the unit.

Table 2 - GVT influent volume ratios modeled for sensitivity analysis and process model results

Scenario	Settled Mixed Liquor:GVT Ratio	Effluent SCVFA mg-COD/g-VS (model result)	Percent Increase (model result)
1	0:1	61	-
2	1:10,000	62	2%
3	1:1,000	63	3%
4	**1:100**	**80**	**33%**

One concern with adding settled mixed liquor to the GVT influent was the potential for it to impact settling/thickening in the unit either in the form of sludge blanket expansion or blanket popping due to excessive biological activity. The concern was assessed through a bench-scale lab simulation performed by filling 4L beakers with GVT influent and dosing with settled mixed liquor at the aforementioned ratios. The beakers were mixed after dosing then left quiescent for observation of blanket settling and any sludge blanket activity/upset events. A camera was set up in a well-lit fume hood to record round-the-clock time-lapse photography of the reactors alongside a digital clock to track the time/date of any blanket popping activity. Time-lapse footage of the reactors was reviewed, and the experiment replicated, to determine blanket popping times of the various concentrations of settled mixed liquor injection. This proof-of-concept experiment displayed a change of blanket activity based on the concentration of settled mixed liquor in each reactor, and was sufficient preliminary testing to proceed with more detailed bench-scale experiments regarding this option. Results from this preliminary testing can be seen in in Table 3. Interestingly, the least amount of settled mixed liquor addition resulted in the most rapid blanket activity, whereas the higher doses of settled mixed liquor caused an increasing delay in blanket activity.

Table 3 – Visual observations from preliminary blanket activity testing with settled mixed liquor dosing to GVT influent

GVT influent : Settled Mixed Liquor	Blanket Rising Begins	Marked Curling of Blanket (lifting edges)	Total Pop (though blanket never reaches surface)
10,000:1	+32 hours	+36 hours	+36.5 hours
1,000:1	+35 hours	+37 hours	+38 hours
100:1	+37 hours	+38 hours	+39 hours

Bench-scale Fermentation Tests

In these tests, the method used for the fermentation bench testing of Option #3 was utilized to provide simulated anaerobic fermentation of the GVT underflow. To change the variable constituent across reactors (in this case, settled mixed liquor volume), the samples were brought to their final volume with filtered secondary clarifier effluent. For this experiment, the volumes of each of the sample constituents are summarized in Table 4.

Table 4 – Experimental setup after dilution ratios summarized in Table 2

Constituent	Scenario 1	Scenario 2	Scenario 3	Scenario 4
GVT Underflow, mL	891	891	891	891
Settled Mixed Liquor, mL	0	0.09	0.9	9
Secondary Effluent, mL	9	8.91	8.1	0

Experimental Method

The method used in this experiment followed the same protocol used in the fermentation testing experiment for Option #3 (previously outlined). The only differing factors in this round of testing were the reactor make-ups, the 'filler' fluid (no distilled water used in this testing), and the inclusion of full COD fractionation analysis (details below).

The GVT underflow, settled mixed liquor and secondary effluent were well mixed via manual shaking, and had small initial aliquots drawn from each vessel for analysis of initial pH, temperature, SCVFA,TCOD, soluble COD (SCOD), floc-filtered COD (FFCOD), and glass-fiber filtered COD (CODG). Temperature and pH were performed immediately on collection. VFA (SCVFA) aliquots were collected directly from prepped collection glass jars into the prepped glass vials and frozen immediately. TCOD was homogenized via laboratory-grade homogenizer for 30 seconds per standard lab protocol. SCOD, CODG and FFCOD had aliquots poured into centrifuge bottles and were centrifuged for 15 minutes at 4000rpm prior to filtration. SCOD was vacuum filtered through a 0.45um membrane filter, using a glass fiber pre-filter above the membrane for expediency. CODG was filtered through a large glass fiber filter (i.e., TSS filter) using vacuum filtration into a glass vessel. FFCOD was treated with 1mL of 50g/L aluminum sulfate solution per 100mL sample, allowed to mix rapidly (200rpm) on magnetic stir plate for 2 minutes, stirred slowly (10rpm) for 15 minutes, then allowed to settle without mixing for 15 minutes before the supernatant was drawn off the top via pipette and filtered through 0.45um membrane filter.

Option #5 – *Route GVT underflow sludge through a small side fermentation tank, then return the tank effluent back to the GVTs for solids separation and elutriation.*

Biowin process modeling was again conducted to develop an array of temperature and solids retention times given the practical size constraints of the existing sludge holding tank. The results also provided guidance for the bench-scale experimental campaigns. The concept with Option #5 also considers that fermentation of primary sludge would release less nitrogen and phosphorus than the APD effluent since there is little or no thickened waste activated sludge (with PAOs) in the primary sludge fermentation tank. Table 5 shows process model output results for estimated SCVFA as a function of temperature and solids retention times in the ranges possible with the existing tank available for fermentation. The model also predicted soluble ammonia and phosphorus concentrations in the range of 31 mg-N/L and 7 mg-P/L, respectively.

Table 5 – Model predicted SCVFA from fermentation of primary sludge (mg-COD/g-VS)

		Solids Retention Time		
		0.5 days	**1.0 days**	**1.5 days**
Temperature, □C	**20**	53	54	63
	25.5	55	56	69
	31	55	70	76

Experimental Method

This bench testing experiment focused on the concept of the standalone fermentation of GVT influent and GVT underflow without the addition of any external carbon sources or heterotroph inoculation.

The methodology for reactor setup, reactor treatment during the fermentation period, sample collection and quality control measures (i.e., duplicate reactors) was mirrored in this experiment as per the method outlined in the previous bench testing campaigns (Options #3 & #4). This testing only deviated from the previously mentioned testing methods in that the reactors contained pure GVT influent or underflow (no external carbon augmentation).

RESULTS AND DISCUSSION

Option #3 evaluated the use of a locally generated recycle stream produced primarily from the crushing and comingling of expired or off-specification beer and soft drink products. The local generator produces in the range of 5,000 to 8,000 gallons per day (19-30,000 L/day) of this stream. From Table 6 below it can be seen that the stream is very high in FFCOD which is typically indicative of a stream that has high potential for SCVFA generation. An analysis of slowly and non-biodegradable soluble COD still needs to be conducted to make this analysis more complete. The stream has a low solids content and does not carry significant nitrogen and phosphorus, making it attractive from an initial screening standpoint. Ethanol contributed 4,389 mg-COD/L or about 4% of the total FFCOD content.

Table 6 – Characterization of locally generated recycle stream

	TSS, mg/L	VSS/TSS Ratio	TCOD, mg/L	SCOD, mg/L	FFCOD, mg/L	Tot. Phos., mg-P/L	Ortho-phos., mg-P/L	TKN, mg-N/L	Ammonia, mg-N/L	Ethanol, mg-COD/L
Average	529	89%	112,422	111,675	116,056	88	76	401	7	4,389[1]
Median	441	91%	113,000	113,500	113,000	85	79	423	5	N/A
Std. Dev.	322	8%	12,951	17,693	16,790	33	24	138	6	N/A

[1] Value from analysis of a single sample (2.09g-COD/g-ethanol).

Figures 2 and 3 summarize the bench-scale test reactor results set up as described in Table 1. The GVT underflow sample showed that SCVFA concentration increased from an initial concentration of 146 mg-COD/L to 648 mg-COD/L over three days. The rest of the trend lines in Figure 2 show how much additional SCVFA was produced beyond what the GVT underflow sample produced. The sample that contained the most locally generated recycle stream (i.e., least diluted recycle stream sample) showed a trend which suggests inhibition or conversion. Over the three-day period, the SCVFA content of the sample with 90% locally generate recycle stream dropped from 1,324 mg-COD/L to 770 mg-COD/L. In this sample, the estimated ethanol concentration was around 4,000 mg-COD/L. The ethanol spike (3,600 mg-COD/L) also did not produce much SCVFA beyond what was produced by the GVT underflow control sample. For comparative purposes between the fermentation bench tests between Options #3 - #5, it must be noted that dilution of the GVT underflow portion of certain reactors (Option #3 in particular) cause different biological and chemical dynamics for fermentation, as well as changes from the standpoint of pure concentration dilution. Tables outlining reactor contents for each option's bench testing should be reviewed for applying context to testing results.

At the higher dilutions (lower percentages) of the locally generated recycle stream there was a substantial increase in SCVFA production. After three days, the sample which contained 60% of produced 1,324 mg-COD/L SCVFA over the control. From Table 2, this sample received 120mL of the stream, so roughly 11 mg-COD SCVFA was produced per mL of the stream. Conversely with the 30% stream, after three days, 1,340 mg/l of SCVFA was produced from a 34 mL sample. Therefore approximately 35 mg-COD SCVFA was produced per mL of stream sample. This three-fold increase between the 30% and 60% waste stream fermentation reactors, from a production per mL standpoint, is significant.

Figure 3 shows the distribution of volatile acids by species in the testing conducted under Option #4. After three days fermentation, the 30% and 60% dilutions produced propionic to acetic acid ratios that were in the range of two-thirds propionic. Conversely the same ratio with the GVT control was roughly at around 50%. This propionic acid dominance within the SCVFA speciation is seen as highly favorable for EBPR performance and PAO selection, and contributes strongly to the merit of utilizing this locally generated recycle stream for EBPR SCVFA augmentation.

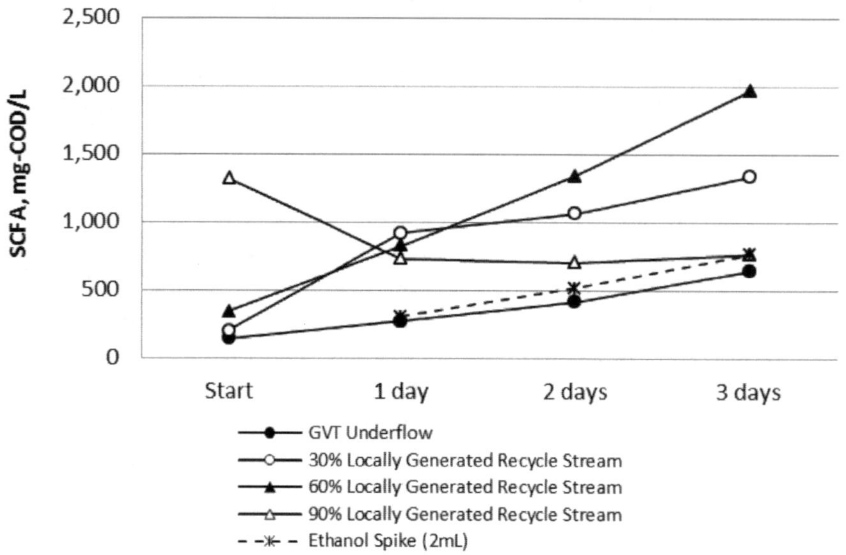

Figure 2 – Trends of SCVFA production as a function of fermentation time for the GVT underflow, 30%-60%-90% locally generated recycle stream and ethanol spike

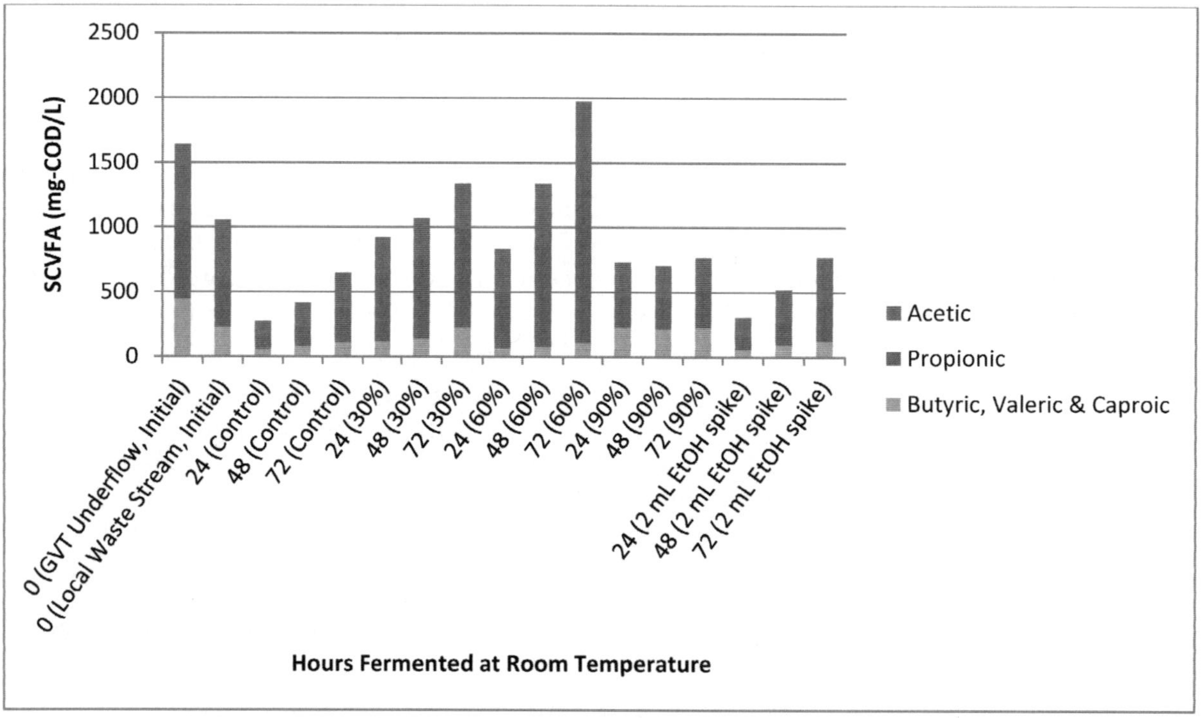

Figure 3 – Fractionated SCVFA production as a function of fermentation time for the GVT underflow, 30%-60%-90% locally generated recycle stream and ethanol spike

The supplier of the recycle stream typically produces between 5,000 and 8,000 gallons per day of the liquid waste stream. Based on the fermentation results summarized on Figure 4, it appears that higher dilutions (less percentage by volume) of the stream will produce more SCVFA, perhaps due to the dilution of suspected inhibitory substances. From Figure 2 and from Figure 4 it can be seen that at 30% of locally generated recycle stream, most of the SCVFA was produced inside of 24 hours. Under the conditions tested, this dilution scaled up would produce about 1,300 lb/day (590 kg/day) of SCVFA. Further work will focus on fermenting higher dilutions (less of the recycle stream) and even direct addition to the anaerobic reactor, to hopefully triple the SCVFA production observed in this initial trial. Also, additional work will investigate how much of the FFCOD as summarized in Table 6 can be practically converted to SCVFA in a reasonable time frame.

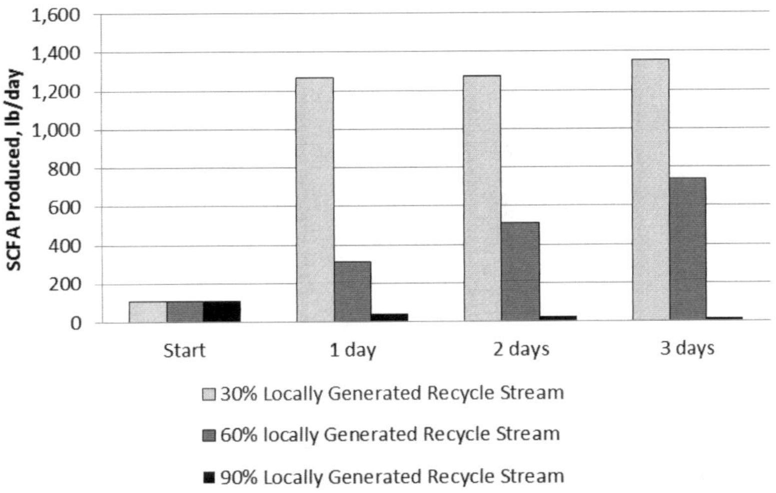

Figure 4 – Daily production of SCVFA from the locally generated recycle stream for the different dilutions over time. The higher dilutions (lower percentages, i.e., 30% locally generated recycle stream) produced more SCVFA while the lower dilutions (higher percentages) showed signs that suggest inhibition. The figure is based on the fermentation performance summarized in Figure 2 and an assumed 8,000 gallons per day (approx. 30,283 L/day) delivery.

In Option #4, the focus is to produce more SCVFA in-situ with GVT by adding settled mixed liquor with the GVT influent. With the experimental method, GVT underflow was mixed with the dilutions of settled mixed liquor as summarized in Table 2. From Figure 5 it can be seen that over a 24 hour period, the settled mixed liquor at the 1:1,000 and 1:100 dilutions did in fact increase the SCVFA produced from 3,716 mg-COD/L to 4,168 mg-COD/L or about 12%. Assuming a 75% elutriation efficiency of the SCVFA into the GVT effluent, 258,000 gallons of recycle is needed to meet the 3 tpd shortfall with GVT underflow alone. Settled mixed liquor addition to the GVT would reduce this recycle amount to 230,000 gallons. The return solids load would be reduced from 53.8 dry tons/day (dton/d) to 48.0 dton/d (48,800 kg/day to 43,505 kg/day) assuming 5% total solids of the GVT underflow. This is a significant increase of the hydraulic and solids loading that would be associated with Option #2. The 1:10,000 dilution did not result in any difference compared to GVT underflow by itself. Further study will focus on mixing the settled mixed liquor with the GVT influent rather than the underflow to assess the interactions of the mixed liquor with the soluble and colloidal fraction of organic matter that is in the water column of the GVT.

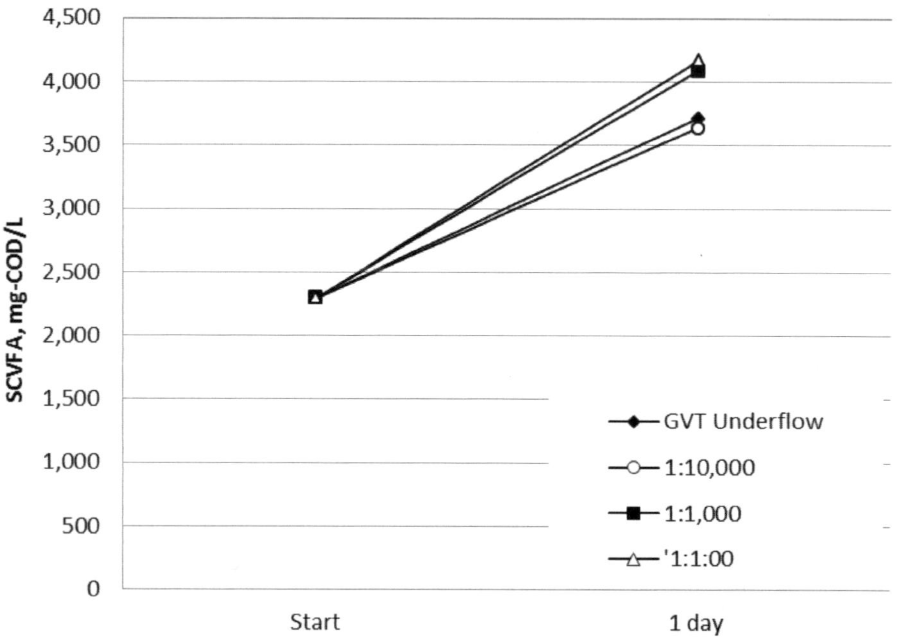

Figure 5 – SCVFA produced over 24 hours from fermentation of GVT underflow alone, and with the addition of settled mixed liquor at 1:10,000, 1:1,000, and 1:100 dilutions. The higher amounts of settled mixed liquor at 1:1,000 and 1:100 exhibit a response in increasing SCVFA production

Figure 6 displays the SCVFA speciation found within each settled mixed liquor addition ratio expressed over the 1, 2, and 3 day retention time period. The prominent finding is that the ratio of acetic acid to propionic acid remains generally equal throughout the different reactor conditions, generating no more or less proportions of acetic or propionic than what is initially found in the GVT underflow stream (compare experimental reactors with 'GVT underflow, initial' data within Figure 6). Some variation between acetic and propionic occurs, but there is no significant shift in relative speciation, which might be worth consideration regarding the influence of propionic:acetic ratios in selecting PAOs over GAOs for optimal EBPR performance.

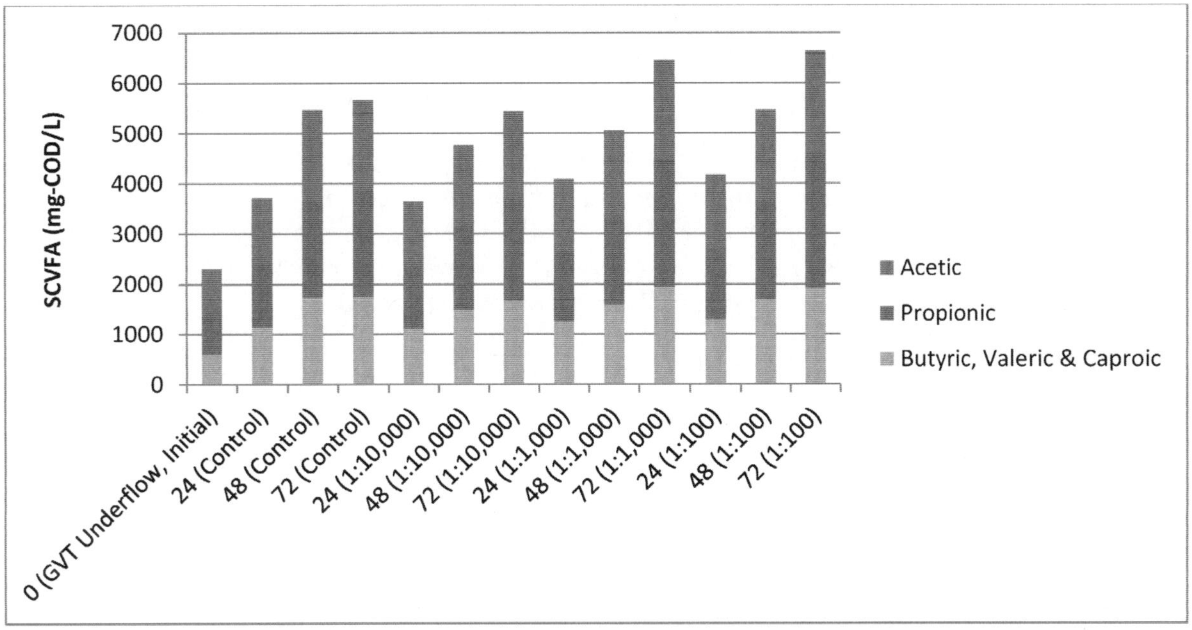

Figure 6 – Fractionated SCVFA produced over 24 hours from fermentation of GVT underflow alone, and with the addition of settled mixed liquor at 1:10,000, 1:1,000, and 1:100 dilutions

Option #5 evaluated SCVFA generation from off-line fermentation of GVT underflow sludge. The results of the bench-scale testing are summarized in Figure 7. Two tests showed that after three days retention, the SCVFA produced per unit volatile solids (VS) fed to the reactor reached 120 mg-SCVFA/gr-VS. This is in-line with 100 to 200 mg-SCVFA/g-VS range typically cited in design guidance resources (*Metcalf and Eddy, 2014*).

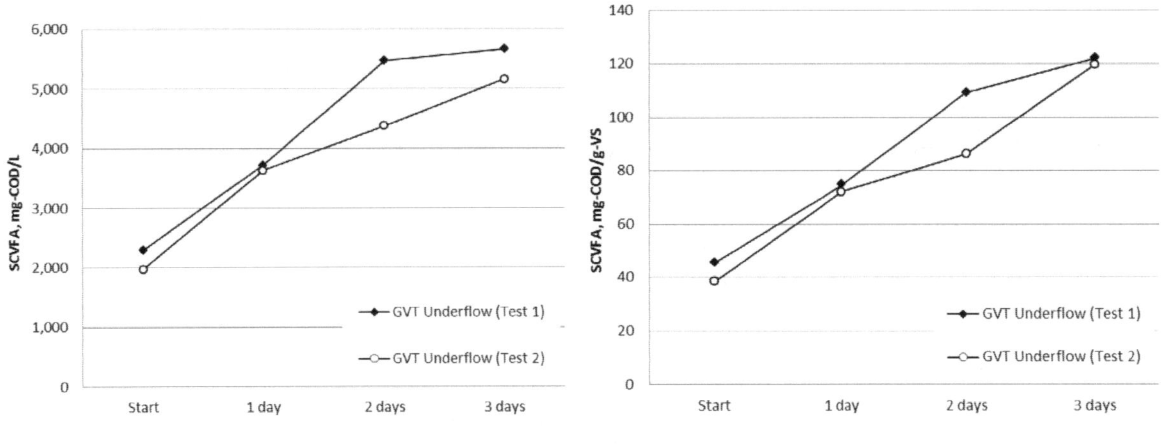

Figure 7 – Total concentration of SCVFA produced (left) and per gram Volatile Solids (right) from fermentation of GVT underflow at room temperature over a 3-day period

In Figure 8, the SCVFA speciation shows interesting results of fermenting GVT influent versus GVT underflow. Fermenting the GVT influent (Figure 8, left) produces an overwhelmingly dominant speciation of acetic acid over propionic and other species. GVT underflow, however, has more of a balanced acetic:propionic ratio. It is also interesting to note that the levels of total SCVFA produced from the GVT influent fermentation at 24 hours closely aligns with full scale characterization of GVT overflow effluent, which normally produces approximately 175 mg-COD/L of SCVFA at similar retention times. This implies that any net gain of SCVFA generated by fermentation of GVT influent may not be realized as this substrate is likely to be in-situ fermented within the anaerobic zone already. Very little is to be gained through fermenting the GVT influent in a separate fermenter, as the regular GVT process provides some portion of this action already.

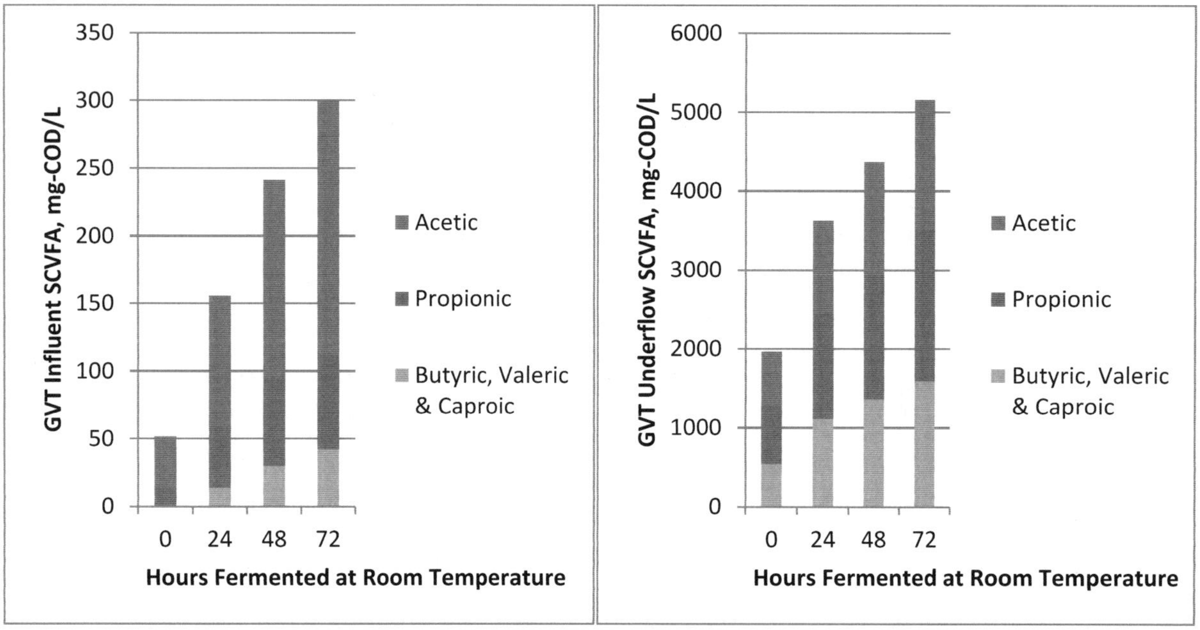

Figure 8 – Fractionated SCVFA (mg-COD/L) produced from GVT Influent (left) and GVT Underflow (right) through fermentation at room temperature over a 3 day period

Figure 9 brings the bench-scale results to full-scale. If in full-scale the fermenter could match the performance of the bench tests, then approximately 15 to 22 tons per day (14 to 20 metric tons per day) would need to be fed to the fermenter for three-day retention period. There is the possibility of heating the contents of the fermenter which could reduce the amount of sludge that needs to be conveyed to the fermenter. However, recirculation of the fermented sludge back to the GVT for liquid-solids separation adds operation complexity.

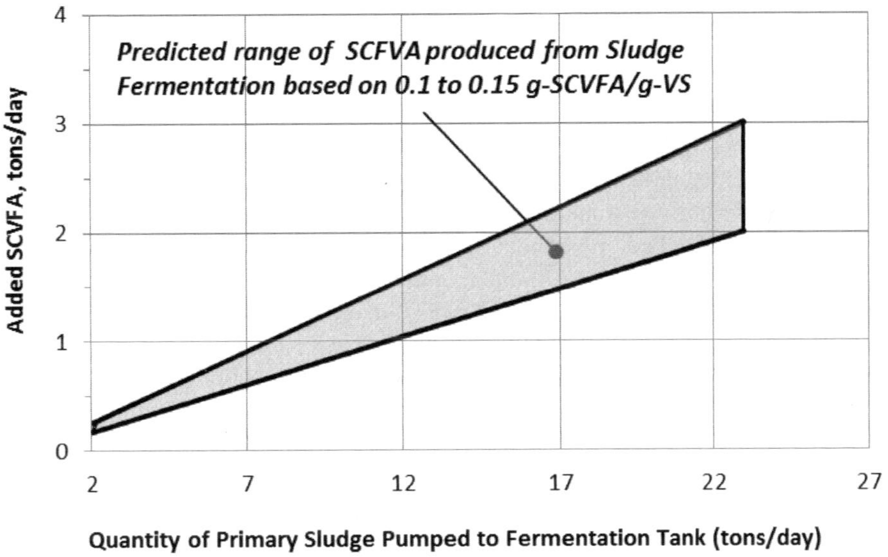

Figure 9 – Predicted range of primary sludge through the GVT that would be diverted to a fermentation holding tank to produce the targeted amount of SCVFA (2 to 3 tons per day)

CONCLUSIONS

In reviewing the process modeling information, laboratory bench scale data, facility cost analyses and waste stream characterizations, the high level take-away information from this initiative at its present state can be summarized as follows.

Option #1 (acetic acid dosing) provides a reliable, high strength carbon source, but comes with a financially burdensome price tag and omits the inclusion of propionic acid (potentially a problem with PAO/GAO selection). The cost of this option in light of other promising, cheaper, and more sustainable options, is the primary drawback of this form of carbon addition. This option will not be considered as a primary solution, but may be used as an emergency backup carbon source.

Option #2 (APD effluent dosing), despite being the highest strength SCVFA source currently available as an in-situ process stream, was deemed unsuitable for the Metro District due to high nitrogen and phosphorus recycle loads. This option is currently no longer under consideration.

Option #3 (fermentation and/or direct dosing of the locally generated recycle stream) showed promising results. During the bench study, the lowest volume ratio (30% recycle stream by volume) produced 1,269 lb/d (576 kg/d) of SCVFA after just 24 hours of fermentation with GVT underflow. Continued investigations aim to double this production rate. Based on fermentation testing results, there appears to be inhibition at higher strength concentrations of the recycle stream, whereas this inhibition didn't appear to have as much of an effect at lower strength concentrations. Additional study regarding proper dosing and possible inhibition sources will prove beneficial. It also must be considered that a consistent and reliable source of this waste stream must be secured, and backup primary sludge fermentation may need to be considered in the event of delivery failure/inconsistencies of the locally generated recycle stream. Fermentation of this stream is likely to be operationally simpler and less time consuming for maintenance than with fermentation of primary sludge. Initial tests with direct dosing (incomplete at the time of this paper) showed encouraging results, but must be repeated for a solid data foundation for this approach. The most noteworthy benefit to this option is that using a waste stream as an unmodified carbon source eliminates high costs associated with purchased carbon sources. Potential drawbacks include the locally generated recycle stream being not quite as strong in propionic acid when used unmodified, as well as the inherent risk of securing a second-party waste stream. This option will be further investigated.

Option #4 (settled mixed liquor injection for GVT underflow fermentation) showed moderate increases in both concentration (mg/L) and by-weight VS rate (mg/gVS) of SCVFA. This option would require minimal equipment upgrades, but may present operational challenges of maintaining GVT blanket density and could cause popped blankets and loss of thickening within the GVTs. Due to the risks of irregular blanket behavior and reduced thickening operations, the concept of settled mixed liquor integration into the GVTs will not likely be used full scale. Furthermore, the concept of elutriating GVT underflow to mix and release SCVFA may not be feasible due to the necessary high rate of recycle within the unit to provide necessary SCVFA levels in the GVT effluent.

Option #5 (GVT underflow and/or influent fermentation) is a promising path towards modifying current equipment to boost SCVFA production using existing process streams. Drawbacks to this option include equipment upgrade costs, operational changes, maintenance impacts, and depriving the anaerobic digesters of a portion of the thickened sludge stream (thereby reducing methane production). This option will be further investigated, perhaps as a backup fermentation stream if external streams are unavailable.

REFERENCES

Carlsson H., Aspegren H. and Hilmer A. (1996) Interactions Between Wastewater Quality and Phosphorus Release in the Anaerobic Reactor of an EBPR Process. Water Res. 1996; 30(6):1517-1527.

Carson K. 2012 Evaluation of Performance for a Novel Sidestream Enhanced Biological Phosphorus Removal Configuration at a Full-scale Wastewater Treatment Plant. Master of Science Thesis, University of Colorado, Boulder, Colorado.

Cavanaugh L., Carson K., Lynch C., Philips H., Barnard J. and McQuarrie J. (2012) A Small Footprint Approach for Enhanced Biological Phosphorus Removal: Results from a 106 mgd Full-scale Demonstration. Proceedings of the 85th Annual Water Environment Federation Technical Exhibition and Conference [CD-ROM], New Orleans, Louisiana, Sep 29 – Oct 3; Water Environment Federation: Alexandria, Virginia.

Lopez-Vazquez CM, Oehman A, Hooijmans CM, Brdjanovic D, Gijzen HJ, Yuan Z, van Loosdrecht MC. (2009) Modeling the PAO-GAO Competition: Effects of Carbon Source, pH and Temperature. Water Res. 2009 Feb; 43(2): 450-62. (Epub 2008 Nov 1).

Metcalf and Eddy. *Wastewater Engineering Treatment and Resource Recovery.* McGraw Hill Education, 2014. Print.

Oehmen A, Lemos PC, Carvalho G, Yuan Z, Keller J, Blackall LL, Reis MAM. Advances in enhanced biological phosphorus removal: from micro to macro scale. Water Res. 2007; 41(11): 2271-2300.

Rossle WH, Pretorius WA. (2001) A Review of Characterisation Requirements for In-line Prefermenters, Paper 1: Wastewater Characterization. Water SA Vol. 27 No. 3 July 2001, ISSN 0378-4738.

Table 7 – Reference for Analytical Methods

Analyte	Analyte Shorthand	Analysis Methods
Ammonia	NH3-N	Determination of Ammonia Nitrogen by Semi-Automated Colorimetry, U.S. EPA, Method 350.1 Revision 2.0, 1993.
		Lachat QuikChem Method 10-107-06-1-K for Ammonia, Lachat Instruments, Loveland, CO. 15 March 2001.
		Lachat QuikChem 8500 Operating Manual, Lachat Instruments, Loveland, CO.
Total Chemical Oxygen Demand (Total, Soluble, Floc-Filtered, Glass-Fiber Filtered)	COD (CODT, CODS, FFCOD, CODG)	Chemical Oxygen Demand, Method 8000, Hach Handbook of Water Analysis, 1979, Hach Chemical Company, P.O. Box 389, Loveland, CO 80537 per Federal Register, March 12, 2007, Table 1B, footnote 14.
Ethanol	EtOH	Method 8015C, Nonhalogenated Organic using GC/FID, Direct Aqueous Injection
Orthophosphate	OP	Phosphorous, All Forms (Colorimetric, Automated, Ascorbic Acid), Method 365.1, Revision 2.0, August 1993.
		Lachat QuikChem 8500 Method 10-115-01-1-A for Orthophosphate in Waters Lachat Instruments, Loveland, CO. 2007.
pH	pH	Standard Methods for the Examination of Water and Wastewater 21st Edition, Method 4500 H+ B, 2000, modified.
Temperature	TEMP	Standard Methods 20th Edition, Method 2550-B
Total Kjeldahl Nitrogen	TKN	ASTM D 3590-02 (06), Test Method B – Semiautomated Colorimetric Berthelot.

		Lachat Quikchem 8500 Method 10-107-06-1-K for Ammonia, Lachat Instruments, Loveland, CO.
		Operation Manual 1000 96301 Rev. 1.0 for the FOSS TECATOR block digesters BD6, BD12, BD20, BD40.
Total Phosphorus	Total P, TP	Standard Methods 4500-P – H-1999, 21st Edition, Manual Digestion and Flow Injection Analysis for Total Phosphorus
		Lachat QuikChem 8500 Method 10-115-01-1-F Determination of Total Phosphorus by Flow Injection Analysis Colorimetry (Acid Persulfate Digestion Method) Lachat Instruments, Loveland, CO. 2007.
		Lachat QuickChem 8500 Operating Manual, Lachat Instruments, Loveland, CO.
Total Solids	TS	Standard Methods for the Examination of Water and Wastewater18th Edition, Method 2540 G, 1991, modified.
Total Suspended Solids	TSS	Fishman, M.J., et al. "Methods for Analysis of Inorganic Substances in Water and Fluvial Sediments," U.S. Department of the Interior, Techniques of Water- Resource Investigations of the U.S. Geological Survey, Denver, CO, Revised 1989, I-3765-85 for total suspended solids, modified.
Total Volatile Solids	VSS	Fishman, M.J., et al. "Methods for Analysis of Inorganic Substances in Water and Fluvial Sediments," U.S. Department of the Interior, Techniques of Water-Resource Investigations of the U.S. Geological Survey, Denver, CO, Revised 1989, I-3753-85 for volatile suspended solids, modified.
Volatile Fatty Acids	VFA, SCVFA	Supelco Application Note 855B
Volatile Solids	VS	Standard Methods for the Examination of Water and Wastewater18th Edition, Method 2540 G, 1991, modified.

Aeration control strategies for nitrogen removal: A pilot and model based investigation

Pusker Regmi[1*], Mike Sadowski[2], Jose Jimenez[1], Bernhard Wett[5], Sudhir Murthy[6], and Charles B. Bott[7]
[1]Brown and Caldwell, 1600 Duke Street, Suite 310, Alexandria, VA 22314
[2]Civil and Environment Engineering Department, Virginia Tech, Blacksburg, VA 24060
[3]ARA Consult, Unterbergerstr.1, 6020 Innsbruck, Austria
[4]DC Water Authority, 5000 Overlook Ave. SW, Washington DC, 20032
[5]Hampton Roads Sanitation District, 1436 Air Rail Ave., Virginia Beach, VA 23455
*Corresponding email: pregm001@gmail.com

ABSTRACT

In an effort to reduce the costs associated with biological nitrogen removal, there has been a huge interest to evaluate the feasibility of innovative aeration control strategies and short-cut nitrogen removal. The use of intermittent aeration control strategy known as ammonia versus NOx (AVN) has shown to promote a very efficient N removal pathway known as nitrite-shunt (Regmi et al., 2014). In a nitrite-shunt process, ammonia is only oxidized to nitrite which is then denitrified to nitrogen gas. This process avoids the oxidation of nitrite and reduces the theoretical oxygen demand and supplemental carbon for N removal.
In this two-pronged study, we will focus on aeration control strategies currently available to optimize N removal performance of an existing WRRF. The main aeration strategies that will be considered are 1) DO setpoint control or conventional aeration control (CAC); 2) Ammonia-based aeration control (ABAC); and 3) AVN control. Process model simulation and pilot experiments will be performed to compare these strategies in terms of effluent N concentrations, aeration demand, alkalinity demand, nitrite-shunt and so on. The results of this study will provide insights on the advantages and limitations of each the aeration control strategy. The results will also aide in the selection of appropriate aeration control strategy based on the specific considerations of WRRFs.

Keywords: Nitrite-shunt, nitrogen removal, nitrification, denitrification, aeration control, AVN

INTRODUCTION

Effluent total nitrogen (N) discharge standards are becoming increasingly stringent and widespread. Therefore, there is a need for efficient and inexpensive N removal technologies. Many biological nitrogen removal (BNR) plants have strict regulations to meet where improved efficiency at the cost of process stability is not justified. Historically, BNR processes are prone to internal and external disturbances, and a limited understanding of the microbial protagonists performing the desired biochemical reactions has impeded selection of specialist microorganisms. In recent years, BNR optimization has been greatly aided by the new knowledge from molecular science and advancements in instrumentation, control and automation (ICA). However, there is a huge void in holistic integration of these discrete optimizations to develop highly efficient BNR systems. Hence, the sustainability of BNR systems in the future relies greatly on the integration of new microbial knowledge and robust process control strategies.

Although ICA was considered in wastewater treatment from the early 1970s (Olsson et al., 2012), ubiquitous implementation of ICA has been late in coming to wastewater treatment plants. The initial implementation of ICA involved aeration control using a DO sensor and programmable logic controller (PLC), resulting in improved biological processes and the efficiency of the overall system (Briggs et al., 1973). Therefore, when online ammonia sensors became more reliable and affordable, control of aeration based ammonia set-points related to the level of ammonia removal required, burgeoned as an effective alternative to meet increasingly stringent ammonia and N limits (Åmand et al., 2013).
Recently, the ammonia-based aeration control using *in-situ* analyzers and controls has gained popularity for significant aeration energy savings and performance improvements of activated sludge processes (Rieger et al., 2014). The ammonia-based aeration control has demonstrated that aerating to nitrify just enough ammonia to meet regulations not only saves aeration energy, but also increases capacity for denitrification and provides alkalinity and supplemental carbon savings (Åmand et al., 2013). Moreover, there has been renewed interest in intermittent aeration control from the perspective of nitrite oxidizing bacteria (NOB) out-selection that allows for the efficient short-cut nitrogen removal *via* nitrite, also referred to as "nitrititation-denitritation (Regmi et al., 2014). The benefits of nitritation-denitritation include decreased aeration demand (assuming carbon is diverted) for nitrification and decreased carbon demand for denitrification allowing efficient N removal.

In this paper we explored the efficacy, simplicity and savings related to a novel aeration control strategies for efficient N removal. We also summarized findings of a pilot study and full scale plant modeling with 1) conventional aeration control 2) ammonia based aeration control 3) a novel ammonia vs NOx control.

EXPERIMENTAL SETUP

The Pilot Plant: A pilot plant receives municipal wastewater from HRSD's Chesapeake Elizabeth treatment facility. The influent water is subject to grit removal, screening, and temperature control before primary treatment using a high rate activated sludge A-stage. Aeration is supplied with motorized valves and membrane disc diffusers.

The B-stage consists of four CSTRs in series followed by a secondary clarifier (Figure 1). The first CSTR was set anoxic to perform nitrogen removal in an MLE configuration. The system was maintained at a 5 hour HRT. Return activated sludge (RAS) from the clarifier and internal mixed liquor recycle (IMLR) from the last CSTR were fed using peristaltic pumps. For control of SRT, wasting occurred directly from the last reactor using a programmable peristaltic pump. Sodium Bicarbonate was fed in order to maintain pH and monitor alkalinity requirements.

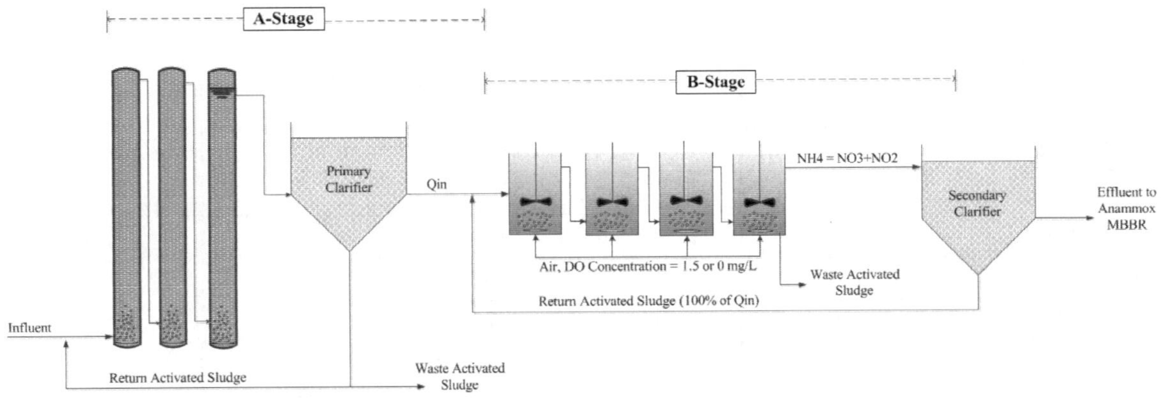

Figure 1 – Pilot plant with AVN Control Configuration

<u>Model Simulation:</u> A calibrated process simulation model for a 15-MGD full-scale WRRF was used to investigate the effect of different aeration control strategies on the performance of the facility. The WRRF consists of pretreatment, primary treatment using two circular primary settling tanks, activated sludge reactor using an anaerobic-anoxic oxic (A^2O) process followed by three secondary settling tanks and UV disinfection. Primary sludge and waste activated sludge are co-thickened using dissolved air floatation, anaerobic digested and dewatered using centrifuges.

Process simulator Biowin^{TM} (Envirosim, Ontario, Canada) was used to assess the N removal performance of the WRRF under different aeration control strategies. The BW controller (add on to the original simulator) was used to run operational scenarios especially for ABAC and AVN controller with intermittent aeration. Other process simulator will also be considered for further investigation based on the availability and applicability.

RESULTS AND DISCUSSION

Three aeration control strategies (CAC, ABAC, AVN) were simulated with same ammonia loading rate, influent COD/NH_4^+-N ratio and total SRT (Table 1). The TIN removal efficiency was progressively better with ABAC and AVN control compared to CAC (Table 1). In fact, AVN aeration control resulted in 20% more TIN removal efficiency compared to ABAC. The COD/TIN removal rate ratio gives the indication of efficiency of influent carbon utilization for N removal. AVN aeration control resulted in the lowest COD/TIN removal rate ratio among three aeration control strategies tested. Under, AVN control nitrification and denitrification are always balanced, which precluded the need for additional alkalinity. This can be considered a major advantage compared to CAC and ABAC (Table 1). AVN control required the least amount of airflow outperforming CAC and ABAC (Table 1). The NOB/AOB ratio was the lowest for AVN control, which indicates NOB out-selection and potential for nitrite-shunt (Table 1). Therefore, nitrite-shunt could be responsible for better N removal performance under AVN control. Figure 2 shows the N species profile with AVN control. The effluent ammonia and NOx were maintained equal in the effluent during the controller model simulation.

Table 1. Comparison of conventional, ammonia based and AVN aeration strategies

Results from Model simulation			
Parameter	*CAC*	*ABAC*	*AVN*
COD:NH3-N Ratio	8.5		
NH3 Loading (kgN/d)	1973		
Total SRT (days)	8		
Target DO (mg/L)	2	0 - 1.5	0 - 1.5
COD:TIN Removal Rate	25	18	8
NOB:AOB Ratio[1]	0.6	0.58	0.4
Effluent NH_3-N (mg/L)	0.2	1.5	2.1
Effluent NOx-N (mg/L)	12.5	8.5	2.1
TIN Removal (%)	57.6	66.7	86
Sodium Hydroxide (mg/L $CaCO_3$)	120	70	NA[2]
Average Airflow Rate (scfm)	16000	12500	10500

Results from Pilot study			
NH_4^+-N loading rate (mg N/L/d)	151.2 ± 5.7	142.9 ± 16.2	214.6 ± 35.7
TIN removal rate (mg N/L/d)	74.2 ± 24.9	103.7 ± 19.1	163.9 ± 24.3
TIN removal Efficiency (%)	49.0 ± 16.3	72.6 ± 10.5	76.9 ± 6.7
COD removal rate/TIN removal rate	17.7 ± 3.9	12.3 ± 1.6	10.6 ± 1.1
Bicarbonate Feed (mg $CaCO_3$/L)	122 ± 22	94 ± 26	NA[2]

NOB:AOB ratio of 0.6 indicates no NOB out-selection and hence, no nitrite-shunt. The lower the NOB: AOB ratio is the higher the degree of NOB out-selection and nitrite-shunt are

[2]*No alkalinity addition*

Similar results are being obtained in the pilot study. Each phase of aeration control comprised a minimum of three weeks of data collection. The aerobic fractions for CAC and ABAC were 0.75 due to MLE configuration (With 25% anoxic volume) whereas the aerobic fraction for AVN operation was 0.53 ± 0.07. Alkalinity addition was shown to decrease from DO Control to ABAC. TIN removal efficiency shows an increasing trend from DO control to ABAC to AVN (Table 1). The COD/TIN removal rate ratio shows a decreasing trend from DO Control to ABAC and AVN (Table 1). Lower values indicate a more efficient use of carbon from the aeration strategy.

This model and pilot-scale study objectively contrasted CAC, ABAC and AVN control in terms N removal performance. The advantages of AVN control were demonstrated clearly over CAC and ABAC. In the final paper, further investigation of SRT control and nitrite-shunt will be presented.

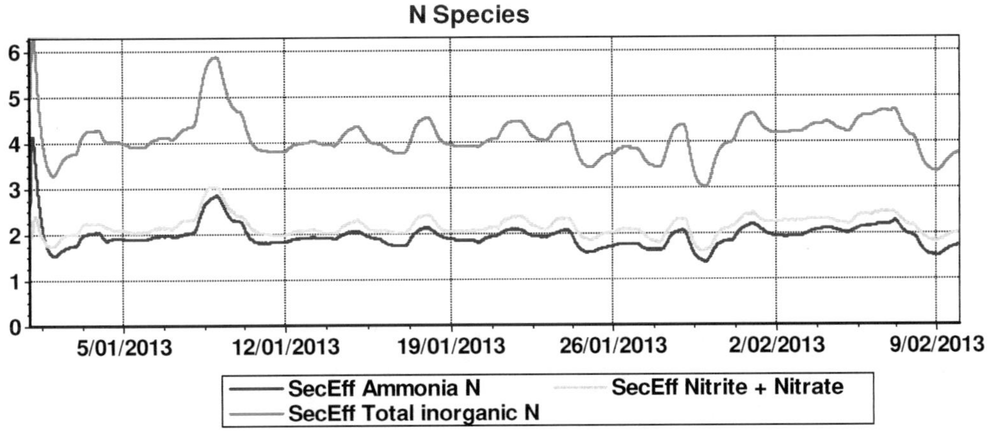

Figure 1. Effluent N species with AVN aeration control

CONCLUSION

The study demonstrates the advantages of advanced aeration control strategy specially the AVN aeration control. It also highlights the possibility of meeting nitrogen removal requirements with lower influent COD/TKN ratio which in turn facilitates upstream carbon capture for energy production.

REFERENCES

Åmand, L., Olsson, G., & Carlsson, B (2013). Aeration control – a review. Water Science & Technology, 67(11), 2374.

Briggs, R., 1973. Instrumentation and control in sewage treatment. In: ICA (1973).

Regmi, P., Miller, M., Holgate, B., Bunce, R., Park, H., Chandran, K., Wett, B., Murthy, S., Bott, C. (2014). Control of aeration, aerobic SRT and COD input for mainstream nitritation/denitritation. Water Research, 57, 162-71.

Rieger, L., Jones, R. M., Dold, P. L., & Bott, C. B (2014). Ammonia-based Feedforward and Feedback Aeration Control in Activated Sludge Processes. Water Environment Research.

Using Step-Feed to Improve Secondary Effluent Ammonia Control

Bruce R. Johnson[1]*, Jennifer Phillips[2], Greg Smith[3], John Sherlock[4]

[1]CH2M HILL, Englewood, Colorado
[2]CH2M HILL, Phoenix, Arizona
[3]CH2M HILL, Tucson, Arizona
[4]Pima County Regional Wastewater Reclamation Department
*Email: bruce.johnson2@ch2m.com

ABSTRACT

Pima County Regional Wastewater Reclamation Department (Tucson, AZ) was facing a mandate from the Arizona Department of Environmental Quality (ADEQ) to reduce the amount of total nitrogen and ammonia in system effluent and to pass the whole effluent toxicity test, as well as increase the treatment capacity to handle the County's growing population. To meet this challenge, the County developed and maintained a Regional Optimization Master Plan (ROMP) for wastewater treatment facilities required through the year 2030. The ROMP Final Report, published in November 2007, recommended the implementation of a number of capital improvements necessary to achieve regulatory compliance and meet the anticipated capacity requirements for the service area. One of these was the construction of a new facility, the Agua Nueva WRF, adjacent to an aging facility that would be decommissioned after the new plant was placed into operation.

The DBO delivery approach used by Pima County to deliver the project encouraged the development of low net-present value approaches, which included the minimization of both energy and chemical usage. Additionally the very stringent disinfection requirements (4 of 7 days non-detect for E. Coli), in combination with a concern over NDMA and THM formation, led to the selection of chloramination for effluent disinfection. In order to minimize operating costs, and minimize the need for supplemental ammonia in disinfection, an innovative secondary effluent ammonia control system was developed. This system uses a combination of the standard approach of adjusting operating dissolved oxygen levels in the aerobic zones with a system that adjusts the step-feed fractions to the various passes. The goal of the step-feed control system is to provide the last pass of the bioreactor with a constant mass of ammonia, i.e. base loading.

The use of primary effluent and bioreactor on-line ammonia monitoring to control the feed splits and DO setpoints for the Aqua Nueva WRF bioreactors was shown to enhance effluent ammonia control, reduce chemical usage and reduce power usage. It was found that a combination of factors in the raw sewage, which include a tripling of ammonia load from minimum to maximum on a diurnal basis, rapid increases in the ammonia load in the morning hours, and influent nitrification inhibition, exceeded the control authority of this control system at ANWRF.

However the ANWRF feed-forward step-feed system was able to significantly improve the control of effluent ammonia and meet the design intent of the DBO project. This shows that step-feeding a bioreactor provides an additional level of control authority above that possible with DO control alone.

KEYWORDS

Ammonia Control, Step-feed, Denitrification, Low Oxygen

INTRODUCTION

Pima County Regional Wastewater Reclamation Department (Tucson, AZ) was facing a mandate from the Arizona Department of Environmental Quality (ADEQ) to reduce the amount of total nitrogen and ammonia in system effluent and to pass the whole effluent toxicity test, as well as increase the treatment capacity to handle the County's growing population. To meet this challenge, the County developed and maintained a Regional Optimization Master Plan (ROMP) for wastewater treatment facilities required through the year 2030. The ROMP Final Report, published in November 2007, recommended the implementation of a number of capital improvements necessary to achieve regulatory compliance and meet the anticipated capacity requirements for the service area. One of these was the construction of a new facility adjacent to an aging facility that would be decommissioned after the new plant was placed into operation.

The County elected to use alternative delivery for the new Agua Nueva Water Reclamation Facility (ANWRF). CH2M HILL was chosen to provide this 120 MLD (32 MGD) facility based upon its design/build/operate (DBO) proposal in November of 2010. The facility saw first wastewater in December of 2013. Table 1 shows the effluent requirements for the facility. In particular, it should be noted the very stringent disinfection and disinfection byproduct requirements of the permit. In addition, while there was no current limit for NDMA (N-Nitrosodimethylamine), it was anticipated that it might be a future requirement.

Table 1: Aqua Nueva WRF Effluent Criteria

Performance Criteria	Maximum Month Design Limit
BOD_5	30 mg/L
TSS	30 mg/L
Total Nitrogen	8 mg N/L (5 mo. Geometric Mean)
NH_3	1.75 mg N/L
Total Phosphorus	1.0 mg P/L
E. Coli	4 of 7 days non-detect
E. Coli	Single Sample Max < 15 / 100 ml
Total THMs	80 ug/L
NDMA	monitor

Figure 1 shows an aerial view of the facility. The secondary treatment system is located in the center of the Figure 1, with the bioreactor being the large round covered tank. The chlorine contact basin is shown on the left of the image in another round tank, but without a cover.

Figure 1: Agua Nueva WRF Aerial View

ANWRF PROCESS DESIGN AND SELECTION

The raw sewage is screened by 10 mm (3/8 inch) bar racks prior to being pumped to a combined aerated grit/flocculation chamber. Effluent from the flocculation chamber goes to the primary dissolved air flotation (DAF) clarifiers, prior to entering the three pass step-feed 5-Stage Bardenpho bioreactor system, shown in Figure 2. The bioreactor is divided into four trains, one in each quadrant. Bioreactor effluent goes to secondary clarifiers and then to a disk filter system prior to disinfection with chloramines. The bioreactor step-feed and DO control system is used to keep a target ammonia concentration in the secondary effluent (nominally 1.5 mg/L) for chloramination.

Figure 2: Agua Nueva WRF Bioreactor Schematic

Waste solids from the end of the first pass of the bioreactor system are returned to the DAF clarifier feed (ahead of the flocculation tanks) for co-thickening with the primary solids. ANWRF pumps the produced biosolids from the DAF float to the nearby regional plant (Tres Rios WRF, formerly Ina Road WRF) for stabilization and dewatering. Figure 3 shows a process flow diagram of the facility.

The DBO delivery approach used by Pima County to deliver the project encouraged the development of low net-present value approaches, which included the minimization of both energy and chemical usage. Additionally the very stringent disinfection requirements (4 of 7 days non-detect for *E. Coli*), in combination with a concern over NDMA and THM formation, led to the selection of chloramination for effluent disinfection. In order to minimize operating costs, and minimize the need for supplemental ammonia in disinfection, an innovative secondary effluent ammonia control system was developed. This system uses a combination of the standard approach of adjusting operating dissolved oxygen levels in the aerobic zones with a strategy that adjusts the step-feed fractions to the various passes. The goal of the step feed control system is to provide the last pass of the bioreactor with a constant mass of ammonia, i.e. base loading.

Achieving controllable and measurable ammonia in the secondary effluent is a challenging design goal, especially in the warm water conditions normally present in Tucson, Arizona, where average summer water temperatures can exceed 30 ^{O}C. In a conventional bioreactor with this long length to width ratio, and at the current aerobic operating SRTs of 6 days, effluent ammonia would be expected to be less than 0.2 mg N/L on average. The design goal of the ammonia control system was to provide a stable secondary effluent ammonia concentration of less than 1.5 mg/L (to stay within the monthly limit of 1.75 mg/L).

Figure 3: Agua Nueva WRF Process Flow Diagram

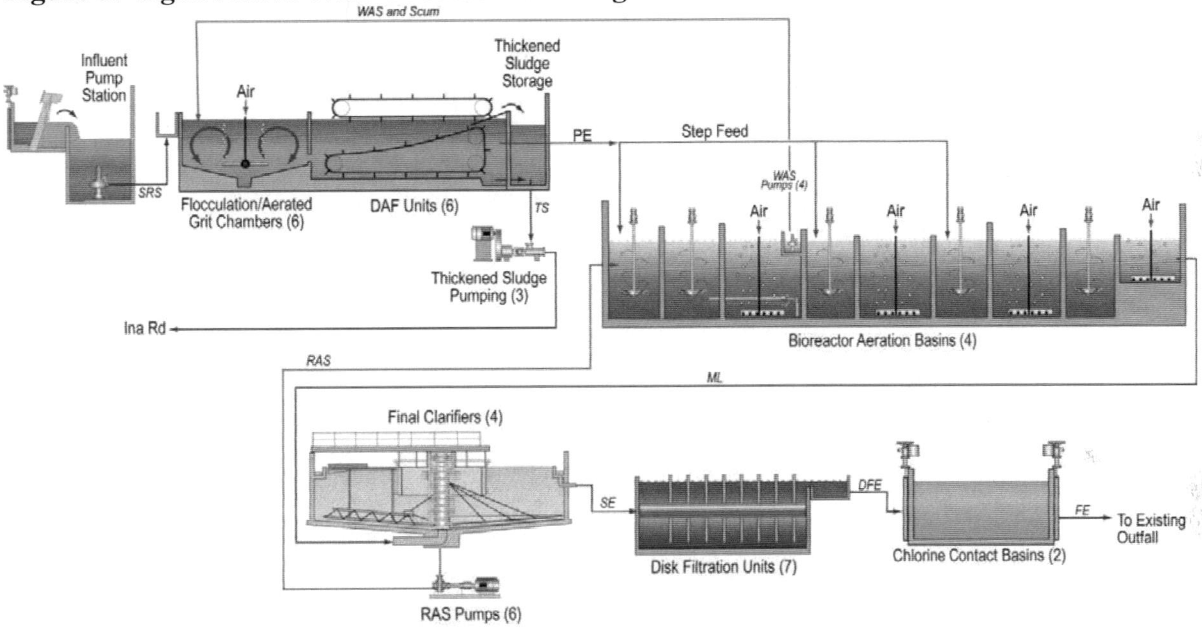

Figure 4 shows the 3-Pass Step-Feed bioreactor process flow diagram at ANWRF. The bioreactor is set up to be 60% aerobic and 40% unaerated (anoxic+anaerobic). Pass 1 has two separately controlled aerobic zones (1A and 1B). Pass 2 has three separately controlled aerobic zones (2A, 2B and 2C). Pass 3's first aerobic section also has three separately controlled aerobic zones (3A, 3B, and 3C), but also includes an aerobic section at the end of the bioreactor (Zone 4). Each aerobic zone described includes its own DO probe and separate fine bubble diffuser system. This adds up to a total of nine separately controlled aerobic zones in each of the four bioreactors.

Figure 4: Aeration Basin Process Configuration

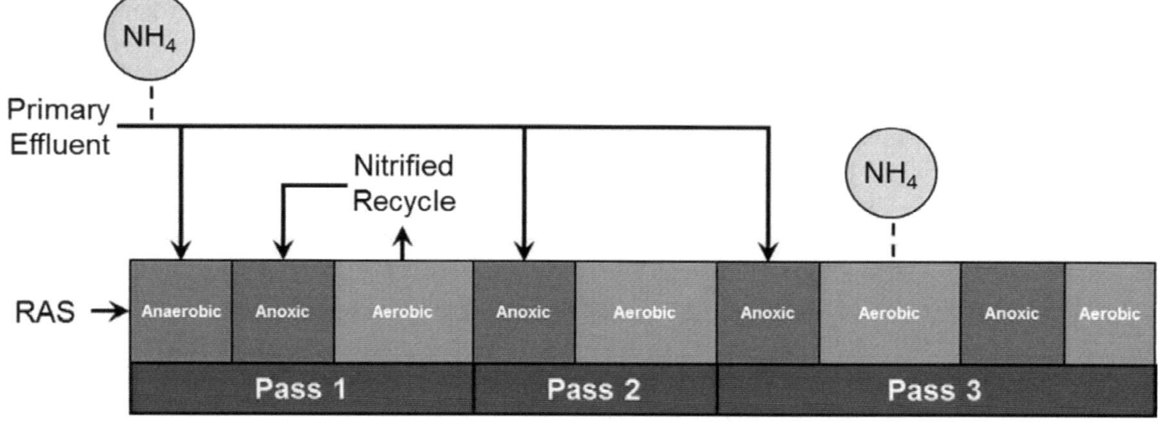

The ammonia control system relies on five in-situ ammonia probes, one at the end of each bioreactor's first Pass 3 aerobic zone (Zone 3c) and a single one in the primary effluent. The primary effluent ammonia probe is used to control the amount of flow sent to Pass 3. The step-feed control system is set up to "base load" Pass 3 at an operator set constant mass of ammonia (primary effluent ammonia concentration times the flow) by adjusting the amount of flow that enters that pass. The remaining flow is directed to the two upstream passes in proportion to the amount of biomass in each pass. The amount of biomass in Passes 1 and 2 are not directly measured, but calculated from the previous feed split. This results in Pass 1 always getting more flow than Pass 2 since it will always have a higher biomass concentration.

The concept behind adding this step-feed control for ammonia control is based on the fact that adjusting DO alone does not give sufficient control authority (ability to control the ammonia within the needed range) to control effluent ammonia to a specific target when considering normal influent nitrogen variability (Rieger, et. al 2012).

In addition, ANWRF has another on-line ammonia sensor located on the filter effluent that has two functions:

1. Controls the liquid sodium hypochlorite dosage to give a constant Chlorine/Ammonia ratio.
2. Monitors for adequate ammonia in the secondary effluent. If it falls below approximately 0.5 mg/L ammonia, supplemental ammonia is added to maintain sufficient disinfection capabilities.

THEORY VERSUS REALITY

The original design went through two phases. In the first phase, the primary effluent ammonia probe was not included in the design. However detailed control system modeling showed that relying on the bioreactor ammonia probes alone did not meet the design intent (Johnson, et. al. 2013). As a result of the control and process simulation, the design went forward with the primary effluent ammonia on-line monitoring for flow split control as described above. The ammonia probes in the bioreactor Zone 3c would then be used to control the DO setpoints as described below:

- Because the last aerobic zone of Passes 1 through 3 (Zones 1B, 2C, and 3C) all have similar oxygen loads (as a result of the step-feed controls), their DO setpoints were identical and set by the Zone 3C ammonia controller. In addition, Zone 4 was also set to operate at the same DO as 3C, with the thought that if that DO was high, Zone 4 also needed to be high to control the secondary effluent ammonia
- Aerobic Zones 2B and 3B DO setpoints were controlled from a PI loop looking at the air rates going to Zone 3c. The concept was that if 3C was needing more air than the design average, the upstream Zone 3C should be adding more oxygen to the system to reduce the load to 3C. This worked in the opposite direction as well, i.e. low air usage in 3C would drive down the DO setpoint in 3B.
- The first aerobic zones in each pass (1A, 2A and 3A) were not to be ammonia controlled, but set to a constant DO (nominally 2.0 to start to minimize high SVI problems). The design concept here is that these zones would normally be going at a very high metabolic rate no matter what, and therefore benefitted very little from automated variable DO control.

In simulations with an estimated influent diurnal profile, this system worked very well. Figure 5 shows the provided influent flow profile. It was necessary to assume that ammonia concentrations were constant during the day since no actual diurnal influent nitrogen data was available at the time of the design.

Figure 5: Design Influent Diurnal Flow Profile

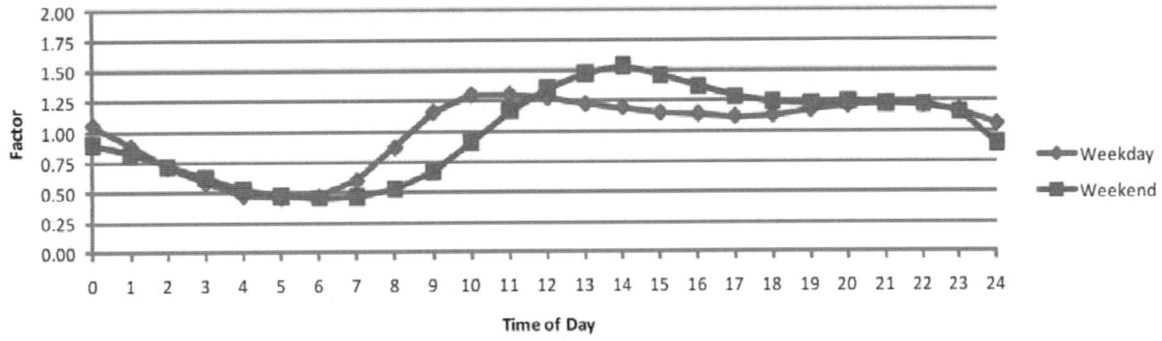

Upon startup of ANWRF, it was found that the diurnal influent flow and ammonia load profile was more variable than expected. Figure 6 shows the average primary effluent flow and load profiles for 2015.

Figure 6: Average Primary Effluent Flow and Ammonia Load Diurnal Profiles

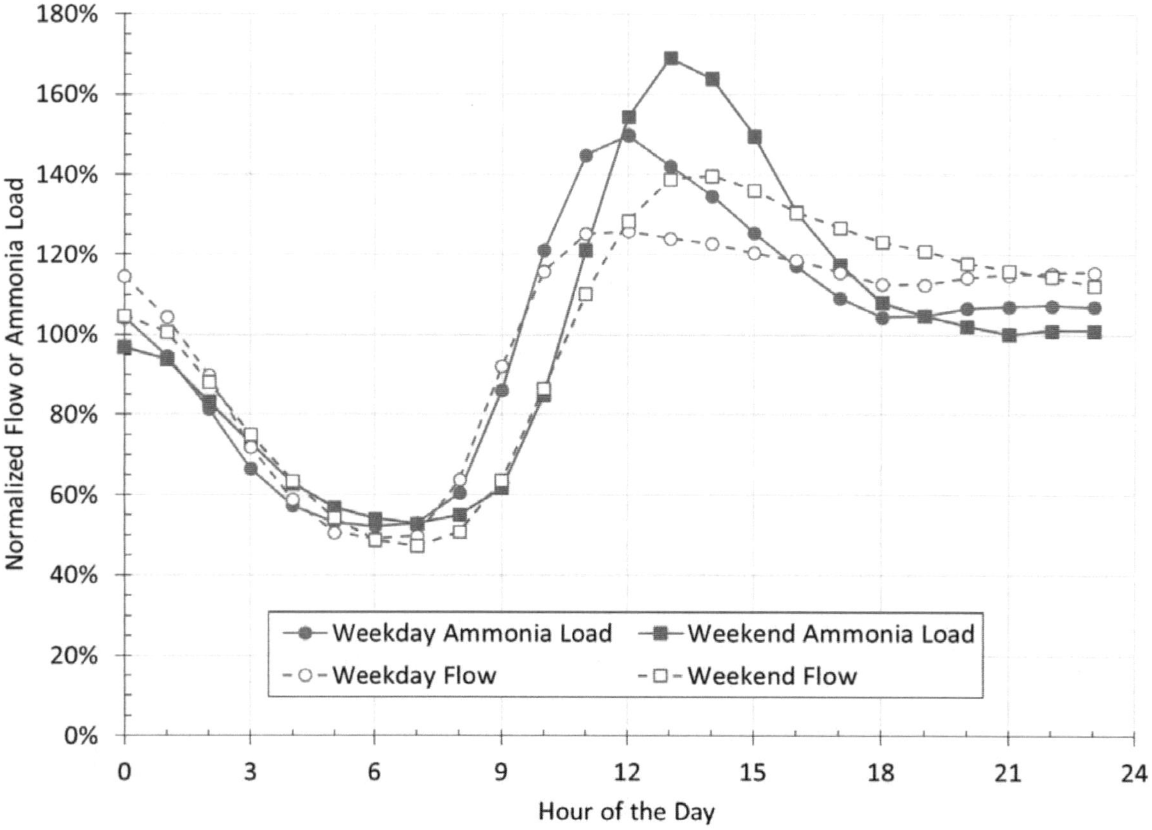

It can be seen that the peak ammonia loads to the facility are actually higher than what was originally estimated. On weekdays, the peak load is actually 150% of the average day value versus a design of 130% and on weekends it is again 20% higher than design, 170% versus 150%. Of note both of the ammonia load peaks are significantly higher than the respective flow peaks (25% higher on weekdays and 30% higher on weekends. Additionally, the profiles shown in Figure 6 have already been attenuated by passing through primary treatment, thus the influent ammonia load profile is actually higher than what is shown here.

The rate of ammonia load increase is also of interest since during weekdays it goes from the minimum of 50% at 7 am to the peak load 5 hours later at 150% (tripling the load). Such a rapid increase is difficult for feedback systems to control.

Additionally, it has been found that the influent to ANWRF is inhibitory to nitrification, to the extent that nitrification is almost fully inhibited in a WERF high F/M test (WERF 2003) during the first two days of the test using primary effluent (Wadhawan 2014) as shown in Figure 7. This inhibition results in an effective maximum nitrification rate of 0.69 1/d (versus 0.83 1/d in the test with BOD water) or a 17% reduction in nitrification rate. As compared to the WERF recommended nitrification rate of 0.9 1/d, this is a 23% reduction in nitrification rates as compared to design.

Figure 7: ANWRF High F/M Nitrification Test Results (Wadhawan, 2014)

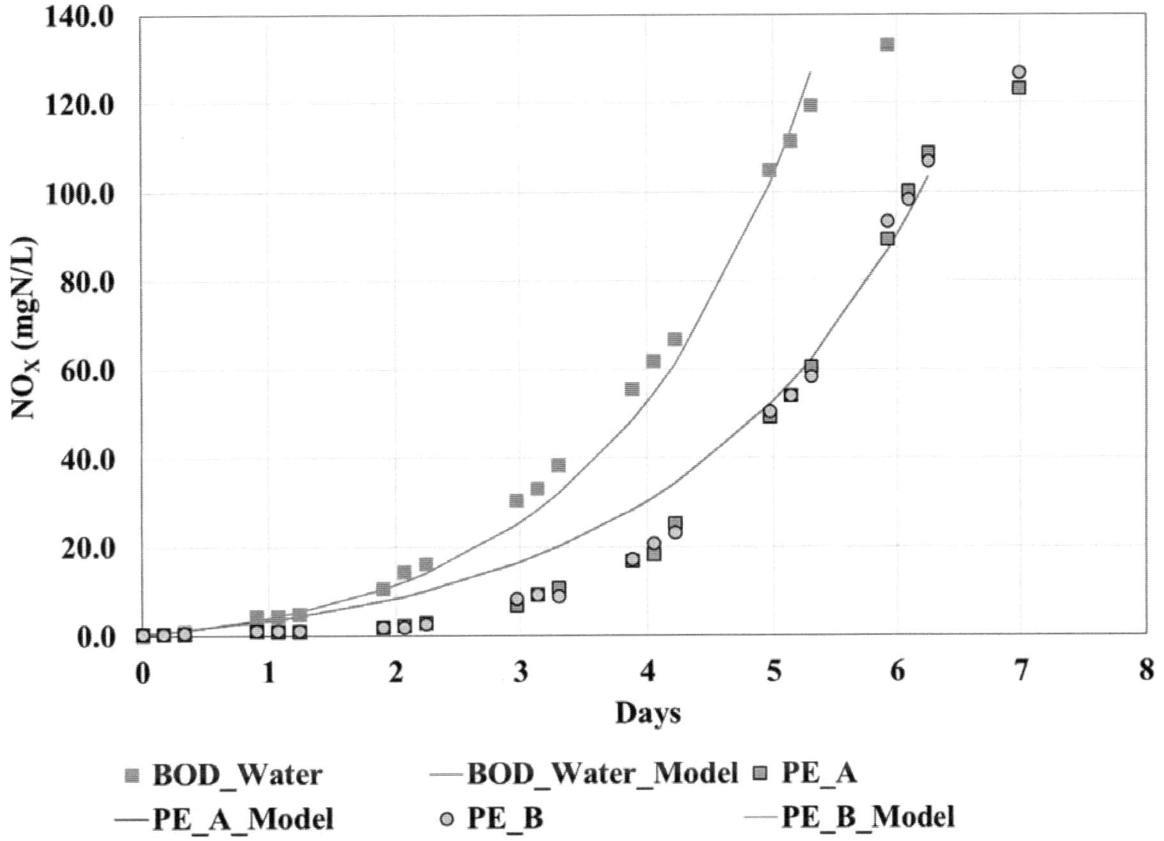

The combination of higher peak loads, rapid load increase, and nitrification inhibition resulted in some very high (unacceptably) effluent ammonia concentrations during the peak loads of the day using the above described control method. The system would eventually catch up within a couple of hours, but a large effluent ammonia peak would have already gone to the effluent. It was therefore decided to change the control system to more actively use feed forward control based on the primary effluent ammonia loading rate. This use of feed forward control would directly address the difficulties of the rapid load increase in the morning.

The current ammonia control system keeps the step-feed flow control approach developed during design, but changes the reactor DO control to a high/low setpoint for all zones except 3C and 4. The high DO setpoint is started when plant primary effluent flows exceed an operator setpoint, normally occurring during the flow increase period in the morning. These settings are adjusted seasonally as wastewater temperatures change, but the current (summer) settings are as described in Table 2.

Table 2: Aqua Nueva WRF DO Control Setpoints

Pass/Zone	Low DO SP	High DO SP
1A	1.5	2.0
1B	1.0	1.5
2A	1.5	2.0
2B	1.0	2.0
2C	0.5	1.5
3A	1.0	1.5
3B	0.5	1.0
3C	Ammonia Control Loop	
4	Same as 3C	

ANWRF AMMONIA REMOVAL PERFORMANCE

The plant operating data is summarized from April 29th, 2015 to June 18th, 2015 as a typical operating period. The ammonia load to the three passes is shown in Figure 8. It can be seen that the Pass 3 loads were very stable during this period, with an operator initiated increase in early June as a result of the higher wastewater temperatures. The load to Passes 1 and 2 are highly variable, as described previously, with the load to Pass 1 always being higher than to Pass 2.

Figure 8: ANWRF Ammonia Loads to Bioreactor Passes

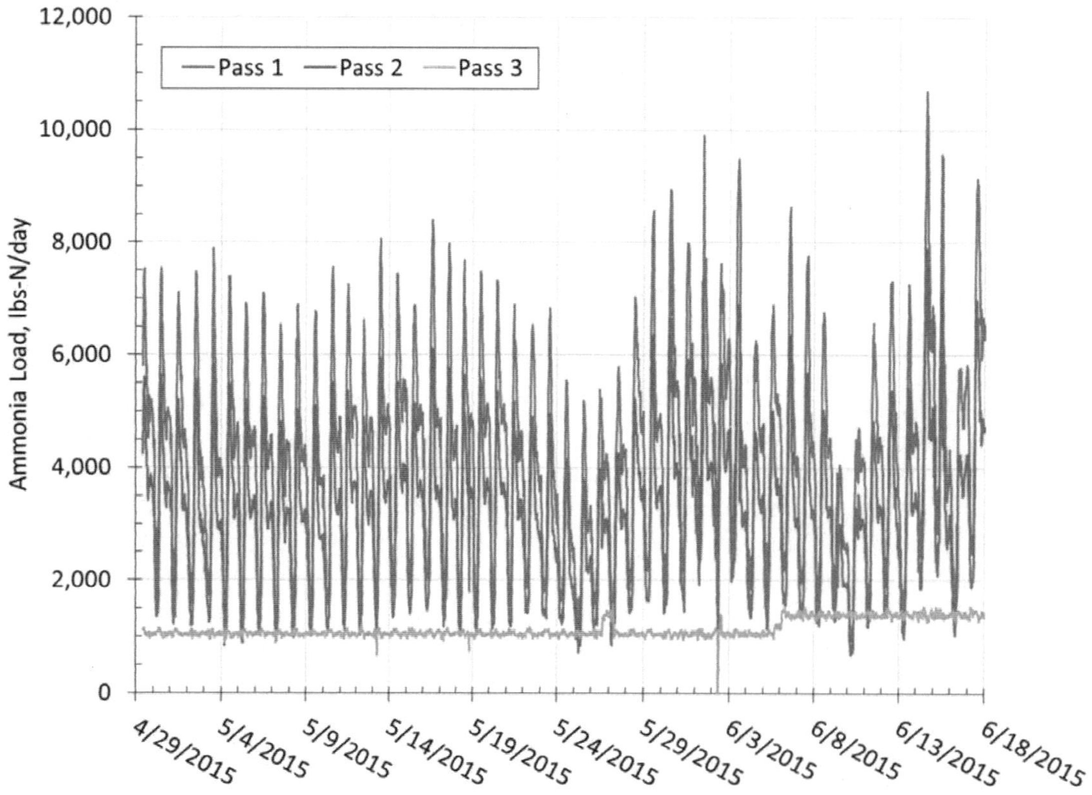

Better Process Control Enables Performance Exceeding Limits of Technology in Nutrient Removal - Truckee Meadows Water Reclamation Facility Experience

Kishora Panda[1*], Michael Drinkwater[1], Todd Saxberg[1]

[1] Truckee Meadows Water Reclamation Facility, Reno, NV.
[*] Email: kpanda@cityofsparks.us

ABSTRACT

Process control and optimization efforts at the Truckee Meadows Water Reclamation Facility (TMWRF), Reno, NV have helped the facility achieve process performance exceeding the design limits of the nutrient removal processes. Compliance with the facility's extremely stringent nitrogen limits on a watershed with one (1) endangered species, one (1) threatened species and a sovereign nation at the end of a desert terminus lake has also been achieved. The facility has an annual and seasonal average total nitrogen (TN) mass discharge limit of 500 lb/d that corresponds to a concentration of 2.2 mg/l and 1.4 mg/l at the 44 mgd permitted flow and 27 mgd current flow, respectively. These TN permit parameters become even more difficult to achieve when one considers the TN number includes a Dissolved Organic Nitrogen (DON) component averaging between 1.3 and 1.8 mg/L that cannot be treated with current technology at TMWRF. The 30-day average effluent total phosphorus (TP) is limited to 0.4 mg/L. The TMWRF nutrient removal system as a whole is an old and unique system that includes enhanced biological phosphorus removal (EBPR), nitrifying trickling filters (NTF), and Envirex upflow fluidized bed denitrification reactors (FBR). Achieving such low nutrient concentrations depends on effective and consistent performance of each of these three (3) sequential biological processes. Extensive process data analysis, field trials, and adjustments to process variables coupled to seasonal operating conditions were performed over a period of one year. Perhaps, of the three aforementioned tertiary processes, most challenging is the operation of NTFs that were designed to an effluent ammonia level of 2 to 4 mg NH_3-N/L. The TMWRF facility was able to meet that challenge and achieved success in maintaining the effluent ammonia below detection limits (0.1 mg/L, 90 percentile) for nearly eight (8) months last year. Among all the process variables studied; ammonia loading management including from the side stream centrate, secondary effluent alkalinity supplement, wetting rate balance, and snail control were found to be the most critical parameters affecting NTF performance. Methanol feed control and fluidization flow precision helped in consistently achieving nitrate below the detection level in the denitrification (FBR) effluent (designed for 0.5 mg/L). For the EBPR, activated sludge process control leading to low nitrate return helped achieve the TP to below 0.4 mg/L while an online chemical polishing step using alum stabilized the effluent phosphorus during biological process upset.

KEYWORDS: Process control, nutrient, nitrifying trickling filter, fluidized bed reactor, denitrification.

INTRODUCTION

The Truckee Meadows Water Reclamation Facility (TMWRF) is a 44-mgd tertiary water reclamation facility serving the cities of Reno and Sparks, Nevada. As shown in Figure 1, the nutrient removal process configuration includes enhanced biological phosphorus removal (EBPR), nitrification trickling filter (NTF), and Envirex fluidized bed reactors (FBRs) for denitrification. The facility is permitted to divert 6,700 acre feet per year (AFY) of reclaimed water for beneficial off-site use and discharge of remaining effluent to the Truckee River. The effluent's eventual destination is a landlocked lake with a threatened and an endangered trout species inhabited by the sovereign Paiute nation. Thus, the discharge permit imposes very stringent discharge limits: 1) annual average total nitrogen (TN) limit of 500 lb/d; 2) 30-day monthly average TN limit of 500 lb/d between May 1 and October 31; and 3) 30-day average total phosphorus (TP) limit of 134 lb/d. At the 44 mgd permitted flow, the nutrient mass discharge limits correspond to 1.4 mg/L and 0.4 mg/L of TN and TP, respectively. At a current average flow of 27 mgd, the TN is then limited to 2.2 mg/L. The plant effluent baseline dissolved organic nitrogen (DON) concentration ranges from 1.3 to 1.8 ppm.

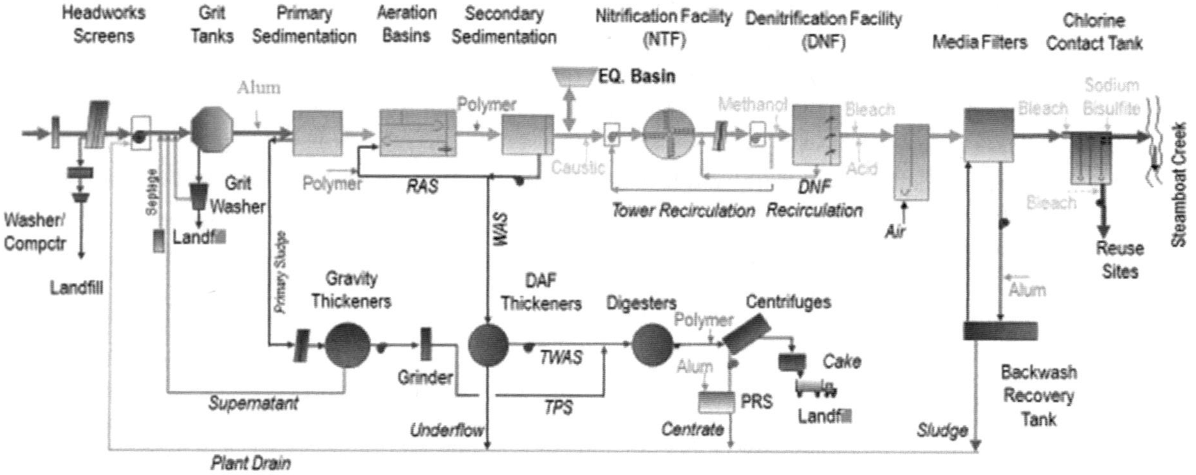

Figure 1. TMWRF Process Flow Schematic

Achieving very low TN concentrations using 30-year old trickling filters (designed for 2 to 4 mg/L NH_3-N in effluent) and fluidized bed reactors is a daunting task. While under such a low nitrogen permit limit, the facility's anaerobic digested sludge dewatering centrate recycle stream comingles with the plant influent at headworks without having any treatment or controlled discharge. While the NTFs perform satisfactorily during the summer months, NTF effluent ammonia concentration during peak winter months rises to as high as 6 mg/L. For example, the poor winter performances in 2013 resulted in an annual average TN load of 535 lb/d to the river and one toxicity violation. Consequently, the facility has been under a strict regulatory watch for nitrogen noncompliance and was imposed regulatory actions in 2013. In addition, these nitrogen capacity issues in the facility became a socio-political issue for the region. A recent study on determining facility's current true treatment capacity found the trickling filters as the limiting process and a true capacity of 27 mgd for the plant. In order to gain the facility's capacity, three alternative treatment options were identified with the lowest alternative costing more than 50 million dollars to treat the dissolved organic fraction in the plant effluent. Implementing any of these alternate options is difficult to fund by the stakeholder cities.

The facility's fluidized bed reactor denitrification (FBR) process is designed for an effluent nitrate concentration of 0.5 mg/L and uses methanol as the external carbon source for the biological denitrification of nitrate to nitrogen gas. To meet the phosphorus requirements in the process during low NTF effluent phosphorus levels, a phosphoric acid feed system is in line to supplement phosphorus in the process. Some issues identified by the current process team include, lack of knowledge transfer from similar processes elsewhere, some unplanned modifications to the reactors over the twenty plus year history of the process, a lack of current and updated process monitoring and control. While addressing these issues, this unit process still performed well consistently due to the simplistic nature of FBR operation. With fluidization and proper carbon dosing the FBR's will reliably produce effluent with nitrate concentrations to non-detect (<0.06 mg NO_3-N /L) levels.

The EBPR process at the TMWRF is a staged configuration of the EBPR activated sludge process with an anaerobic zone preceding three aerobic zones. Gravity thickeners for the primary sludge thickening supply volatile fatty acids through its overflow that is recycled and split evenly back to the aeration basin influent. To aid the EBPR process, soluble phosphorus in the centrate is chemically precipitated in a Phosphorous Rich Supernatant (PRS) process using alum before the recycle stream is blended with the raw sewage at the headworks. The PRS process returns precipitated phosphorous laden sludge to the primary clarifiers which is settled out rapidly and sent to anaerobic digestion. Under very widely varying temperatures, control of nitrification inside the activated sludge and prevention of nitrate recycle to the anaerobic zone are very crucial for effective performance of the EBPR process. Poor process control, especially solids inventory control, resulted in significant nitrification during the spring and summer months. Consequently, it was found an excess nitrate/nitrite return to the anaerobic zone through RAS return would negatively affect or shut down the EBPR process. The EBPR process is difficult to control during these events due to all volatile fatty acids being from gravity thickening are consumed by the nitrate return. . For example, there were two 30-day average TP permit violations in 2013. In retrospect the 2013 emergency headworks bypass utilizing miles of black HDPE pipe through the spring and summer may have begun the nitrification process early than anticipated.

The plant has planned for and began design work on a series of CIP projects to assist in meeting nutrient goals in the long term. However, since a short term solution was absolutely needed to prevent further permit excursions, plant staff implemented carefully planned process control and optimization activities starting April, 2014. The three nutrient removal systems were studied by the plant staff to determine the factors that were impacting the process performances, identify the problems, and possibly implement solutions to the previous problems through control and optimization with an aim to enhance process capabilities.

METHODOLOGY

Each nutrient removal system was studies by reviewing historical operations data, analyzing design conditions vs. current operating states, conducting laboratory tests, collecting additional data, and reviewing literature. The details of the activities are described below.

Nitrification Trickling Filter

Factors that were identified for detail evaluation of the trickling filter were grouped into four major categories; 1) ammonia loading, 2) hydraulics e.g., recirculation rate and wetting rate, 3) water quality impacting nitrification process, e.g., secondary effluent pH and alkalinity, and 4) environmental factors such as snail control, air flow, ambient temperature, etc. In addition, new instruments were installed to gather more data and monitor the process when optimization efforts were implemented.

Ammonia loading to the NTF. Impact of ammonia loading on NTF performance was studied by reviewing NTF performance at various ammonia loading states, conducting additional sampling and monitoring, and exploring options to reduce or equalize the ammonia loading to the NTF.

NTF design ammonia loading data were compared with the current operating states. Historical ammonia loading patterns were analyzed to see any relationship between performance and loading patterns and the time when the load appears at the filters, e.g., daytime vs. night time. Additional filter effluent testing was conducted to track performance of filters over 24 hour period. This was studied in light of changes in diurnal secondary effluent ammonia loading and concentration variations. The variations in secondary effluent ammonia loading and concentration including daily, weekly, and seasonal were reviewed to see if the load can be equalized or managed in a way to reduce influent variations to the nitrifying trickling filters. The ammonia loading was also analyzed to evaluate the effect of centrate recycle on both loading and concentration of ammonia in secondary effluent. Since the NTF effluent recycle dilutes the net ammonia concentration going to NTFs, the impact of NTF effluent recycle was also evaluated.

Secondary effluent ammonia is monitored using an online ammonia analyzer. However, due to NTF effluent recycle, the net ammonia load/concentration that enters into the NTF is less than that in the secondary effluent. In order to track the real time performance of the NTF by comparing the ammonia in NTF influent to nitrate formed (measured by a nitrate analyzer at the NTF effluent), a new ammonia analyzer was installed at the inlet to the NTF. A mass balance model using net ammonia concentration in the NTF influent and the nitrate concentration in the NTF effluent provided real time performance of the NTFs. Operators watched the Active Factory trends of these three (3) parameters to monitor performance of the NTFs and troubleshoot when the performances deteriorated.

Various options including side stream centrate treatment were analyzed to reduce, equalize, and regulate the centrate ammonia contribution to the main process. Available centrate treatment options and existing unused infrastructures in the facility were evaluated to see if it was feasible to set up a temporary centrate treatment process to reduce centrate ammonia load to some extent possible. The facility had four abandoned concrete tanks (1 MGal each, used for Phostrip process) and one offline aeration tank that were found to be feasible to setup an ammonia stripping process using the aeration tank's fine bubble diffusers. The Washoe County Air Quality Management District was contacted to see if there were any limitations with regard to ammonia emissions from the process. Figure 2 is a schematic of the temporary centrate treatment system setup.

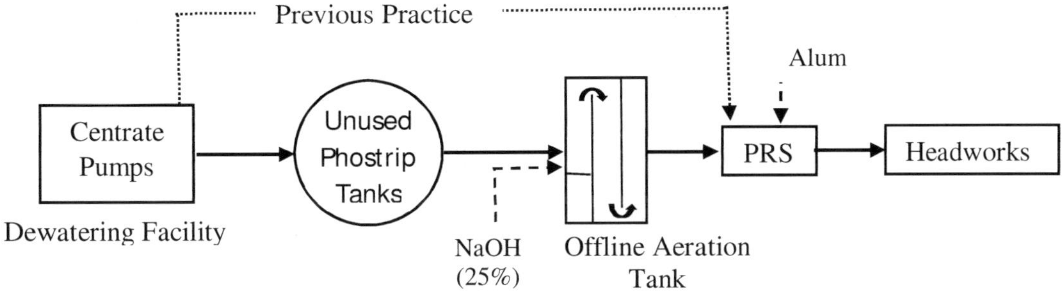

Figure 2. Temporary Centrate Ammonia Stripping Process Schematic

As shown in figure 2, prior to temporary centrate ammonia treatment, the centrate was pumped from dewatering facility to a phosphorus removal system (PRS) tank where alum is fed to precipitate phosphorus partially and then overflowed out of the PRS to headworks. The centrate pumping was continuous. In the temporary centrate ammonia treatment setup, the centrate is pumped to the four unused Phostrip tanks and then pumped from the Phostrip tanks to the offline aeration tank only between 1PM and 4AM. This pumping time window was determined after experimenting different centrate pumping windows to find the time when the centrate load will coincide with plant low ammonia load period. Centrate was pumped into first aerobic zone of the aeration tank where 25% NaOH was fed to raise the pH. The chemical metering pump was controlled to maintain a pH between 8 and 9 at the downstream of the chemical mixing. Airflow was regulated between 400 and 2,100 SCFM. Centrate overflowed from the aeration tank to the PRS. The treatment system was operated only during the winter months when the NTF performance is affected.

Hydraulics. NTF design documents, available literature on NTF, and the current operating hydraulic characteristics of NTF were studied in detail to determine if any hydraulic parameters and conditions had any correlation with poor performance of the NTF. Most important ones included in this study were recirculation rate, wetting rate, distributor arm speed, and SK (Spulkraft) value. Recirculation rates were varied to provide proper wetting rate. All distributor arm nozzles were balanced to provide equal arm speed in all NTFs.

Water quality. Since secondary effluent water quality besides ammonia has to be conducive for nitrification process, available literature and NTF design data were reviewed to find any correlation between poor NTF performance and secondary effluent water quality parameters such as BOD, TSS, alkalinity, and pH. Therefore, COD and BOD measurements were started in secondary effluent. In addition, a set of two pH and dissolved oxygen meters were installed at the common NTF influent and effluent.

Since alkalinity plays a major role in in the nitrification process, secondary effluent alkalinity data were analyzed to see if the secondary effluent alkalinity is not limiting and affecting the process. About 7.14 mg/L as $CaCO_3$ alkalinity is consumed per mg/L ammonia nitrified to nitrate and a residual alkalinity of minimum 50 mg/L as $CaCO_3$ is recommended in the nitrification process effluent. Biesterfeld et al., (2003) found from their study Littleton/Englewood plant NTFs that nitrification rates were inhibited when alkalinity was at 40 mg/L as CaCO3 and lower, while 45 mg/L as $CaCO_3$ and above was not inhibitory. NTF effluent 24 hour alkalinity testing and historical alkalinity data analysis indicated alkalinity deficiency in the nitrification process, particularly during the winter months. A temporary caustic feed system was set up to feed caustic (25% NaOH) to secondary effluent. A pH meter and daily laboratory alkalinity check provided basis for controlling the caustic feed.

Environmental factors. Design and operating data review and additional testing and sampling were conducted to verify capacity of fan air supply, NTF influent and effluent temperature and dissolved oxygen, and NTF inlet and outlet air temperatures. Snail and other biofilm grazing control was continued using the centrate.

FBR Denitrification Process

The process data, unit design data, and current state of the reactors were reviewed along with the information gathered from long time operators on the history of changes made to the reactors. Since the FBR has a recirculation stream that also partly mixes with the NTF influent, a possibility of excess methanol dosing leading to high BOD in the NTF influent was suspected. A TOC analyzer was installed at the FBR effluent to track excess methanol in FBR effluent. A field testing of BOD and TOC in FBR effluent was performed to develop a BOD/TOC ratio for the FBR effluent. Operator tracked the FBR effluent TOC and adjusted methanol dosing to FBR accordingly. Other major modifications made to the FBR were effluent piping relocation, influent flow increase, and recirculation flow reduction.

Enhanced Biological Phosphorus Removal System

Historical operating and performance data was reviewed to reveal any evidence of correlation between previous operating states and EBPR process upsets. The activated sludge process including RAS/WAS pumping methodology, SRT, and F/M data were reviewed. Some field modifications and control changes were made to the RAS/WAS pumping. In addition, previous solids wasting based on maintaining a target MLSS concentration was changed to maintaining a target SRT. All activated sludge process parameters such as F/M ratio, MLSS concentration, SRT, RAS and WAS flow rates were monitored and adjusted closely to minimize nitrification in the activated sludge process. Also, the centrate phosphorus precipitation rate in PRS was adjusted to align with the EBPR process phosphorus treatment capacity and performance capability.

RESULTS AND DISCUSSIONS

Results of the process control and optimization efforts are presented below for each nutrient removal system followed by a review of the effect on the facility's overall nutrient management.

Nitrification Trickling Filter

Ammonia loading to the NTF. Figure 3 is a schematic of the nitrogen removal systems (nitrification and denitrification) showing process flows, design loadings, and current performances. The NTFs (total of 6 units) are designed for a maximum ammonia load of 12,000 lb of which plant influent contributes about 10,000 lb and centrate up to 2,650 lb (about 26% of plant influent). Prior to the optimization efforts, NTF effluent NH_3-N concentration stayed between non-detect (<0.1 mg/L) and 6 mg/L.

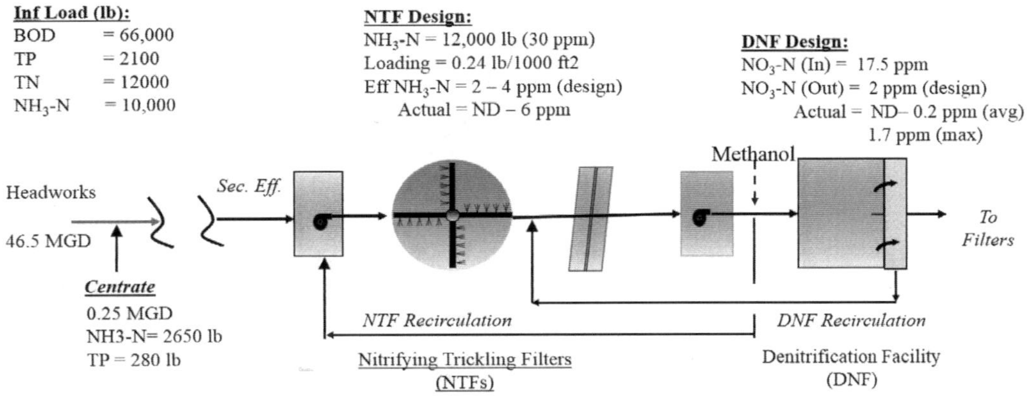

Figure 3. TMWRF Nitrogen (Nitrification and Denitrification) Removal System Design Criteria and Operating Performances

Seasonal variations in secondary effluent ammonia concentrations were studied. As shown in Figure 4, the secondary effluent ammonia concentration exceeded the design value of 30 mg/L most of the time during winter months. Whereas, the secondary effluent ammonia concentration in the summer months stayed between 7 and 12 mg NH_3-N/L. Figure 4 also shows profiles of secondary effluent nitrite and nitrate concentration. It may be noted from the profiles of ammonia, nitrite, and nitrate that amount of ammonia drop in summer months does not represent equal amount of nitrite and nitrate formation. This finding clearly shows that the nitrification is not the only mechanism for ammonia concentration drop in summer months. Partial volatilization of ammonia under summer high water temperature and fine bubble ammonia stripping in aeration tanks may be occurring resulting in additional ammonia drops beyond nitrification. This loss is minimum during cold winter months resulting in a secondary effluent ammonia concentration drop equivalent to the extent of nitrification in the activated sludge process (ASP). This observation may be a phenomena typical to plants with activated sludge for BOD removal followed by separate nitrification processes.

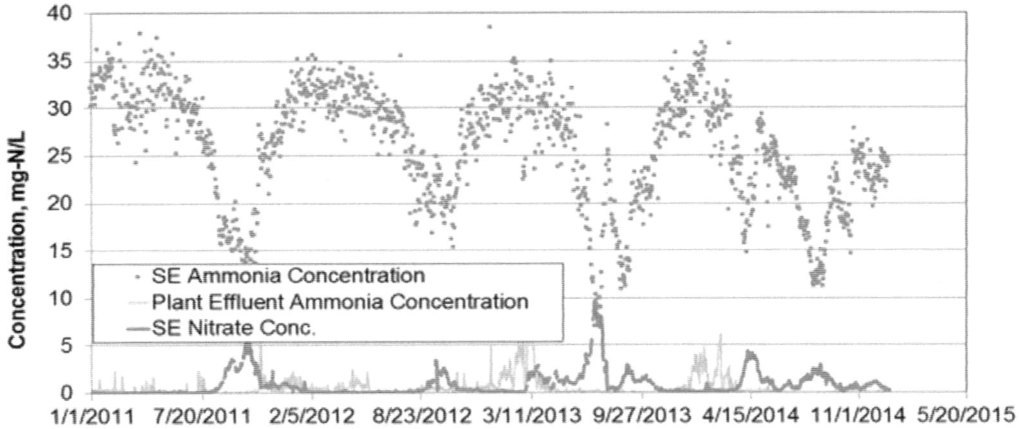

Figure 4. Plant Effluent Ammonia and Secondary Effluent (SE) Nitrate and Ammonia Profiles

Review of the 2011-14 diurnal secondary effluent ammonia loading data showed that the average daily ammonia loading mostly stayed below the design load of 12,000 lb though instantaneous ammonia loading (lb) and concentration exceeded the design values for about 4-6 hours during late afternoon through midnight of the winter months (December through March). However, except 2011 winter, the ammonia loading rate remained below the design loading rate of 0.24 lb NH_3-N/d/ 1000 ft^2 (Figure 5). Also, additional NTF effluent hourly grab sample data showed poor night time NTF performance compared to day time. High ammonia loading and lower night time temperature possibly had a synergetic effect on poor night time performance of the NTF. Therefore, during winter months, efforts were focused toward reducing net ammonia load and concentration peaks during night time and shifting them more to day time. With no control on ammonia load from plant influent, operations staff used the temporary centrate treatment system and controlled discharge to headworks as a means to regulate the ammonia loading and concentration in the secondary effluent.

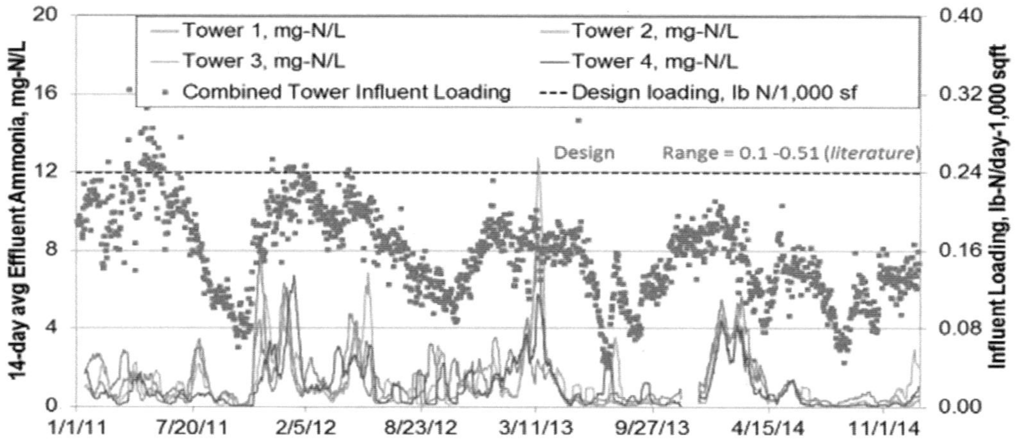

Figure 5. NTF Influent Design and Actual Loading (lb N/d/1000 ft^2 and NTFs 1-4 Effluent Ammonia Concentrations (Grab Samples)

The temporary centrate treatment system performance and operating conditions are shown in Figure 6. It may be noted that this was not an engineered system, rather a crude method that the operations staff devised to treat the centrate to whatever extent possible. Figure 6 shows centrate temperature and pH and percent centrate ammonia reduction in the temporary centrate treatment system. The centrate from the dewatering system has an average ammonia concentration of 1,275 mg NH_3-N/L. An average of 30 to 40 percent of the centrate ammonia concentration reduction was achieved. This reduction in the treatment system contributed to about 11 percent ammonia load reduction to the headworks. Though the percent ammonia removal was observed to go higher at higher air flow rates, the air flow rate was restricted to about 500 SCFM per zone maximum to avoid excess foam formation. The excess foam level, particularly during night time, was an issue for operator access around the treatment system.

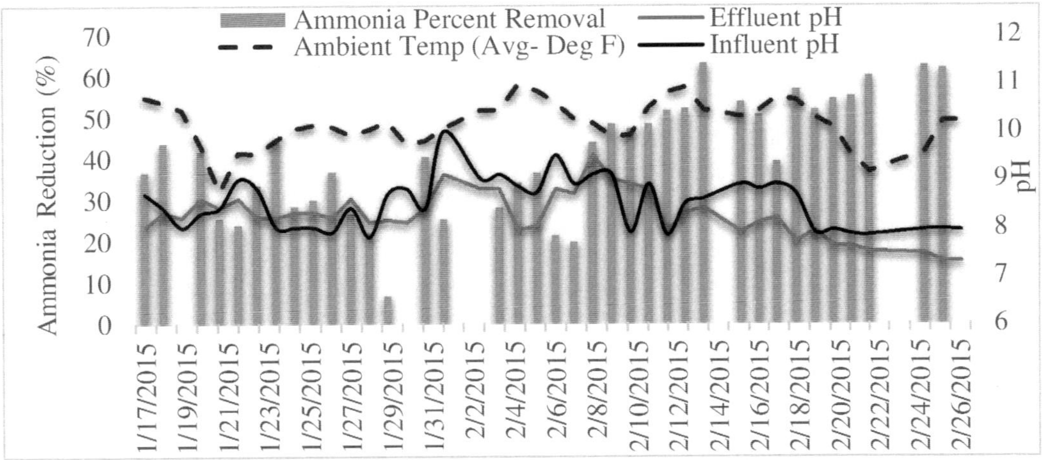

Figure 6. Temporary Side Stream Centrate Treatment Performance (Water Temperature, pH In, pH Out, and Percent Ammonia Removal)

All the above efforts including side stream centrate treatment, controlled centrate discharge, plant flow restrictions and some plant influent equalization in headworks, and additional monitoring and control resulted in instantaneous secondary effluent ammonia load and concentrations below design values. As shown in Figure 5, secondary effluent ammonia concentrations were mostly below 28 mg NH_3-N/L during 2014-15 winter months (Dec – Mar).

Hydraulics. Recycle of NTF effluent is an effective way of keeping all areas of the media biologically active when the influent flow is low to provide proper media wetting. The NTF wetting rates are typically maintained in the range of 0.4 – 1.0 L/s. m^2 or 0.59 – 1.46 gpm/ft^2 (Tchobanoglous et al., 2010). Prior to the optimization efforts that began in April, 2014, the TMWRF NTF recirculation rates were maintained between 0 and 0.4 that was resulting in a wetting rate below the design wetting rate of 0.88 gpm/ft^2 (Figure 7). As shown in the figure, low NTF recirculation flows were also causing lower wetting rates prior to April, 2014. NTF influent pumping rates were increased to maintain a minimum recirculation rate of 0.5 and a minimum wetting rate of 0.88 gpm/ft^2.

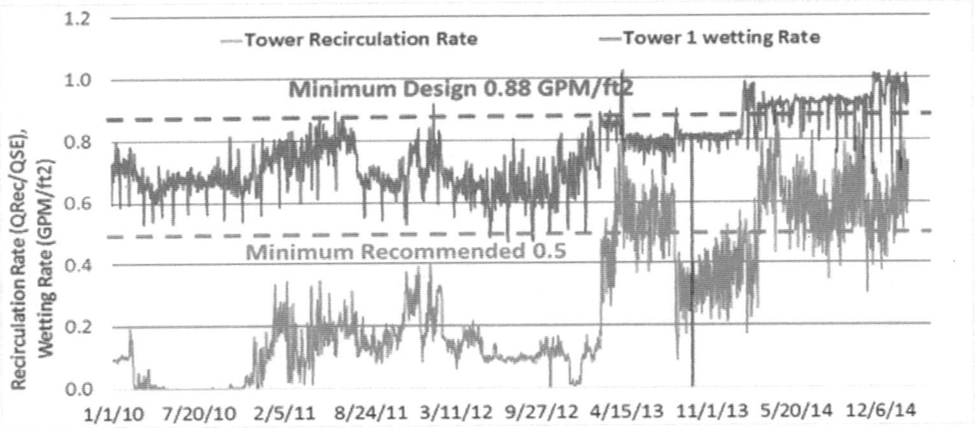

Figure 7. NTF Recirculation and Wetting Rates

The effect of higher recirculation indirectly helped reduce the NTF influent ammonia concentration (Figure 8). Prior to the optimization efforts, due to the low recirculation rates, the secondary effluent and NTF influent ammonia concentration were almost same. However, after increasing the recirculation rates, the net ammonia concentration in NTF influent started dropping below the secondary effluent ammonia concentrations. Though the increased recirculation was successful in reducing the ammonia load, operations staff is further studying the impact of very low ammonia concentrations and widely diurnal variations on NTF performance.

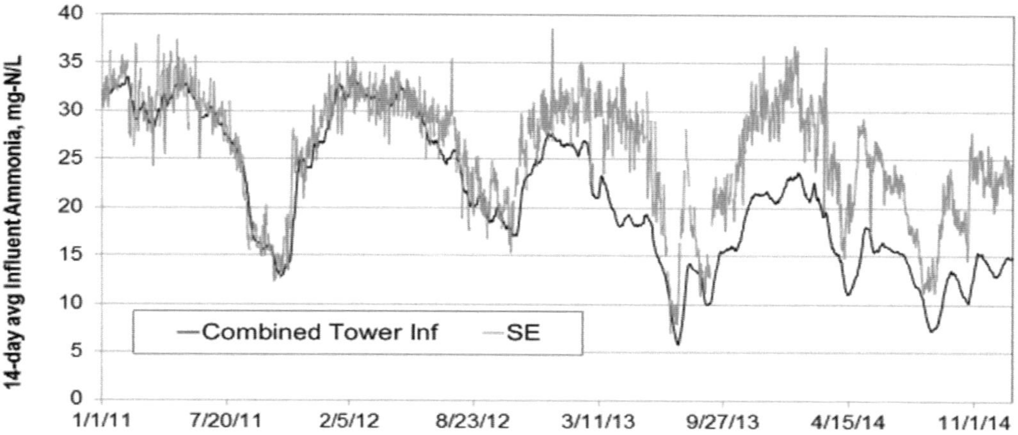

Figure 8. Effect of Increasing NTF Recirculation Flow on NTF Influent Ammonia

Water quality. Secondary effluent water quality parameters other than ammonia such as BOD, TSS, alkalinity, pH, temperature, and dissolved oxygen are critical to developing efficient nitrification process. The historical data analysis results revealed the following and necessary corrections made to each are also described in the corresponding paragraph.

Nitrification performance is greatly impacted by the NTF influent BOD concentration. A high amount of organics to ammonia in NTF influent promotes growth of heterotrophic bacteria that competes with nitrifiers on occupying filter media, thus affecting nitrification rate. Harremoes (1982) found that nitrification could occur at a maximum soluble BOD concentrations below 5 mg/L, be inhibited proportionately with soluble BOD exceeding 5 mg/L, and be insignificant when soluble BOD is 30 mg/L or more. Figueroa and Silverstein (1991) found that nitrification rates are inhibited at BOD values above 10 mg/L in agreement with observation made by Parker and Richards (1986). TMWRF secondary effluent process optimization resulted in an average BOD of 20 mg/L (average COD of 40 mg/L) in the secondary effluent causing no significant impact on the nitrification performance. Weekly bumping of the aeration tank Sanitaire diffusers also played a significant role in keeping the diffusers in good condition for optimum oxygen transfer. For any upset in the activated sludge process leading to high effluent COD seen in the daily laboratory results, operation staff will flush the NTF's and it has been observed to immediately improve performance by knocking off the heterotrophic growth. We have seen significant COD measurement results in the flush water when testing NTF effluent during an hour long flush cycle.

Secondary effluent with high TSS concentrations can also impact the performance of an NTF. Literature indicates various limiting TSS concentrations (in the range of 15 to 20 mg/L) in the secondary effluent above which it affects the NTF performance. TMWRF secondary effluent TSS generally stays below 10 mg/L thereby posing no threat to NTF performance.

Figure 9 shows the profiles of plant effluent ammonia and temperature, NTF influent ammonia and alkalinity, and calculated alkalinity requirement per the NTF ammonia concentration. As shown in the figure, from Dec, 13 to Mar, 14, the calculated alkalinity requirement was higher than alkalinity present in the NTF influent matching with high ammonia concentration in plant effluent. Later between Mar, 14 and Jul, 14, alkalinity deficiencies can be noticed resulting in ammonia appearance in plant effluent though in small amounts compared to the episode in Dec, 13 through Mar, 14. Due to the direct impact of alkalinity on nitrification performance, it was concluded that poor NTF performances in winter months was most likely due to alkalinity deficiency in the NTF influent. This was also field verified by conducting grab alkalinity tests in NTF effluent during peak winter months. It was found that NTF effluent alkalinity reached as low as 20 mg/L as $CaCO_3$ often in night times. Thehe recommended minimum residual alkalinity in nitrification effluent is 50 mg/L as $CaCO_3$ for effective nitrification (EPA 2000). The temporary caustic (25% NaOH) feed system was able to feed up to 1,500 gallon of caustic per day and maintain a pH around 7.5 in NTF influent. The temporary system was operated in the winter months of 2014-15. Further study and installation of a permanent caustic feed system with better control is underway.

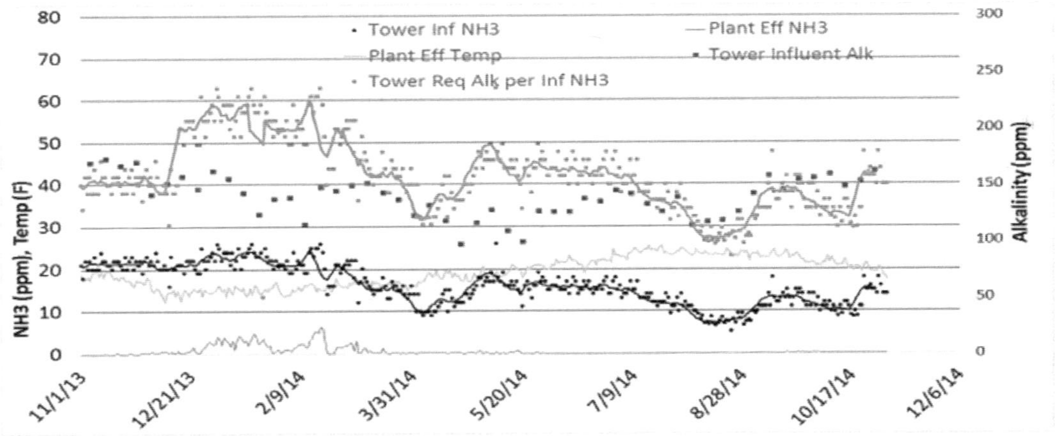

Figure 9. Available and Required Alkalinity for NTF Process and Plant Effluent Ammonia

The NTF influent and effluent pH monitoring started after the beginning of the optimization efforts. There was not enough data available to correlate any impact of NTF influent pH on NTF performance. The nitrification process is pH sensitive and optimal nitrification rates occur at pH values in the 7.5 to 8.0 ((Tchobanoglous et al., 2010). The caustic addition to secondary effluent helped maintain NTF influent pH between 7.2 and 7.5.

Monitoring data on NTF influent and effluent temperature indicated a very little difference in water temperature, with an average of 0.4 degree Celsius during the winter months. So due to this low temperature drop, it was concluded that the temperature was not a big factor in NTF performance, particularly during the winter months. Albertson et al., (2001) noted that the rate of nitrification is unchanged over a temperature range of 9- 20^{0}C and decrease above and below this temperature range. TMWRF influent temperature has been to 12^{0}C and the performance of NTFs is not affected any more than the performance affected from the other wintertime operating conditions.

Effective oxygen transfer from air to water within the NTF media is necessary to maintain an aerobic condition and meet nitrification process oxygen demand. Okey and Albertson (WEF, 2000) from the analysis of data from four different studies involving trickling filters designed for both BOD removal and nitrification found that the bulk liquid dissolved oxygen concentration had more effect on the nitrification rate than temperature and a DO concentration of 2.8 mg/L is required for nitrification. The two dissolved oxygen analyzers at the NTF influent and effluent provided real time DO levels in water. The data showed that NTF influent dissolved oxygen concentrations stayed above the recommended values most of the time.

Environmental factors. Dissolved oxygen in NTF influent and effluent remained higher than 2.5 mg/L and the installed fan capacity was calculated to be sized properly for design supply of air. NTF air temperature data did not show a big change in temperature between the NTF inlet and outlet indicating no possibility impact of temperature on the NTF performances. The NTFs are designed for a minimum ambient air temperature of 13^{0}C and during 2014-15, the ambient temperature stayed at 14^{0}C.

NTF snail control is performed using dewatering facility centrate. The pH of the centrate is raised to 9.1 and then recirculated for one hour and then flushed 8 hours with secondary effluent sending the flushed water to headworks. This predator eradication treatment successfully eliminates the adult snails but TMWRF has not found an answer to the egg pouches which show resilience to most chemicals and still hatch approximately every three (3) weeks. By this time the tower being treated is generally beginning to show poor performance from the expected hatch and subsequent biofilm grazing and it comes up again for cycled predator control treatment. Currently must treat two (2) NTFs must be treated per week to stay ahead of the hatch cycle and accompanying biofilm grazing. During the optimization activities, a centrate recirculation pipe bypass valve was also found to have been partially open leading to short circuiting of some centrate into the NTF effluent. This short circuiting issue was raising the plant effluent consistently during NTF snail control treatment days. After this finding, the centrate bypass through this valve was stopped eliminating further occurrence of higher ammonia concentration in NTF effluent.

FBR Denitrification Process

Currently methanol is dosed proportional to either flow or mass of NO3 entering the FBR. Since the amount of methanol required for completion of denitrification process also depends on dissolved oxygen concentration in FBR influent, an effective control was needed to minimize the methanol overdose. Methanol dose was adjusted by looking at the daily laboratory NO_3 results and real time TOC readings on SCADA. TOC readings within a range of 9 to 18 mg/L was found to be effective in optimizing the methanol feed. A BOD to TOC ratio of 1.48 was determined from separate laboratory analysis.

Enhanced Biological Phosphorus Removal System

Previous upsets to the EBPR process were attributed primarily to excess nitrate recycled to the anaerobic zone of the EBPR process through RAS and subsequent to lack of VFAs. The

RAS/WAS pumping rate analysis showed that the WAS rate was not enough to lower the target SRT value particularly during the summer months. In the activated sludge process, a set of two (2) RAS pumps distribute secondary clarifier sludge to a common header from which a flow control valve diverts a set amount of WAS to sludge thickening process while the remainder goes to the front of the aeration tanks or EBPR zones as RAS. The RAS pumping was capped at 0.5 of aeration influent flow and the hydraulics of the RAS pumping and WAS diversion pipe were not allowing the system to meet the target WAS flow. The RAS pumping was raised to as high as 0.75 and that corrected WAS flow deficiencies temporarily. A separate project is underway evaluating the RAS/WAS pumping system capacity and capability further. These changes, along with increased activated sludge process monitoring and updated control changes minimized nitrification in the activated sludge process and helped the EBPR process performance.

Polymer is also fed to the aeration tanks for filament control and to the secondary clarifiers for chemically enhanced settling/phosphorus removal when the biological system is not individually effective to keep total phosphorus below 0.4 mg/L. This can be noted by gauging the difference between Ortho Phosphate and total Phosphorus indicating a settling and solids capture problem.

Phosphorus precipitation rate in PRS and the effluent PRS phosphorus level is maintained in correlation with EBPR efficiency. Prior to the optimization activities, alum feed control was based on PRS effluent pH which was changed to target PRS effluent total phosphorus concentration. Simply stated, there are times when the PRS system can be ran with less Alum as the EBPR system will take up any phosphorous coming back through headworks recycle to the primary clarifiers. After the optimization efforts, total phosphorus concentrations in plant effluent remained consistently below 0.4 mg/L.

Effect on Facility's Overall Nutrient Management

Figure 10 shows the trend of TMWRF plant effluent ammonia and TKN before and after the optimization efforts. The difference between TKN and ammonia is the organic nitrogen fraction that contains mostly dissolved organic nitrogen which lies in the range of 1.3 to 1.8 mg/L. The inorganic fractions that includes ammonia, nitrite and nitrate mostly contribute to the varying total nitrogen concentration in plant effluent. In addition, TKN and ammonia profiles being parallel as shown in figure 11 and nitrate and nitrite being below or close to non-detection levels, the total nitrogen in plant effluent is mostly impacted by the ammonia concentrations. The high TKN profile peaks of previous four winter seasons (2011 thru 2014) were above the target value of 2.2 mg/L TN (corresponding to 500 lb/d TN and plant's average flow of 27 mgd) indicating poor NTF performance and overall nitrogen management by the facility. After implementation of the optimization efforts in April, 2014, the TKN and so as the ammonia peak stayed below mostly the target 2.2 mg/L TN limit during 2014-15 winter months. During this period, ammonia concentrations mostly stayed below detection limits (<0.1 mg NH_3-N/L).

The monthly and annual daily average total nitrogen load to the river was calculated for 2013, 2014 and 2015 and are shown in Figure 11. As shown in the figure, extremely poor winter performances often exceeding 600 lb/d TN resulted in a total annual average of 535 lb/d TN in 2013. In 2014, effects from the optimization efforts in April and December, the annual total nitrogen load was dropped to 474 lb/d and in compliance for the year. In 2015, optimization efforts resulted in a very good winter performances. TMWRF estimates the total nitrogen load to the river at or below 425 lb/d provided the summer reuse demand remains unchanged from previous years.

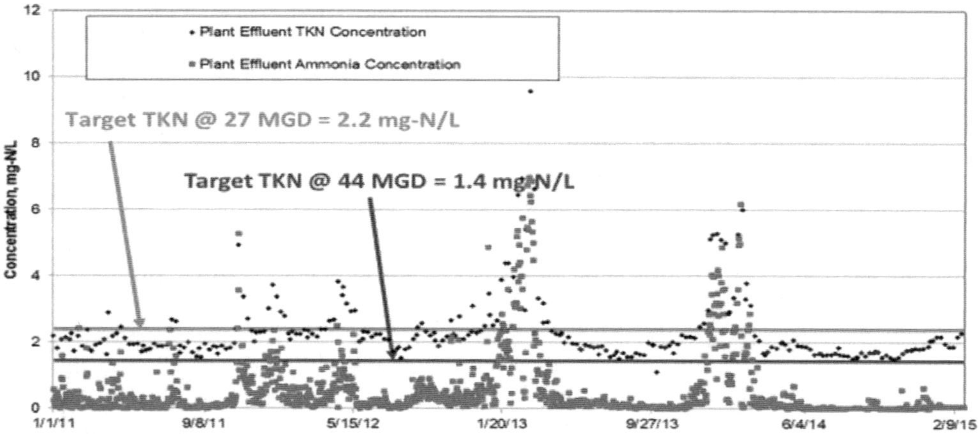

Figure 10. Plant Effluent TKN and Ammonia Concentration Before and After the Optimization Activities

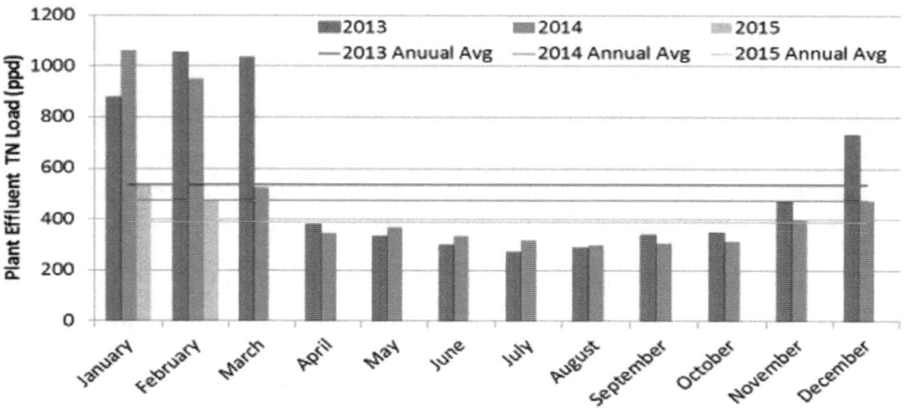

Figure 11. Monthly and Annual Daily Average Nitrogen Discharge to Truckee River (2013 to 2015)

CONCLUSIONS

The plant effluent permit still demands an efficient nutrient treatment using the facility's unique and aging treatment train which had been performing poorly, leading to recent permit excursions. Poor winter performance of NTFs and biological phosphorus removal in summer months have been particularly challenging to plant operations.

While management was belatedly forced into considering options on a 50- 250 million dollar investment for an advanced nitrogen treatment system installation in the facility to bring the total nitrogen load below the permit limit of average 500 lb/d, operations staff accepted the challenge to explore all possibilities to enhance the performance of the current nutrient treatment systems. The optimization efforts that included better process control, operational changes, and temporary treatment systems were successful in achieving the NPDES permit limits for Nitrogen in 2014.

The total nitrogen load to the river was dropped from 535 lb/d to 474 lb/d in 2014. With the level of performance achieved in the plant so far in 2015, TMWRF is confident in maintaining the total nitrogen load at or below 425 lb/d in 2015. There were no phosphorus permit violations in 2014 and the same is expected in 2015.

The optimization efforts undertaken by TMWRF staff helped them understand the process and treatment system capabilities better, and saved the facility, for now, from having to address major capital project plan changes in a hurried manner in addition to mitigating further permit excursions. The results gave TMWRF time to consider additional improvements and helped them plan for related projects and additional upgrades to centrate sidestream methodology for example.

REFERENCES

Albertson, O. E.; Okey, R. W.; Pearce, P.; Parker, D.S. (2001) *Trickling Filter Mythology*. Journal of Environmental Engineering, 127(1), 83-91.

Biesterfeld, S.; Farmer, G.; Russell, R.; Figueroa, L. (2003) *Effect of Alkalinity Type and Concentration on Nitrifying Biofilm Activity*. Water Environ. Res., 75, 196.

EPA (2000) *Wastewater Technology Factsheets nitrification Trickling Filter.* EPA 832-F-00-015,

Figueroa, L.; Silverstein, J. (1991) *Pilot Scale Trickling Filter Nitrification at the Longmont WWTP*. Proceedings Environmental Engineering Specialty Conference, American Society of Civil Engineers, Reno, NV.

Harremoes, P. (1982) *Criteria for Nitrification in Fixed-Film Reactors*. Water Science and Technology, vol. 14. P.167.

Lacan, I.; Gray, R.; Ritland, G.; Jenkins, D.; Resh, V.; Chan, R. (2000) *The Use of Ammonia to Control Snails in Trickling Filters*. Proceedings of the Water Environment Federation, WEFTEC 2000: pp. 79-96(18).

Parker, D.S.; Richards, T. (1986) *Nitrification in Trickling Filters*. Journal Water Pollution Control Federation, vol. 58, p. 896.

Tchobanoglous, G.; Burton, F. L.; Stensel, H. D. (2010) *Wastewater Engineering Treatment and Reuse*, Metcalf & Eddy, Inc. McGraw Hill, 4th Edition.

Aeration Considerations in the Design and Operation of Biological Nutrient Removal Processes

Joe Rohrbacher[1]*, PE, Paul Pitt[1], PhD, PE, Jim Cramer[1], PE, Diego Rosso[2], PhD

[1]Hazen and Sawyer
[2]University of California-Irvine
*Email: jrohrbacher@hazenandsawyer.com

ABSTRACT
Aeration system design and control is a critical consideration in the performance of biological nutrient removal (BNR) systems. Excessive aeration not only increases energy costs, but it can limit the reliability of nitrogen and phosphorus removal processes and decrease process stability and sludge settleability, compromising compliance and leading to increased chemical usage to meet effluent permit limits. Important considerations in aeration system design of nutrient removal systems include:

- Providing an effective taper to control dissolved oxygen (DO) concentrations
- Limiting the return of DO in RAS and internal recycle streams
- Decreasing DO prior to supplemental carbon addition
- Eliminating the potential of back mixing into anoxic or anaerobic zones
- Providing access for maintenance of diffused aeration equipment
- Maintaining aerobic conditions to avoid secondary phosphorus release
- Providing zone dissolved oxygen control
- The use of instrumentation to control aeration

This paper presents case studies illustrating the importance of aeration system design in achieving reliable, cost-effective BNR.

KEYWORDS: Aeration, Energy Optimization, Nutrient Removal, Nitrogen, Phosphorus, Sustainability

INTRODUCTION
Electricity comprises a significant portion of operating costs for municipal wastewater utilities in the United States. Approximately 3 percent of energy consumed in the United States is by water treatment plants and water resource recovery facilities (WRRFs) (Krause et al., 2010). It is well known that the aeration process is typically the largest consumer of energy at conventional wastewater treatment plants. Seventy percent of WRRFs in the United States exceeding 2.5 million gallons per day (MGD) utilize activated sludge secondary treatment, where 45 to 75 percent of electricity use is consumed in the aeration process (Rosso and Stenstrom, 2006). Because the aeration treatment process consumes the majority of energy in WRRFs utilizing secondary treatment, improving the efficiency of aeration can result in significant cost and energy savings to utilities.

Energy efficient aeration is key to moving WRRFs from net energy consumers to producers. In addition, aeration system design and control is a critical consideration in the performance of biological nutrient removal (BNR) and enhanced biological nutrient removal (ENR) systems. Excessive aeration not only increases energy costs, but it can limit the reliability of nitrogen and phosphorus removal processes and decrease process stability and sludge settleability, compromising compliance and leading to increased chemical usage to meet effluent permit limits. Important considerations in aeration system design of nutrient removal systems include:

- Providing an effective taper to control dissolved oxygen (DO) concentrations in various zones
- Limiting the return of DO in RAS and internal recycle streams
- Decreasing DO prior to supplemental carbon addition
- Eliminating the potential of back mixing into anoxic or anaerobic zones
- Providing access for maintenance of diffused aeration equipment
- Maintaining aerobic conditions to avoid secondary phosphorus release
- Providing zone dissolved oxygen control
- The use of instrumentation (i.e. ammonia analyzers) to control aeration

METHODOLOGY

The impacts of aeration system design and operation on nutrient removal performance were evaluated at three WRRFs. Case studies were selected to illustrate the importance of careful attention to detail and process flexibility when designing aeration systems for nutrient removal facilities. These case studies are provided to demonstrate several of the important considerations listed above and provide examples of designs which have facilitated reliable, cost-effective nutrient removal and those designs which have hindered and increased the cost of reliable nutrient removal.

RESULTS

Case Study No.1 – Provide a Deoxygenation Zone to Limit Dissolved Oxygen in Internal Recycles Flows and Reduce Carbon Feed

A 57 million liter per day (ML/day) (15 mgd) WRRF was converted from a nitrifying conventional activated sludge facility to enhanced nutrient removal (ENR) to meet annual average effluent TN and TP concentrations of 5 mg/L and 0.3 mg/L, respectively. The ENR process has the flexibility to be operated in numerous configurations including as four-stage and five-stage ENR processes. The last portion of the aerobic zone prior to the post-anoxic zone was provided with surface mixers to allow air to be reduced or shut off while keeping solids in suspension prior to internal recycle withdrawal or carbon feed as shown in **Figure 1**.

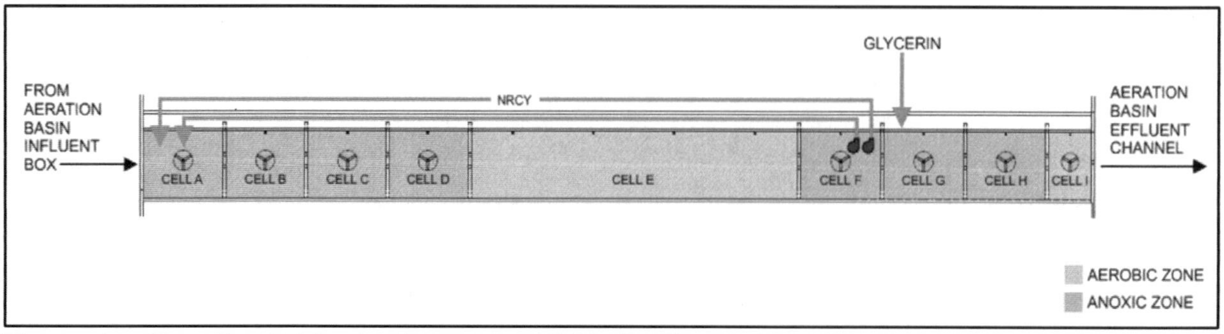

Figure 1 – Case Study No. 1 - Four-stage ENR Configuration with Deoxygenation Zone prior to Post-Anoxic Zone

This design feature has limited dissolved oxygen return to the pre-and post-anoxic zones, improving denitrification performance and reducing carbon addition. The mixed zone at the end of the aerobic zone also contributes a substantial amount of additional denitrification as shown in **Figure 2**.

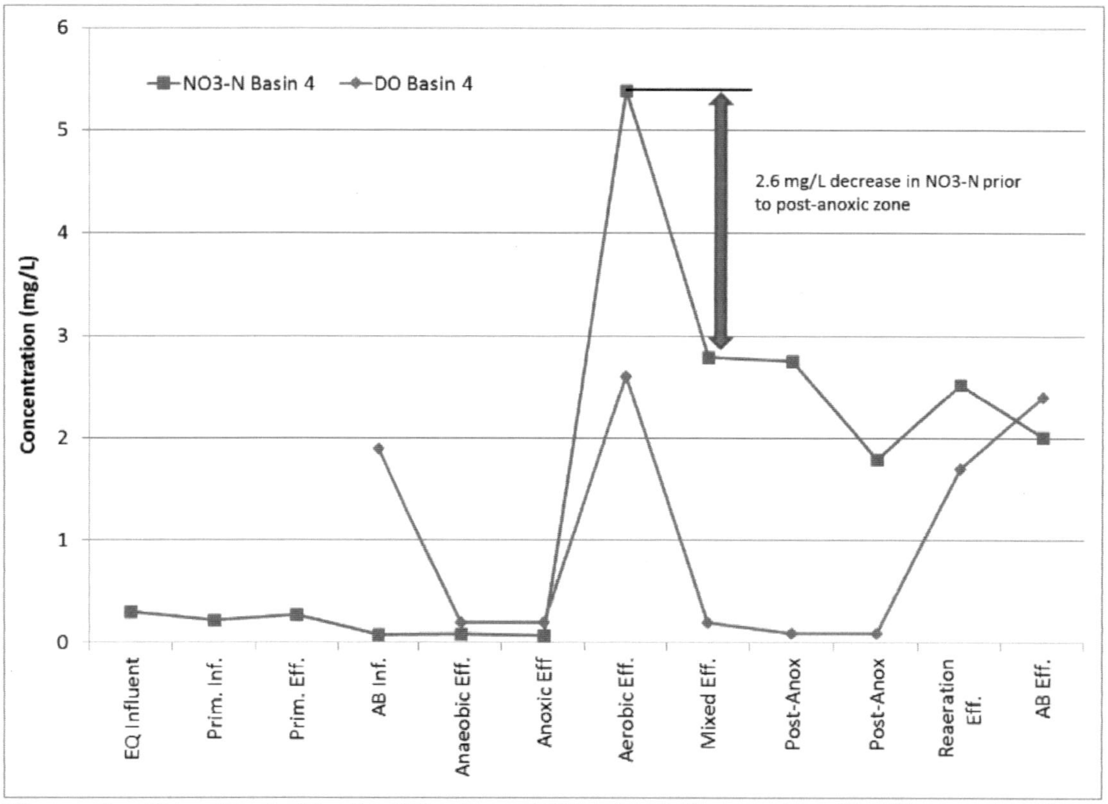

Figure 2 – Case Study No. 1 - Nitrate and Dissolved Oxygen Profile Illustrating Decrease in Nitrate Prior to Post-Anoxic Zone

Process simulations were performed using a calibrated process model previously developed using BioWin 4.1 (Envirosim, LTD.) to assess the impacts of the deoxygenation zone. The process simulations indicated operation of Cell F without air resulted in an additional 1.5 mg/L of TN removal compared to if Cell F were aerated. The process simulations predicted that the additional TN would require approximately 230 gpd of additional glycerin product at the current 38 ML/d (10 mgd) design flow. The resulting annual savings in supplemental carbon addition is estimated to be approximately $150,000, whereas the additional power cost associated with operating the mixing equipment in Cell F is estimated at $17,000/year. The construction cost of providing the additional mixers in Cell F in each of the five ENR basins was estimated to be $300,000. Installation of mixers in Cell F to allow for operation as a deoxygenation zone had a simple payback of just over two years.

Case Study No.2 – Avoid Submerged Baffle Walls between Unaerated and Downstream Aerobic Zones

A 110 ML/d (29 mgd) WRRF operated in an anaerobic-anoxic-oxic (A2O) configuration currently needs to meet an effluent TP limit of less than 1 mg/L at design flow. The facility is facing a reduction in the permitted effluent TP concentration to below 0.5 mg/L and is facing potential TN limits as well. A process evaluation including detailed wastewater characterization, process profiling and calibrated process model development was performed to determine current performance and optimization strategies for meeting more stringent effluent nutrient limits.

Process profile sampling of the WRRF was performed over three days in September 2014. The process profiles focused on profiling the ammonia, nitrate, and ortho-phosphate concentrations through the bioreactors. In addition, dissolved oxygen concentration, temperature, and pH were recorded during sample collection. The first portion (roughly 25%) of the basins were operated without aeration, and NRCY was returned to Cell C of each basin. Operation of each cell in the bioreactors during the process profiles was as follows:

- Cells A and B (12.5% of the total volume) – Anaerobic
- Cells C and D (12.5% of the total volume) – Anoxic
- Cells E and F (75% of the total volume) - Aerobic

The submerged baffle wall between Cells C and D permits the back-mixing of aerated MLSS into the anoxic zone due to the "air-lift" effect of diffused aeration as shown in **Figure 3**. The anoxic zone is operated with approximately 0.5 mg/L dissolved oxygen due to the back-mixing, and nitrate removal across the anoxic zone is minimal (**Figure 4**). Reduced denitrification decreases the reliability of the existing biological phosphorus removal process (BPR) due to increased nitrate return to the anaerobic zone. There is very little taper to the installed diffusers, and full nitrification was observed within the first 1/3 of the aerobic zone.

Figure 3 – Case Study No. 2 - Back-mixing of Aerated MLSS Across Submerged Anoxic Zone Effluent Weir

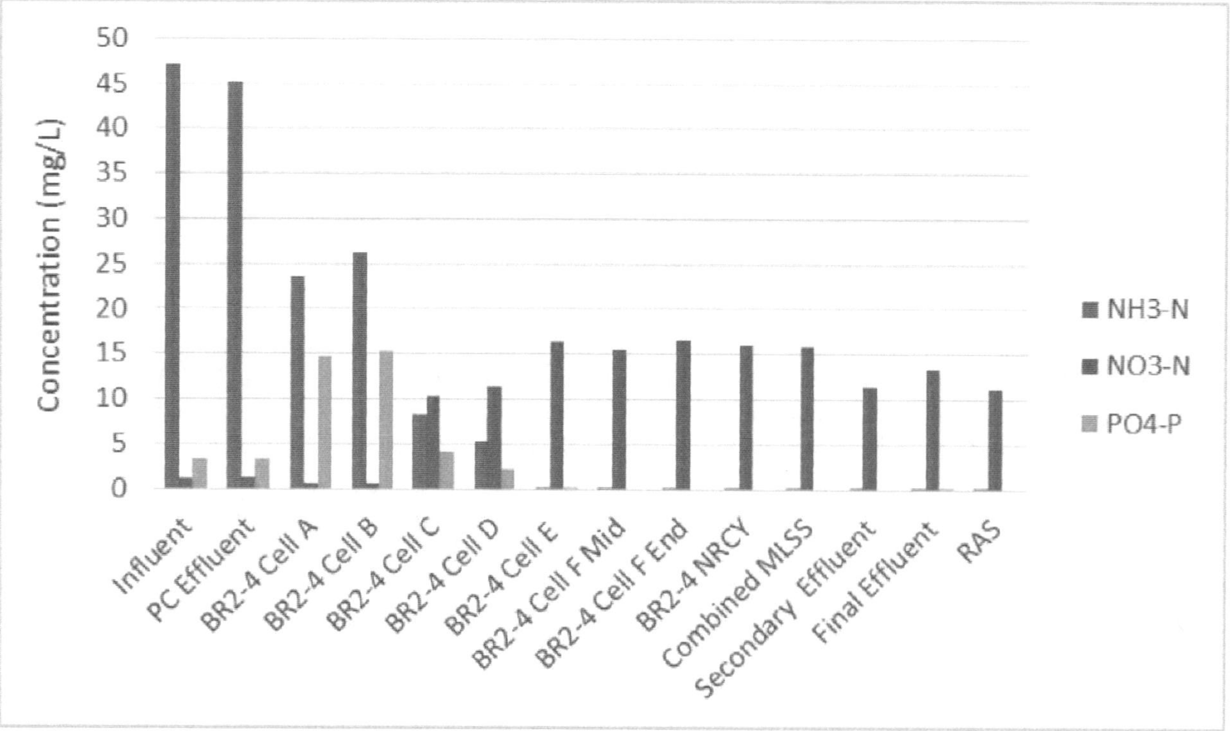

Figure 4 – Case Study No. 2 - Nutrient Profile Indicating Backmixing in Cell E and Full Nitrification in First 1/3 of Aerobic Zone

Figure 5 presents the typical dissolved oxygen profile observed during the sampling event. High DO concentrations were observed in Cell F during most of the profiles, with DO concentrations exceeding 4 mg/L. The DO was also high in the NRCY channel, indicating potentially high DO input to the anoxic zone Cell C. The impacts of back-mixing from Cell E to Cell D was evident in the DO data as DO was present in Cell D in several profiles. The DO profiles indicate that DO return is adversely impacting denitrification performance and may be a contributing factor to the variability in historical effluent nitrate and phosphorus concentrations. The process profiles also indicate the accumulation of excess dissolved oxygen through the aerated zones due to full nitrification occurring by the end of Cell E and the poorly tapered aeration system design.

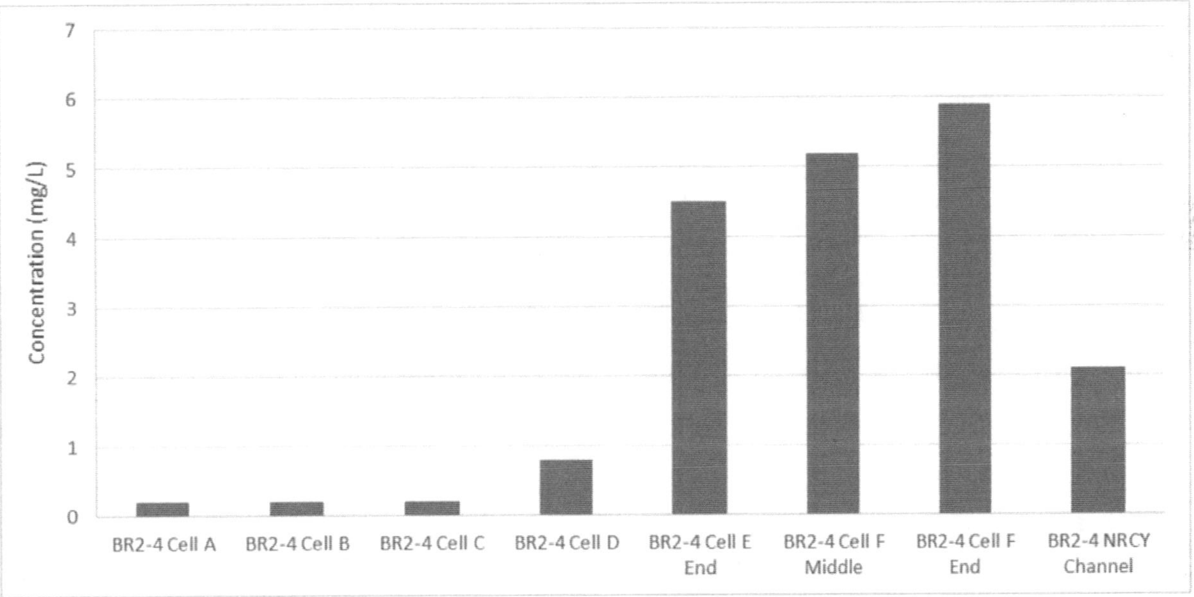

Figure 5 – Case Study No. 2 - Dissolved Oxygen Profile Exhibiting Excess DO due to Back-mixing in Anoxic Zone (Cell D)

Case Study No. 3 – Inadequate Dissolved Oxygen Distribution Results in Reduced Nutrient Removal Efficiency

A 55 ML/day (14.5 mgd) WRRF is operated in a step feed BNR configuration. The step feed configuration consists of three trains of four cells each, which discharge to a common effluent channel followed by additional aerobic volume prior to secondary clarification. Only two of the three step feed trains (Trains 1 and 3) had been operated since the facility was modified to step feed in 2010. The third step feed train (Train 2) was placed in service in January 2012 to provide additional cold weather nitrification volume, and **Figure 6** illustrates the activated sludge process operation once the third train was put online. Putting the third train online was expected to maintain current levels of nutrient removal performance during cold weather. However, as **Figures 7 and 8** demonstrate, respectively, effluent TN and TP concentrations increased once the third train was put online.

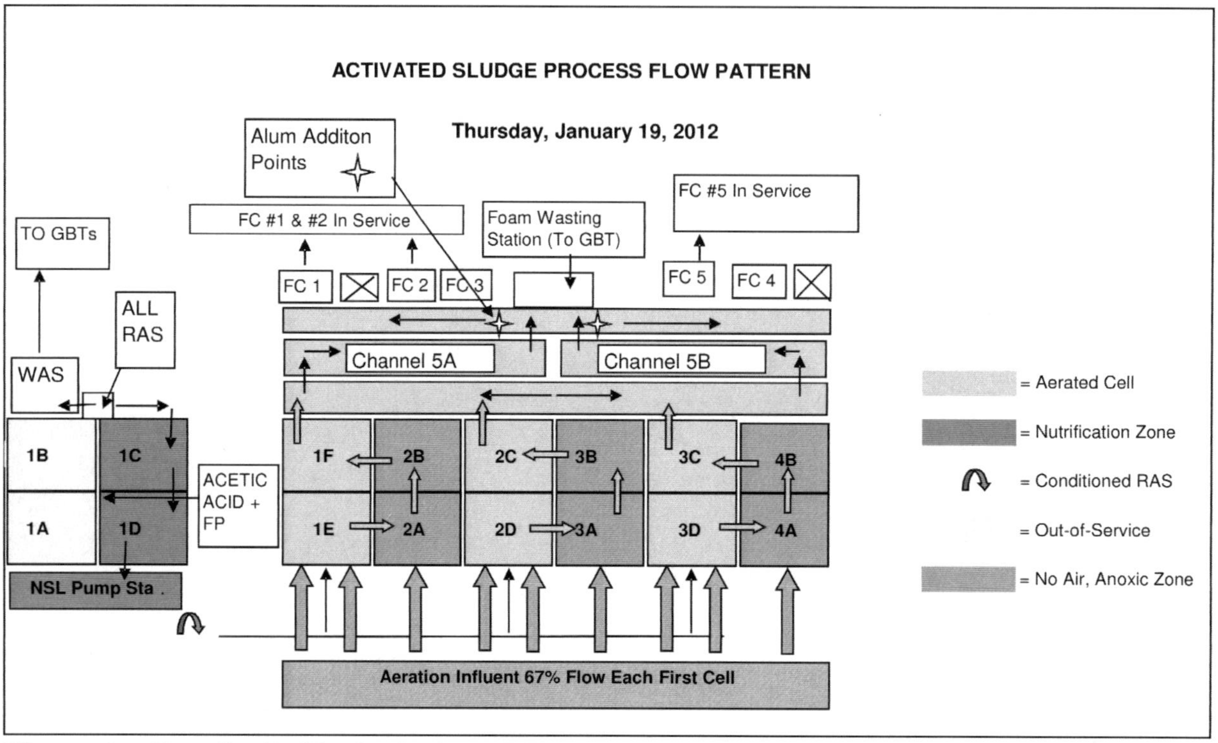

Figure 6 – Case Study No. 3 - Activated Sludge Process Flow Pattern

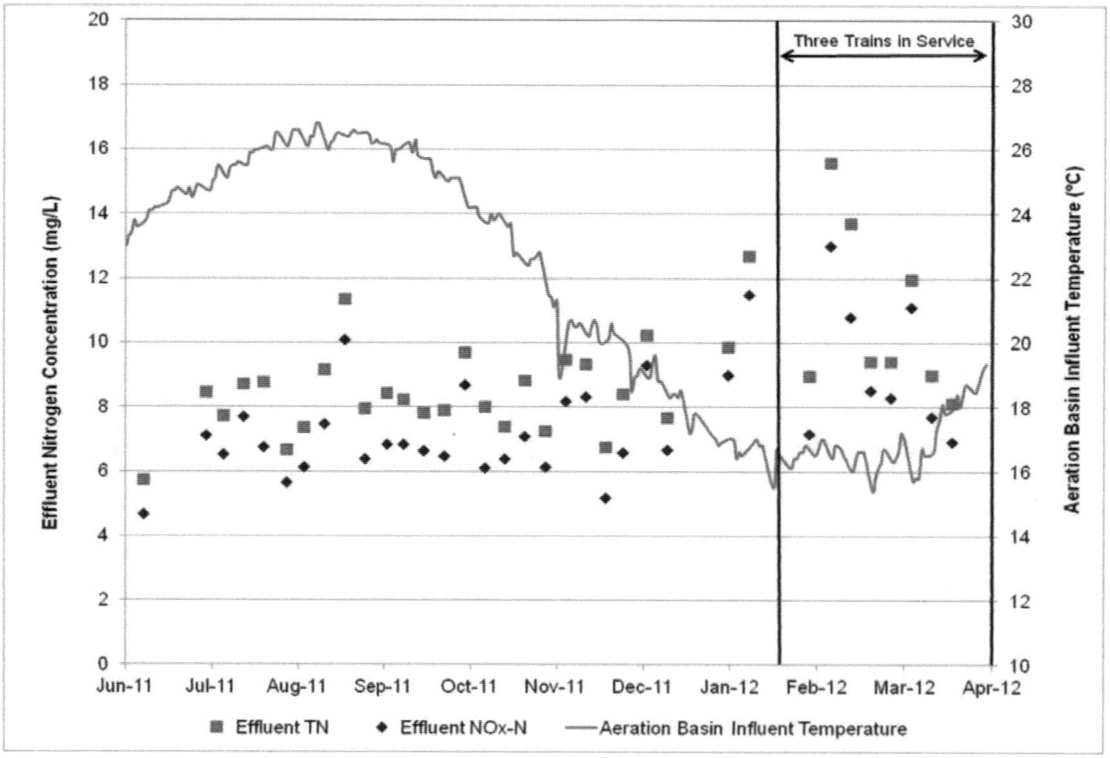

Figure 7 – Case Study No. 3 – Effluent Total Nitrogen Concentrations

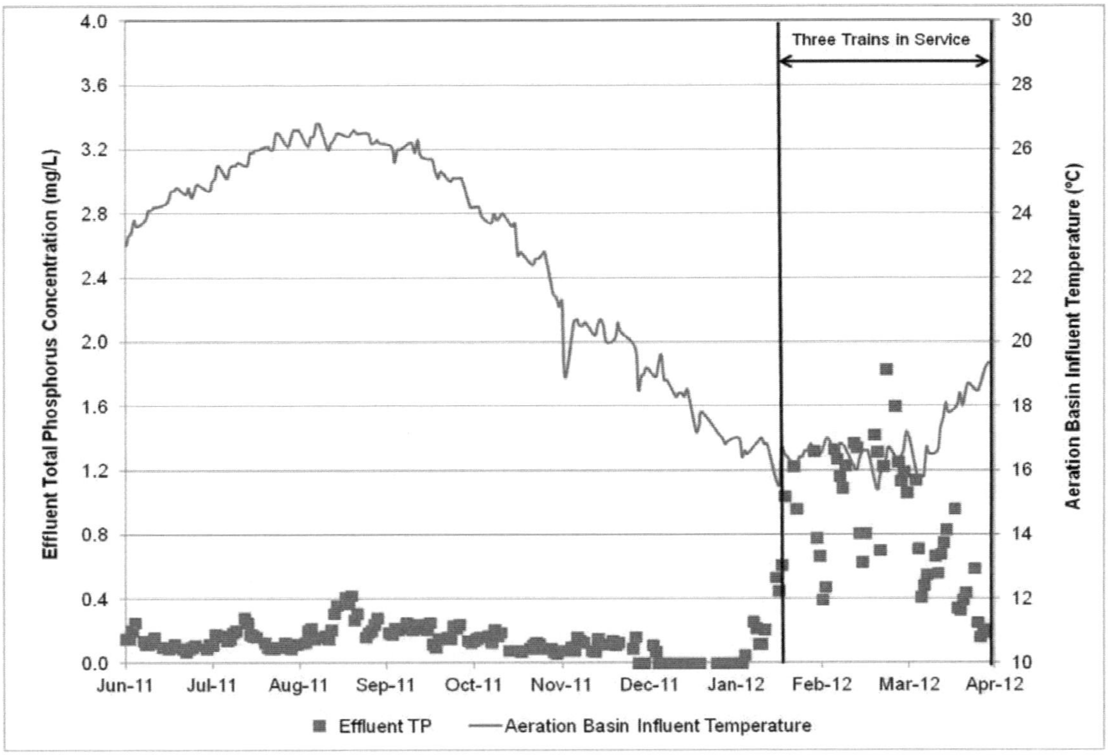

Figure 8 – Case Study No. 3 – Effluent Total Phosphorus Concentrations

Process aeration at the time was provided by jet aeration. Control of the jet aeration system was minimal and the amount of air provided to each cell varied among basins. The lack of control led to differential dissolved oxygen concentrations and subsequent nutrient removal performance between trains. **Figure 9** presents the dissolved oxygen profiles through the bioreactors after the third train was put online. Very low dissolved oxygen concentrations were observed in the Train 2 aerobic zones compared to the other two trains, resulting in incomplete nitrification and insufficient phosphorus uptake in Train 2.

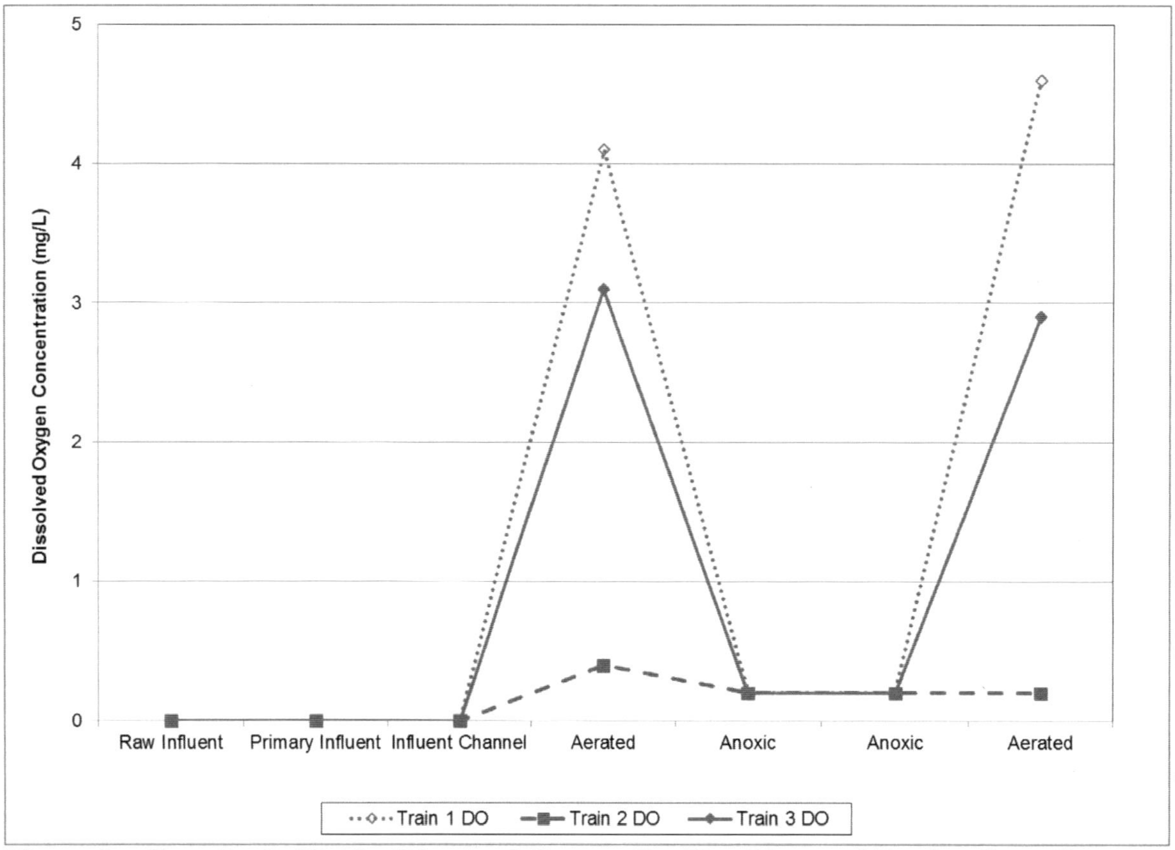

Figure 9 – Case Study No. 3 – Dissolved Oxygen Profiles Observed During Sampling

Figure 10 profiles the ammonia and nitrate concentrations observed through the process during supplemental sampling. The profiles indicate an effluent ammonia concentration of approximately 8 mg/L in the second aerobic zone of Train 2, whereas ammonia was almost completely removed in Trains 1 and 3. Subsequent nitrification of the additional ammonia in the effluent channels and downstream aerated zones resulted in additional nitrate production.

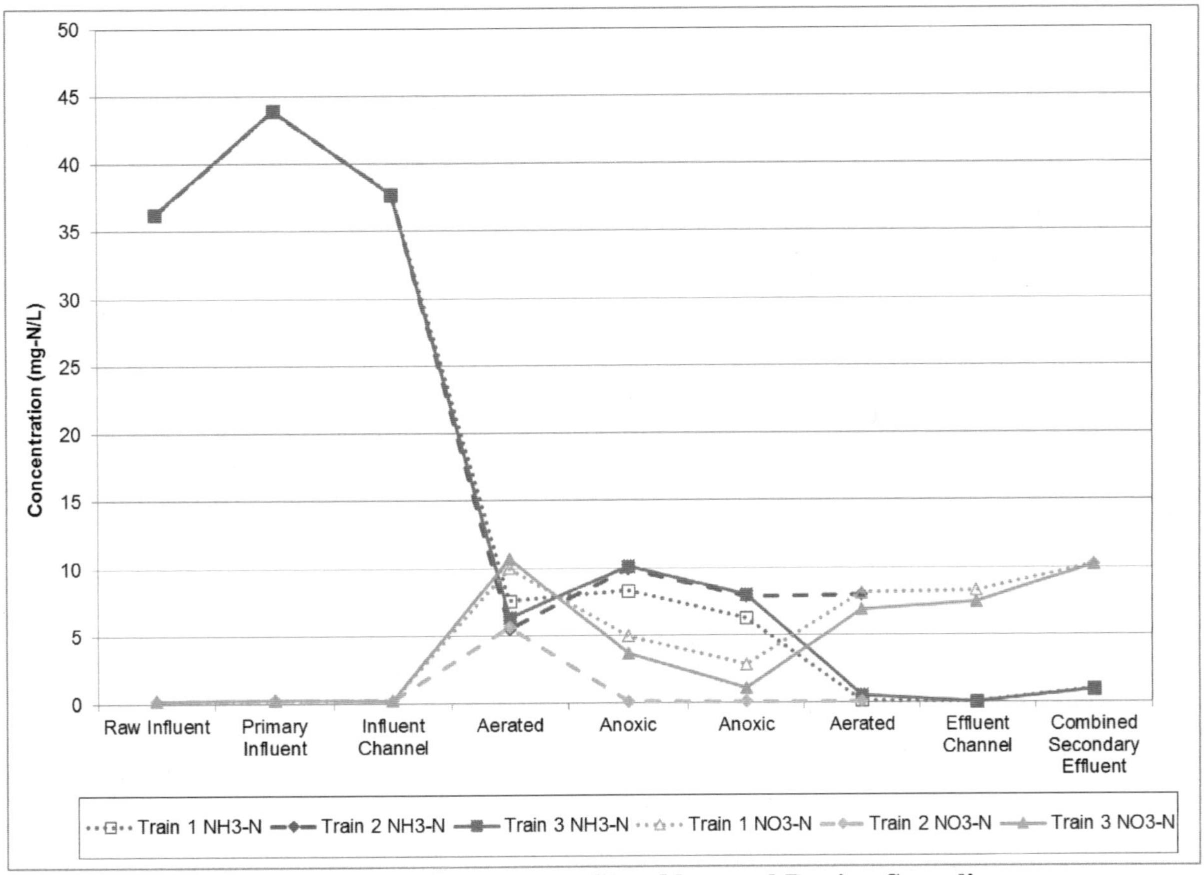

Figure 10 – Case Study No. 3 – Nitrogen Profiles Observed During Sampling

Figure 11 presents the ortho-phosphate concentration profiles. Biological phosphorus removal in Train 2 is particularly poor due to the lack of dissolved oxygen in the final aerobic cell, which prevented luxury phosphorus uptake. The Train 2 effluent ortho-phosphate concentration was nearly 10 mg/L during sampling. Train 1 and Train 3 performed well from a biological phosphorus removal perspective with ortho-phosphate concentrations of 0.4 mg/L and 0.7 mg/L, respectively. The poor performance in Train 2 severely limited biological phosphorus removal efficacy, and increased alum addition was necessary to sufficiently reduce effluent TP concentrations.

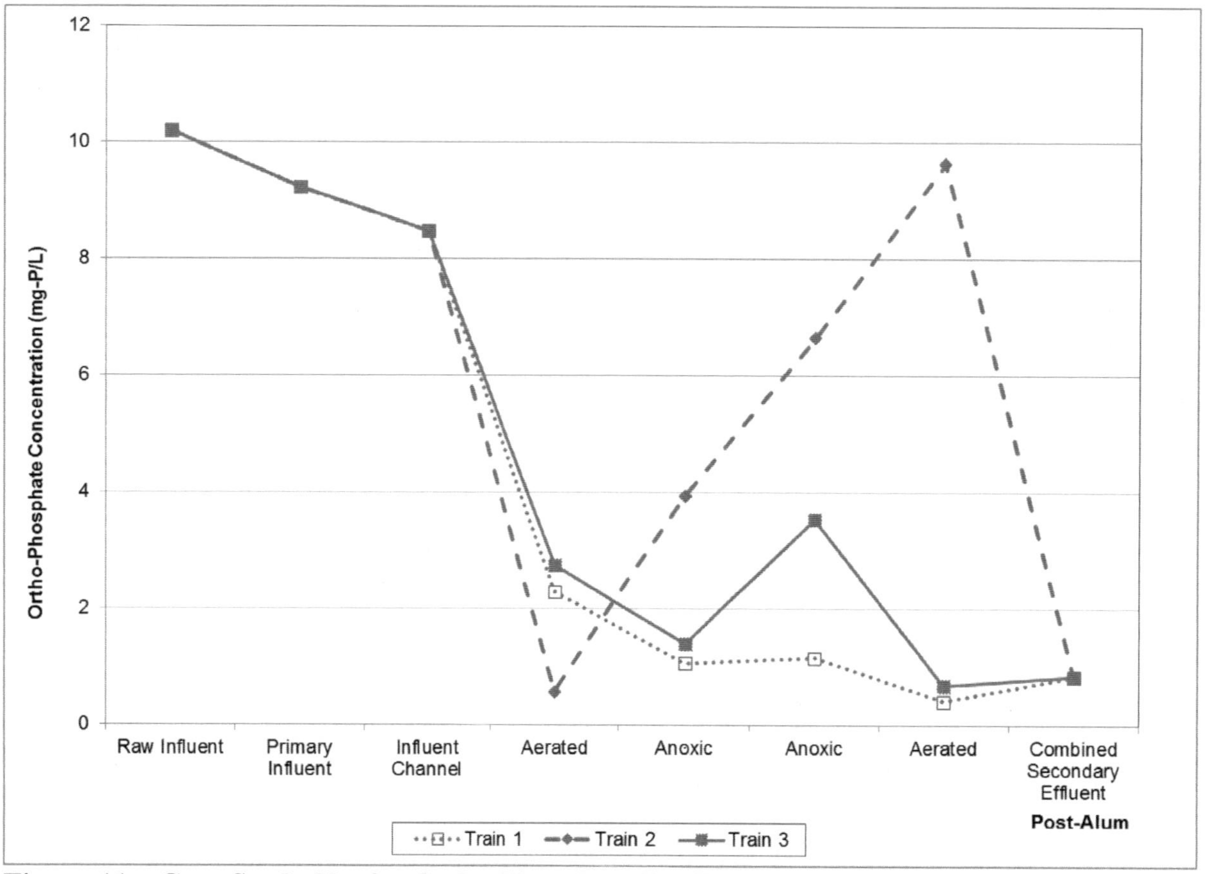

Figure 11 – Case Study No. 3 – Ortho-Phosphate Profiles Observed During Sampling

In conclusion. nitrogen and phosphorus removal efficiency decreased once Train 2 was online due to the inability to maintain sufficient aerobic zone dissolved oxygen concentrations, which adversely impacted nitrification and denitrification performance as well as prevented luxury phosphorus uptake from occurring in Train 2. It was recommended that Train 2 be taken back offline to improve performance. The facility has since replaced its existing jet aeration system with a new fine bubble diffused aeration system, which should improve overall nutrient removal performance year round regardless of which trains are in service.

CONCLUSIONS

The case studies illustrate the importance of aeration system design and operation in providing reliable, cost-effective nutrient removal. Lessons that can be learned from the success and shortcomings of existing facilities include the following:

- <u>Provide an effective diffuser taper to facilitate dissolved oxygen control and avoid excessive dissolved oxygen concentrations</u> - The oxygen demand varies due to changes in organic and nitrogen load down a plug-flow biological reactor. A tapered aeration diffuser layout configures the diffusers to follow the oxygen demand profile in the reactor. Thus, more diffusers should be located at the influent end (i.e. highest load and oxygen demand) and be tapered or reduced towards the effluent end of the reactor.

Reducing the density of diffusers towards the end of the aeration basin reduces the risk of excessive aeration to provide adequate mixing and reduces the potential for excess DO concentrations in internal recycle streams. Process modeling can be used to simulate oxygen uptake rates for each cell to determine the oxygen distribution among the zone of the aeration basin.

- Decrease dissolved oxygen prior to internal recycle flow collection points and supplemental carbon addition – Providing a separate DO control zone and operating at low DO concentrations or providing an unaerated, mixed zone prior to nitrified recycle collection and/or supplemental carbon addition to a post-anoxic zone improves denitrification performance and reduces supplemental carbon requirements.

- Eliminate the potential of back-mixing into anoxic or anaerobic zones - Baffle walls between unaerated and aerobic zones should allow for a hydraulic drop from the unaerated zone to the aerated zone to prevent back-mixing of aerated MLSS. Baffles should also be carefully designed to allow passage of scum and foam through the basin.

- Maintain aerobic conditions to avoid secondary phosphorus release – Operating biological nutrient removal systems at low DO and nitrate concentrations results in anaerobic conditions. If these conditions occur in the absence of sufficient volatile fatty acids (VFAs), the phosphorus accumulating organisms responsible for luxury phosphorus uptake will release stored phosphorus and will be unable to perform luxury phosphorous uptake in subsequent aerobic zones. Well-designed BNR and aeration systems provide adequate dissolved oxygen control and anoxic zone flexibility to reduce the chance of anaerobic conditions occurring downstream of the initial anaerobic zone commonly provided in BNR and ENR systems.

- Provide automated dissolved oxygen control systems suitable for the process objectives and WRRF staff capabilities - The use of instrumentation (i.e. dissolved oxygen probes and ammonia analyzers) to further automate aeration system control is gaining traction. It is important that these systems are properly designed and maintained in order to maximize the benefit gained from these systems. Engineers should be cognizant of WRRF staffing limitations when it comes to operation and maintenance of instrumentation and design systems accordingly.

REFERENCES

Krause, Terry L. et. al. 2010. Design of Muncipal Wastewater Treatment Plants: WEF Manual of Practice

Rosso, Diego and Stenstrom, Michael K. 2006. Economic Implications of Fine-Pore Diffuser Aging. Water Environment Research, Volume 78, No. 8 (August): 810-815

NUTRIENT RECOVERY AT THE F. WAYNE HILL WATER RESOURCES CENTER: DESIGN AND FULL SCALE START UP

Ron Latimer, Vivi Nguyen, Eddie McCallum, Paul Pitt: Hazen and Sawyer
Robert Harris, Tyler Richards: Gwinnett County DWR

The F. Wayne Hill Water Resources Center (FWHWRC) is Gwinnett County's largest and most advanced wastewater treatment facility with a capacity of 60 mgd. The FWHWRC uses enhanced biological phosphorus removal and chemical trim to meet a stringent TP limit of 0.08 mg/L. Solids handling consists of anaerobic digestion of combined primary sludge and WAS in egg-shaped digesters.

Over the past three years, struvite precipitation has increasingly become a major issue at the FWHWRC. Plant staff have used high pressure blasting of centrate lines in the past; however, struvite deposits are now being found in the dewatering centrifuges and upstream of dewatering. Struvite control, beyond reactive blasting of centrate lines, was needed moving forward. In addition, combined primary and secondary sludge from Gwinnett County's 22 mgd Yellow River WRF is transferred to the FWHWRC resulting in significant additional phosphorus load, and will result in more severe phosphorus recycle loads in the future.

To address the struvite issues and decrease the impacts of phosphorus recycle loads on the main liquid stream, while at the same time creating a sustainable fertilizer through nutrient recovery, Gwinnett County evaluated the OSTARA Pearl® and the Multiform Harvest nutrient recovery processes. The OSTARA Pearl® process creates controlled conditions of struvite precipitation for P and N recovery as a high quality fertilizer (Crystal Green). The Multiform Harvest process also creates a controlled struvite recovery process. Implementation of a Pearl® type process or Multiform Harvest process will provide a significant reduction of nuisance struvite formation. In addition, each process can reduce the recycle phosphorus load by up to 80-90%.

The controlled struvite precipitation in the Pearl® process can be enhanced with the addition the patented process called WASSTRIP (wasted activated sludge stripping). The WASSTRIP process combines dewatered centrate with phosphorus and magnesium-rich WAS filtrate (from a WAS P-Release Reactor) into the Pearl® process, increasing Crystal Green production and reducing struvite in solids handling.

The evaluation included bench scale testing, pilot scale testing, and a business case evaluation. From the results of the evaluation and a competitive RFP process, Gwinnett County opted to move forward with a Nutrient Recovery project utilizing Ostara's WASSTRIP process. Design commenced in 2013, construction began the following year and is scheduled to be completed in June 2015. Once complete, this facility will be only the second full scale WASSTRIP installation in the United States. This case study will present the design for the facility as well as startup and operating data for the WASSTRIP and Nutrient Recovery processes.

Pilot Testing

Prior to the design phase, the WASSTRIP process was piloted at FWHWRC. Figure 1 shows the process schematic and sampling points of the pilot reactor setup. WAS and PS were pumped into a tank on a 50/50 mass basis, mixed for 3 hours, then dosed with polymer and fed to a screen filter for solids separation. The P-Release filtrate stream was fed into another

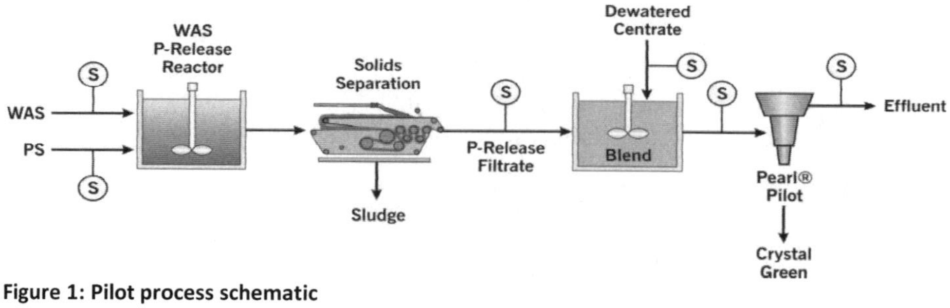

Figure 1: Pilot process schematic

tank where it was blended with dewatering centrate on an 80/20 volume basis and fed to the Pearl® reactor. The pilot scale process was able to achieve PO4-P concentrations in the P-Release filtrate comparable to bench scale values.

Figure 2 shows influent, effluent and P-removal % across the OSTARA Pearl® reactor. These results show the majority of the releasable PO4-P from the Bio-P WAS can be removed prior to digestion, producing a green fertilizer and reducing struvite potential significantly.

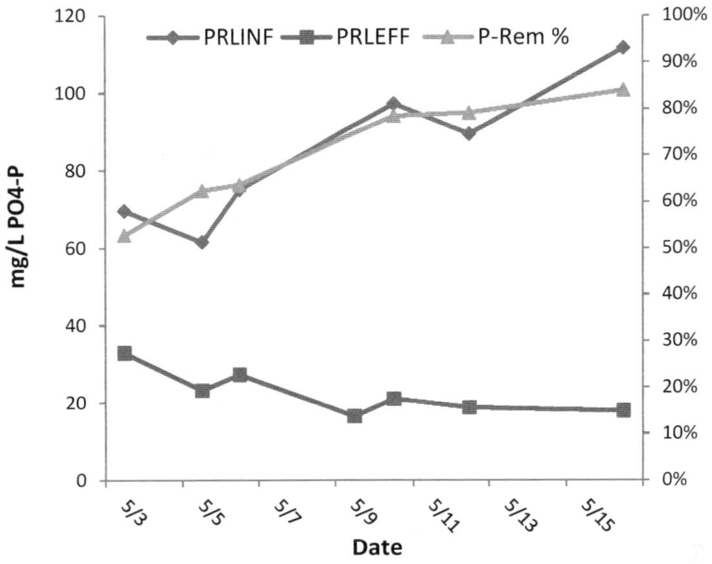

Figure 2: Influent and effluent PO4-P concentrations of Pearl® reactor along with % P-removal.

Full Scale Design

The Nutrient Recovery System consists of the following elements: nutrient recovery reactors, centrate and filtrate storage tanks, transfer pumps, fertilizer product handling system, and chemical feed systems. The WASSTRIP process consists of a holding tank where primary sludge and WAS are allowed to react anaerobically for approximately two to four hours. The typical WAS concentration at the FWH facility averages 12,000 to 15,000 mg/L and primary sludge at 30,000 mg/L, so pre-thickening prior to the P release tank was not included. The combined sludge is then thickened by rotary drum thickeners and the filtrate which is phosphate and magnesium rich is stored as feed to the nutrient recovery reactors. Filtrate and centrate equalization tanks, plus one swing tank are 500,000 gallons each. The facility has two nutrient recovery reactors with space for a third one in the future. Each reactor has a nominal capacity of 4,400 pounds of struvite production daily. Figure 3 shows the construction of part of the nutrient recovery facility.

Figure 3: Construction of the Pearl® reactor

The design also includes PVC centrate pipes with removal sections in the dewatering building. Parallel HDPE centrate pipes into the recovery facility were used including an acid feed loop to allow periodic cleaning of the feed pipes. The equalization tanks were designed to allow heavy solids to settle and include automatic draining and washdown.

The FWHWRC nutrient recovery process is unique in that the utility adds $Mg(OH)_2$ to the collection system for odor and corrosion control and therefore Mg does not need to be added to the Ostara process. Data from the first month of operation of the WASSTRIP and nutrient recovery process will be provided including reactor influent flows, reactor influent phosphorus and ammonia loads, pounds of struvite produced, percent removal of phosphorus and ammonia over the reactors and chemical usage. Additional discussion to be presented include startup and operational issues. As this is only the second WASSTRIP installation in the U.S., the operational data to be presented will be a valuable to the industry data base on the potential of this technology.

Figure 4: P Release tank with combined primary and WAS

Figure 5: Top of centrate EQ tanks with retractable covers

Prove It! Demonstrating Performance of a Combined Carbon and Ammonia Removal BAF

Jon R. Pearson[1*], Terry L. Desmarais[2], Paula Anania[2], Erik N. Meserve[1], Erik J. Grotton[3]

[1] AECOM, Wakefield, Massachusetts
[2] City of Portsmouth, New Hampshire
[3] Blueleaf Incorporated, Charlton, Massachusetts
* Email: Jon.Pearson@aecom.com

ABSTRACT

The City of Portsmouth, New Hampshire conducted an initial piloting program that recommended the use of a two-stage Veolia Water Technologies BIOSTYR™ Biological Aerated Filter (BAF) to upgrade the City's existing wastewater treatment facility. A subsequent pilot study, referred to as the Demonstration Study, was conducted at the pilot scale to test the first stage BAF at, or close to, the maximum month loading rates proposed for the full-scale design and at wastewater temperatures close to the design minimum temperature. The Demonstration Study also provided the opportunity to determine BAF backwash solids characteristics and their impact on other unit processes. The Demonstration Study confirmed the full-scale design loading rates and provided information used in the design of chemical feed and solids handling processes. The methodology, results, and conclusions of the Demonstration Study are reviewed in this paper.

KEYWORDS: BAF, nitrification, pilot study, co-settling, co-thickening, solids thickening, dewatering

INTRODUCTION

The City of Portsmouth, New Hampshire has been issued a Consent Decree by the United States Environmental Protection Agency (EPA) to upgrade the City's existing chemically enhanced primary treatment (CEPT) Peirce Island Wastewater Treatment Facility (WWTF) to provide secondary treatment. The EPA has also indicated that removal of total nitrogen (TN) to 8 mg/L with the potential for lower future TN limits is likely to be required in the near future. As such, the City has decided that the WWTF upgrade to comply with the Consent Decree will also include TN removal. The WWTF's current design flow is 18.2 ML/day (4.8 MGD) with a peak flow of 88.3 ML/day (22 MGD). The City's Final Wastewater Master Plan (WWMP) (November 2010) recommended high rate, small footprint treatment technologies to provide secondary treatment due in large part to extreme site limitations on Peirce Island. Because these high rate technologies have limited full scale installations, it was recommended that the technologies be piloted to determine the most applicable technology for upgrading the WWTF. The piloting program was then undertaken in phases as described in Pearson and Desmarais (2013). Based on the results of the initial piloting program, the use of a two-stage Veolia Water Technologies BIOSTYR™ Biological Aerated Filter (BAF) was recommended. The first BAF stage accomplishes both carbon oxidation and nitrification while the second stage is designed for denitrification.

The initial pilot study in 2012 focused mainly on secondary treatment because the EPA had not stated its intent to impose a total nitrogen requirement at the time. As a result, results from the initial pilot study did not include extensive data on ammonia removal from the combined carbon oxidation and nitrification BAF at maximum month loading rates. No data were obtained in the initial pilot study at cold temperatures while the pilot operated in this configuration. A subsequent pilot study, referred to as the Demonstration Study, was conducted at the pilot scale between January and June 2014 to test the first (Carbon-Nitrogen or CN) stage BAF at, or close to, the maximum month loading rates proposed for the full-scale design and at wastewater temperatures close to the design minimum temperature of 10° C to verify performance over the range of loadings expected. Additionally, the Demonstration Study provided the opportunity to determine BAF backwash solids characteristics such as solids coagulation, settling, thickening and dewatering and for the existing WWTF's operations staff to become more familiar with the BAF process.

Testing began in early January 2014 and throughout January, February, and March extremely cold temperatures resulted in a number of freezing and breakage challenges. Additionally, the cold wastewater temperature limited development of nitrification during the startup period. By the end of March, nitrification began to take hold in earnest. Testing continued until the beginning of June, by which time wastewater temperatures had risen significantly. Loading rates were adjusted for this rise in temperature to maintain close to the maximum month loading rates proposed by Veolia Water Technologies in the full-scale design.

METHODOLOGY

The primary objective of the Demonstration Study was to evaluate the ability of the first stage BAF to remove BOD and oxidize ammonia at or near the proposed full-scale maximum month loading rates and at wastewater temperatures close to the design minimum temperature of 10°C. Additionally, this study sought to assess BAF backwash solids characteristics such as solids coagulation, settling, thickening and dewatering. The data gained in this study was compared to the basis of design for the full-scale process.

Pilot Equipment

The pilot treatment systems used in the initial phase of piloting were used again in the Demonstration Study. The major components of these systems are described below.

Primary Clarifier. The use of the CEPT process during dry weather will be discontinued with the upgrade of the Peirce Island WWTF. As such, the influent to the two-stage BAF process will be conventional primary effluent, or non-chemically enhanced primary effluent, during dry weather conditions. Non-chemically enhanced primary effluent was therefore used as the influent to the BAF pilot. Since the existing WWTF must operate the CEPT process to meet the existing NPDES permitted effluent limits, the pilot system included a pilot scale primary clarifier. The rectangular pilot primary clarifier was sized to mimic the hydraulic overflow rate of the existing primary settling tanks at the dry day average flow and was fabricated onsite. Raw wastewater flow to the pilot primary clarifier was provided by a pilot influent pump (Pearson and Desmarais, 2013). Effluent from the pilot primary clarifier was then pumped to the BAF. Figure 1 shows a photograph of the pilot primary clarifier.

Figure 1. Pilot Primary Clarifier

BAF. The BAF pilot system consisted of a 0.91 m (3 foot) diameter pilot scale upflow filter column as well as air and backwash systems. The pilot filter cell was a vertical column containing the polystyrene bead media and air feed grids. The BAF pilot unit was provided by Veolia Water Technologies (formerly Kruger) and included feed pumps, backwash pumps, pneumatic valves, compressor, and a programmable logic controller and associated instrumentation for automated operation (Pearson and Desmarais, 2013).

Primary effluent was pumped to the inlet of the upflow filter cell. Air was added to provide oxygen for the aerobic biomass. The effluent from the BAF pilot was directed into a storage tank to provide water to backwash the filter media. During backwash of the BAF, the forward flow was stopped and backwash pumps provided flow to the top of the cell and downwards through the cell to flush out accumulated solids and excess biomass (Pearson and Desmarais, 2013). Figure 2 presents a schematic of the pilot BAF unit, and Figures 3 and 4 present photographs of the pilot BAF unit.

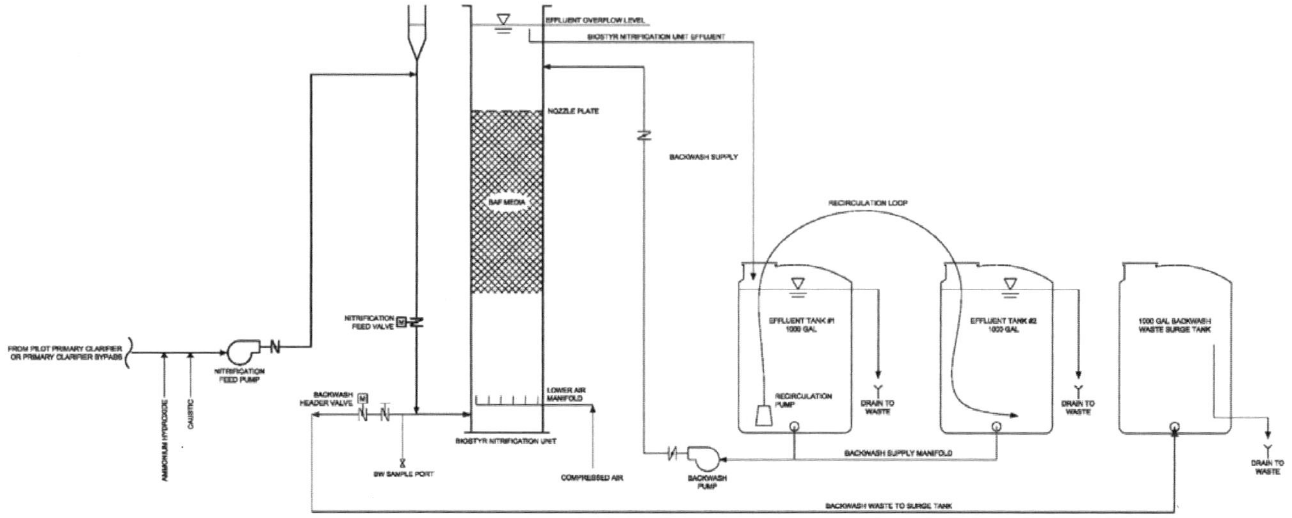

Figure 2. BAF Pilot Unit Schematic

Figure 3. BAF Pilot Unit

Figure 4. Enclosed BAF Pilot Unit

Demonstration Study Protocol
The Demonstration Study included a number of experimental trials to provide the desired data. Each of the experiments conducted are described below.

Jar Tests. The full-scale design of the two-stage BAF system includes the return of the BAF backwash waste to the primary clarifiers to be co-settled with the raw influent wastewater. While the settling of raw influent wastewater will not be chemically enhanced, the backwash waste sludge will be dosed with ferric chloride and polymer to assist in settling. Backwash waste sludge from the pilot unit was used to conduct a series of twelve jar tests with the goal of assessing ferric chloride and polymer doses that would effectively coagulate the backwash solids. A response surface experiment was designed to test a range of ferric chloride and polymer doses. Upon completion of the jar tests, the supernatant was tested for pH, COD, soluble COD, TSS, TN, NOx, and TKN.

Settling Tests. Three settling column tests were conducted to simulate primary clarifier performance while co-settling BAF backwash with the influent wastewater after the BAF is installed. Settling column tests were conducted as described in Metcalf & Eddy, 2014. A schematic is shown in Figure 5 below. The depth of water in the settling column was 3.66 meters (12 feet) in order to match the existing primary clarifiers. The first test was used to compare the settling column to the primary clarifier while the second two tests were used to simulate future conditions with backwash waste mixed with raw wastewater. Ferric chloride and polymer were mixed with the backwash waste using the doses determined in the jar testing prior to combination with the raw wastewater. TSS was measured at regular height and time intervals along the test settling column to generate settling curves.

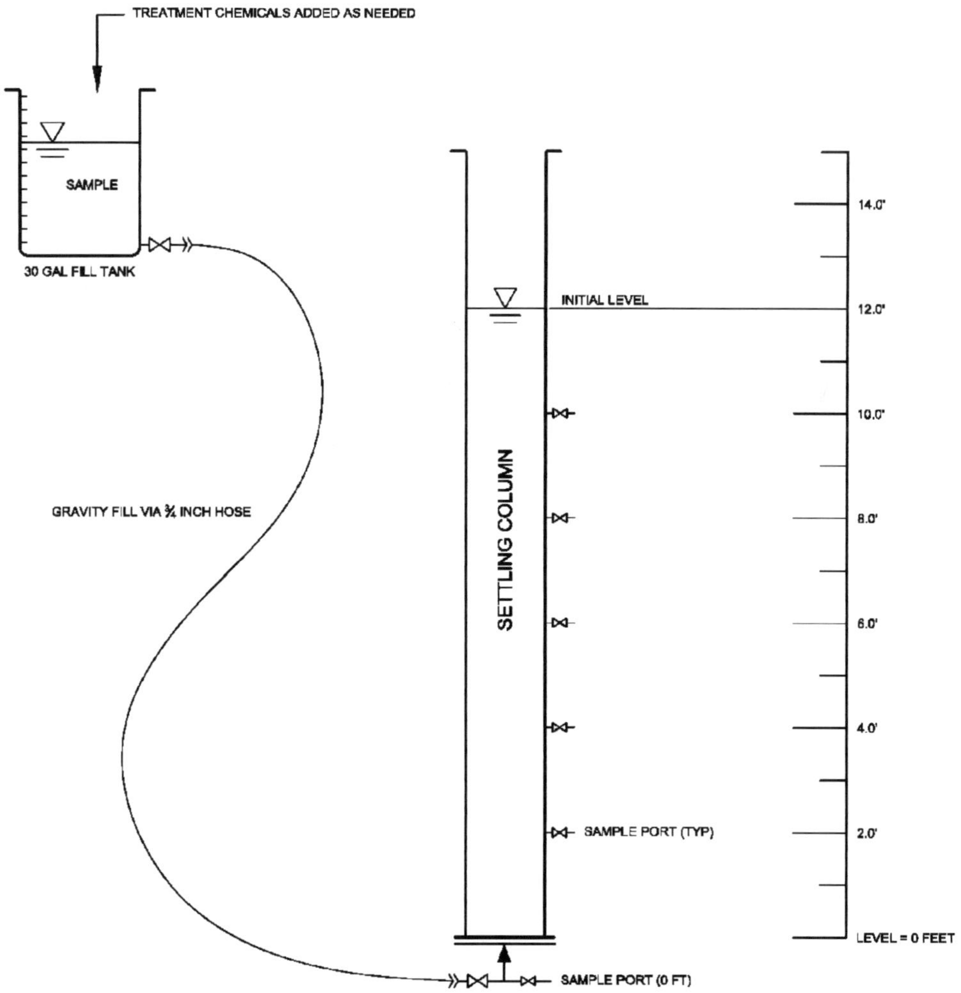

Figure 5. Settling Test Schematic

BAF Pilot Unit. After re-starting the pilot unit that remained on-site after the initial piloting, the objective of the Demonstration Study was to operate continuously at or near the maximum month loading rates proposed for full-scale and at or near the minimum design temperature of 10°C. Table 1 below presents these loading rates.

Table 1. First Stage BAF Maximum Month Hydraulic and Pollutant Loading Rates

Parameter	Loading Rate Objective
Hydraulic Loading Rate	2.5 $m^3/m^2/hr$ (1.04 gpm/sf)
BOD_5 Loading Rate	2.35 $kg/m^3/d$ (147 lb BOD_5 per 1,000 ft^3/d)
NH_3-N Loading Rate	0.22 $kg/m^3/d$ (13.7 lb BOD_5 per 1,000 ft^3/d)

Initially, the Demonstration Study was intended to be operated during the winter in order to evaluate the maximum month loading rates at or near the minimum design temperature. Extremely cold temperatures during January, February and March 2014 however caused numerous problems and delays with the operation of the pilot equipment. On several occasions valves froze and became inoperable. Freezing caused breakages of various pipes, valves, and the BAF feed pump. The pilot primary clarifier also became partially frozen and the waste valves were frozen. Broken pilot components were repaired or replaced, heat tracing and insulation were installed, and a new plastic enclosure around the pilot unit and space heaters were installed. In an effort to encourage nitrification, the pilot was operated on CEPT effluent from the existing primary clarifiers. Eventually nitrification began to occur and the feed to the unit was returned to the pilot primary effluent. However, by this time it was mid- to late-April and wastewater temperatures had begun to rise above 10°C.

In order to compensate for the higher influent temperatures, a temperature correction was applied to the design loading rates. The hydraulic loading rate was controlled on a daily basis to obtain the temperature-adjusted design loading rate for BOD_5. Influent BOD_5 concentrations were estimated based upon field analyses of influent COD, using a ratio of 1.0 mg/L BOD_5 per 1.9 mg/L of COD determined for Portsmouth wastewater during the initial piloting. Ammonium hydroxide was added to adjust the influent ammonia concentration in order to obtain the temperature-adjusted design loading rate for NH_3-N, based upon the field analyses of influent ammonia. Once nitrification took hold, the pilot operated for 6 weeks at or near the temperature-adjusted design loading rates for BOD_5 and NH_3-N.

Compression Tests. As discussed above, the full-scale design includes the return of the BAF backwash waste sludge to the primary clarifiers where it will co-settle with the raw influent wastewater. The BAF backwash waste will be chemically treated to enhance settling, but the raw influent wastewater will not normally be chemically treated. After co-settling in the primary clarifier, the primary sludge/backwash waste sludge mixture will be pumped to two gravity thickeners for co-thickening. No additional chemicals will be added to the primary sludge/backwash waste sludge mixture.

Three compression tests were conducted to simulate gravity thickener operation after the BAF is installed. The mass balance for the final design of the two-stage BAF system indicates that the backwash waste sludge comprises approximately 60% of the primary sludge/backwash waste sludge mixture. The primary sludge used in these experiments was collected from the pilot primary clarifier. Backwash waste sludge from the BAF pilot was treated with ferric chloride and polymer as a backwash was occurring and directed to a surge tank as shown in Figure 6 below. After settling for three hours to simulate co-settling in the primary clarifier, the surge tank was decanted and the backwash waste sludge collected, mixed with the primary sludge and the compression tests were completed. The primary sludge/backwash waste sludge mixture was allowed to thicken for three hours. The settled solids volume was periodically measured, TSS measurements were made at the beginning and conclusion of the test and the BOD_5 of the decant layer was measured at the conclusion of the test.

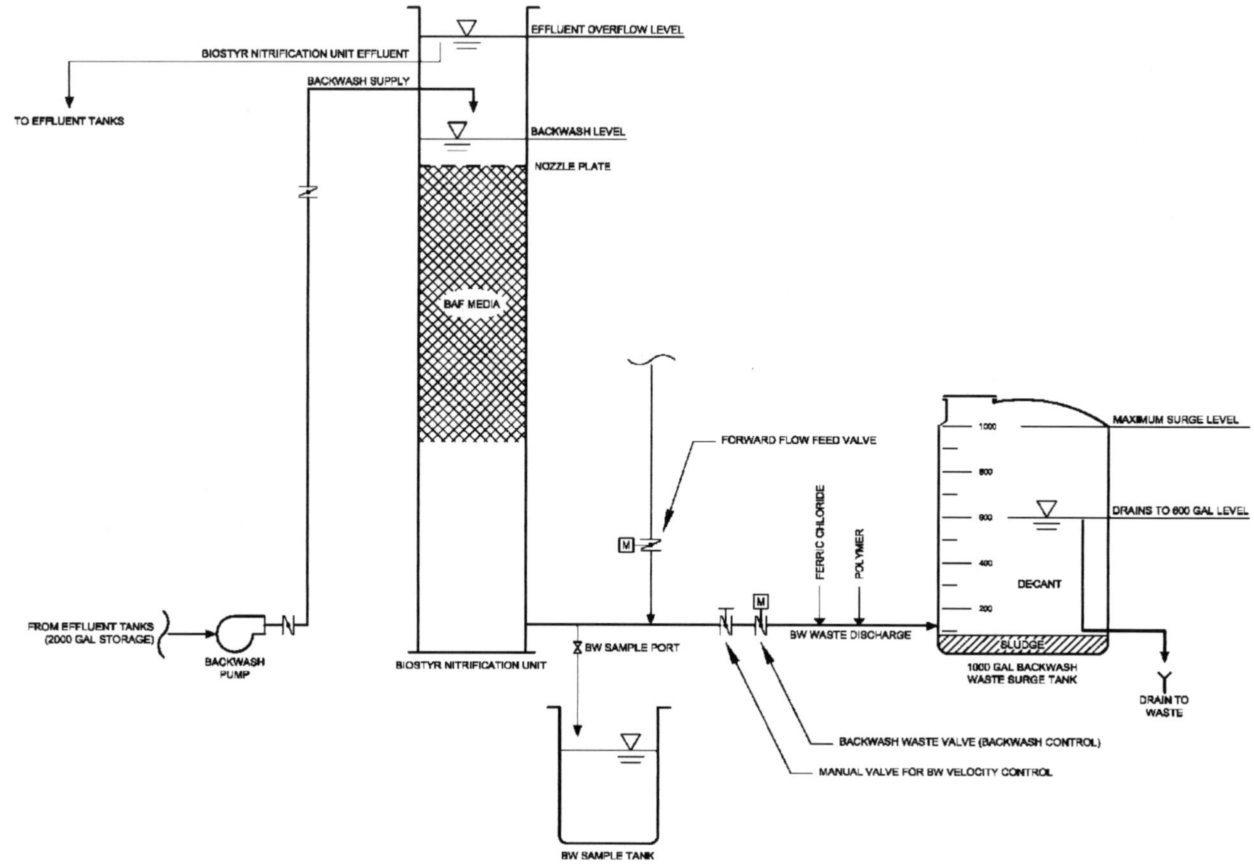

Figure 6. BAF Backwash Schematic

Dewatering Tests. The primary sludge and settled backwash solids mixture used in the compression tests was sent to two different screw press vendors to test at the bench scale the ability of a screw press to dewater the combined solids mixture.

RESULTS

The BAF pilot unit was operated from January to June 2014. Primary effluent was fed to the pilot unit at flow rates ranging from 0.14 L/s to 0.46 L/s (2.2 gpm to 7.3 gpm), BOD_5 loading rates ranging from 1.12 kg $BOD_5/m^3/d$ to 3.52 kg $BOD_5/m^3/d$ (70 to 220 lb $BOD_5/1000\ ft^3/day$) and ammonia loading rates ranging from 0.22 kg NH_3-$N/m^3/d$ to 0.44 kg NH_3-$N/m^3/d$ (13.6 to 27.6 lb NH_3-$N/1000\ ft^3/day$). The results of the piloting effort are summarized below.

Jar Tests

Backwash waste sludge from the pilot unit was used to conduct a series of jar tests with the goal of assessing ferric chloride and polymer doses that would effectively coagulate the backwash solids. A response surface experiment was designed to test a range of ferric chloride and polymer doses. Response surface designs are used to evaluate the effect of two or more independent variables on a dependent variable. They are used when the two independent variables are thought to interact with each other. In water treatment processes, a good example of interacting dependent variables is coagulant dose and pH, where it is well-established that coagulation performance is dependent on pH, while most coagulants change the pH of the water. Since both variables are interactive, and both affect treatment performance, it is difficult to assess performance using a one-factor-at-a-time approach. The response surface design evaluates a range of both variables at the same time. The results of the response surface experiment are displayed below in Table 2.

Table 2. Response Surface Experiment Results

Jar No.	Chemical Doses		Analytical Results						
	Ferric (mg/L)	Polymer (mg/L)	pH	COD (mg/L)	sCOD (mg/L)	TSS (mg/L)	TN (mg/L)	NOx (mg/L)	TKN (mg/L)
0[1]	-	-	7.11	-	-	625	-	-	-
1	0	0	7.25	200	184	21	17.3	6.19	11.1
2	10	1	7.08	203	59.6	30	16	5.94	10
3	20	1.1	7.03	144	50.8	38	16.9	6.82	10.1
4	10	0.5	7.07	154	109	31	16	6.3	9.68
5	30	1	7.00	123	38.3	47	3.76	3.76	0
6	34.1	0.75	6.92	128	33.4	45	13.6	4.68	8.95
7	20	0.4	6.98	137	32.4	44	15.1	5.1	10
8	20	0.75	6.96	137	41.2	53	14.7	4.41	10.3
9	20	0.75	7.00	122	30.3	52	12.7	2.67	10.1
10	30	0.5	6.92	134	28.2	53	11.2	2.55	8.68
11	5.9	0.75	7.10	156	32.4	47	13.8	2.44	11.3
12	20	0.75	6.97	157	32.5	46	13.7	3.7	9.98

[1]Backwash Bulk Sample

Response surface results normally include the preparation of contour plots, showing the expected value of the dependent variable across the range of independent variables. A contour plot is a technique used to compare different combinations of two variables (dose, pH) to estimate a third variable (i.e.COD and TN). This graphical approach can be used in this situation to optimize pH values and coagulant doses to get a minimal total nitrogen concentration. An overlaid contour plot can be created to show the operating conditions to meet multiple objectives.

The overlaid contour plot showing the combined results of Total COD, soluble COD, TSS and TKN is shown below in Figure 7. A high and low limit were selected for each parameter. The high limits were selected to be well below typical primary clarifier effluent values for these parameters (the low limits were set to -100 for all parameters). The shaded gray areas indicate that one or more parameters would be greater than the high limits. The unshaded white area indicates that all parameters would be below the high limits. The white area indicates the potential operating area within which any point would theoretically provide adequate contaminant removal, and any coordinated dose of ferric chloride and polymer could be selected from this area. The overlaid contour plot generally indicates that ferric chloride doses greater than 20 mg/L (as $FeCl_3$) produced acceptable settled water, based upon the selected contaminant limits. Any non-zero polymer dose appears to have been effective (zero polymer dose was not included in the experiment design). The gray area in the upper right hand corner indicating unacceptable sCOD removal at higher ferric/polymer doses is probably an anomaly that would be eliminated if there were more data.

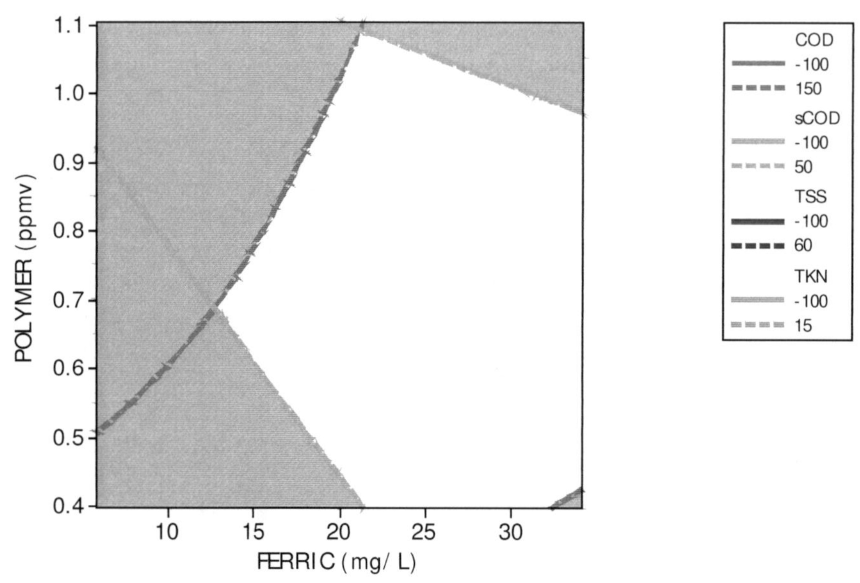

Figure 7. Overlaid Contour Plot of Total COD, Soluble COD, TSS and TKN

TSS removal without coagulant or polymer resulted in a removal rate of 96%. Adding coagulant and polymer slightly increased the effluent TSS because the ferric chloride slightly increased the influent TSS. Nonetheless, the removal rates under all conditions tested were higher than the 74% assumed in the BAF process design. Although effluent TSS may be slightly higher than if chemicals were not added, the coagulant and polymer are expected to remove soluble COD, which will reduce aeration costs associated with the BAF.

Settling Tests

Settling tests were conducted to simulate primary clarifier performance while co-settling BAF backwash with the influent wastewater after the BAF is installed. Prior to completing settling column tests with BAF backwash and influent wastewater together, a settling column test was conducted using only influent wastewater in order to compare the settling column to the full-scale primary clarifier. During this test, the water introduced into the settling column received the same ferric chloride and polymer dose as did the raw wastewater that was introduced into the primary clarifiers.

The results of this test are shown in Table 3 and show that the settling column performed similarly to the existing primary clarifier in regards to effluent TKN and BOD_5. However, the settling column did not match the performance of the full-scale primary clarifier in regards to effluent TSS. At the conclusion of the test, the TSS at the top of the test column was approximately 160-170 mg/L. The effluent TSS from the WWTF is typically 40-60 mg/L. It is believed that the increased wall effects in the small diameter settling column as compared to a large diameter clarifier caused this difference.

Table 3. Comparison of Primary Clarifier to Settling Column

Parameter	Settling Column	Primary Clarifier
TSS, initial (mg/L)	269	233
TSS, final (mg/L)	173	40 – 60 (typical)
TKN, final (mg/L)	30	24
BOD_5, final (mg/L)	103	90

With regards to sludge, the first settling column test measured an underflow solids concentration of 0.6%, which is greater than the plant's typical primary sludge concentration of less than 0.1%. However, it is difficult to compare the results of this test to the current full-scale operation because the plant does not allow a sludge blanket to develop and compact, but rather operates the primary sludge pumps continuously which results in the low primary sludge concentration.

The next two settling column tests simulated the proposed full-scale process. Raw wastewater was combined in the predicted proportion with BAF backwash waste. The raw wastewater was not chemically pre-treated with ferric chloride and polymer but the BAF backwash waste was dosed with ferric chloride at 20 mg/L and polymer at 1 mg/L. The results from these two tests were comparable to each other and are shown graphically in Figures 8 and 9. Sludge TSS ranged between approximately 1.8 and 2 percent solids and overall TSS removal rates were between 45 and 65 percent.

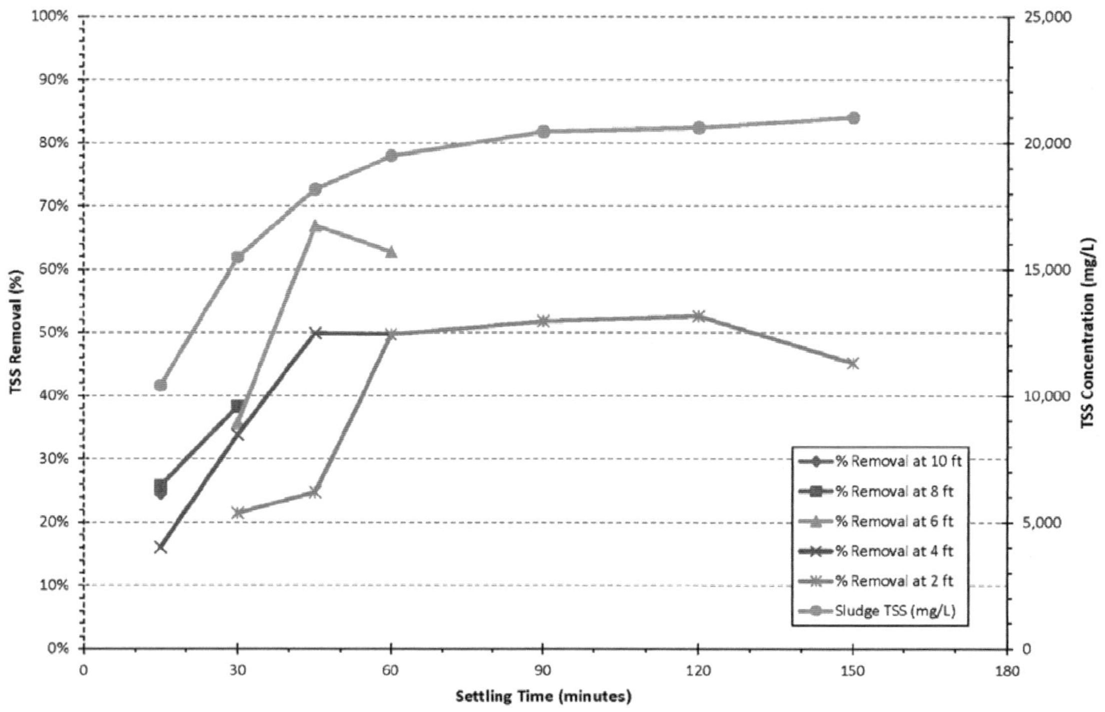

Figure 8. Settling Column Test No.2

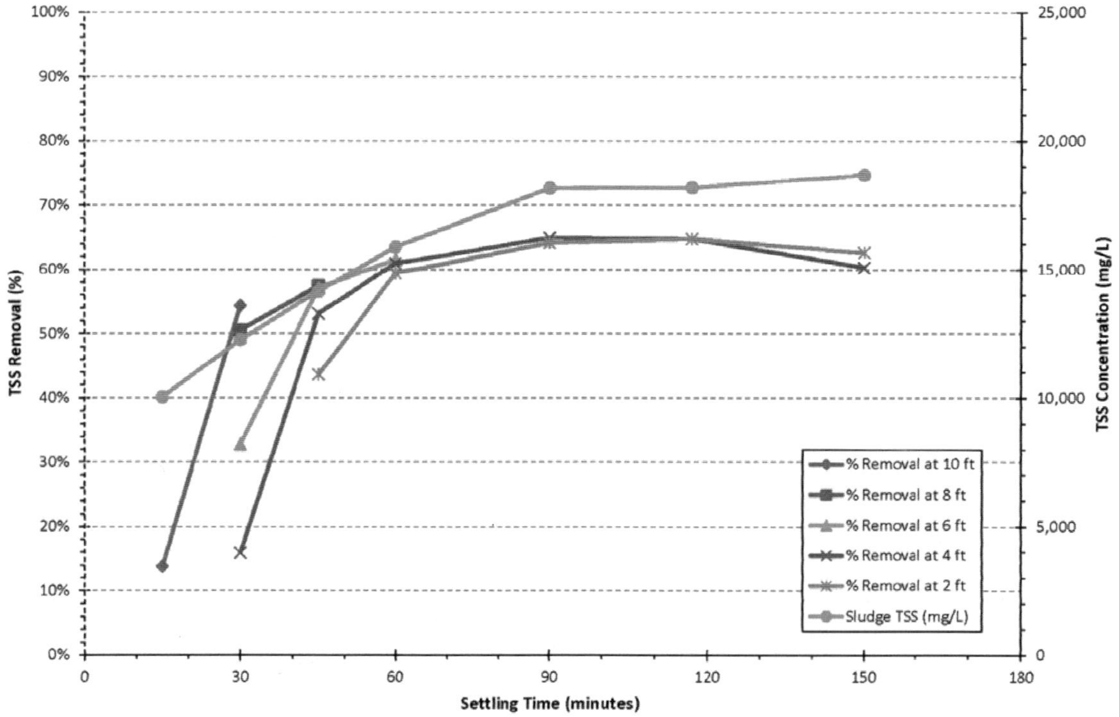

Figure 9. Settling Column Test No.3

The predicted hydraulic retention time in the primary clarifiers after the BAF upgrade is approximately 1.75 to 2.4 hours. At this detention time, the second and third settling column tests show between 53 and 65 percent TSS removal and effluent TSS of approximately 230 mg/L. After applying the ratio of full-scale to bench-scale effluent TSS measured in the first test, the effluent TSS from the second and third tests is estimated to be approximately 80 mg/L, which is well under the maximum concentration of 120 mg/L that the BAF vendor requested.

BAF Pilot Unit

As discussed previously, nitrification did not fully develop in the pilot unit until after wastewater temperatures had begun to warm in the spring. In order to encourage the growth of nitrifying bacteria, the pilot unit was fed with CEPT effluent from the full-scale primary clarifiers and a combination of CEPT effluent and primary effluent from the pilot primary clarifier. The BAF vendor recommended this change in order to reduce the BOD_5 loading rate and encourage nitrifying bacteria. Figure 10 displays the hydraulic data from the Demonstration Study from approximately the time the feed flow was changed to CEPT effluent until the end of the test. Nitrification began to take hold as the flow was changed to a combination of CEPT effluent and primary effluent (referred to as RAW on the figure). Data reported in the sections below is from the time period after only primary effluent was fed to the pilot unit.

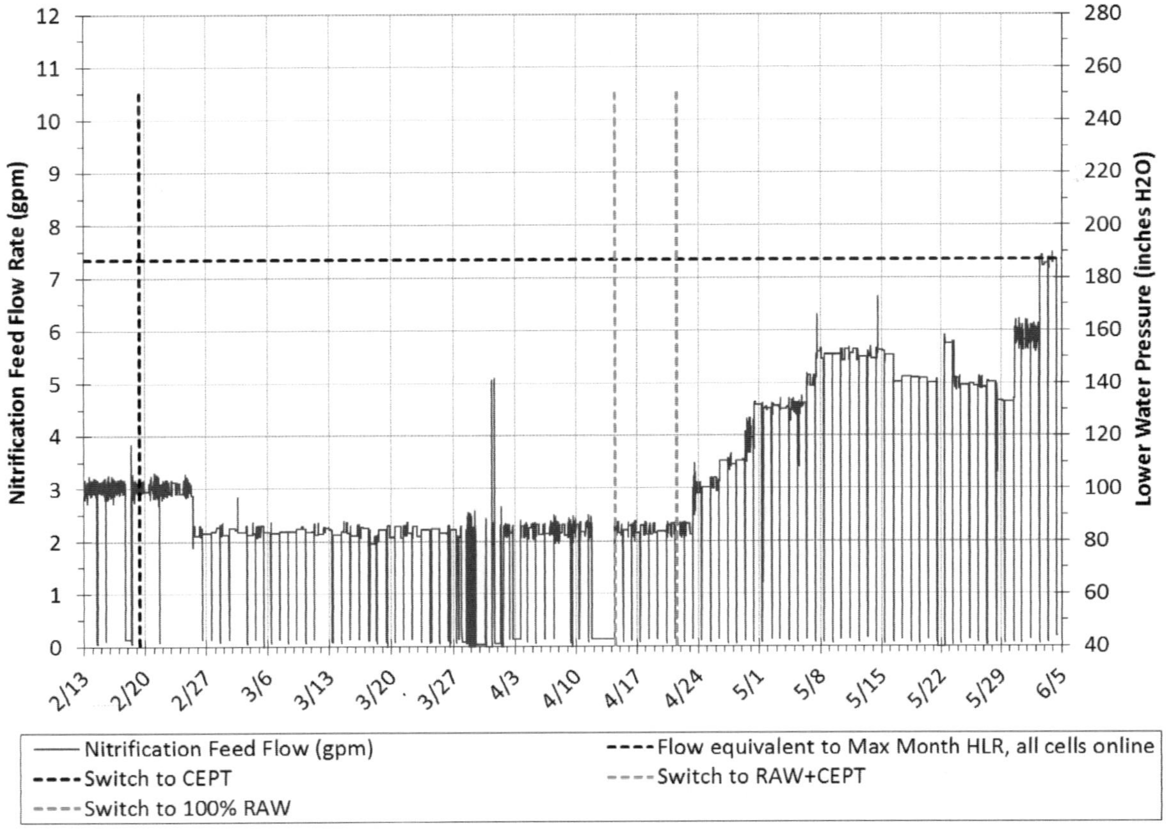

Figure 10. Demonstration Study Hydraulic Data

Table 4 summarizes the influent and effluent data collected by the composite samplers after nitrification had taken hold and the feed flow had been changed to primary effluent from the pilot primary clarifier. Figures 11 through 14 display this data in box plot form. Where applicable, grab data is also included.

Table 4. Laboratory Results of Influent and Effluent Composite Samples

Sample Round	Sample Date[1]	BAF Influent Analytical Results (mg/L)					BAF Effluent Analytical Results (mg/L)							
		BOD_5	TSS	TKN	NH_3-N	PO_4-P	BOD_5	TSS	TKN	NH_3-N	NO_3-N	NO_2-N	PO_4-P	ALK
001	5/5	160	140				11	12						
002	5/6	99					12	18	0.8	0.24	16	< 0.5 [2]	0.93	140
003	5/7			28	20				1.9	0.8				
004	5/8			44	21				0.5	3.0	16	< 0.5 [2]	0.16	120
005	5/9	160	130				16	14						
006	5/11	130	110				8	13						
007	5/12	130	110				10	13						
008	5/13	170	150	31	18	1.10	12	21	1.1	0.67	16	< 0.5 [2]		120
009	5/14	160	210	34	21		14	22	3.7	1.5				
010	5/15	170	150	34	20	1.30	17	21	0.9	1.2	17	< 0.5 [2]		120
011	5/16	200	150				11	10						
012	5/18	140	110				14	16						
013	5/19	120	120				13	15						
014	5/20	150	130	35	18	1.80	13	16	0.7	0.31	17	< 0.5 [2]		110
015	5/21	160	170	30	19		17	19	2.2	0.23				
016	5/23	230	210	44	22	3.00	23	27	1.2	1.2	16	< 0.5 [2]		130
017														
018	5/27						12	19	1.2	0.32	20	< 0.5 [2]		89
019	5/28	200	220	35	17		15	24	2.3	0.23				
020	5/29	150	130	30	18	1.40	11	16	1.7	0.16	18	< 0.5 [2]		95
021	5/30	150	120				13	13						
022	6/1	130	130				13	20						
023	6/2	140	120				15	15						
024	6/3	200	220	40	23	1.90		28	6.1	4.6	15	0.7		140

(1) Date indicates sample collection; auto sampling of composite samples began 1 day prior to collection.
(2) Below reporting limit for parameter; the value shown indicates the limit.

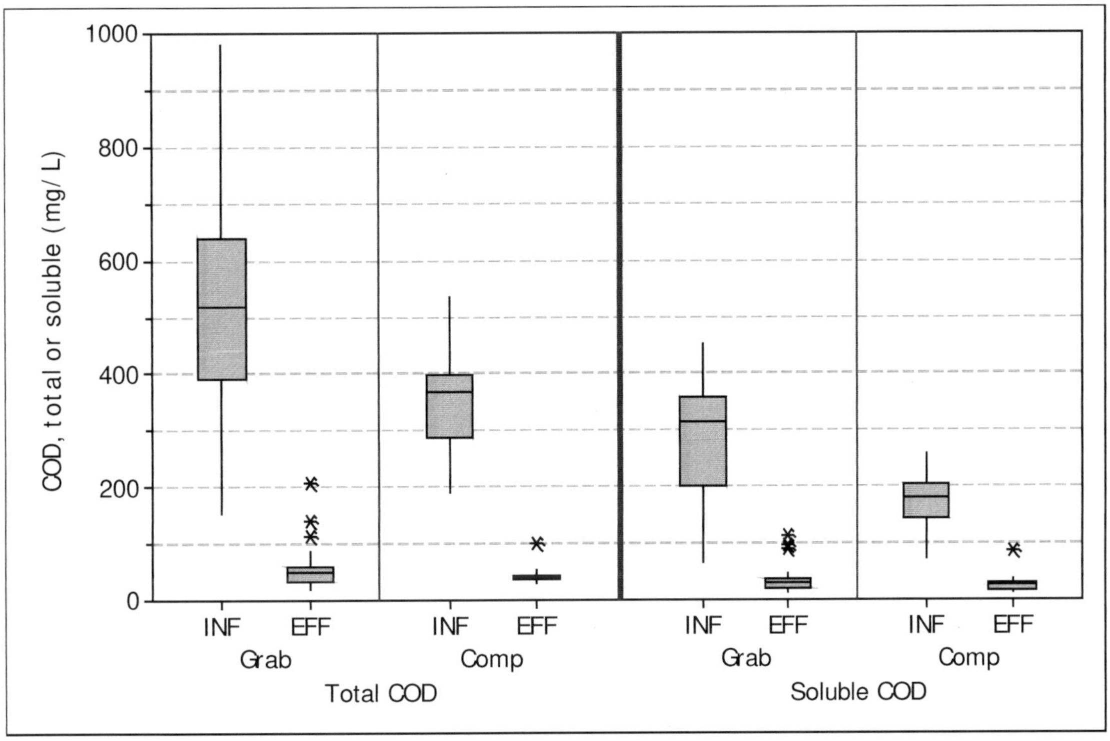

Figure 11. Boxplots of Influent and Effluent COD Data

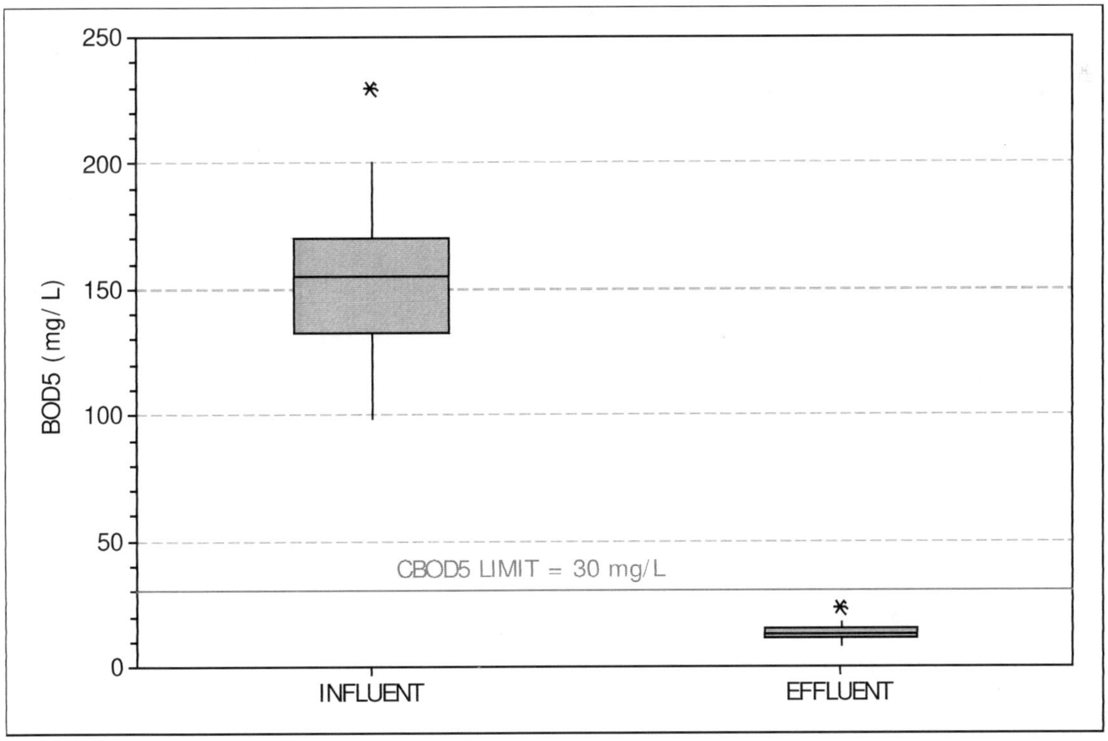

Figure 12. Boxplots of Influent and Effluent BOD_5 Data

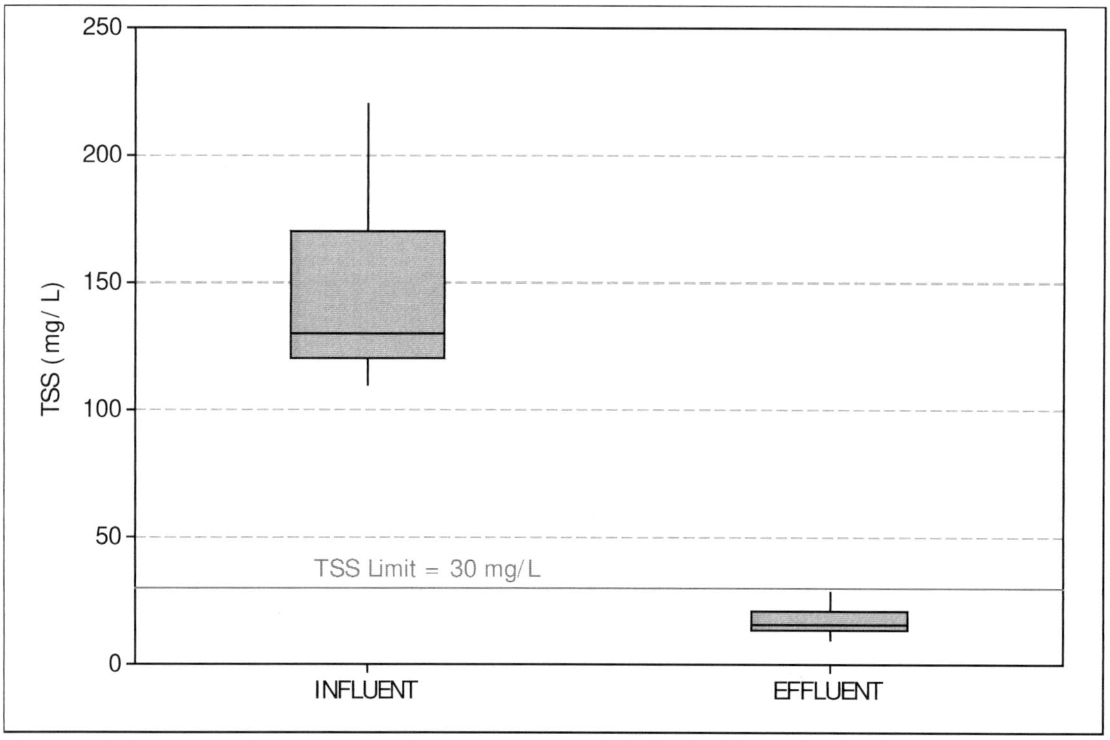

Figure 13. Boxplots of Influent and Effluent TSS Data

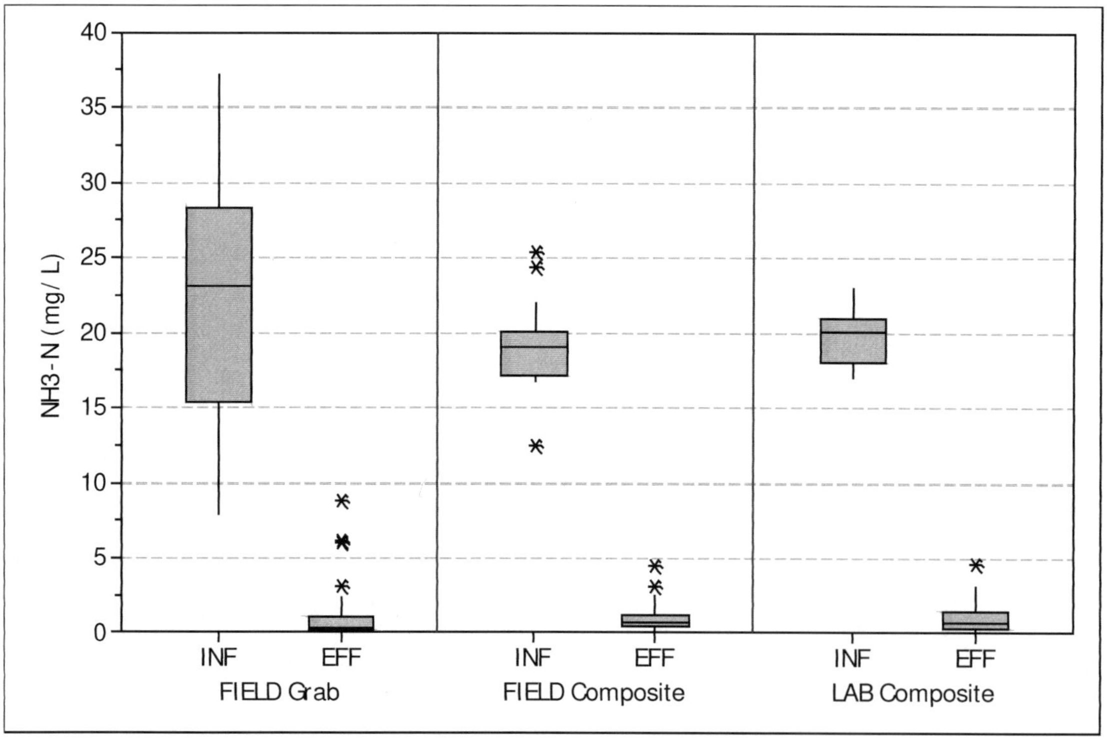

Figure 14. Boxplots of Influent and Effluent NH_3-N Data

Contaminant Loading Rates. The actual loading rates applied to the BAF pilot unit during the Demonstration Study can be compared to the contaminant loading rates proposed by the BAF vendor for the full-scale design at 10°C using the data collected during the Demonstration Study. The loading rates proposed for the full-scale design are shown below in Table 1. By the time the unit was nitrifying reliably, warmer wastewater temperatures resulted in the need to adjust the target loading rates to account for the warmer wastewater temperatures. Based on the average influent wastewater temperature of 13.3°C, the design BOD_5 loading rate of 2.35 $kg/m^3/d$ (147 $lb/1000\ ft^3/d$) at 10°C was temperature corrected to 2.64 $kg/m^3/d$ (165 $lb/1000\ ft^3/d$) and the NH3-N design loading rate of 0.22 $kg/m^3/d$ (13.7 $lb/1000\ ft^3/d$) at 10°C was temperature corrected to 0.25 $kg/m^3/d$ (15.3 $lb/1000\ ft^3/d$).

Figures 15 and 16 present the BOD_5 loading rate versus BOD_5 removal. In these figures, and subsequent similar figures, the following items should be noted:

- The X-axis presents the contaminant loading rate per day on a unit volume while the Y-axis presents the contaminant removal across the CN BAF on the same basis.
- The red squares present the data (loading applied per day per volume vs. loading removed).
- The red line represents a linear regression best fit line of the data which has been forced through the graph's origin. This line was expected to curve and the slope decrease (flatten) as load is increased. Where the beginning of this inflection point was observed, the data that no longer fit the straight line portion of the curve were removed to provide a more accurate linear regression line. The horizontal portion of the curve represents the maximum or rate limiting loading rate.

Figure 15 shows the data collected during grab samples and that this loading rate was achieved frequently. The slope of the linear regression line in this figure is 0.93, signifying 93% removal of BOD_5 across the CN BAF. Additionally, this figure shows a linear relationship between load applied and load removed both above and below the temperature corrected design loading rate indicating that for the range of loads tested, the performance of the CN BAF was not impacted by increased loads, and the maximum loading rate was not approached. The linear portion of the line appears to end at approximately 4 to 4.81 $kg/m^3/d$ (250 to 300 $lb/1000\ ft^3/d$), well above the design loading rate. Figure 16 displays data collected from composite samples and shows similar results. Based on these figures, 91 to 93% removal of BOD_5 is expected through the CN BAF.

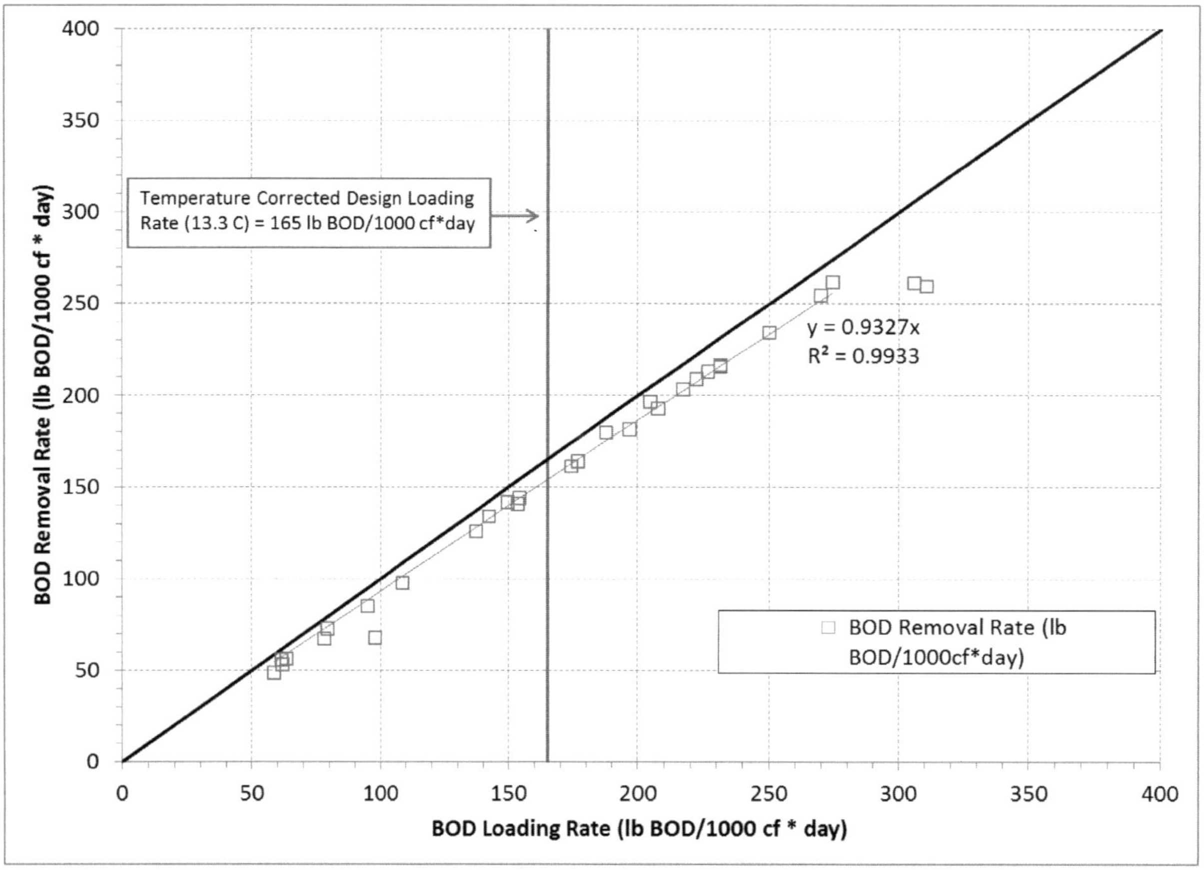

Figure 15. BOD_5 Loading Rate vs. Removal through First Stage BAF Reactor – Grab Data

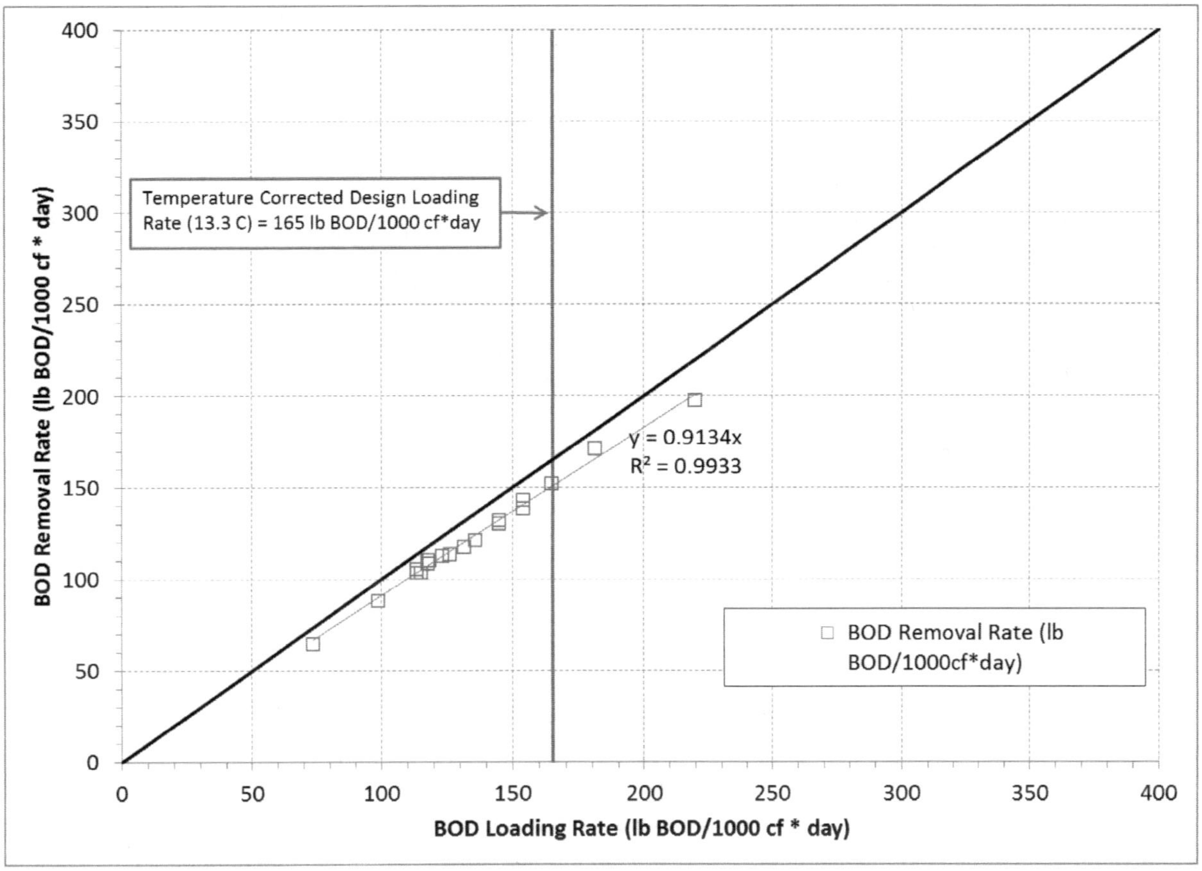

Figure 16. BOD_5 Loading Rate vs. Removal through First Stage BAF Reactor – Composite Data

Figures 17 and 18 display the NH_3-N loading rates vs. NH_3-N removal. Figure 17 shows the data collected during grab samples and that this loading rate was achieved frequently. In Figure 15 the slope of the linear regression line is 0.98, signifying 98 percent removal of NH_3-N across the Stage 1 BAF. Figure 18 shows data collected in the composite samples. The slope of this line is 0.95, signifying 95 percent removal. Both of these figures show a linear relationship between load applied and load removed both above and below the temperature corrected design loading rate indicating that for the range of loads tested, the performance of the CN BAF was not impacted by increased loads, and the maximum loading rate was not approached. The linear portion of the line appears to end at approximately 0.32 to 0.40 kg/m^3/d (20 to 25 lb/1000 ft^3/d), above the design loading rate. Based on this figures, 95 to 98 percent removal of NH_3-N is expected through the CN BAF.

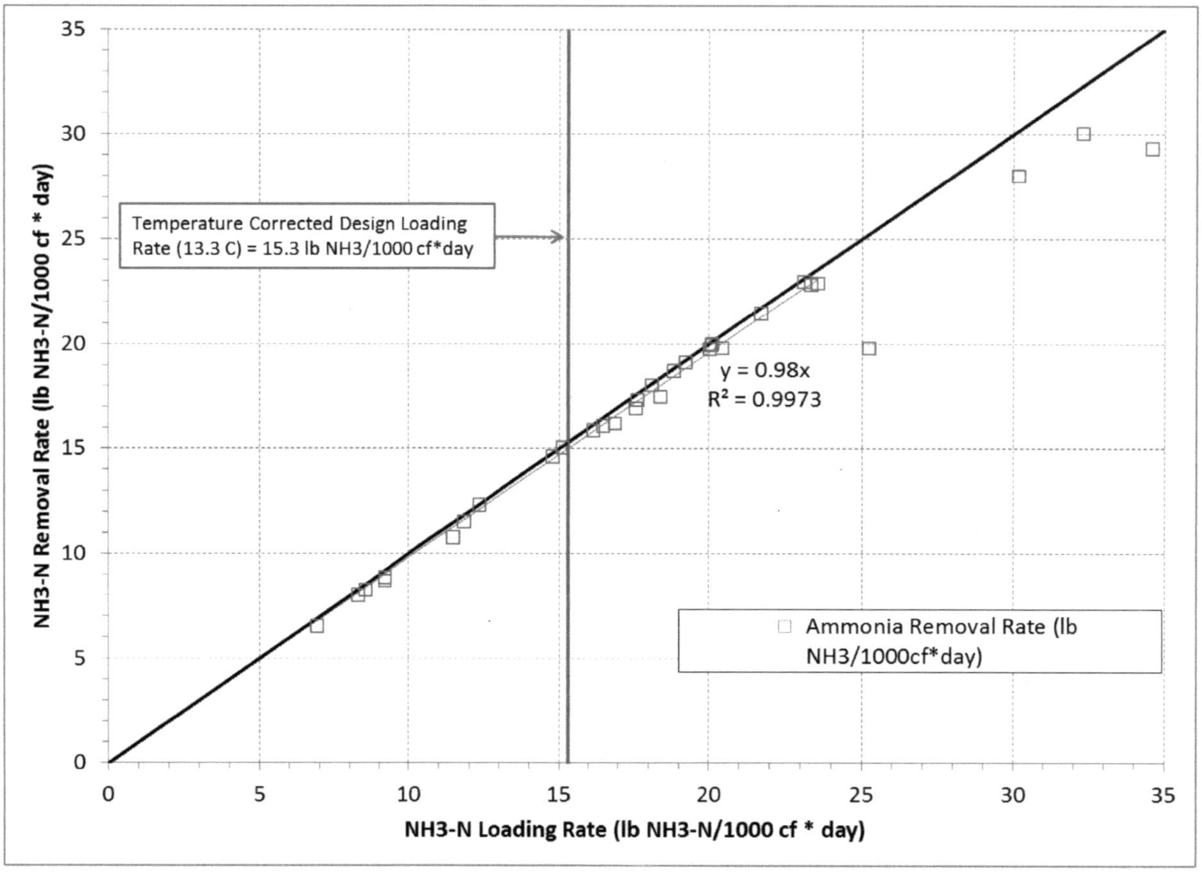

Figure 17. NH_3-N Loading Rate vs. Removal through First Stage BAF Reactor – Grab Data

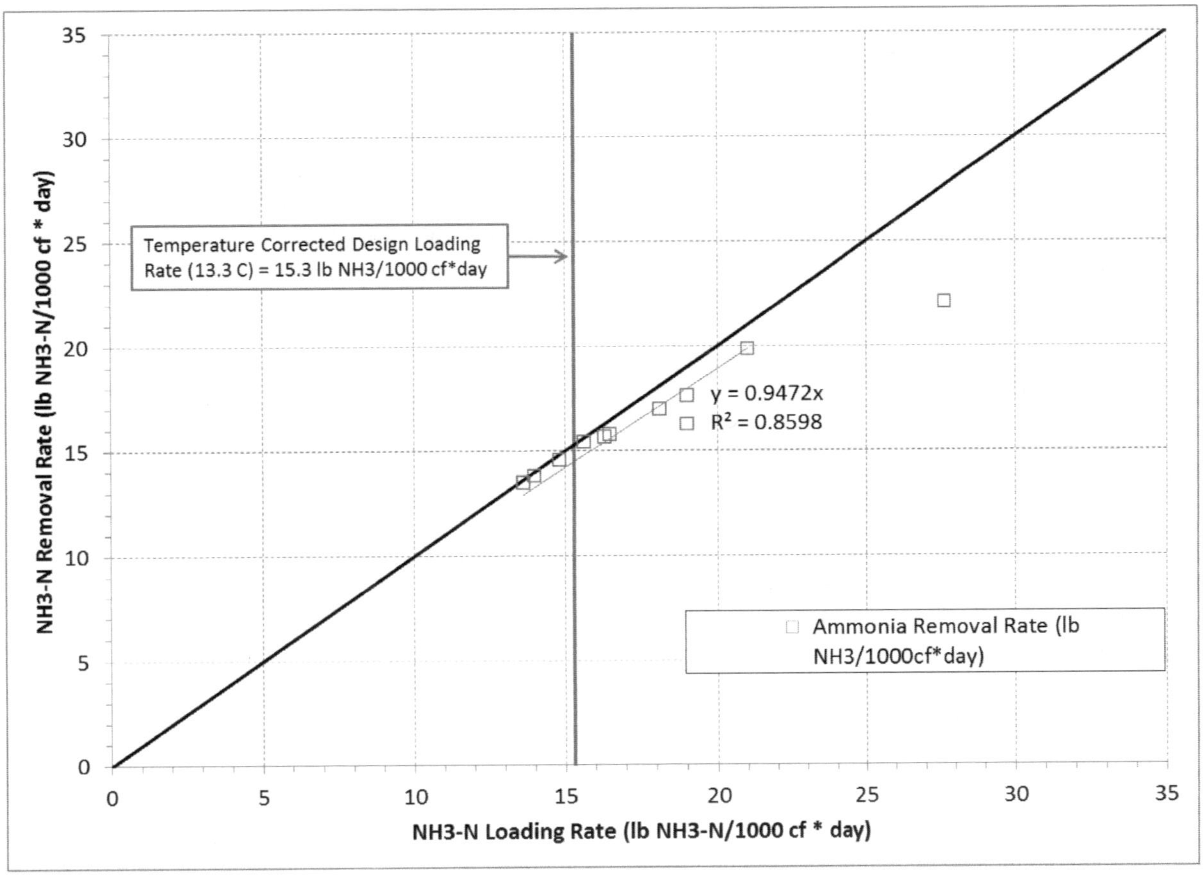

Figure 18. NH_3-N Loading Rate vs. Removal through First Stage BAF Reactor – Composite Data

Sodium Hydroxide Dose. Sodium hydroxide (caustic soda) was added to the pilot primary effluent during the Demonstration Study to maintain pH and provide improved growth conditions for ammonia oxidizing bacteria (AOBs). Doses ranged from approximately 40 mg/L to 50 mg/L in order to maintain influent alkalinity of approximately 200 mg/L when raw wastewater was run through the pilot primary and from approximately 70 mg/L to 90 mg/L when CEPT effluent was used to feed the BAF.

The mass balance calculations show that a caustic soda dose of approximately 7 to 15 mg/L should provide sufficient alkalinity to buffer the primary effluent during nitrification during times when CEPT is not being used. When CEPT is in use, the caustic soda dose rises significantly to approximately 50 mg/L because of the alkalinity demand created by the ferric chloride.

The mass balance calculations assume a primary effluent TKN of approximately 22 to 24 mg/L whereas the primary effluent TKN measured during the Demonstration Study was approximately 33 mg/L. Supplemental ammonia was being added to the process during this time to encourage nitrification and test the system at higher loading rates. If the mass balance calculations are modified to use a primary effluent TKN of 33 mg/L, the calculated caustic soda dose rises to approximately 60 mg/L when CEPT is not running and 105 mg/L when CEPT is running. Both of these are very similar to the doses measured during the Demonstration Study, verifying the dose.

It is suspected that the difference in primary effluent TKN between the mass balance and the TKN in the BAF feed water can be attributed to the use of supplemental ammonia and that the difference in TKN causes the disagreement between the Demonstration Study results and original mass balance calculations. As a result, the full-scale design will provide caustic soda pumps that are able to deliver high doses should it be necessary.

Compression Tests

Three compression tests were conducted to simulate gravity thickener operation with co-settled primary sludge and BAF backwash solids after the BAF is installed. Settled backwash waste was combined with primary sludge in settling columns. At the completion of all three tests, the thickened sludge was approximately 2.5 – 3 percent solids after 2.5 to 3 hours of settling time. As shown in Figure 19, much of the compression was completed during this length of time in all three tests. Final sludge volumes increased with the initial TSS of the samples. The third compression test (Test 6 (6/3/14)) was extended overnight to almost 19 hours in length. The additional time resulted in sludge that was 0.8% thicker.

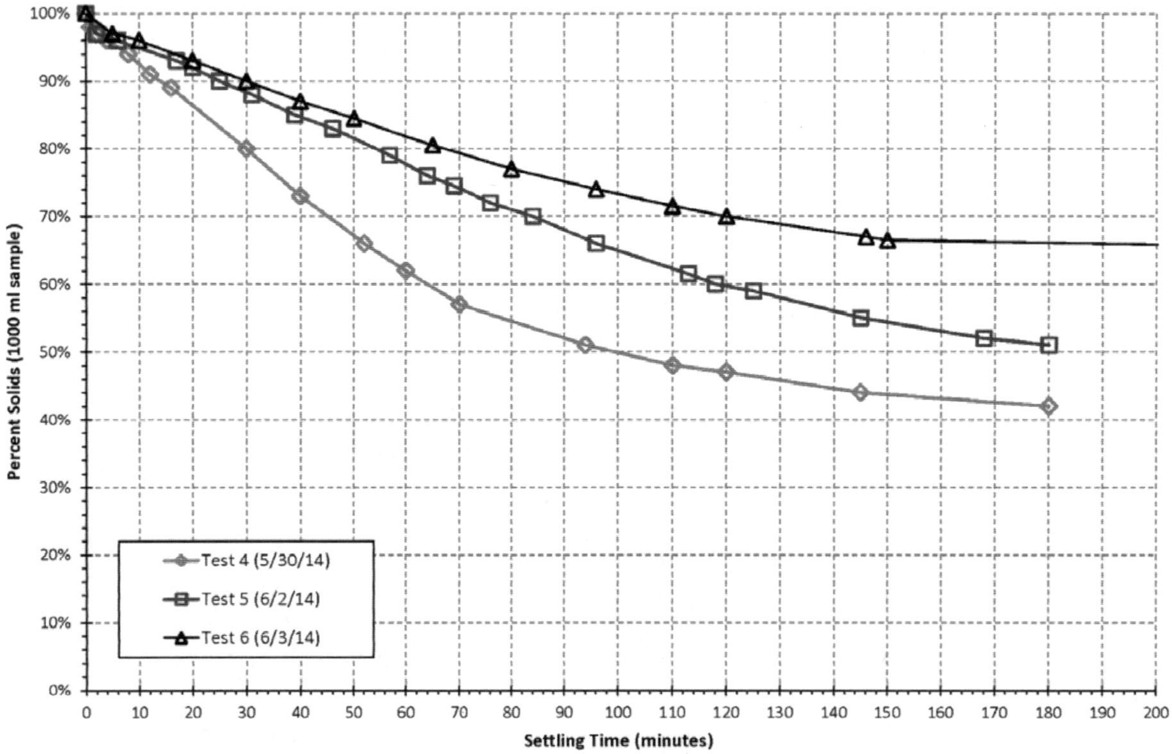

Figure 19. Compression Settling Data

The process design for the gravity thickeners has used standard solids and hydraulic loading rates. The resulting detention time at full-scale is approximately 14 hours, so full compression of the solids is expected. The mass balance is based on an assumed thickened sludge percentage of 7 percent based on standard design parameters for thickeners accepting solids from a primary clarifier and fixed film process such as the BAF. No changes to the design basis were made because it is not possible to calibrate these bench-scale results to current performance and determine if and how much the bench-scale results differ from the full-scale. In the case of the gravity thickener, not only does the bench-scale test have increased wall effects but it does not replicate the high surface area to volume ratio, slow stirring, and rake arms which will all benefit settling.

Dewatering Tests
Sludge from the compression tests was gathered into two samples and sent to two screw press vendors, FKC and Huber, for laboratory testing to determine the effectiveness of screw presses on this sludge for dewatering. As will be done at full-scale, both companies mixed polymer into the sludge sample to assist in dewatering prior to running it through the screw press. Table 6 below displays results from the two vendors.

Table 6. Screw Press Lab Test Results

Parameter	FKC	Huber
Predicted polymer usage	11 g/kg (22 lb/dry ton)	12.5 – 17.5 g/kg (25 – 35 lb/dry ton)
Predicted dewatered cake solids	27.65%	>27%

As shown in the table above, both vendors achieved very similar results that indicate that the predicted sludge can be dewatered to a concentration of 27 percent or greater with polymer usage between approximately 12.5 and 17.5 g/kg (25 and 35 lb/dry ton). Due to the fact that screw presses have not been used in the US to dewater BAF sludge prior to this project, a conservative assumption of 20 percent dewatered cake solids had been used. The two bench-scale tests completed confirmed that this assumption was indeed conservative.

CONCLUSIONS

The Demonstration Study tested the ability of the first-stage BAF to operate at the maximum month loading rates during cold temperatures, but also provided information related to the settling, thickening and dewatering of the BAF backwash solids. This information was compared to the process design calculations that had already been completed for the full-scale design. This section briefly summarizes the conclusions presented in the sections above.

Backwash Coagulation
The jar testing on the backwash found that a ferric chloride dose of 20 mg/L or greater and a non-zero polymer dose was effective. TSS removal was greater than the values used in the process design calculations, supporting the assumptions made in these calculations.

Primary Clarification
Settling column tests of combined raw wastewater and chemically treated BAF backwash show that the raw wastewater and BAF backwash solids can be co-settled successfully. Tests with raw wastewater and chemically treated BAF backwash solids were used to predict a primary effluent TSS of 80 mg/L, well below the maximum concentration of 120 mg/L requested by Veolia Water Technologies.

BAF
The BOD and NH_3-N loading rates used during the Demonstration Study were temperature corrected to close to, or exceeding, those proposed by Veolia Water Technologies. Once nitrification was established, the effluent quality during the study showed that BOD_5 and ammonia were fully oxidized the majority of the test period. The Demonstration Study provided confidence that the first stage BAF is able to remove BOD_5 and fully nitrify at loading rates of approximately 2.35 $kg/m^3/d$ (147 lb/1000 ft^3/d) for BOD_5 and up to 0.22 $kg/m^3/d$ (13.7 lb/1000 ft^3/d) for NH3-N.

The caustic soda dose required during the Demonstration Study varied from the calculated dose in the mass balance, however supplemental ammonia was being added to the system to mimic the high loading rates during a max month situation. Because the mass balance calculations were validated and typical ammonia concentrations are much lower, no changes to the caustic storage and feed system other than providing capability for high doses were made as a result of the Demonstration Study.

Full-scale start-up should occur outside of the winter, if possible. If it must occur during winter, start-up may be accelerated by encouraging AOB growth through continued use of the CEPT process to lower the BOD_5 loading rate and through addition of enough caustic soda to maintain a neutral pH.

Gravity Thickener
The compression tests conducted simulated gravity thickener operation after the BAF is installed, show that the combined primary sludge and BAF backwash solids can be effectively co-thickened and suggest that there is sufficient hydraulic retention time to achieve full compression of the solids.

Solids Dewatering
Bench-scale tests evaluating the ability of screw presses to dewater co-settled, co-thickened sludge from the primary clarifier and BAF show that the sludge is able to dewatered to approximately 27 percent solids using between 22 and 35 lb/dry ton of polymer. The testing confirmed that assumption used in the process design calculations was conservative and did not need to be modified.

The upgrade of the Peirce Island WWTF is currently in final design. Construction is expected to begin in 2016 and to be completed within three to four years. The project encompasses a complete upgrade of the WWTF, including a new Headworks Building, two-stage BAF for nitrogen removal, gravity thickener and solids dewatering, and a new Operations/Laboratory Building.

ACKNOWLEDGEMENTS

The authors acknowledge and thank the wastewater treatment facility operators of the City of Portsmouth, New Hampshire, Blueleaf Incorporated, and Veolia Water Technologies for their assistance in completing the Demonstration Study.

REFERENCES

Pearson, J., Desmarais, T., Rice, P., Anania, P., Formica, M., Grotton, E. (2013) Testing the Waters – Piloting High Rate Treatment Technologies for Secondary Treatment and Total Nitrogen Removal Process Selection. *Proceedings of the 2013 WEF Technical Exhibition and Conference*; Chicago, Illinois.

Metcalf & Eddy. *Wastewater Engineering: Treatment and Resource Recovery,* Fifth Edition; McGraw-Hill Education: New York, 2014; pp 354-355.

Startup of a Low Phosphorus Nitrifying Tertiary Membrane System

Mario Benisch, HDR Portland
JB Neethling, HDR Folsom
Mike Zeltner, HDR Boise
Casey Fisher, City of Coeur d'Alene

ABSTRACT

The first stage of the tertiary membrane bioreactor facility for low level phosphorus removal and nitrification has been commissioned at the City of Coeur d'Alene (the City, CDA, Figure 1) in January of 2015. This marks the conclusion of significant investments made by the City over a span of eight years to develop a cost efficient technical solution for compliance with effluent phosphorus limits, which at the time (0.01 mg/L TP monthly average, 2006) where unprecedented.

Because this tertiary membrane process with chemical sludge inventory or tertiary membrane bioreactor (TMBR) has not previously been proven at full scale, the City decided to proceed with caution and only install 1/5 of the capacity, required in the future, but to do so earlier than required for compliance in order to prove and fine tune the process before fully committing to the TMBR.

The performance of the TMBR system itself is only part of the greater picture. Even operating at only 1.0 mgd, the TMBR is expected to impact the main plant in three ways; (1) equalized ammonia rich dewatering centrate will be routed directly to the TMBR to grow additional nitrifiers which will be wasted (seeded) to the contact stabilization basins, (2) along with the nitrifiers chemical sludge will be sent to the contact stabilization, which is expected to contribute significantly to the secondary phosphorus removal due to its ability to removal PO_4-P through adsorption and complication, and (3) the seeding is expected to reduce the minimum solids retention time (SRT) and MLSS in the contact stabilization basin as well as reduce the secondary alum dose.

BACKGROUND

The startup of the TMBR at the City at the start of 2015 completes nearly 10 years of applied research and development as well as negotiation with and educating of regulatory agencies and other stakeholders. The original marching orders of 10 microgram (0.01 mg/L) effluent total phosphors came at a time when limits below 0.05 mg/L where unprecedented and beyond the limit of technology.

Between 2006 and 2012 different technologies and processes were piloted along with research focusing on phosphorus fractionation, bioavailability, or the intricacies of low level phosphorus analysis ([1] – [9]). The selected technology offered the advantage of simultaneously producing a very low effluent total phosphorus concentration and nitrifying the residual ammonia concentration in the effluent from the trickling filter solids contact facility. Presently, effluent ammonia concentration from the trickling filter solids contact TFSC plant ranged between 25 and 7 mg/L depending on the season and year and lower ammonia limits (<4 mg/L) are anticipated for the near future. Figure 1 shows photos of the City's TMBR facility.

Previous nitrification improvements [19] through the addition of fixed media had reduced the summer effluent ammonia concentrations but the anticipated year around ammonia limits would otherwise require an expansion of the secondary treatment system via a new parallel treatment train. By achieving nitrification in the tertiary process the parallel train is no longer needed, reducing the overall program cost on the order of $20 million.

Figure 2 shows the process schematic of the CDA WWTP with the full TMBR. Figure 3 shows the current process with only partial TMBR. Because of lingering uncertainties about the full scale performance of a TMBR for nitrification and low level phosphorus removal it was decided to proceed with caution and only construct a 1 mgd worth of membrane filtration capacity while providing the additional aeration (solids contact) volume that would be needed ultimately to meet the future ammonia limits.

In the future, the treatment process with the TMBR at build-out, the majority of the effluent from the secondary clarifier will enter the TMBR except flows in excess of a maximum hydraulic capacity that is yet to be determined. Presently, only about ¼ of the plant flow enters the TMBR. In addition, since not all of the mixing tank volume is needed this volume is utilized as an extension of the current solids contact tank, effectively doubling its volume. When the TMBR receives the remaining membrane modules the volume presently used as the solids contact expansion will be repurposed as the TMBR mixing tank volume.

Since overcoming some of the startup issues the TMBR is producing the desired effluent quality. An extensive sampling and monitoring program captures the relevant water quality and process control parameters to fully asses the TMBR performance and its impact on the total plant operation.

Figure 1: TMBR Facility

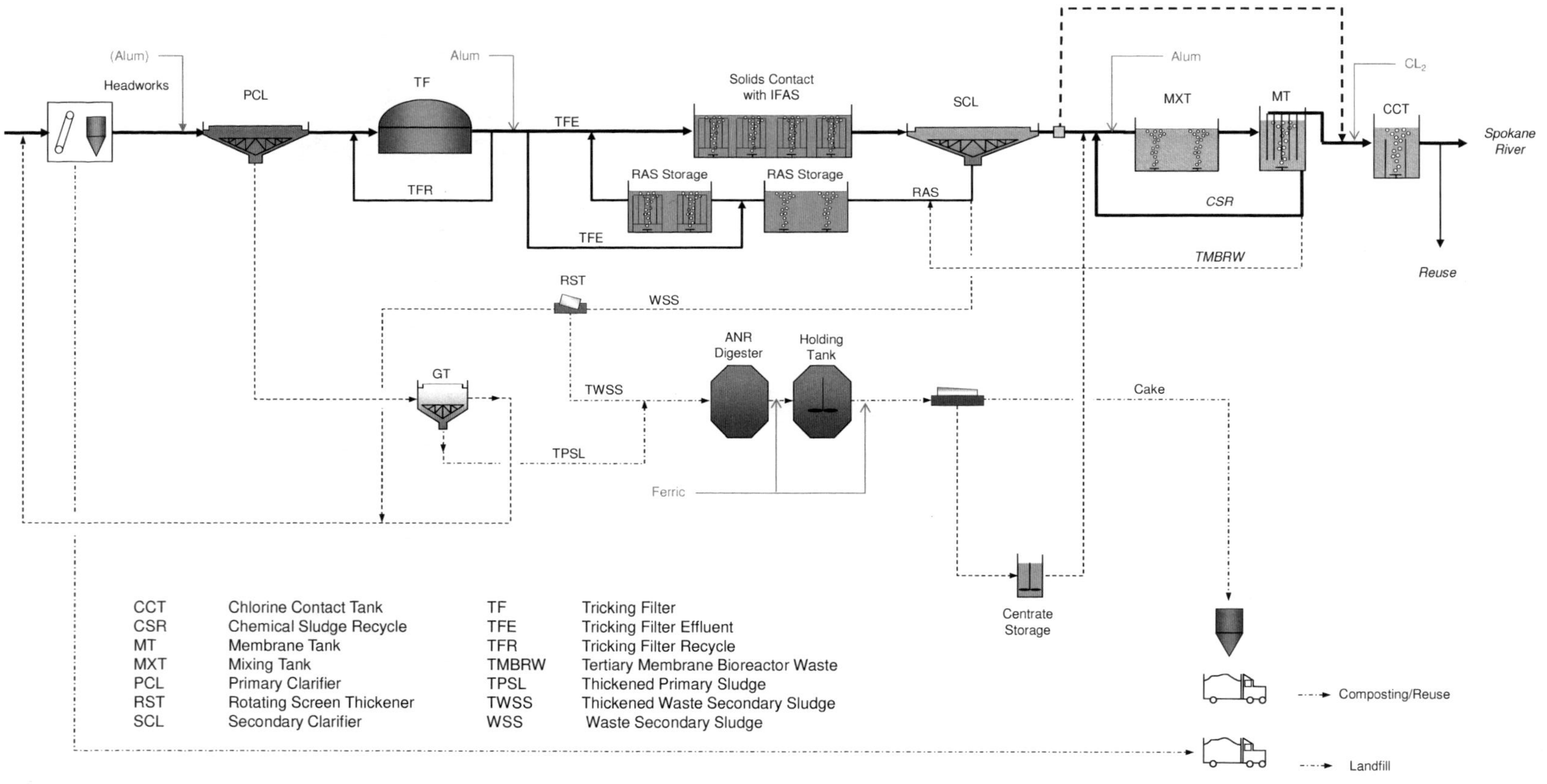

Figure 2: Schematic of Future Process with TMBR

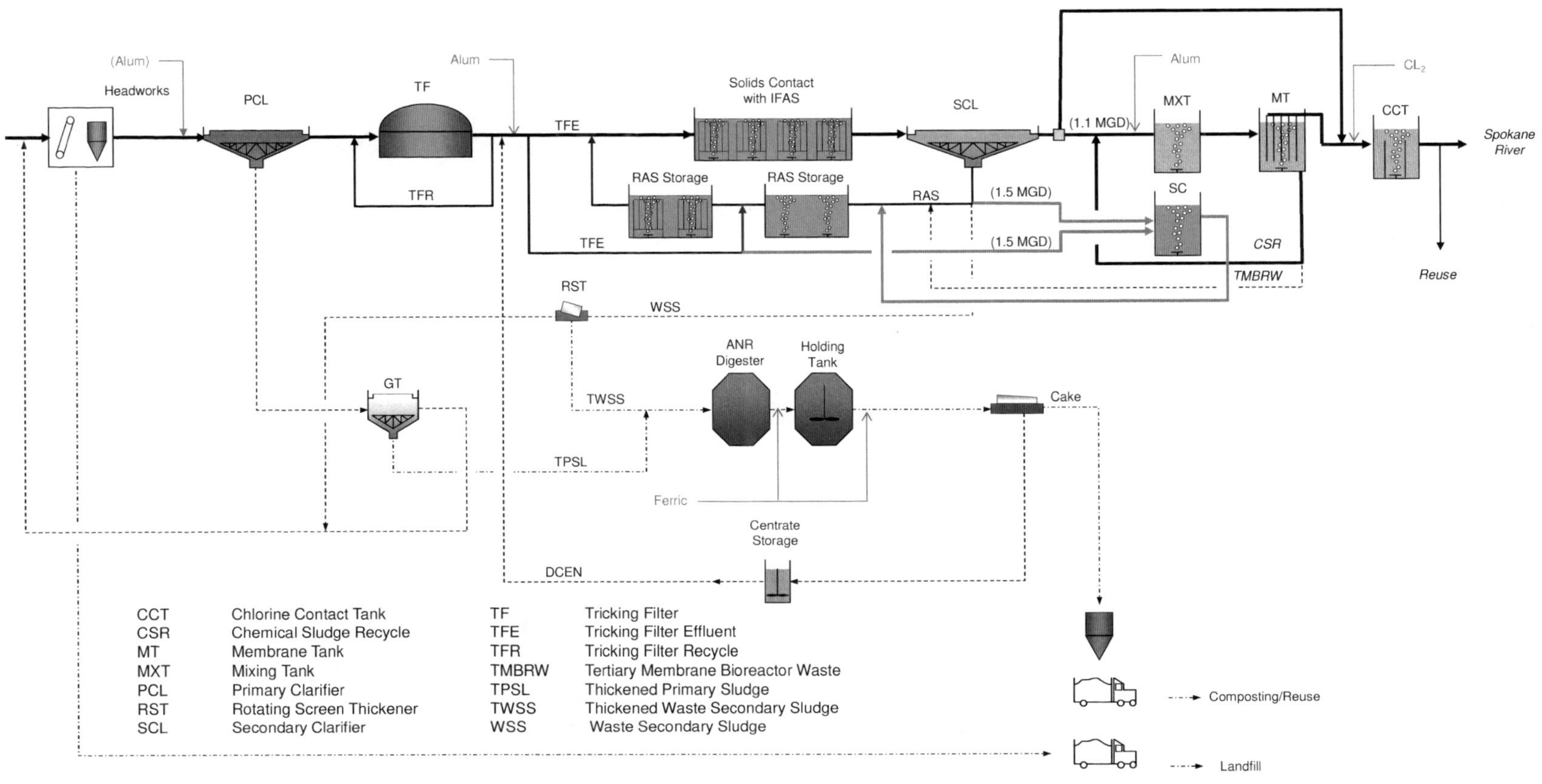

Figure 3: Schematic of Current (2015) Process with partial TMBR

Results and Discussion

Startup

The startup sequence included seeding the TMBR with active biomass from a nearby MBR facility to speed up commissioning; with mixed results. While full nitrification was reached immediately it was also quickly lost due to what was determined to be an excessive clarifier launder chorine addition; the TMBR was effectively chlorinated. Since the chorine dosing issue has been resolved and pH control (alkalinity addition) was fine tuned, the TMBR nitrification performed as expected with effluent ammonia in the 0.1 mg/L range.

The startup period was impacted by unique gelatinous scaling that was not observed during the pilot phase (Figure 4 and Figure 5). Chemical analysis showed the scale was organic, very greasy, had some aluminum content. It did dissolve rapidly in hypochlorite solution. While the exact cause of the scaling was not fully established two possible causes were identified based on the operational difference compared to the pilot operation; (1) the seeding with active biomass and (2) the recently installed scour air upgrade to the BioWeb media in the solids contact basins. This scour air may have produced excessive biofilm slough that was not present during the two year pilot operation. The scale appeared to grow from the outside of the membrane modules where the Leap scour system provided less kinetic? energy due to the placement of the air supply. Once the scale started on the outside it continued to grow inwards and or "fusing" the individual membrane fibers together

The membrane modules were cleaned with a combination of chemical soaking and pressure washing. After cleaning the TMBR was restarted. This time without seeding but by adding a 30 mg/L alum dose and running the scour air continuously. Over time it is the intent to bring the scour air gradually down while carefully monitoring for scale buildup. Five months into the startup and commission the TMBR is now running without major issue though the fine tuning of the operation strategy and process control remains and ongoing process.

Figure 4: Scaled Membrane Module

Figure 5: Collected Scale after Manual Removal

Nitrification performance

The nitrification improvements of the first stage of the TMBR upgrade had two components – increasing the solids contact volume and fully nitrifying ¼ of the flow in the TMBR. Figure 6 shows the overall nitrification improvements for the most recent years, with ammonia effluent concentrations which are in line with what was expected. Even without the TMBR in full

Planning, Design, and Operation of a BNR Plant with Air Mixed Selectors

Don Esping[1*], Carol Mordorski[2], Christine Voigt[2], George Sprouse[2]

[1] Brown and Caldwell, St Paul, Minnesota
[2] Metropolitan Council Environmental Services, St. Paul, Minnesota
*Email: DEsping@brwncald.com

ABSTRACT

The Metropolitan Council Environmental Services in Saint Paul, Minnesota was mandated to reduce effluent total phosphorus (TP) discharges at its Blue Lake WWTP. To meet the TP standards, MCES retrofitted the plant's existing bioreactors with air mixed selectors. Stress testing and two years of full-scale plant operations has shown the air mixed selector configuration reliably reduces effluent TP annual discharges below 0.5 mg/L and there is no quantifiable difference in phosphorus removal performance between a bioreactor configured with a baffled, mechanically mixed selector and an air mixed selector configuration. Air mixed selectors can also be cost effective solutions for selectors aimed at improving sludge quality. This paper presents the Blue Lake WWTP case history of meeting the reduced TP levels from planning to design and subsequent operations under planned and unforeseen changes to the plant's system.

KEYWORDS

Phosphorus Removal, Selectors, Mixing, Sludge Quality

INTRODUCTION

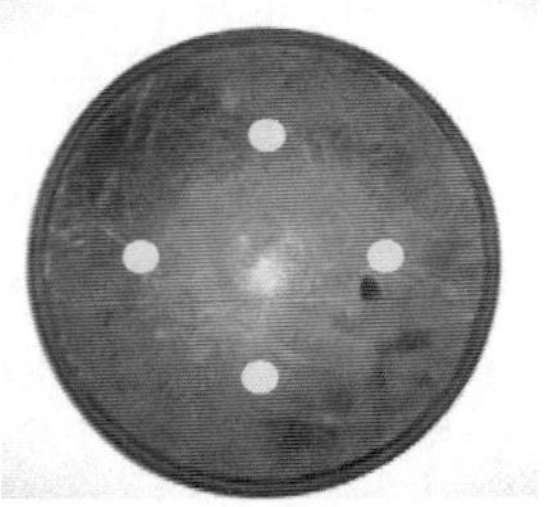

Metropolitan Council Environmental Services (MCES) currently owns and operates the Blue Lake wastewater treatment plant (WWTP) which treats an annual average flow of 32 mgd. Historical operation in the early 2000's at another MCES WWTP showed reduced phosphorus discharges if the airflow to the bioreactors aerated selector zone was reduced. Based upon this observation, MCES staff modified the Blue Lake bioreactors in 2003 to include a pseudo-air mixed selector as shown in Figure 1 in an effort to reduce total phosphorus (TP) discharges. The pseudo-selector retrofit consisted of "plugging" all existing fine pore diffusers in the selector zone except for a total of 38 diffusers along the tank sides which were retrofitted with custom

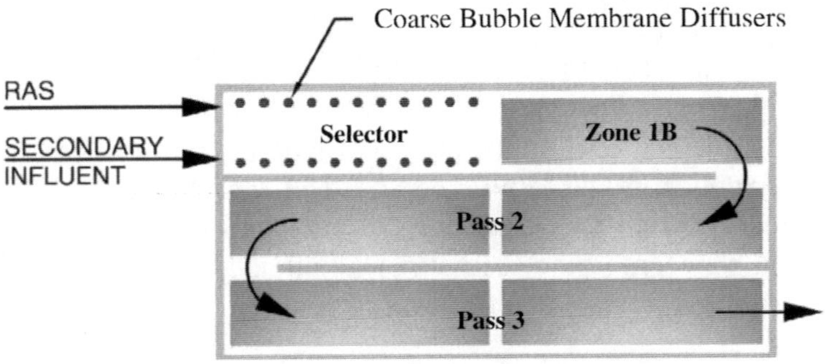

Figure 1. Blue Lake WWTP Bioreactor Configuration.

fabricated coarse bubble diffusers to deliver air for mixing while minimizing air transfer efficiency. The custom diffusers were fabricated by punching four ¼-inch holes in a solid EDPM membrane diffuser as shown in Figure 1. The diffuser openings were sized to provide approximately 4.3 scfm/diffuser with 38 diffusers providing roughly 165 scfm or 0.05 scfm/sf. The selector retrofit did not include a baffle wall to separate the selector zone from the downstream fine pore aeration grid; therefore, back mixing from the fine pore grid zone to the selector zone occurred. Dedicated airflow control to the selector zone and Zone 1B fine pore aeration grid was not provided in the retrofit. To prevent the selector zone airflow from varying while controlling the downstream Zone 1B dissolved oxygen (D.O.), the Pass 1 airflow was set to provide constant airflow resulting in a constant airflow to the selector zone with varying D.O. levels in Zone 1B. Operation with the pseudo-selector reduced annual TP discharges below 1 mg/L at existing flows and loadings, but it was unclear how the air mixed selector configurations would perform under increased future loading conditions.

PLANNING

In 2004, MCES began planning efforts to evaluate improvements required to reduce TP discharges below 1 mg/L on a 12-month rolling average basis which was proposed for inclusion in Blue Lakes National Pollutant Discharge Elimination System permit to be renewed in 2008. To remove uncertainty about the performance of air mixed selectors at future loadings, phosphorus removal planning efforts incorporated a process proving program to determine:

- Can the 1.0 mg/L effluent TP goal be met under projected future loading conditions using the pseudo-air mixed selector configuration?
- Is there a quantifiable difference in phosphorus removal performance between a bioreactor configured with a baffled, mechanically mixed selector versus the pseudo-air mixed selector configurations?

A four month systematic stress testing program was conducted to evaluate phosphorus removal under future loading conditions with the existing solids processing flow scheme: gravity thickening of primary sludge with subsequent sludge screening, gravity belt thickening of waste activated sludge, and combined raw sludge dewatering followed by heat drying. Systematic stress testing was completed by removing aeration basins from service to simulate increased future loading conditions.

Stress testing of projected Year 2005, 2010, 2015, 2020 and 2025 flows and loadings conditions was conducted. Testing showed the air mixed selector configuration reliably produced effluent TP concentrations averaging 0.1 to 0.34 mg/L under current and simulated Year 2020 influent loadings while maintaining complete nitrification. Elevated effluent TP was observed under Year 2020 projected conditions due to one event of high effluent suspended solids. If this data point is eliminated from the data set, the effluent TP averaged 0.18 mg/L. Testing at simulated condition at Year 2025 projected loadings was stopped when oxygen transfer limitations in the existing aeration system suppressed nitrification.

Table 1. Blue Lake Process Proving Effluent Total Phosphorus.

Simulated Loading Condition	Effluent Total Phosphorus, mg/L	
	Average	Range
Year 2005	0.17	0.0 – 0.4
Year 2010	0.13	0.0 - 0.3
Year 2015	0.15	0.0 - 0.45
Year 2020	0.34	0.18 – 2.9

Observed phosphorus removal performance with the pseudo-air mixed selector configuration was compared to simulated plant operation with a mechanically mixed selector configuration using the BioWin™ simulator. Simulation results presented in Figure 2 show the measured plant performance with the pseudo-air mixed selector was typically 0.1 mg/L less than simulated effluent predictions. Simulated and measured effluent TP levels greater than 1 mg/L represented conditions of solids carryover in the secondary effluent.

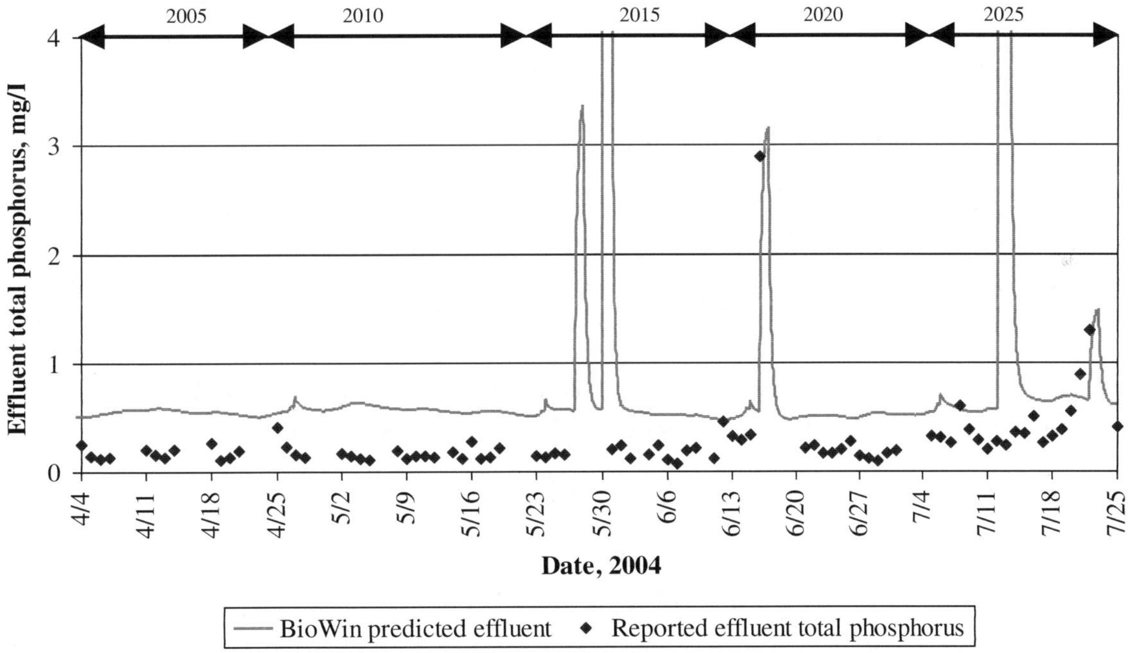

Figure 2. Comparison of Blue Lake Process Proving Reported and Predicted Effluent Total Phosphorus Discharges.

DESIGN

Successful demonstration of the pseudo-air mixed selector configuration led to the detailed design of the air mixed selector configuration shown in Figure 3. The selector zones comprise 17 percent of the total bioreactor volume and measure 115 feet long by 27 or 32 feet wide with 15 foot side water depths. The air mixed selector design incorporates 42 custom fabricated coarse bubble diffusers located along the tank side walls evenly spaced at 5 foot intervals. The diffusers are designed to provide 4 scfm/diffuser or roughly 0.05 scfm/sf basin. Additional design features were added to address limitations observed during stress testing and plant operation. A baffle wall was added to separate the selector zone and Zone 1B to minimize back mixing to the selector zone observed during stress testing. A dedicated selector zone airflow control valve and meter were also added to each bioreactor providing the ability to control the selector zone and Zone 1B airflows separately.

Baffle wall design accommodated two considerations: limited head loss at peak flows and providing the ability to remove the baffle wall. The fiberglass reinforced plastic (FRP) baffle is primarily an underflow baffle to minimize head loss and back-mixing with a slight submergence to allow passage of scum/floatable material to minimize its accumulation in the selector zone. The ability to remove the baffle wall addressed concerns that the back mixing between the selector zone and Zone 1B may have deposited solids in the selector zone generating volatile fatty acids in the deposited solids with subsequent sequestration from back-mixing.

Initial construction and operations of the air mixed selector configuration proceeded as planned until the plant placed the first bioreactor in service and conducted routine diffuser "air bumping" in the Zone 1B to clean the diffusers. When the Zone 1B air diffuser grid was "air bumped", the liquid level in Zone 1B increased, creating a backward flow from Zone 1B over the baffle wall to the selector zone. This backward overflow and resulting circular internal flow pattern in Zone 1B applied excessive pressure to the baffle wall causing the baffle wall to flex ultimately causing the center support column to fail at its base. (Figure 4) To address this issue, the baffle wall design was modified to include an FRP support beam spanning across the tank side walls and attached to the top of the center support column providing additional structural support. (Figure 5b)

Figure 3. Blue Lake Air-Mixed Selector.

Figure 4. Selector Baffle Wall After "Air-Bumping" Zone 1B Fine Pore Diffusers.

OPERATIONS

Figure 5 shows the air mixed selector zone under current operations. Secondary influent (primary effluent) enters the selector through an overflow weir gate and return sludge enters at the head of the tank. The plant currently introduces 80 to 100 scfm of air into each selector zone for mixing the selector. This flow rate is roughly ½ the selector zone design airflow rate. Visual observation of the selector shows that the first 1/3 to 1/2 of the selector zone is extremely well mixed as a result of the energy from the influent flow and mixing air. Visual observations at the end of the selector zone show the bioreactor to be less homogenously mixed with some floating biomass but the mixed liquor is still in suspension. (Figure 5b)

Figure 5. Air Mixed Selector in Operation.

Figure 6 shows side-by-side phosphate profiles along two existing bioreactors: one with the air mixed selector (red) representative eight of the plants bioreactors and the other equipped with floating mixers (blue) which represents the remaining plant bioreactor. Both profiles are consistent with stable enhanced biological phosphorus removal (Bio-P) systems with high phosphate concentrations in the selector zone and subsequent phosphate uptake in downstream aerated zones. The morning profile selector phosphate concentration might suggest the air mixed selector configuration is less efficient than the mechanically mixed; however the afternoon profiles show the selector phosphate concentrations are essentially the same. The phosphate concentration at the end of Pass 3 in both bioreactors is less than 0.05 mg/L. Similar phosphate release and subsequent uptake patterns were observed in multiple profiles taken side-by-side from the two bioreactors.

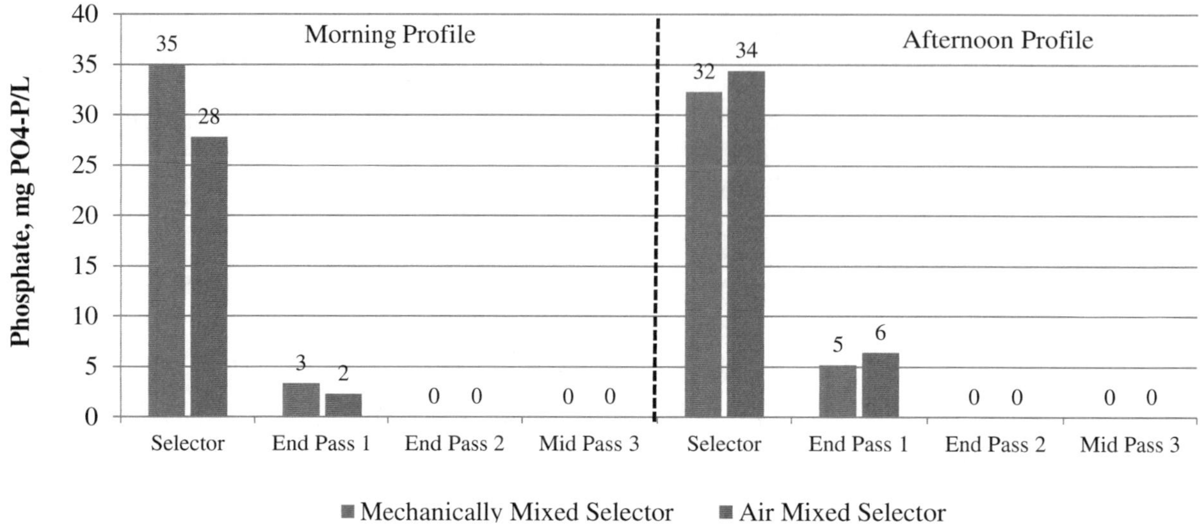

Figure 6. Comparison of Air Mixed and Mechanically Mixed Selector Bioreactor Phosphate Profiles. (March 11, 2015)

PERFORMANCE

Figure 7 shows a history of the effluent TP discharges beginning with operations prior to pseudo-selector installation, through three major construction projects, and up to current operations with anaerobic digestion. Table 1 and Figure 7 show the effluent TP discharges have decreased during the plants three key operational changes, even with higher nitrogen and phosphorus loadings from anaerobic digestion recycles and higher influent TP loadings from a new industry beginning in Year 2010. Sludge quality has also improved in each operational scenario with current sludge quality being excellent with the 90th percentile SVI less than 75 mL/g. Both effluent TP discharges and SVI variability have decreased under current operations compared to pre-selector conditions.

Dewatered cake solids did not change when Bio-P was implemented with the pseudo-air mixed selector, but significantly decreased when anaerobic digestion was implemented as a result of: (1) dewatering digested biosolids and (2) dewatering a Bio-P sludge that has been anaerobically digested. The combination of these two factors has resulted in a decrease in dewatered cake solids from 26 percent total solids to 19 percent total solids.

To reduce operating costs and meet agency energy goals, in 2014, the Blue Lake plant began experimenting with reducing the bioreactor D.O. concentrations to minimize blower energy usage. Figure 8 shows the plant effluent ammonia and TP discharges from November 11, 2014 through June 9, 2015. Effluent TP discharges with Zone 1B/Pass 2/Pass 3 D.O. levels of 1.5/1.4/1.1 mg/L are typical of operations at historical D.O. levels of 1.5/1.8/2.3 mg/L. Reducing the D.O. levels to 1.2/1.2/0.6 mg/L increased the effluent TP slightly, but the larger impact on effluent quality was the increase in effluent ammonia due to reduced nitrification rates when operating at lower D.O. in cold weather. Limited testing showed reducing the Zone 1B/Pass 2/Pass 3 D.O. set points to levels of 1.2/1.2/0.6 are acceptable for the desired TP reduction and effluent ammonia criteria will govern the D.O. levels to achieve effluent ammonia discharge requirements.

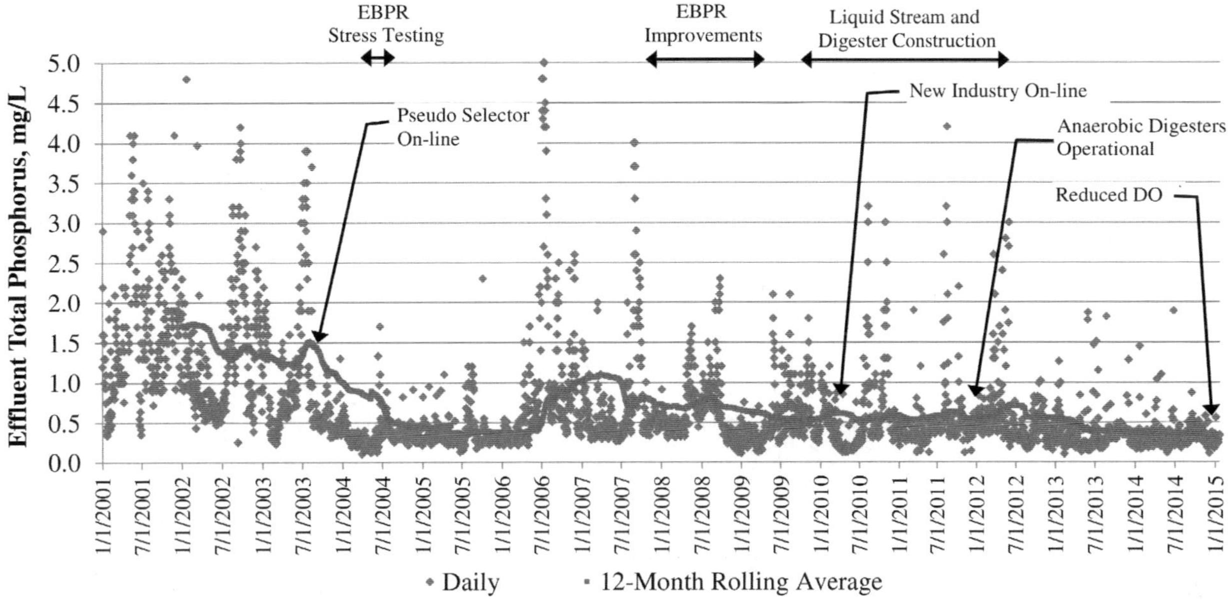

Figure 7. Blue Lake Effluent Total Phosphorus.

Table 1. Comparison of Blue Lake Performance.

Parameter	**Units**	**Pre-selector**	**Pseudo Air Mixed Selector**	**Air Mixed Selector with Anaerobic Digestion**
Period	--	1/1/01-12/31/03	1/1/04-1/1/08	1/1/13-1/1/15
Influent TP				
Average	mg P/L	6.8	6.3	8.4
Effluent TP				
Average	mg P/L	1.34	0.64	0.36
90th Percentile	mg P/L	2.50	1.10	0.50
Effluent TIN				
Average	mg N/L	9.9	11.0	22.5
90th Percentile	mg N/L	10.0	12.2	24.2
Sludge Volume Index				
Average	mL/g	116	86	64
90th Percentile	mL/g	157	108	72
Dewatered Cake Solids	% TS	26	26	19

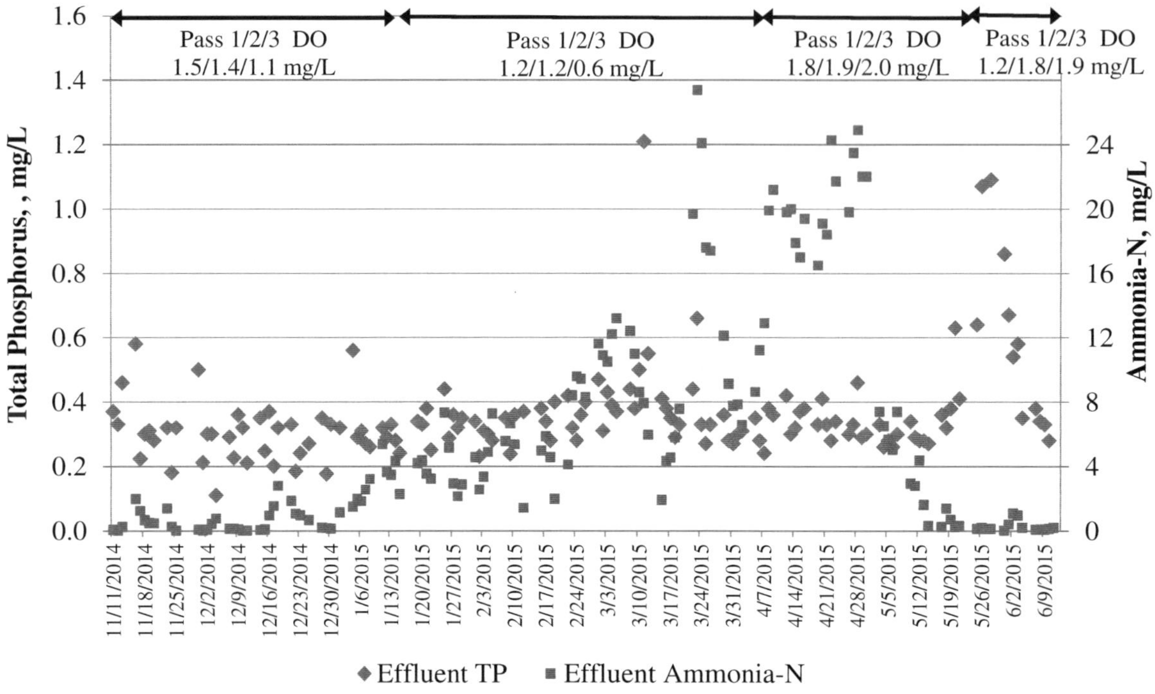

Figure 8. Impact of Dissolved Oxygen on Effluent Total Phosphorus.

CONCLUSION

Based upon full-scale plant operations over the past two years, air mixed selector configurations can reliably reduce effluent TP annual discharges below 0.5 mg/L. WWTP process modeling and side-by-side bioreactor profiling showed there was no quantifiable difference in phosphorus removal performance between a bioreactor configured with a baffled, mechanically mixed selector and an air mixed selector configuration. Based upon successful operation at these facilities, air mixed selectors could be more commonly used to eliminate the complexity of mechanical mixers and reduce capital and maintenance costs. Air mixed selectors can also be cost effective solutions for selectors aimed at improving sludge quality. Pilot testing and understanding of air mixing in individual anaerobic selectors will assist in defining when air mixing is a viable alternative to mechanical mixed selector zones.

REFERENCES

Esping, D., Bratby, J., Schierholz, E., Mordorski, C., Rogacki,L., Swaggart, C. (2007) "Process Proving for Phosphorus Removal with Air Mixed Anaerobic Selectors at Two Large Plants." Proceedings of the Water Environment Federation 80th Annual Conference & Exposition (WEFTEC), San Diego, CA, 2007.

Low Cost Biological Nutrient Removal for Treatment Plants in the Upper Long Island Sound Watershed

Jeanette Brown[1], Emily Bird[2], Stacy Passaro[3]
[1*]JJ Environmental, LLC, Darien, CT
[2]NEIWPCC, Lowell, MA
[3]Passaro Engineering, Mount Airy, MD
*email: jjenvironmental@gmail.com

ABSTRACT

Long Island Sound (LIS) suffers from low dissolved oxygen caused by excess nitrogen accelerating algae growth. Nitrogen is delivered from the five-state LIS watershed including areas of Connecticut, Massachusetts, New Hampshire, New York, and Vermont. To address the dissolved oxygen impairment, Connecticut Department of Energy and Environmental Protection and New York State Department of Environmental Conservation completed a Total Maximum Daily Load Analysis (TMDL) to Achieve Water Quality Standards for Dissolved Oxygen in LIS which assigned wasteload allocations for treatment plants in Connecticut and New York and to reduce nitrogen from plants in those states. This interstate project was undertaken to recommend ways of reducing nitrogen from treatment plant effluent in the upper LIS watershed states: Massachusetts, New Hampshire, and Vermont. The objective was to explore ways to sustainably remove the largest quantity of nitrogen from the LIS watershed at the lowest unit cost to protect water quality and minimize financial impacts on communities. The results of that study are presented.

INTRODUCTION

During the early 1980s, Long Island Sound showed significant water quality degradation, mostly in the form of hypoxia (low dissolved oxygen). In response to hypoxic conditions, Federal and state (New York and Connecticut) legislators supported monitoring, research and action plans to preserve and protect the waters of Long Island Sound through the Long Island Sound Study (LISS) National Estuary Program partnership. Their support resulted in the following:

- In 1985, Congress appropriated funds to both Connecticut and New York to monitor and research water quality.
- The Clean Water Act reauthorization in 1987 established a National Estuaries Program and Long Island Sound was designated an Estuary of National Significance.
- In 1994 the LISS Comprehensive Conservation Management Plan (CCMP) was developed by a management conference consisting of federal, state, interstate and local agencies, universities, environmental agencies, industry and the public.

The LISS CCMP is a significant document that identified several problems associated with the degradation of water quality in Long Island Sound, including:

- Hypoxia, or low dissolved oxygen (DO)
- Toxic contamination
- Pathogen contamination
- Floatable debris
- Habitat loss and its impact on living marine resources
- Land use and development practices which have contributed to degradation of natural habitat and water quality

The LISS Management Conference has focused its resources on the most pressing problem, hypoxia. Hypoxia is caused when excess amounts of nitrogen enter Long Island Sound by way of discharges from wastewater treatment plants, surface runoff and atmospheric deposition. Nitrogen is typically the limiting nutrient in marine waters for phytoplankton, a microscopic plant. When phytoplankton die, they sink to the bottom waters where the decay process uses oxygen in the water column which reduces the available amount of DO below critical levels. Hypoxia is defined as DO concentrations below 3.0 mg/L. Severe hypoxia, also known as anoxia, occurs when DO concentrations decrease to less than 2.0 mg/L. The CCMP concluded that the main objective for Long Island Sound corrective actions should focus on reducing the amount of nitrogen entering the Sound to increase DO concentrations.

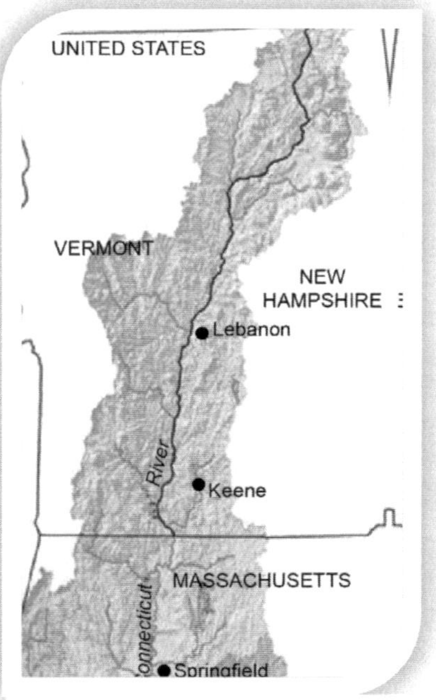

Figure 1 Upper Long Island Sound Watershed

A Long Island Sound Total Maximum Daily Load (LIS TMDL) for nitrogen was approved by U.S. Environmental Protection Agency (EPA) in 2001. This TMDL specifies a 58.5% reduction in total nitrogen (TN) by 2014. Connecticut proposed to remove about 6,056 metric ton (6,670 US tons) per year of TN and New York proposed to remove about 15,570 metric tons (17,150 US tons) per year. Beginning in 2000, the waste load allocation for each treatment plant that discharges to LIS decreased annually through 2014. Wastewater treatment plants throughout Connecticut and New York have adopted denitrification processes to meet the TMDL target load for nitrogen.

The LIS TMDL included a provision to re-evaluate and revise the TMDL at a later date. That process is currently underway by a workgroup comprised of representatives of the five watershed states (Connecticut, Massachusetts, New Hampshire, New York and Vermont), EPA, and NEIWPCC. The workgroup is considering many source categories for possible nitrogen reductions and is evaluating management strategies for achieving those reductions.

There are three major river basins which contribute nitrogen to Long Island Sound; Connecticut, Housatonic and Thames. Of them, the Connecticut River is the largest. The Connecticut River begins near the Canadian border and passes through Vermont, New Hampshire, Massachusetts and Connecticut. The Housatonic River begins in western Massachusetts and the Thames River watershed includes areas in southern Massachusetts. About 37-50% of the nitrogen load (due to wastewater discharge, runoff and atmospheric deposition) from these three rivers to LIS was attributed to areas north of the Connecticut border according to the report "Estimated Nitrogen Loads from Selected Tributaries in Connecticut Draining to Long Island Sound, 1999–2009" (Mullaney and Schwartz, 2013). The drainage areas for these rivers north of the Connecticut/Massachusetts border constitute the Upper Long Island Sound Watershed (LISW) (Figure 1).

As a component of the evaluation and management strategies to reduce nitrogen in LIS, this Upper Basin (Massachusetts, New Hampshire, and Vermont) project was conducted to assess the feasibility and financial impact of installing low-cost biological nitrogen removal retrofits at select facilities in the Upper Connecticut River basin in Massachusetts (MA), New Hampshire (NH), and Vermont (VT) as well as facilities in the Housatonic and Thames River basins in Massachusetts.

PROJECT DETAILS

Objectives

The objectives of the project were to:

1. Perform a detailed and accurate evaluation of the treatment plants including but not limited to existing and design capacity, expected near term future flows, seasonal flow and load variation, capacity of bioreactors and clarifiers and wastewater characteristics;
2. Evaluate ability to configure existing tankage and pumps for nitrogen removal;
3. Determine impact of nitrogen removal on operation and maintenance budgets;
4. Determine nitrogen removal training needs for plant staff;
5. Recommend whether operational and/or low cost modifications will be practical; and
6. Quantify the achievable reduction in effluent nitrogen concentrations and mass.

To achieve these objectives, the project was divided into several tasks including:

- Preparation and approval of the Quality Assurance Project Plan (QAPP)
- Kick-off meetings
- Initial site visits and data collection from 29 treatment plants
- Special sampling and testing program
- Preliminary modeling using an EXCEL-based nitrogen removal model
- Technical memorandum documenting WWTPs selected for further study (BioWin modeling and second site visits
- Development of recommended modifications to achieve low cost nitrogen removal
- Comparative cost analysis based on dollars per pound of nitrogen removed

Special Sampling Program

Although several of the treatment plants included in this project had effluent nitrogen data, few had influent nitrogen data. Additionally, for the computer modeling phase, parameters such as soluble chemical oxygen demand (SCOD), soluble Kjeldahl nitrogen (SKN) and influent alkalinity were necessary. So a special sampling and testing program was established using a laboratory contracted directly to NEIWPCC. This program was explained to plant personnel at each initial site visit including a review of proper sampling procedures and identification of sample locations to ensure these samples would be representative and reflect the influent characteristics entering the biological system. Furthermore, plant personnel were instructed to take samples after sidestreams such as digester supernatant, septic waste, filtrate or centrate from thickening or dewatering operations entered the treatment plant. Additionally, if plants only dewatered on certain days or received septic on certain days, they were asked to include at least one of those days as part of the sampling period to ensure any increased loadings were captured in the data. Because of limited funding the program only allowed for three consecutive days of sampling.

Plant personnel transferred the samples in pre-labeled bottles provided by the laboratory. Plant personnel filtered samples for both the soluble COD and soluble Kjeldahl Nitrogen determinations. Samples were taken of influent or primary effluent for six analytes (BOD_5, sCOD, NH_4-N, TKN, TSS/VSS and Alkalinity and final effluent for five analytes (BOD_5, sCOD, NH_4-N, sKN and NO_x). The results were used, in part, for initial selection of plants for further study and for the modeling phase.

Facility Selection

During the first site visit, various plant records were collected at the 29 plants including two years of operating data, plant schematics and plant design criteria. Additional data from the site visits included wet weather-cold weather operational problems, number and capacity of various pumps and blowers or aerators, sizes of tanks and types of sidestreams and the locations where they enter the process. Prior to the site visit, each plant was sent a survey form asking various data including: average of all influent, primary effluent and final effluent monitoring parameters for a two-year period, operating parameters, chemicals used, January-March average temperature, and other pertinent information. This information along with the results of the sampling program was used to determine which plants would be included for further study which includes BioWin modeling for all activated sludge plants, conceptual designs for all plants including those plants using Rotating Biological Contactors (RBCs), and cost estimation for all plants.

Various graphical and statistical methods were used in analyzing data from each plant. Flow and BOD loading variability, temperature, and influent/effluent nitrogen species, if available, were analyzed. Flow data were analyzed using frequency distribution graphs to compare current average flow to design flow and 80% of the design flow. This is important because if the plant is currently receiving flows close to 80% of design flow, it might be an indicator that low-cost nitrogen removal was not appropriate for that facility since typically that is the threshold for communities to limit future sewer connections. Additional statistical analysis used to make assessments of the treatment plants included mean, standard deviation, linear regression and correlation for all important parameters such as flow, wastewater temperature, BOD, TSS and nitrogen concentrations (if available). In addition to this analysis, a desk top nitrogen removal model was used as a screening tool to evaluate the plants. In selecting facilities for further study, several things were taken into consideration such as current flow as compared to design, current organic loading, temperature, COD to N ratio, and current ability to nitrify.

Of the twenty nine facilities, twenty were selected for further study and evaluation: eleven in Massachusetts, five in Vermont and four in New Hampshire. The further study program included:

- BioWin models: at current average conditions (average temperature and three-month low temperature), at 80% design flow and at low temperature and 80% design flow
- Preliminary conceptual design base on BioWin model
- Second site visit to discuss conceptual design and get operator input
- Finalizing conceptual design
- Cost estimate (capital plus some operating)

RESULTS AND DISCUSSION

The project was not intended to produce detailed designs, but only to evaluate the potential of low cost TN removal strategies and produce a preliminary estimate of the associated costs of the retrofit for planning purposes. The conceptual design for each facility was based on the model output or calculated nitrogen removal potential and the equipment or modifications necessary to achieve that removal. These modifications (mostly MLE-type processes) are relatively easy to operate and control and allow for interim removal of nitrogen while decisions are being made for a long-term nitrogen removal strategy. Depending on the existing plant configuration and process, the conceptual design and cost estimation included addition of nitrate recycle pumps, creation of anoxic and or swing zones mixers. In some cases, the design and cost estimation includes the addition of variable frequency drives (VFDs) and air piping modifications and/ or addition of isolation valves. In all cases, conceptual designs and cost estimations included instrumentation: pH, DO, nitrate/nitrite, ammonia.

Table 1. Predicted TN Removal and Capital Cost

STATE	lbs TN/day	lbs TN/year	Capital Cost, $
MA	**1934**	**704,529**	**$3,252,492**
NH	292	106,479	$877,332
VT	**92**	**33,517**	**$606,414**
TOTAL	2317	844,525	$4,736,238

Table 2. Predicted TN Removal and Capital Cost by

WATERSHED	Design, MGD	ADF, MGD	lbs TN/day	lbs TN/year	Capital Cost, $
Connecticut River	**36.2**	**16.1**	**1,150**	**418,177**	**$3,327,885**
Houstonic River	20.2	13.0	917	334,965	$1,042,546
Thames River	6.00	3.0	250	91,383	$365,807
TOTAL	62.4	32.1	2,317	844,525	$4,736,238

Some of the twenty plants were already removing nitrogen to varying degrees so the estimates of nitrogen removal predicted by the model were adjusted so that the values reported represented the net mass of nitrogen removed rather than the total amount predicted. The study determined that if these retrofits are made, as much as 2,313 lbs/d or 844,525 lbs/year of additional nitrogen could be removed from the watershed for a total capital investment of $4,736,238 (Table 1). The cost per pound of nitrogen removed ranged from $2.18 to 4.70 over the 10-year term and from

$1.71 to $3.50 over the 20-year term. Massachusetts had the potential of removing the largest mass of nitrogen due to the larger number and size of plants in that state.

There were three different watersheds represented by the twenty plants: Connecticut River (17 plants), Housatonic River (2 plants) and Thames River (1 plant). The capital cost and TN removal for each watershed is shown in Table 2.

The results of this study demonstrated that with a relatively small capital investment, the nitrogen load to Long Island Sound can be reduced through process control adjustments coupled with low cost retrofits.

ACKNOWLEDGEMENTS

This project was funded by an agreement awarded by the Environmental Protection Agency to the New England Interstate Water Pollution Control Commission in partnership with the Long Island Sound Study. The authors are indebted to the treatment plant staff at each facility for their cooperation on this project. They supplied information on a timely basis, gave freely of their time to supply data, take samples, answer questions and give plant tours.

REFERENCES

Mullaney, J.R., and Schwarz, G.E., 2013, Estimated Nitrogen Loads from Selected Tributaries in Connecticut Draining to Long Island Sound, 1999–2009: U.S. Geological Survey Scientific Investigations Report 2013–5171, 65 p., http://dx.doi.org/10.3133/sir20135171. ISSN 2328–0328 (online)

IFAS Systems with Structured Sheet Media for Simultaneous Carbon Oxidation, Nitrification, Denitrification and Implications

Hua Li*, Jia Zhu*, Michael Faeth*,

*Brentwood Industries, Inc., 610 Morgantown Road, Reading, PA, USA (Emaillli@brentw.com);

ABSTRACT

Submerged fixed-bed media has been utilized in wastewater treatment either as a hybrid system in conjunction with return activated sludge (i.e. integrated fixed-film activated sludge or IFAS) or as a fixed-film only system without RAS since the 1960s. The benefits of such systems include, but are not limited to, plant capacity increase without requiring new construction, enhanced nitrification at both normal and low temperatures, treatment capacity protection during loading surges and improved sludge settling characteristics, etc. Most of these systems were designed to specifically target carbon oxidation and/or nitrification. The simultaneous nitrification and denitrification (SND) capability of an IFAS system (especially with PVC structured sheet media) and its impact to both secondary and biological nutrient removal systems have not been adequately explored even evidence of significant SND exists. A pilot study was conducted to evaluate SND of various processes at the City of Reading Wastewater Treatment Plant utilizing PVC SSM as the fixed-film carrier. The pilot study consisted of five phases, each simulating a different process. Various HRTs were tested. This paper presents the details of the pilot configuration, results, data analyses, discussion and conclusions.

KEYWORDS

Biofilm; Integrated Fixed-Film Activated Sludge; IFAS; Structured Sheet Media; Fixed-bed; Simultaneous Nitrification and Denitrification; SND; Biological Nutrient Removal; BNR

INTRODUCTION

Submerged fixed-bed media has been utilized in wastewater treatment either as an integrated fixed-film activated sludge (IFAS) system in conjunction with return activated sludge (RAS) or as a fixed-film only system without RAS since the 1960s (Lessel, 1991, 1994; Muller 1998; Schlegel and Koeser 2007, Flamming *et al.* 2011, Ye *et al.* 2010, Zhu *et al.* 2012, Li *et al.* 2014). The benefits of submerged fixed-bed media systems include, but are not limited to, plant capacity increase without requiring new construction, enhanced nitrification at both normal and low temperatures, treatment capacity protection during loading surges and improved sludge settling characteristics, etc. While most of these systems were designed to specifically target carbon oxidation and/or nitrification, the simultaneous nitrification and denitrification (SND) capability of a fixed-bed IFAS system compared to an activated sludge system have not been adequately explored.

Fixed-bed media includes a few types such as the rope-type, perforated polyethylene tubes, PVC structured sheet media (SSM) and a knit-web like fabric media. The various fixed-bed media systems share common and general benefits and/or characteristics due to the similar working mechanisms. At the same time, each system has its own distinct characteristics because of the difference in media type. Significant SND has been observed in full-scale and/or pilot aerobic fixed-bed media IFAS systems where consistent nitrification is achieved (e.g., Randall and Sen 1996, Muller 1998, Sen *et al.* 2010, Ye *et al.* 2010, Zhu *et al.* 2012). Measurement of denitrification rates shows that an average of 30 percent of the nitrates generated are denitrified in a media segment located farther away from a feed point, and can increase to above 80 percent for media located near a feed point (Sen *et al.* 1995). In the Calw-Hirsau full-scale plant experiments, TN was simultaneously removed to an extent where no denitrification reactor was needed (Muller 1998). While some researchers interpret the observed SND as being a mechanism taking place in deeper layers of the biofilm (Revsbech *et al.* 1989, Hagedorn *et al.* 1994), others believe that nitrification takes place mainly on the media and denitrification takes place mainly in the suspended growth (Hibiya *et al.* 2003).

The objectives of this pilot study were to further explore the SND capabilities of IFAS systems with SSM and its effect on the design of both secondary and biological nutrient removal (BNR) systems; to develop design parameters for future IFAS system design; and to provide a technology alternative for retrofit plants with limited land availabilities and capacity increase pressures.

MATERIALS AND METHODS

Pilot Configuration and Design

The study was carried out at the City of Reading Wastewater Treatment Plant. Circular plastic tanks were placed in a truck trailer and used as anoxic and/or aerobic reactors (Figure 1). The dimensions of the tanks are shown in Table 1. Brentwood vertical-flow media VF3000 and cross-flow distribution media CF3000DM media were used as the film carrier. The vertical-flow (VF) media has vertical flutes as reflected in its name. The cross-flow (CF) distribution media (DM) is a proprietary media type that has been designed to be placed underneath the VF media and to provide uniform liquid/air distribution to the bottom of the VF media. Additional media information is provided in Table 2. The pilot drew wastewater from the plant primary effluent. Fine bubble diffusers were used in the aerobic stages.

The pilot study consisted of five phases. Phase I was a two stage aerobic activated sludge process (ASP) without media. Phase II was a two stage aerobic IFAS process with media in both stages. Phase III was a two stage aerobic fixed-film only process without RAS. Phase IV was a three stage (anoxic for stage 1, aerobic for stages 2 and 3) fixed-film process with media in all three stages but without RAS. Phase V was a three stage (anoxic for stage 1, aerobic for stages 2 and 3) IFAS process with media in all three stages and RAS. Various HRTs were tested.

Table 1. Reactor Information

	Anoxic	Aerobic #1 Or Aerobic #2	Clarifier
Total Volume, L (Gallons)	473.1 (125)	662.4 (175)	788.4 (208)
Total Surface Area, m2 (ft2)	0.47 (5.03)	0.63 (6.77)	0.66 (7.06)
Total Tank Depth, m (ft)	1.0 (3.33)	1.09 (3.57)	1.2 (4.00)

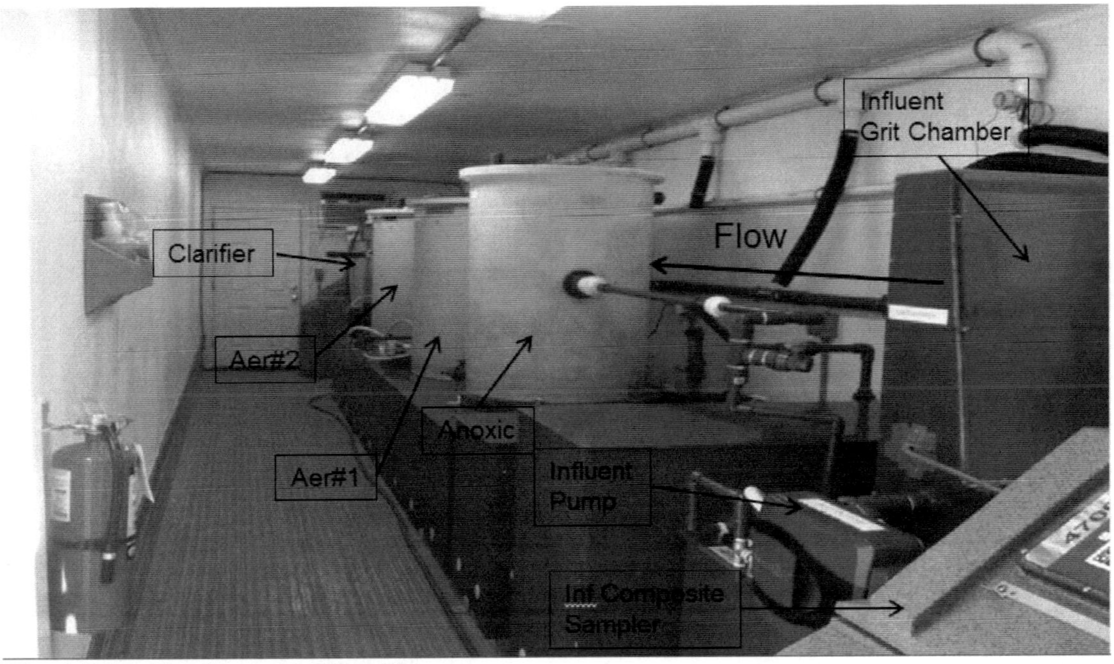

Fig. 1 Pilot Trailer and Pilot Setup

Table 2. Media Information for Anoxic and Aerobic Reactors

	Anoxic (Phase IV & V)	Aerobic #1 OR #2
Media	VF3000	VF3000 & CF3000DM
Media Specific Surface Area, m^2/m^3 (ft^2/ft^3)	196(60)	158(48)
Media Fill Volume, m^3, (ft^3)	0.29 (10.1)	0.33(11.53)
Media Fill Percentage, %	60	49
VF3000 Media Dimension, m (ft)	Concentric Ring OD 0.77(2.52), ID 0.28(0.92)	0.38(2.52) High, 0.51(1.67) Wide, 0.89(2.92) Long
CF3000DM Distribution Media	N/A	0.38(2.52) High, 0.51(1.67) Wide, 0.89(2.92) Long
Total Media Height, m, (ft)	0.71 (2.33)	0.76 (2.5)
Total Media Plan Area, m^2(sqft)	0.40 (4.33)	0.43(4.61)
Total Specific Surface Area, m^2 (sqft)	56.3(606)	51.4(553)

Table 3 Pilot Testing Conditions

	Phase I AS		Phase II IFAS		Phase III FBBR	
	Aer#1	Aer#2	Aer#1	Aer#2	Aer#1	Aer#2
Test Date	Nov 2011 – Feb 2012		Feb – Apr 2012		Apr – May 2012	
F/M Ratio, kg SCOD/kg MLVSS-day (SCOD Removal Rate, g SCOD/m^2 media-day)	0.9-1.6	0.08-0.19	0.9-2.5	0.04-0.4	(12-25)	(2.1-6.3)
Volumetric Loading {Removal} Rate, kg SCOD/m^3-day (lbs SCOD/kcf-day)	1.5-3.7 (92-230)	0.8-2.7 (48-167)	1.2-3.8 (76-238)	0.7-2.6 (46-162)	{59-118}	{10-30}
MLSS, mg/L	1,482-2,058	1,642-2,020	1,044-2,968	1,256-2,838	83-288	91-290
Air Flux, SCFM/ft^2-media	1.7	1.2	2.1	1.3	2.1	1.9
Average DO, mg/L	5.9	7.4	3.2	6.0	4.9	7.3
RAS, % INF	100-158		105-180		0	

RESULTS AND DISCUSSION

Two-Stage Secondary Systems – Phase I, II & III

The primary effluent wastewater from the plant had high soluble organic matter. Since SCOD was used as the monitoring parameter for organic loading calculations in the study, SCOD/SBOD ratio was evaluated at the beginning of the pilot to provide a tool if SBOD needed to be examined. Table 4 shows the SCOD/SBOD ratios for the pilot influent and effluent wastewater, respectively. The SCOD/TN ratio is approximately 10. Table 5 shows some of the performance results of the pilot study.

Table 4. SCOD, SBOD and BOD

Samples	SCOD	BOD	SCOD:BOD	SBOD	SCOD:SBOD
Influent (Composite, mg/L)	384	245	1.6	230	1.7
Effluent (Composite, mg/L)	84.7	--	--	40	2.1

Table 5. Pilot Results

	HRT, hr		Inf	SCOD, mg/L			TN, mg/L			NH_3-N, mg/L			NO_3-N, mg/L			Alk, mg/L		
	An	Aer# 1 & 2	Tem °C	Inf	Eff	Rem, %	Inf	Eff	Rem, %	Inf	Eff	Rem, %	Inf	Eff	%	Inf	Eff	%
Phase I: ASP	0	9	15.0	328	59.7	81.8	33.9	19.8	41.5	24.3	1.3	94.5	0.8	11.3	1,297	236	161	32.0
		6.1	14.6	303	54.2	82.1	36.5	18.8	48.6	22.9	0.9	96.0	0.8	9.6	1,108	266	182	31.8
Phase II: SSM IFAS	0	9*	16.0	353	53.3	84.9	34.0	11.1	67.3	24.6	0.7	97.0	0.6	1.5	132	296	245	17.1
		7.3	17.2	396	56.2	85.8	39.9	8.1	79.7	28.8	0.7	97.6	0.4	1.3	213	314	258	17.9
		6.1	17.9	376	49.0	87.0	33.6	8.1	75.9	26.5	0.3	98.9	0.7	4.3	523	294	248	15.8
Phase III: FBBR	0	9	18.7	355	73.9	79.2	36.5	26.5	27.4	29	6.3	78.1	0.5	13.1	2,677	320	246	23.3
		7.3	18.9	384	81.4	78.8	39.2	23.3	40.6	27.4	0.9	96.9	0.5	15.8	3,050	318	216	32.2
		6.1	20.0	361	72.9	79.8	33.4	21.5	35.7	26	2.9	89.0	0.5	10.3	2,129	299	243	18.9
Phase IV: FBBR-BNR (IMLR, No RAS)	2.78	7.8	24.6	377	76.2	79.8	34.1	10.7	68.5	26.2	1.5	94.3	0.6	3.7	481	336	277	17.4
Phase V: IFAS-BNR (IMLR & RAS)	2.3	6.5	24.2	358	57.4	83.9	34.5	7.0	79.6	27.3	0.2	99.2	0.5	3.3	531	334	260	22.2
	1.9	5.3	24.7	371	68.8	81.5	36.4	6.9	80.9	30.1	0.6	97.9	0.4	3.8	757	342	272	20.6

Phase I of the study was a two-stage aerobic activated sludge process with two aeration reactors in series and served as a baseline for comparison with the results of subsequent phases. Phase II was a two-stage aerobic IFAS system with media in both aeration reactors. A comparison of the results (Table 5, Figures 2 and 3) from Phase I and Phase II indicates that with the same HRT and similar influent conditions, the IFAS system performance was slightly better in terms of SCOD and NH_3-N removal efficiencies. The performance of the IFAS system in terms of nitrate and therefore total nitrogen (TN) removal, however, was much better than the activated sludge process in Phase I. The IFAS system was not only able to reduce the NH_3-N concentration to a very low level (under 1.0 mg/L), it also reduced the NO_3-N concentration to a low level (under 2.0 mg/L with longer than 6.1-hour HRTs and under a 5 mg/L with a 6.1-hour HRT). The low effluent NH_3-N and NO_3-N concentrations resulted in an average total effluent TN concentration of 8.1 mg/L for the IFAS system without an anoxic reactor, which is significantly lower than the effluent TN concentration (just below 20.0 mg/L) from the activated sludge process in Phase I. Moreover, The IFAS system consumed approximately half of the alkalinity that was consumed by the activated sludge process under similar operating and influent conditions. The lower effluent NO_3-N and TN concentrations, together with the significantly lower consumption of alkalinity clearly indicate the presence of simultaneous denitrification in the IFAS system. Considering the main difference in the configuration and operation of the two phases was the presence of the media, it is logical to assume that the media caused the simultaneous nitrification and denitrification to take place, even though the exact mechanism has yet to be identified. One interesting finding worth noting was that the first stage of the activated sludge process in Phase I actually performed better in removing ammonia than the first stage of the Phase II SSM IFAS system (Figures 2 and 3), but the first stage of Phase I had more nitrite and nitrate in it than the first stage of Phase II. This appears to indicate a different nitrification and denitrification mechanism in the IFAS system than in the activated sludge system.

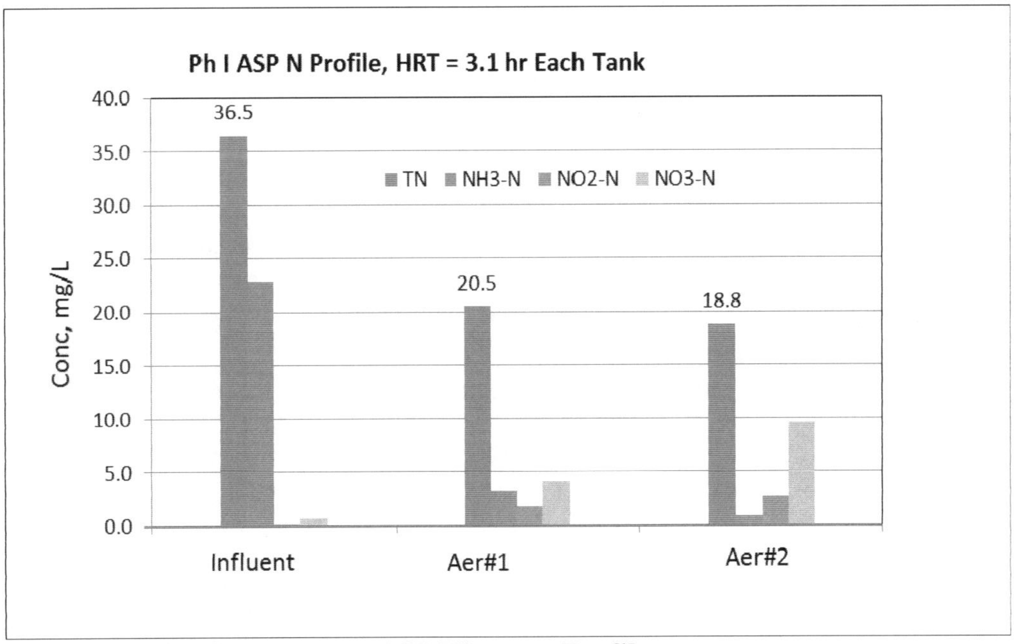

Figure 2. Phase I Two-Stage ASP Nitrogen Profile

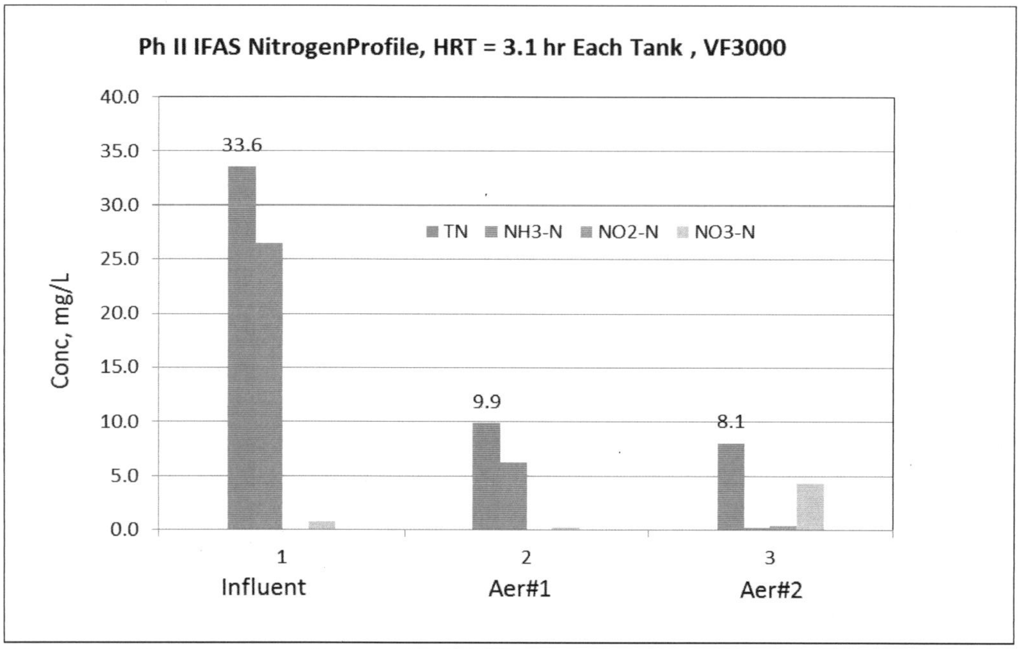

Figure 3. Phase II Two-Stage IFAS Nitrogen Profile

Similar SND was observed in the first aeration reactor of a full-scale aerobic SSM IFAS system where TN removal efficiencies ranged from 38 to 73% (averaging 60%) including contribution from nutrient uptake (Zhu *et al.* 2012). A linear relation was found to exist between soluble SCOD removed and TN removed. In addition, a few conditions that existed in the system were believed to have been conducive to SND. Firstly, the soluble COD (SCOD) concentration ranged from 423 to 760 mg/L and was high in the influent. The SCOD/TN ratio ranged between 7 and 10. Secondly, the mixed liquor suspended solids concentration was high with a range between 6,000 and 9,000 mg/L. Thirdly, the DO concentration was mostly under 1.1 mg/L and low.

From the operating conditions of Phase II, it appears that some of the conditions that existed in the full-scale plant in Zhu *et al.* (2012) were different from those in the pilot. For instance, the DO concentration in the pilot was mostly high. It was between 1.9 and 3.8 mg/L on average in the pilot while it was mostly under 1.1 mg/L in the full-scale plant. This seems to indicate that SND in a SSM IFAS system is more tolerant to higher DO concentrations, which contradicts with the understanding that high DO concentrations inhibit SND in a traditional activated sludge system. Additionally, the mixed liquor suspended solids (MLSS) concentration in the pilot was mostly low. It was between 1,400 mg/L and 2,800 mg/L. This seems to indicate that a high MLSS concentration is not necessary for SND to take place in a SSM IFAS system.

Phase III of the study was a two-stage aerobic media-only system without RAS. The purpose of the phase was to evaluate the impact of RAS on the system performance. As seen from the results (Figure 4), the overall performance was inferior compared to that of Phase I and II with similar other operating and influent conditions. In particular, the TN removal efficiency was worse than Phase I and II. The removal efficiency ranged from 27% to 41%. This seems to indicate that RAS and therefore a certain MLSS concentration are critical for elevated denitrification to take place even though the MLSS concentration does not have to be higher than what an activated sludge system typically requires. The media-only system, or the fixed-bed bioreactor (FBBR), reduced SCOD first to a low concentration in the first aeration reactor and then converted most ammonia to nitrate in the second aeration reactor. The lack of bioflocculation and conditioning from MLSS and animal grazing seemed to have resulted in higher

WERF Addresses Challenges of Minimizing Effluent Nutrient Concentrations

J.B. Neethling[1*], H.D. Stensel[2], J. Sandino[3], R. Tsuchihashi[4], D. Clark[1], A. Pramanik[5]
[1] HDR Inc., [2] Univ. Washington, [3] CH2M Hill, [4] AECOM, [5] WERF
*Email: JB.Neethling@hdrinc.com

ABSTRACT

The WERF Nutrient Removal Challenge has investigated the nutrient removal in wastewater treatment plants. Over 7 years the Challenge completed more than 30 research project with collaborators from universities, utilities, government, consultants and equipment suppliers. Utilities provided data to assess treatment performance and efficiencies valued and millions of dollars.

EPA's proposed nutrient criteria is approaching and often below the concentrations achievable with current technologies. Data from operating plants indicate that soluble organic nutrients become the limiting factor for reaching low effluent values. Inorganic and particulate nutrients can be removed to low concentrations. Variability in performance contributes to the risk and ability to reliably meet permits. Longer averaging periods in permits provide relief from short term effluent excursions from this variability. As nutrient limits decrease to 3 mg TN/L and 0.1 mg TP/L energy and chemical use increase the environmental cost with only modest improvement in water quality.

KEYWORDS: WERF Nutrient Removal Challenge, nutrient species, nitrogen, phosphorus, limit of technology, sustainable operation

INTRODUCTION

The WERF Nutrient Removal Challenge is a multi-year collaborative research initiative established in 2007 to develop and provide current information about wastewater treatment nutrients (specifically nitrogen and phosphorus in wastewater), their characteristics, and bioavailability in aquatic environments to help regulators make informed decisions. This paper presents and overview of the approach and key findings of the Challenge regarding the nutrient species in wastewater effluents, the ability of treatment processes to remove these compounds, the reliability of performance, and the implications for water quality and sustainability.

Findings from studies from various collaborators and investigators documented in WERF reports form the basis of the paper. Companion papers to for this Nutrient Symposium session provide additional details on nitrogen and phosphorus species (Stensel et al., 2015); design issues (Sandino et al. 2015), operational considerations (Tsuchihashi et al. 2015), and regulatory aspects (Clark et al. 2015).

GOALS AND OBJECTICES

The goals for the WERF Nutrient Removal Challenge are to provide scientific information regarding the fate, removal and operational requirements to reduce nutrient in wastewater effluent to low levels. The needs range from assessing the technologies available or emerging for nutrient removal, their cost, operational requirements, and reliability of performance. In addition, the implications of nutrient reduction on operating costs, green house gas (GHG) emissions and the public's ability to pay, needs to be addressed for a range of permit conditions.

There is no uniform approach or strategy by EPA and states to control nutrients. EPA provided guidelines and proposed nutrient criteria for various ecoregions. The nutrient criteria are generally very low. For example, ecoregional criteria for rivers and streams for total nitrogen (TN) are between 0.12 and 2.18 mg TN/L and for total phosphorus between 10 and 76 ug TP/L. Translating these criteria into site specific in-stream concentrations would consider environmental factors, including dilution from the receiving water.

However, since many waters already contain or exceed the nutrient criteria concentrations, there is no capability of dilution and consequently the nutrient criteria becomes the in-stream standard.

The goal of the Challenge is to determine the ability of technologies to reach these low nutrient concentrations and identify the limiting factor for preventing further removal. The focus of the work is therefore on technologies that are able to achieve "low" nutrient concentrations. The term "low" is not defined, but in general, means concentrations that approach the ecoregional criteria: TN below 2.18 mg/L and TP below 76 ug/L.

Nutrient reductions from water resources recovery facilities can also be achieved by means other than mechanical treatment. Measures such as water reuse, trading, offsets and others can be used to reduce the nutrient loading on the water body. These measures are extremely important for utilities to consider when possible since they may offer low cost avenues for nutrient reduction. This Challenge did not research these strategies except for inclusion in the general discussion. The degree of nutrient reduction achievable using these strategies was not researched. The Challenge focused nutrient removal "inside the fence" as the primary means.

Sustainable nutrient removal is another central theme in this work. Since this Challenge focus on "inside the fence" reduction means, it only address part of the classic sustainability factors. Economic impacts from capital and operating costs are considered in the research. Environmental factors are partially considered by determining the GHG impacts associated with nutrient reduction technologies. Other environmental impacts, such as in stream impacts following discharge or emissions in collection systems are considered outside the bounds. These factors are also very site specific and require consideration for local conditions. Social impacts are not accounted for since these are considered to be specific to the local community and their own social values. In short, the sustainability assessment focused mainly on economic impacts and GHG emissions as an indicator of the environmental impacts. There is one notable exception: for regulatory and permitting work, the receiving water impacts are addressed.

Resource recovery implications and costs were not researched since the WERF Nutrient Recovery Challenge focus on this aspect. In our work, technologies for resource recovery are acknowledged and included as part of a potential solution without researching all options and alternatives.

APPROACH

Prioritization and collaboration formed the base for the research under the Challenge. Insufficient funding was available to research every topic independently. The research team therefore approached the Challenge in a collaborative way, inviting other stakeholders such as universities, researchers, utilities, regulators, consultants, equipment suppliers, and others to participate in the work. Stakeholder meeting following the Nutrient Conference in Baltimore (2007), Washington (2009), and Miami (2011) provided a forum for identifying research needs and prioritizing them. These meetings also serve to identify potential collaborators. Unsolicited proposals were entertained and considered when they addressed key concerns identified in the highly ranked research needs.

The Challenge followed five broad research topics: regulations, nitrogen, phosphorus, design, and operation. This provided a focus and balance to the work. All research projects included elements or more than one focus area, and almost all were a balance between fundamental, applied, and field research.

The 6 authors of this paper form the core team that worked together on all aspects of the Challenge to develop plans, select research teams, review documents, and conduct research. An independent Issue Advisory Team (IAT) administered by WERF provided oversight to the team. More than 30 research projects were conducted by different research teams. Collaborators leading projects came from universities, consultants, utilities and included generally stakeholders from regulators, equipment suppliers, and others. The ability to manage the WERF funding and solicit in-kind funding to leveraging other ongoing work was extremely valuable in directing effort to teams where synergy with ongoing work provided much broader coverage of the research topic.

FINDINGS

Five general findings from the research program are discussed below. Details on individual research projects are documented in report (see list of reports in reference section). Results from the various technical areas are discussed in the companion papers. In this paper, the overall findings of significance are presented. The six areas are:

- Nutrient species matters
- Treatment performance variability impacts reliability
- Economic and GHG emission impacts of nutrient removal
- Achievable nutrient level
- Regulatory compliance with adaptive management

Nutrient Species Matters

Not all nutrient species are equal from a treatment and environmental impact perspective. Treatment technologies are generally more efficient to reduce one species better than others; in some cases one species is reduced and new species produced. The environmental impacts in lakes and rivers of the different species are also not the same. Table 1 shows the general classification of the nutrient species.

Table 1. Classification of Nutrient Species

Classification	Technology Implication	Water Quality Implication
Particulate and soluble	Determine the ability to remove the species by solids separation. Impact biological kinetics	Particulate species will settle to bottom Impact rate of nutrient uptake
Organic and inorganic	Inorganic species are generally biological removed Organic species requires transformation for biological uptake; allowed sorption technologies. Some organic species is very slowly degraded. Soluble organic nutrients species can be produced during biological treatment and during solids processing.	Inorganic species are bioavailable Organic species require transformation to be bioavailable and requires time
Bioavailability/Biodegradability	The term biodegradability applies to treatment. It measures the ability of a biological process to degrade (remove) the nutrient. Biodegradation must occur in the timeframe of the treatment technology – typically within hours. Residual soluble organic nutrients leaving treatment plant are considered essentially non-biodegradable (produced or slowly degradable).	The term bioavailability applies natural waters. It indicates the ability to grow algae using the nutrient species in the environment. Bioavailable nutrient uptake extend over long periods (months) and is impacted by environmental factors such as salinity, temperature, sunlight, synergies with other organisms, etc.

Soluble vs. Particulate. The analytical definition of particulate (or filterable) is dependant on the pore size of the filter. Typically, a 0.45 um filter size is used to distinguish between particulate (filterable) and soluble (dissolved or non-filterable) species. Colloidal particles can sometimes pass through the filtration process and be measured as a soluble nutrients. Sattyatewa et al. (2010) used two filter sizes (0.45 um and 1.2 um) as well as flocculated-filtered samples to identify the "truly soluble" organic nitrogen and colloidal nitrogen. While this further differentiation is helpful, little field data is available about its occurrence. Standard Methods and most WRRFs use 0.45 um filtration remains a useful practical differentiation point between particulate and soluble species.

Particulates can be removed with filtration; and specifically with coagulation/flocculation/ sedimentation/filtration processes. This fraction can therefore be removed with existing technologies to very high degree. Membrane or filtered water typically contains near zero TSS; and a fraction of a milligram fine solids that will pass the 0.45 um filter.

Particulates play a complex role in natural waters. During quiescent conditions, fine particles settle to the bottom. Stable colloidal particles (i.e. very small particles that won't flocculate and settle) will remain in suspension for long periods. Particles can also be resuspended with slight turbulent conditions and return to the liquid column. The particulate form will not be readily available for algal growth and exert a slow release/availability to support algal growth.

Soluble inorganic compounds are generally reactive and can be removed biologically or chemically. Soluble organic compounds remaining following biological treatment is typically slowly or non-degradable organics of complex structure. These could be humic or fulvic molecules. These naturally occurring soluble organic compounds resist conventional treatment but can be broken into smaller molecules with oxidation, typically at high doses or prolonged periods (for example months of sunlight) and become biodegradable and bioavailable. Li and Brett (2015) showed that phosphorus contained in certain soluble organic compounds can support algal growth. However, not all soluble organic phosphorus can support algal growth. Gu et al. (2015b) evaluated oxidation and biodegradation of phosphorus containing organic compounds and found that even advanced oxidation only partially oxidize the organic compounds. Sedlak et al. (2013) used XAD resin to adsorb remove hydrophobic organics from treated effluent. This same hydrophobic fraction was then shown to be non-bioavailable in bioassays.

The hydrophobic soluble organics appears both recalcitrant towards treatment and also toward algal growth. These compounds not only pose a problem for removal during wastewater treatment but are also less available for algal growth.

Treatment performance variability impacts reliability

Treatment reliability refers to the ability to consistently produce the desired effluent. Neethling et al. (2009) and Bott and Parker (2011) proposed using Treatment Performance Statistics (TPS) to describe treatment performance in terms of the statistical distribution of the effluent. TPS values are defined as the percentile achieved in the performance and can be used to assess the ability to meet a given target/permit limit.

Three TPS levels provide insight into the reliability of treatment:

- TPS-50% - Represents the median performance (or average), the value that is achieved and also exceeded 50% of the time. Since this value is exceeded half the time, it's not an satisfactorily performance level to meet an annual limit, since it will be exceeded every other (50% of the time).
- TPS-80% - Represents a treatment level that is achieved 80% of the time, or exceeded 20% of the time. An annual permit will be exceeded once in 5 years; a monthly permit 12 times in 5 years.
- TPS-95% - Represents a treatment level that is achieved 80% of the time, or exceeded 20% of the time. An annual permit will be exceeded once in 5 years; a monthly permit 12 times in 5 years.

Bott and Parker (2011) studied variability and reliability performance of 22 BNR exemplary treatment plants. All plants showed variability in performance is inevitable for all treatment plants and should be considered in the evaluation of engineering design and in the ability to meet permits. The variability in performance is attributed to both uncontrolled factors such as variable influent, wet weather conditions, or industrial influences, and controllable factors such as construction, mechanical failure, or disrupted chemical supplies.

In order to meet permits on reliable basis, WRRFs must consider the variability in performance of treatment plants. Neethling and Stensel (2012) analyzed the data from various WERF studies to calculate TPS values. Based on these values they proposed to use the TPS-80% for a reliable annual average permit and the TPS-95% for a reliable monthly permit limit.

Economic and GHG emission impacts of nutrient removal

Nutrient removal requires additional facilities and for lower limits, while energy and chemical use increase. Falk et al. (2011) determined the GHG emissions associated with 5 levels of nutrient removal: from Level 1 at secondary treatment without nutrient removal to Level 5 at 2 mg TN/L and <0.05 mg TP/L. The investigation determined that as the degree of nutrient removal increased, the impact on operational sustainability (function of capital and O&M costs, pumping, mixing, aeration, N_2O emissions, chemicals manufacturing/hauling, and biosolids hauling) increased significantly. Results are presented in Table 2.

The GHG emissions increase about 1,000 CO_2 eq mt/yr with each treatment level step until the transition from Level 4 to 5. For Level 3 and Level 4, methanol is required to provide sufficient BOD for denitrification and chemical addition with filtration for phosphorus removal are required following the secondary clarifiers. The 70 percent increase from Level 4 to 5 is in part due to the energy/chemical demands associated with MF/RO and brine reject handling by deep well injection (DWI) for 50% of the flow.

Table 2. Greenhouse Gas Emission Results (10 mgd; 100,000 PE for 100 gal/d/capita)

Treatment	Level 1 No N/P Removal (CO_2 eq mt/yr)	Level 2 8 mg N/L 1 mg P/L (CO_2 eq mt/yr)	Level 3 4-8 mg N/L 0.1-0.3 mg P/L (CO_2 eq mt/yr)	Level 4 3 mg N/L 0.1 mg P/L (CO_2 eq mt/yr)	Level 5 2 mg N/L <0.02 mg P/L (CO_2 eq mt/yr)
Primary	240	240	240	240	240
Activated Sludge	2,980	3,820	4,430	4,430	4,430
Tertiary	200	200	650	1,530	6,990
Solids *	-150	-10	-40	50	-30
Subtotal	3,270	4,250	5,380	6,250	11,630
Miscellaneous **	1,320	1,320	1,320	1,320	1,320
Total	4,590	5,570	6,600	7,570	12,950

* The solids treatment stream includes cogeneration of biogas, which results in negative values for most treatment levels

** Miscellaneous results include odor control, non-potable water, high pressure air system, miscellaneous pumping/mixing/sludge storage/transfer pumps, HVAC, administrative buildings energy, and miscellaneous equipment.

The above results are placed in better perspective when the GHG emissions impacts are expressed in terms of the incremental increase in GHG emissions per additional lb of N or P removed as shown in Figure 1 (log-scale). For nitrogen, the incremental increase in GHG emissions increases exponentially beyond Level 4. As for P, the increase in GHG emissions with treatment per additional pound of P removed increases exponentially after Level 3. Based on this figure, an argument could be made that a point of diminishing returns is reached following Level 3 where the GHG emissions required to remove the last few percentile of N or P is significantly higher than the initial benefit.

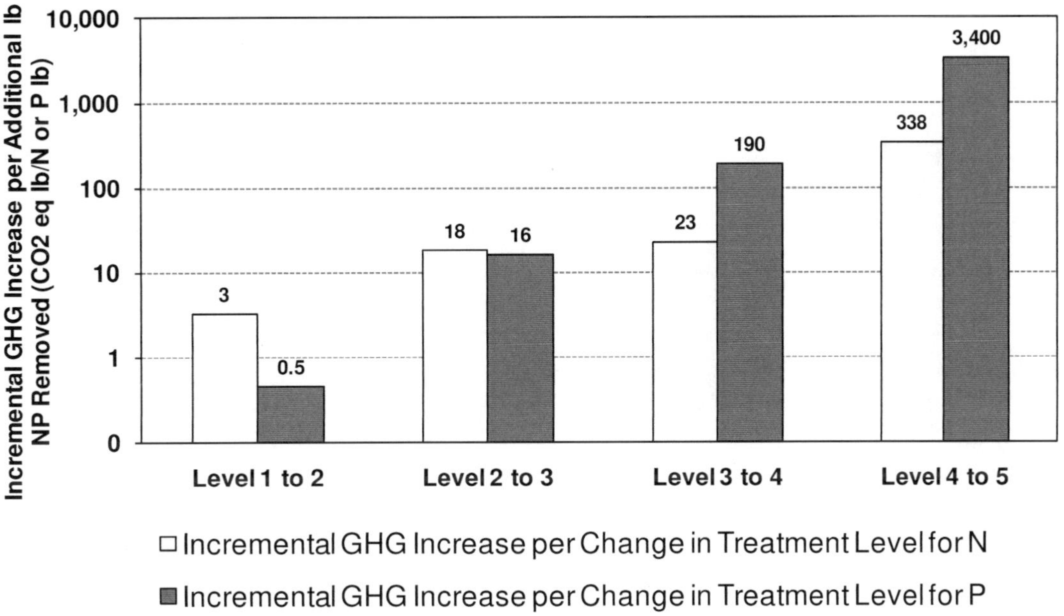

Figure 1. Incremental GHG Emissions Increase per Additional Pound N or P Removed (10 mgd; 100,000 PE for 100 gal/d/capita)

Because the rationale behind nutrient removal is to improve water quality, the benefit of nutrient removal was quantified using a water quality surrogate that reflects potential algal growth. The plot of GHG emissions along with potential algal production for the five treatment levels is shown in Figure 2. Nearly 95 percent of the potential algae production is reduced by changing from Level 1 to Level 3 treatment. An additional 4 percent of potential algae production is reduced from Level 3 to 5, but requires nearly twice as much GHG emissions (6,590 to 12,950 CO_2 eq mt/year).

Capital and operating costs also increase with more advanced treatment as summarized in Table 3. It is important to highlight that total project capital cost are for a greenfield plant.

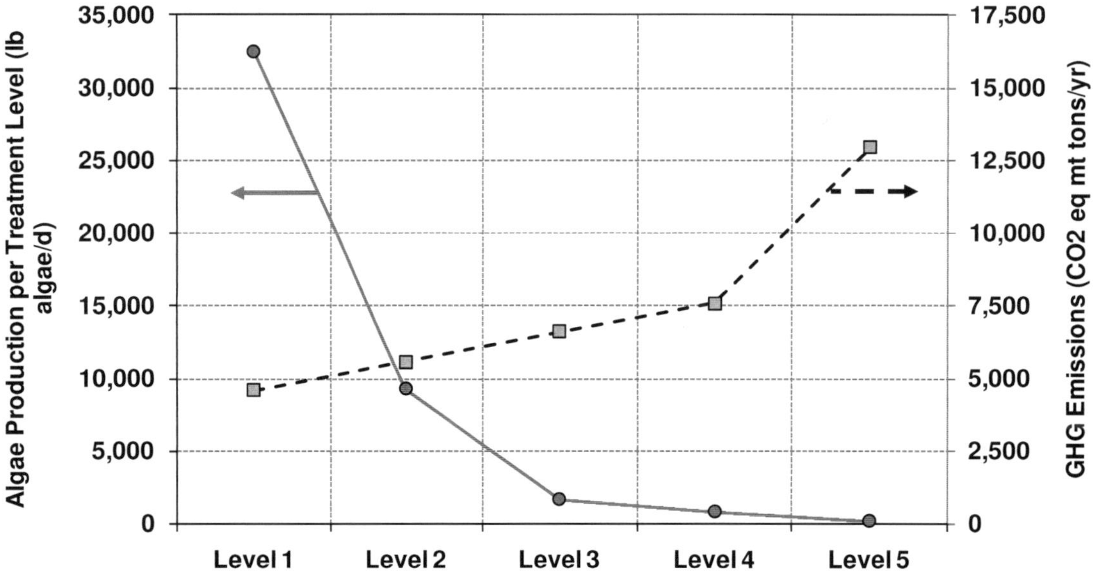

Figure 2. GHG Emissions and Potential Algal Production per Treatment Level (10 mgd; 100,000 PE for 100 gal/d/capita).

Table 3. Total Project Capital and Operations Costs for a 10 mgd Facility [i]

Level	Total Project Capital Cost ($/gpd) [ii]	Total Project Capital Cost (Million $) [ii]	Operations Cost ($/MG Treated) [iii]	Operations Cost ($1,000/yr) [iii]	Total Present Worth Project Costs (Million $) [iv]
1 (No N/P Removal)	9.3	93	250	910	110
2 (8 mg N/L; 1 mg P/L)	12.7	127	350	1,260	150
3 (4-8 mg N/L; 0.1-0.3 mg P/L)	14.4	144	640	2,350	180
4 (3 mg N/L; <0.1 mg P/L)	15.3	153	880	3,200	210
5 (1 mg N/L; <0.02 mg P/L)	21.8	218	1,370	4,990 **	300

i The total project capital cost are for a Greenfield plant

ii The total project capital cost are the equipment cost, construction, and "soft costs"

iii Operations cost = energy and chemical cost. Labor and maintenance costs are excluded

iv The assumed discount rate was 5 percent at an escalation rate of 3.5 percent (capital, energy, non-energy)

In addition to the increased GHG emissions values and reduced potential algal production, there are some ancillary benefits associated with transitioning from secondary treatment to nutrient removal. These include but are not limited to lower nutrient, BOD and SS discharge loads, improved water quality and lower potential algal growth, higher removal of heavy metals and trace organic compounds, and improved process stability.

Achievable Nutrient Level

It is possible to nearly completely eliminate nutrients from effluent using ion exclusion technologies such as reverse osmosis. As shown above, the cost and GHG impacts for that level of treatment increases dramatically and the benefits small. The best achievable nutrient level would be one where the benefits (reduction in algal growth) remains significantly high while the costs (including capital and operational costs, plus environmental costs such as GHG emissions) acceptable.

The achievable effluent nutrient concentrations recorded at full scale facilities was presented by Neethling and Stensel (2012). They summarized data generated by several investigations to estimate the nutrient levels achievable for 80-95% of the time. The nutrient levels in Table 4 are based on reported data under full scale operating conditions. The large datasets containing 3 years' data were used to calculate the 80 and 95th percentiles. Data from other studies (such as Gu et al. 2012) provided detailed speciation data but for a limited basis.

The achievable nutrient ranges in Table 4 represent the performance reported in long term data at the 80th and 95th percentiles. These values are indicative of annual and monthly reliable compliance values. However, some individual data points were lower. These lower values, are highlighted in the comments section and indicated opportunities to reduce the effluent concentrations more by changing operation, adding chemicals, or additional facilities. Treatment plant operators will generally not push nutrients significantly below the permit requirements since chemical addition is required to reach the ultimately low nutrient concentrations. The comments in Table 4 address opportunities where lower effluent limits might be possible.

Regulatory compliance with adaptive management

Current Permit Approach. WERF research show a great variety of approaches to establishing effluent limits for nitrogen and phosphorus have been used across the country and some have resulted in very restrictive conditions. Overly restrictive effluent discharge permits generally result from the application of permitting approaches designed to protect receiving waters from short term effects in the near field effluent mixing zone, such as the approach taken to control toxics.

In other cases, effluent nutrient limits have been tailored to the site specific circumstances necessary to achieve receiving water quality objectives. This often results in more appropriate permit structures that reflect an understanding of both the impact of nutrient discharges on receiving waters and the capabilities of nutrient removal treatment.

Tailoring nutrient permits to site specific circumstances often results in the need to deviate from conventional effluent limit structures applied to other parameters and combine receiving water objectives with technically feasible treatment limits.

Table 4. Achievable Nutrient Levels Based on Reported Data[a]

Nutrient Species	Range[b]	Comment	Reliability
Ammonia	0.25 – 0.5	Ammonia levels below 0.1 mg/L has been reported.	Nitrification is susceptible to influent and other process variations. Variable temperature, flow, loading, toxicants impacts reliability
NOx	1.5 – 3.0	Secondary plus tertiary denitrification with carbon addition. Values below 1 mg/L reported.	Online instrumentation for chemical dosing required. Carbon usage increase as effluent concentration decline
Soluble organic nitrogen	1.5 – 2.5	Based on survey of plants using limited data. Effluent highly specific to influent and presence of industrial discharges that contain SON. Solids and biological processing can produce SON. Half of surveyed plants had SON below 1.25 mg/L.	Concentrations remain relatively stable with low variability.
Particulate nitrogen	0.08 – 0.12	Require filtration to reduce effluent below 1 mg TSS/L. Calculated value.	Dependent on filtration process. Generally good.
Total nitrogen	3-6	Sum of N species. Controlled mainly by SON and NOx.	See individual component comments above.
Soluble reactive phosphorus (SRP)	5 – 15	SRP below 5 ug/L reported with high chemical dose.	Reliability improve with solids inventory or high chemical dose.
Soluble non-reactive phosphorus (SNRP)	15 – 25	Based on survey of plants using limited data. Effluent highly specific to influent and presence of industrial discharges that contain SNRP. Solids and biological processes may produce SNRP. Half of surveyed plants had SON below 10 ug/L.	Concentrations remain relatively stable with low variability.
Particulate phosphorus	10 – 20	Require filtration to reduce effluent below 0.5 mg TSS/L. P content of solids lower for chemical P removal. Calculated value.	Dependent on filtration process. Generally good.
Total phosphorus	30 – 50	Sum of P species. Values below 50 ug/L in full scale operation requires significant chemical dose and membrane-equivalent filtration.	See individual component comments above.

Notes:

a Based on plant performance and other samples; achieved 80-95% of the time (Neethling and Stensel, 2013)

b Units: mg N/L or ug P/L

c Permit limits for plants: TN range from 2 to 11 mg/L; TP range from 36 to 2000 ug/L

Flexible Permits Foster Compliance. Emphasis in nutrient discharge permitting should focus on providing the greatest amount flexibility possible in the structure of nutrient limits in order to preserve the opportunity for the most creative and economical approaches to managing nutrients. Traditional permit structures for publically owned treatment works generally include both monthly and weekly limits on both a concentration and mass basis. This may inadvertently eliminate the most effective watershed solutions to nutrient management by creating disincentives to wastewater dischargers to explore combinations of advanced wastewater treatment and other watershed management practices. It is preferable to structure discharge permits in such a way that receiving water quality objectives are met with the greatest flexibility that can be provided to the treatment processes. This is important in order to avoid unnecessarily restrictive effluent discharge conditions that result in little additional water quality protection but consume inordinate amounts of energy and chemicals that result in other deleterious environmental impacts.

Adaptive Management approaches provide step wise improvement. In this approach progressively more stringent requirements over time allow technically achievable effluent limits step towards point source nutrient control and at the same time evaluate the overall progress towards meeting water quality objectives in the watershed. Finding the best combination of advanced treatment for nutrient removal and other watershed management practices presents a challenge for utility managers and regulators. Adopting technically achievable and cost effective levels of advanced wastewater treatment provides utilities with flexibility to foster creative solutions and stepwise improve treatment to meet overall water quality objectives.

Preserving an opportunity for adaptive management approaches to guide the process of nutrient management over time may improve water quality incrementally, without overly restrictive discharge permits that result in over investment in advanced treatment. Permits structured around no net increase in existing loadings, or simple seasonal or annual loading reductions, may provide a foundation for adaptive management. In some cases, states in the process of adopting numeric nutrient criteria have used technology based effluent limits to achieve some level of point source nutrient reduction while preserving the opportunity for adaptive management approaches. This has been found to be especially important where numeric nutrient criteria are very low concentrations of nitrogen and phosphorus that have the potential to result in water quality effluent limits beyond the capabilities of advanced nutrient removal treatment.

Technology Performance Statistics. As discussed above, technology performance statistics can quantify variability and reliability of effluent performance from advanced nutrient removal facilities, especially those operating at low or very low levels. Short-term limitations, such as maximum daily and maximum weekly, should not be imposed for nutrients. Technology performance statistics provide a science based approach to characterize feasible effluent limits within the capabilities of advanced nutrient removal treatment and also characterize the variability in effluent performance and reliability of treatment.

Predictive Water Quality Models. When water quality models are available to simulate the water quality response to nutrient loadings, discharge permit scenarios can be simulated to develop the basis for the most flexible and sustainable permit structure possible. These models must include the various nutrient species in order to account for the differential kinetics associated with different nutrient species.

Probabilistic Analysis. Where there is recognition that variability exists in receiving water flows and water quality, consider the application of probabilistic approaches to define levels of effluent performance to meet performance objectives and at what frequency. Extremely low receiving water flow conditions are not likely to coincide with maximum effluent discharge conditions. Likewise, aquatic life and recreational beneficial uses would generally not be thought to be impaired if a single rock, pool or riffle, or even a short reach of river had benthic algae higher than a target value. Probabilistic analysis can provide a tool to analyze the frequency at which specific conditions may occur in receiving waters based on variability in both effluent and receiving water.

Watershed Permitting and Water Quality Trading. Since nutrients are often a broad watershed scale issue in terms of water quality, consideration should be given to watershed permitting. Watershed permitting provides a structure that allows for collaboration among point source dischargers, nonpoint sources, and other stakeholders to achieve watershed nutrient management objectives. Approaching nutrient management considerations from the watershed scale, as opposed to individual permits, may reveal the opportunity for watershed permits to result in effective collaborations.

CONCLUSION

Advanced nutrient removal is required to meet nutrient limits based on EPA's ecoregional criteria. These limits require that removal of individual nutrient species be optimized. Often the removal of two species are in conflict with one another (example nitrification versus denitrification). It some processes competition for a common substrate can limit reduction of a nutrient; for example biological phosphorus removal and biological denitrification both require a carbon source.

Variability in treatment plant effluents is unavoidable. Both controllable (e.g., equipment failures) and uncontrollable factors (e.g., wet or cold weather) contribute to performance variation. In order to meet permits consistently, the designer must accommodate process fluctuations, which can lead to higher construction and also higher operational costs. Permit structures that accommodate these variations can reduce the cost of compliance without reducing the environmental protections. Using longer averaging periods can provide relief.

A point of diminishing returns appears to exist as the cost of operation increase, while the net nutrient reduction becomes small. At the same time, the soluble organic residual nutrient species dominates the effluent. These organics are also less bioavailable to algae in the receiving stream.

REFERENCES

Belia, L. and Johnson, B. (2013) "Uncertainty Evaluations in Model-Based WRRF Design for High Level Nutrient Removal - Literature Review and Research Needs" WERF Nutrient Removal Challenge Project NUTR1R06q.

Bott, C.B, and Parker, D.S. (2011), "Nutrient Management Volume II: Removal Technology Performance & Reliability" WERF Nutrient Removal Challenge project NUTR1R06k.

Clark, D., Neethling, JB, Stensel, H.D., Sandino, J., Tsuchihashi, R., Pramanik, A. (2015) "Developing Attainable and Protective Permits for Nutrients" WEF Nutrient Symposium

2015, July 26-28, 2015, San Jose, CA.
Clark, D.L. Hunt, G., Kasch, M.S., Lemonds, P.J., Moen, G.M., Neethling, J.B. (2010) "Nutrient Management Regulatory Approaches To Protect Water Quality - Volume Review Of Existing Practices" WERF Nutrient Removal Challenge project NUTR1R06i.
Clark, D.L., Dupuis, T., Falconer, H., Hatch, L., Kasch, M.S., Lemonds, P.J., and Neethling, J.B. (2015 in press) "Nutrient Management Volume III: Development of Appropriate Permitting Frameworks" WERF Nutrient Removal Challenge Project NUTR1R06z.
DeBarbadillo, C., Benisch, M., Barnard, J.L., Falk, M.F. (2015 in press) "Evaluation of Performance and Greenhouse Gas Emissions for Plants Achieving Low Phosphorus Effluents" WERF Nutrient Removal Challenge Project NUTR1R06v.
Eleuterio, L., and Neethling J.B. (2009) "Low Phosphorus Analytical Measurement Study" WERF Nutrient Removal Challenge Project NUTR1R06F.
Falk, M.W.; Neethling, J.B.; Reardon, D.J. (2011) "Striking the Balance between Nutrient Removal and Sustainability. WERF Research Project under Nutrient Challenge" WERF Nutrient Removal Challenge Project NUTR1RO6n.
Gray, H., Smith, S, and Parker W. (2015) "State Of Knowledge Of The Use Of Sorption Technologies For Nutrient Recovery From Municipal Wastewaters" WERF Nutrient Removal Challenge Project NUTR1R06x
Gu, A, Rahman, S.M., Eckelman, M., and Onnis-Hayden, A. (2015a in press) "Sustainability Evaluation Of Nutrient Removal Technologies Using Comprehensive Life Cycle Assessment" WERF Nutrient Removal Challenge Project NUTR1R14f
Gu, A. and Tooker, N. (2015b in press) "Impact Of Advanced Oxidation Processes On The Composition And Biodegradability Of Soluble Organic Nutrients In Wastewater Effluents" WERF Nutrient Removal Challenge Project NUTR1R14e
Gu, A., Onnis-Hayden, A., Laquidara, M., Stinson, B., and Tsuchihashi, R. (2010) "Protocol to Evaluate Alternative External Carbon Sources for Denitrification at Full-Scale Wastewater Treatment Plants" WERF Nutrient Removal Challenge Project NUTR1R06b.
Gu, A.Z., Lei Liu, Annalisa Onnis-Hayden, Scott Smith, Holly Gray, Dwight Houweling, Imre Takács, (2012) "Phosphorus Fractionation And Removal In Wastewater Treatment-Implications For Minimizing Effluent Phosphorus", WERF Nutrient Removal Challenge report NUTR1R06l.
Khunjar, W., Strahota, M., Pitt, P., and Gellner, W.J. (2014) "Evaluating the Impacts of Cold and Wet Weather Events on Biological Nutrient Removal in Water Resource Reclamation Facilities" WERF Nutrient Removal Challenge Project NUTR1R06s
Li, B. and Brett, M.T. (2011) "Spokane Regional Wastewater Phosphorus Bio-Availability Study" WERF Nutrient Removal Challenge project NUTR4C09.
Li, B. and Brett, M.T. (2015) "The Bioavailable Phosphorus (BAP) Fraction in BNR System Effluents" WERF Nutrient Removal Challenge project NUTR1R06m.
Li, B., Fan, L., and Brett, M.T. (2015) "Mineralization Kinetics Of Soluble Phosphorus And Soluble Organic Nitrogen In Advanced Nutrient Removal Effluents" WERF Nutrient Removal Challenge project NUTR1R06p.
Maher, C., Neethling, J.B., and Pagilla, K. (2014) "Solids Role In Tertiary Chemical Phosphorus Removal By Alum" WERF Nutrient Removal Challenge Project NUTR1R06t
Neethling, J.B., Dave Stensel, Denny Parker, Charles Bott, Sudhir Murthy, Amit Pramanik, Dave Clark (2009) "What is the Limit of Technology (LOT)? A Rational and Quantitative Approach" WEFTEC 2009.

Neethling, J.B., David Stensel (2012) Nutrient Species Implications for Technology Performance and Water Quality Impacts. WEF/IWA Nutrient 2013 Conference, Vancouver, BC.

Pellegrin, M-L., Minetti, A., Stensel, H.D., Sandino, J., and Neethling, J.B. (2014) "Application Membrane Bioreactor Processes for Achieving Low Effluent Nutrient Concentrations" WERF Nutrient Removal Challenge Project NUTR1R06u.

Sandino, J., Tsuchihashi, R., Neethling, JB, Stensel, H.D., Clark, D., Pramanik, A. (2015) "Advances in the State of the Practice for the Design of Nutrient Removal Facilities" WEF Nutrient Symposium 2015, July 26-28, 2015, San Jose, CA.

Sattyatewa, C., Pagilla, K., Sharp, R., and Pitt, P. (2010) "Fate of Organic Nitrogen in Four Biological Nutrient Removal Wastewater Treatment Plants" Water Environment Research, Vol. 82, No. 12, 2306-2315 (2010).

Sedlak, D.L, Jeong, J., and Liu, H. (2013) "Uptake By Algae Of Dissolved Organic Nitrogen From BNR Treatment Plant Effluents" WERF Nutrient Removal Challenge Project NUTR1R06e

Smith, D.S. and Gray, H.E. (2014) "Surface Complexation Modelling And Aluminum Mediated Phosphorus" WERF Nutrient Removal Challenge Project NUTR1R06r

Stensel, H.D. (2008). Dissolved Organic Nitrogen (DON) in Biological Nutrient Removal Wastewater Treatment Processes. Compendium from WERF Nutrient Removal Challenge project NUTR1R06 (2008)

Stensel, H.D., Neethling, JB, Clark, D., Sandino, J., Tsuchihashi, R., Pramanik, A. (2015) "The Importance of Nutrient Species for Treatment Performance." WEF Nutrient Symposium 2015, July 26-28, 2015, San Jose, CA.

Tsuchihashi, R., Sandino, J., Neethling, JB, Stensel, H.D., Clark, D., Pramanik, A. (2015) "Advances in the State of the Practice for the Operation of Nutrient Removal Facilities" WEF Nutrient Symposium 2015, July 26-28, 2015, San Jose, CA.

Tsuchihashi, R.T., Canterbury, B., and Neethling, J.B. (2015, In press) "BNR Process Monitoring And Control With Online Nitrogen Analyzers For Nitrogen Credit Exchange Program In Connecticut" WERF Nutrient Removal Challenge Project NUTR1R06y

The importance of nutrient species for treatment performance and surface water impacts

H.D. Stensel[1*], J.B. Neethling[2], D. Clark[2], J. Sandino[3], R. Tsuchihashi[4], A. Pramanik[5]
[1] Univ. Washington, [2] HDR Inc., [3] CH2M Hill, [4] AECOM, [5] WERF
*Email: stensel@uw.edu

ABSTRACT

Surface water quality impairment has led to more stringent effluent total N (TN) and total P (TP) concentration limits. The WERF Nutrient Challenge Program has determined the role of nitrogen and phosphorus species within TN and TP with regard to their removal in treatment facilities and their availability for algal consumption. A large fraction of effluent TN and TP persists as soluble organic nitrogen (SON) and soluble nonreactive phosphorus (SNRP) in effluents from systems designed to maximize nutrient removal. These compounds are less bioavailable for algae growth compared to inorganic the inorganic compounds. The bioavailability of SON and soluble phosphorus was studied with algal bioassay on effluent samples from full-scale nutrient removal facilities and a model was developed that accounted for an initial rapid consumption rate followed by ever decreasing rates as less N and P remained. This type of nutrient uptake model could improve water quality model predictions of surface water impacts of effluent nutrients.

KEYWORDS: WERF, bioavailable, algae, soluble organic nitrogen, nonreactive phosphorus, nitrogen species, phosphorus species, bioassay

INTRODUCTION

Phosphorus and nitrogen enrichment of surface waters from point and nonpoint sources can accelerate eutrophication with impaired water quality due to the photosynthetic activity of algae or periphyton and diurnal changes in the water dissolved oxygen (DO) content. This has led to more stringent effluent nutrient concentration limits. A common effluent discharge limit for the Chesapeake Bay area and coastal estuaries with impaired surface waters from eutrophication is an effluent total phosphorus (TP) concentration of 0.10 mg/L and effluent total nitrogen (TN) limit of 3.0 mg/L. In a few locations in the U.S. and Canada discharge limits for TP and TN have been set at much lower values in the range of 0.05 mg/L for TP and 1.0 - 2.0 mg/L for TN.

The ability to meet very low effluent limits (typically TN and TP <3.0 and < 0.10 mg/L) has been demonstrated for advanced nutrient removal (ANR) water resource recovery facilities (WRRFs), which commonly include a variety of designs for biological nutrient removal (BNR) processes, chemical phosphorus precipitation, and effluent suspended solids removal. The Water Environment Research Foundation (WERF) Nutrient Challenge Program has engaged in projects to determine the nutrient removal capability of ANR facilities and the ultimate effluent concentrations that could be achieved. Evaluation of plant performance and effluent quality has shown the importance of understanding the nitrogen and phosphorus species contained in effluent TP and TN to better determine how the wastewater characteristics and WRRF design and operation can impact the ability to minimize effluent nutrient concentration. These studies also showed that some of the species were much more resistant to removal in the WRRFs, and thus their relative impact on water quality and availability for algae growth was of interest. A better understanding of the role of different P and N species is needed to provide a more scientific approach to assessing water quality and to help regulators make scientifically-based decisions, which is one of the goals of the WERF Nutrient Challenge Program. This paper summarizes the WERF Nutrient Challenge Program with regard to nutrient speciation, species removal processes, and assessment of nutrient species impact on surface waters.

NITROGEN AND PHOSPHORUS SPECIES AND REMOVAL IN WRRFS

Nutrient species are broadly categorized according being in soluble or particulate form, and then the soluble form is further defined in terms of inorganic or organic constituents. Table 1 shows a summary of commonly used parameters used to describe nitrogen and phosphorus species. The analytical definition of particulate (or filterable) is dependent on the pore size of the filter. Typically, a 0.45 um filter size is used to distinguish between filterable (particulate) and non-filterable (soluble) species. Systems designed to achieve a minimal effluent nitrogen and phosphorus concentration include tertiary filtration, tertiary membrane separation, or membrane bioreactors (MBRs) to remove most all of the particulate nutrients.

Table 1 – Species Commonly used to Characterize Nutrients

Nutrient Species	Abbreviation	Category	Comment
Nitrogen			
Ammonia	NH4	s	Direct Measure
Nitrate	NO3	s	Direct Measure
Nitrite	NO2	s	
Total oxidized Nitrogen	NOx	s	Sum of NO2 and NO3
Soluble Organic N	SON	s	Direct Measure
Particulate N	pN	p	Non-filterable, separated by 0.45 um filter; mainly organic species
Total N	TN	s&p	Sum of all N species
Phosphorus			
Reactive P, mainly ortho-P	SRP	s	Particulate reactive P also found
Soluble Non-reactive P	SNRP	s	Hydrolyzable and organic P species
Soluble Organic P	SOP	s	
Particulate P	pP	p	Non-filterable, separated by 0.45 um filter; organic species and chemical precipitants
Total P	TP	s&p	Sum of all P species

Note: s=soluble; p=particulate

Effluent Nitrogen Species in Nutrient Removal Facilities

About 60-70% of the influent wastewater nitrogen is in the form of ammonia with the rest as soluble, colloidal and particulate organic nitrogen (Tchobanoglous et al. 2014). Biological treatment methods are the most common and cost effective means to remove the ammonia-nitrogen. Effluent TN consists of ammonia-N (NH_4-N), nitrate-N (NO_3-N) plus nitrite-N (NO_2-N), soluble organic nitrogen (SON), and particulate organic nitrogen (Figure 1). Note that the ammonia-N consists of free ammonia (NH_3) plus ionized ammonia (NH^+_4) but at typical pH conditions over 98% of it is as ionized ammonia.

Systems designed to achieve a minimal effluent TN concentration include tertiary filtration, tertiary membrane separation, or membrane bioreactors (MBRs) so that most of the particulate nitrogen is removed. To meet the goal of achieving minimal effluent TN concentrations a low concentration of t-N, NOx-N and SON is needed. The NOx-N concentration is minimized by using an external carbon source to maximize denitrification efficiency in a postanoxic reactor in the BNR system. The NH_4-N concentration is minimized by having a long enough aerobic reactor SRT in the BNR system to maximize ammonia oxidation. Most of the influent organic nitrogen is degraded, but a residual soluble organic nitrogen (SON) persists in BNR processes. Studies on the fate of nitrogen removal in full-scale BNR facilities by Czerwionka et al. (2012) found that SON was minimized by the end of an anoxic zone in anaerobic-anoxic-aerobic (A^2O) systems and that in some systems the SON concentration increased with aeration time.

At present at appears that there is no operating strategy to minimize the effluent SON concentration, and it appears to be greatly affected by the influent wastewater and return flows characteristics. Effluent SON concentrations from BNR processes have varied widely ranging from 0.50 to 2.50 mg/L as shown by the results of WRRF effluent sampling in Figure 2 (Stensel, 2008).

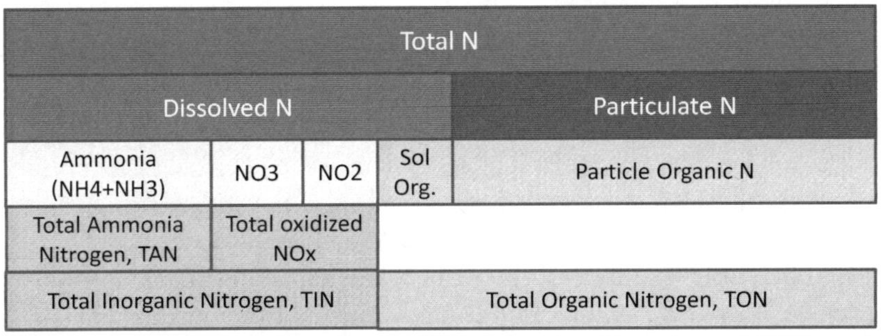

Figure 1. Nitrogen Species in Wastewater Treatment Effluent

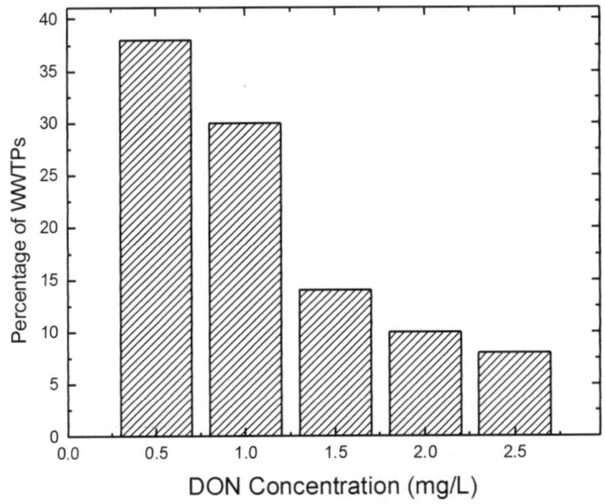

Figure 2. Summary of Effluent Soluble Organic Nitrogen (SON) Concentration (0.45 μm Filtration) from 188 Maryland and Virginia Wastewater Treatment Plants (Pagilla, 2007)

The SON concentration can represent a large fraction of effluent TN in ANR facilities designed to maximize nitrogen removal. An example of effluent NH_4-N, NOx-N and SON concentrations is shown in Table 2 for 5 facilities designed and operated to minimize effluent TN concentration. The effluent SON concentration ranges from 47 to 90% of the effluent TN. Because of the large impact of effluent SON and the unpredictability of effluent SON, the WERF Nutrient Challenge Program carried out projects to investigate the availability of effluent TN to algae by use of bioassay methods and by sampling and analysis of a number of BNR facilities. This work will be discussed in the section of this paper on bioavailability of WRRF effluent residual TN and TP.

Table 2. Comparison of Effluent Nitrogen Species Concentrations at the 50 Percentile Statistic for WWRFs Achieving Low Effluent TN Concentrations (The sum of the N species concentrations may not equal the TN concentration due to not all components sampled on the same day) (Stensel et al. 2013)

	Effluent concentration, mg/L			
WWRF	NH_4-N	NOx-N	SON	TN
Fiesta Village	0.01	0.03	0.9	1.0
Truckee Meadows	0.05	0.11	1.3	1.6
Western Branch	0.04	0.64	0.7	1.5
Tahoe Truckee	0.28	0.43	1.7	2.5
B-MBR	0.20	0.70	1.2	2.2

Effluent Phosphorus Species in Nutrient Removal Facilities

The phosphorus species are identified according to the results of analytical methods for phosphorus measurements. As shown in Table 1 and in Figure 3 below, the main effluent soluble phosphorus species are soluble reactive P (SRP), soluble acid hydrolyzable phosphorus (SAHP), and soluble organic P (SOP). The SRP generally represents orthophosphate (PO_4^{-3}) that is readily available for uptake by algae and bacteria and some colloidal P that reacts in the test procedure. The SAHP includes condensed phosphates and phosphorus precipitated or adsorbed with metal hydroxides or contained in metal complexes. The total soluble P (TSP) minus the SRP is equal to the soluble nonreactive P (SNRP). The SOP is calculated by subtracting the SAHP from the SNRP and includes phosphorus of biological origin, such as nucleic acids, inositol, and proteins. Effluent soluble organic P may also be from soluble organic compounds in the influent wastewater that pass through the system undegraded.

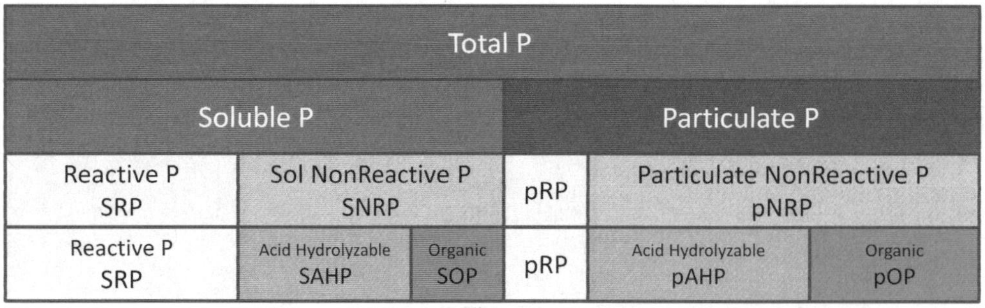

Figure 3. Phosphorus Species in Wastewater Treatment Effluent

Two methods for removal of phosphorus in ANR facilities are enhanced biological phosphorus removal (EBPR) and chemical precipitation. In EBPR the soluble orthophosphorus is removed from solution by phosphorus accumulating organisms (PAOs) and removed via waste sludge. In chemical precipitation alum or ferric metal salts are commonly added to form metal hydroxides, which sorb the soluble P and remove it from the system in particulate form. Some of the SAHP and SOP may also be sorbed. To achieve an extremely low effluent soluble phosphorus (SP) concentration chemical addition at high dose has been used alone or after EBPR with effluent granular or membrane filtration. A WERF Nutrient Challenge study by Gu et al. (2014) evaluated effluent phosphorus species from a number of full-scale facilities operating with secondary and tertiary phosphorus removal methods. An example of effluent phosphorus species after tertiary filtration with high chemical dose to minimize effluent TP concentration is given in Table 3. With high chemical dose the effluent SRP was typically below 0.020 mg/L and about 80% of the remaining effluent SP was SNRP in the form of SAHP and SOP. While the effluent SRP is considered readily available to algae the bioavailability of SAHP and SOP is not known. Thus, further studies were carried out under the WERF Nutrient Challenge Program to assess the bioavailability of effluent P from full-scale tertiary phosphorus removal systems.

Table 3. Example of Tertiary Effluent Soluble Phosphorus Species After High Chemical Dose and Granular Media or Membrane Filtration (Gu et al., 2014).

	Effluent soluble P, mg/L				
Filter Process	SRP	SNRP	SAHP	SOP	Total SP
Mono Media-1	0.005	0.008	0.001	0.007	0.021
Mono Media-2	0.006	0.011	0.001	0.010	0.028
Dual Media-1	0.005	0.013	0.001	0.012	0.031
Dual Media-2	0.100	0.006	0.001	0.005	0.112
Dual Media-3	0.013	0.029	0.028	0.001	0.071
Membrane-1	0.019	0.016	0.033	0.013	0.081

BIOAVAILABILITY OF WRRF EFFLUENT RESIDUAL TN AND TP

Laboratory bioassay procedures were used to measure the bioavailability of effluent nitrogen and phosphorus to algae for a number of full-scale ANR facilities under studies conducted by the WERF Nutrient Challenge Program. This section describes the procedures developed in work on the nutrient bioavailability assays and the results observed from WRRF effluent samples and its implications with regard to the impact of nutrient species on surface waters.

Bioavailability bioassay Methods

The bioavailability of full-scale ANR facilities effluent TN and effluent TP was measured in a number of WERF Nutrient Challenge Program studies (Li and Brett, 2011; Sedlak et al., 2013; Li and Brett, 2015; Li et al., 2015) using the algal bioassay method outlined in Standard Methods (APHA, 2005) with some modifications. Briefly, the bioassay technique involves the growth of *Pseudokirchneriella subcapitata* (formerly *Selenastrum capricornutum*) algal cultures in whole or fractionated wastewater effluent samples to which all necessary nutrients are added. No nitrogen was added in bioassay tests for the bioavailable SON studies and no phosphorus was added for the bioavailable phosphorus (BAP) studies. For the SON studies, the bioassay was carried out in sterilized glass flasks on shaker tables at 25°C using a 12-hour light/dark cycle for

2 weeks. In the BAP studies, glass flasks were incubated for a two-week period at 24°C on a shaker table. In the SON study, the amount of nitrogen used for algal growth was based chlorophyll-a measurements by fluorescence as described previously (Pehlivanoglu and Sedlak, 2004). Standards with known nitrogen concentrations were set up in separate incubation flasks to correlate the amount of nitrogen consumption and Chlorophyll-a production. In later work by Li et al. (2015) the SON concentration was measured with time to assess algal consumption directly instead of indirectly by the chlorophyll-a production. Algal growth from bioavailable phosphorus in the BAP study was determined by algal cell density measurements using a Coulter Multisizer III particle size analyzer by passing the samples through a 100 mm aperture. A correlation function between the phosphorus consumed and algal cell density growth was obtained from flasks set up with standard solution concentrations.

Based on previous work (Pehlivanoglu and Sedlak, 2004; Sattayatewa et al., 2008) an inoculum of bacteria was added to enhance the release of labile forms of organic nitrogen from the SON in studies by Sedlak et al. (2013), but not by Li and Brett (2011, 2015) and Li et al. (2015). Microscopic observations at the end of the bioassay tests by Sedlak et al. (2013) confirmed that the bacteria seed did not contaminate the samples with growth of other types of algae. Control bioassays employed deionized water (negative control) and 0.70 mg N/L of nitrate added to the hydrophobic fraction (positive control). A small amount of algal growth typically was observed in the controls due to the presence of bioavailable forms of nitrogen in the inoculum.

The SON study by Sedlak et al. (2013) developed procedures to minimize the effect of inorganic nitrogen on the bioassay test accuracy. Effluent samples from biological nutrient removal WRRFs typically have low effluent ammonia-N concentrations but can have a high enough effluent NOx-N concentration to impact the measurement accuracy for bioavailable SON, considering that effluent SON concentrations can be as low as 0.50 to 1.0 mg/L. A procedure to remove the NOx-N by an anion exchange resin was investigated to provide a sample for the bioassay that would consist of mainly SON and ammonia-N. However, losses of SON on the ion exchange resin used to remove NOx-N led to a technique that first separated hydrophobic SON on a Rohm and Haas amberlite XAD-8, resin followed by separation of the NOx-N with an anion exchange resin. The hydrophobic SON removed on the first resin was then extracted for use in bioassay testing. In this way the bioavailability of the hydrophobic and hydrophilic fractions of the effluent SON could be determined separately. More recently, the XAD-8 resin has been discontinued and replaced by a similarly effective product called Supelite DAX-8 resin (Sigma-Aldrich, Inc.).

Effluent Nutrient Species and Bioavailability

Effluent Bioavailable SON. Effluent NOx-N and NH_4-N are considered readily available for algal consumption and growth. The WERF studies were intended to determine if the effluent SON was as easily bioavailable as these inorganic nitrogen species.

The protocol for separating the hydrophobic and hydrophilic fractions of the effluent SON was applied to ten full-scale WRRFs with effluent TN, NOx-N and SON concentrations ranging from 2.0 to 12 mg/L, 0.10 to 11.2 mg/L, and 0.6 to 1.6 mg/L, respectively. All of the facilities had full nitrification. The bioavailability of the separated hydrophobic and hydrophilic fractions was evaluated with the described bioassay method (Sedlak et al., 2013). Most of the effluent SON was hydrophilic, which was nearly completely bioavailable. The hydrophobic SON fraction was not bioavailable over the 14-day incubation period and ranged from a concentration of 0.15 to 0.37 mg/L (Figures 4), which was 12 to 28 percent of the effluent SON concentration. Evaluation of the carbon to nitrogen ratios for the hydrophobic SON suggested that it may be associated with humic substances in the effluent.

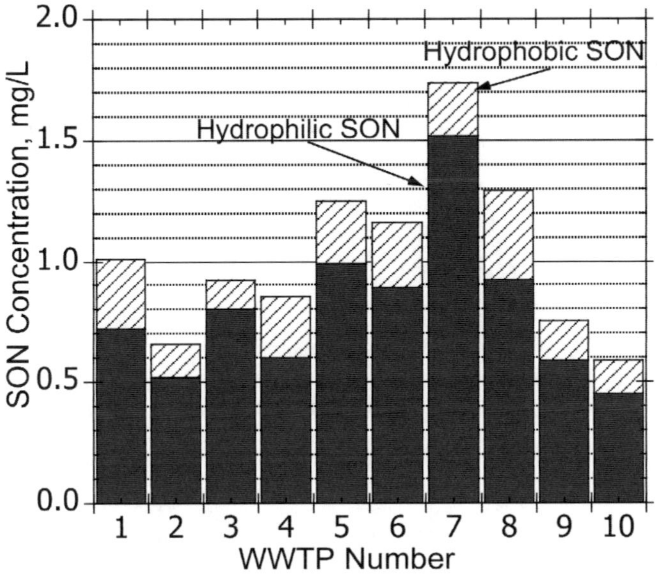

Figure 4. Hydrophobic and Hydrophilic SON Concentrations in Effluent SON from Ten WRRFs Studied (Sedlak et al., 2013).

Effluent BAP. Effluent samples from seventeen full-scale biological nutrient removal facilities were evaluated to determine the BAP content of the effluent phosphorus (Li and Brett, 2013). The facilities had tertiary chemical addition to achieve low effluent phosphorus concentrations. The results in Figure 5 compares the effluent soluble BAP (sBAP) and soluble reactive P (SRP) concentrations to the effluent total soluble P concentrations for the plant samples. The lower effluent soluble P concentrations are the result of increased chemical dose for tertiary phosphorus removal. The portion of sBAP in the effluent soluble P concentration decreased from 40% at an effluent soluble P concentration of 0.10 mg/L to 25% at an effluent soluble P concentration of 0.02 mg/L. There is also a lower fraction of SRP at the lower effluent soluble P concentration suggesting a larger fraction of the effluent soluble P is made up of soluble hydrolyzable P and soluble organic P. Based on sBAP the higher chemical dose is providing more effective phosphorus removal than if just based on total effluent soluble P concentrations.

Relative rates of bioavailability for effluent SON and soluble phosphorus

Though a portion of the effluent SON and soluble phosphorus was found to be bioavailable in the bioassay studies, it could not be concluded that the reminder was inert to biodegradation. It was more likely that some of the substances were more slowly degraded. Having a better understanding of the different biodegradability rates of the effluent nutrients is important for a better understanding of their impacts upon surface waters and how those impacts change with time or distance in receiving streams.

Thus, a WERF Nutrient Challenge Program study by Li et al. (2015) was aimed at evaluating the bioavailability of the effluent SON and soluble phosphorus in longer-term incubation studies and with time to observe changes in conversion rate as the easily available compounds are used

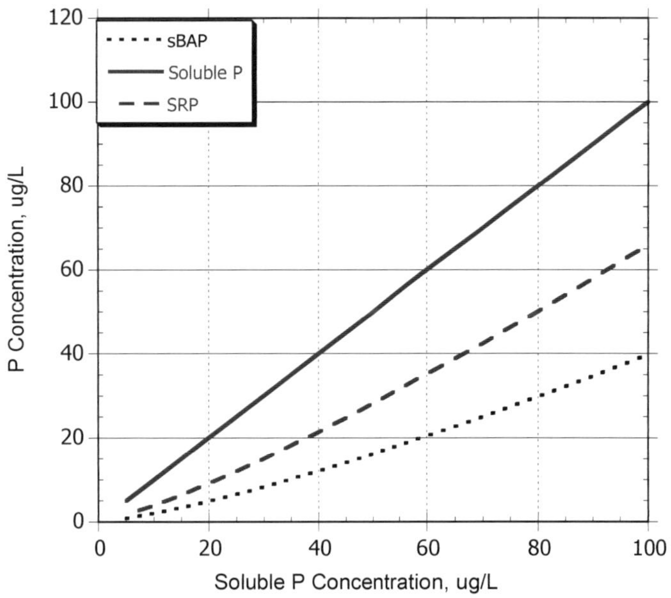

Figure 5. - Soluble Effluent Reactive P and Soluble BAP Concentrations versus Soluble P Concentration Based on Data Correlation for Effluent Samples from 17 Full-Scale WRRFs with Tertiary Chemical Treatment (Li And Brett, 2013).

first. As in the other studies, effluent from a number of full-scale facilities were used for long-term bioassays using the same methods described above. In the case of SON bioavailability, its concentration was measured with incubation time to observe its consumption for algal growth instead of basing it on chlorophyll-a production. For both nitrogen and phosphorus, a high initial algal nutrient consumption rate was observed followed by slower consumption rates with time. Different decay models were evaluated for their compatibility with the characteristic changes in SON and soluble P concentrations with incubation time. The best fit was found when using a Gamma model, which was described by Boudreau and Ruddick (1991) to represent changes in decay kinetics for waters containing some portion of more recalcitrant substances. The Gamma model is defined by the following equations.

$$\frac{C_t}{C_o} = \left[a(a+t)^{-1} \right]^v \qquad (1)$$

$$K = v(a+t)^{-1} \qquad (2)$$

Where,
Co = initial concentration, Ct = concentration at time t, t = time in days, K = first order substances decay rate at time t, and a and v are model coefficients.

An example of the model fit for a WRRF designed with biological nutrient removal and effluent tertiary treatment with chemical addition to achieve low effluent TP concentration is shown in Figure 6, in which the fraction remaining of the initial soluble P concentration is plotted versus incubation time. The data shown represents average values from 5 effluent samples, which had similar soluble phosphorus removal characteristics. Within the first 2 days almost 70% of the soluble phosphorus was removed. By the 14th day 20% of the initial soluble phosphorus remained and the model fit indicates a slower algal consumption rate of the SP over time. Changes in the first-order phosphorus removal rate coefficient (Figure 7) for this model fit show the K values decrease with time as a greater fraction of the soluble phosphorus remains as more recalcitrant.

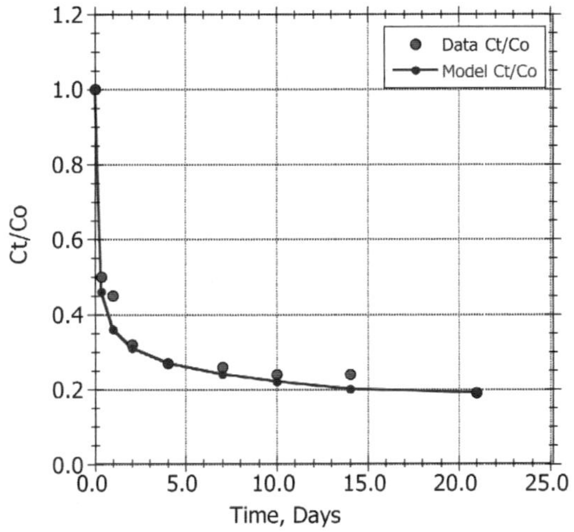

Figure 6. Comparison of Fraction of Initial Soluble P Concentration versus Time to Model Fit for Bioassay Test Data from Tertiary Chemical Treatment Effluent Samples (Initial soluble P concentration equal 45 ug/L).

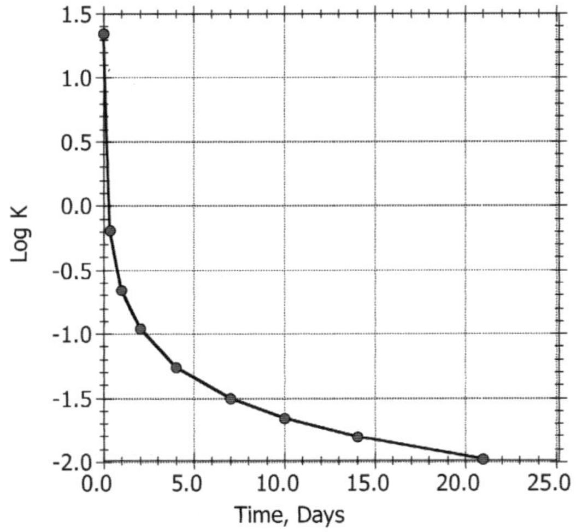

Figure 7. Changes in First Order Coefficient, K, for Phosphorus Uptake Rate Versus Time for Bioassay Test Data from Tertiary Chemical Treatment Effluent Samples.

A similar approach was taken to observe the biodegradability of effluent SON with effluent samples from a number of full-scale ANR facilities. An example of a facility effluent sampling and testing result is shown in Figure 8 in which the average fraction remaining from 5 effluent samples is plotted with time along with the Gamma model fit. Again the SON bioavailability decreases with time as well as its first-order consumption coefficient, K as shown in Figure 9.

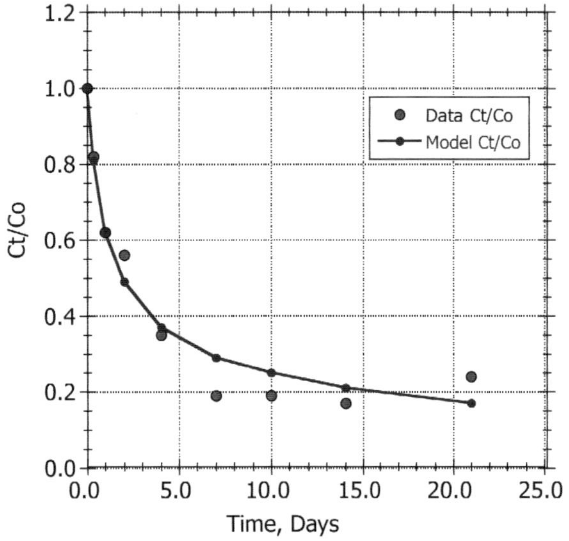

Figure 8. Comparison of Fraction of Initial SON Concentration versus Time to Model Fit for Bioassay Test Data from Bardenpho MBR WRRF Effluent Samples (Initial SON concentration equal 0.80 mg/L).

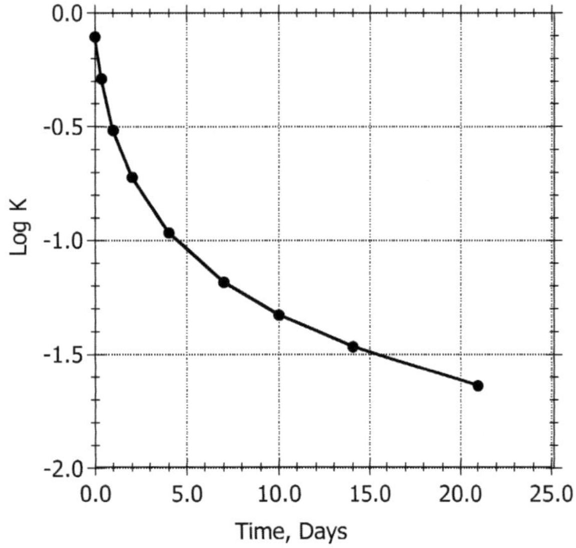

Figure 9. Changes in First Order Coefficient, K, for SON Uptake Rate Versus Time for Bioassay Test Data from Bardenpho MBR WRRF Effluent Samples.

These results show that the bioavailability of effluent SON and soluble P concentration would vary with time in a surface water with a portion that is readily available and a portion for which the biodegradability rate deceases as the more easily degradable components are consumed. It suggests that changes in water quality models used to assess the impact of effluent nutrients should considered the effect of effluent SON and soluble phosphorus in discharges from ANR facilities. These results further suggest that effluent nutrient discharge permits should not be based on only TN and TP concentrations but should consider the effluent N and P species. For effluent TP the algal consumption rates will vary with the proportion of effluent SRP and SNRP in the total SP and time in the receiving surface water. For TN the algal consumption rates will vary with the proportion of inorganic and SON, the portion of hydrophobic and hydrophilic SON, and time in the receiving surface water.

CONCLUSIONS

Studies under the WERF Nutrient Challenge Program have evaluated the treatment performance and effect of nutrient speciation for a number of full-scale WRRFs to better understand factors that impact the WRRF effluent nutrient concentration and the impact of the effluent nutrient concentration on algae growth in surface waters. The following conclusions are drawn from the collective studies on the fate of nitrogen and phosphorus in the ANR WRRFs and in the treated effluent.

1) For nitrogen the effluent TN concentration is mainly soluble and consists of low concentrations of NH_4-N, NOx-N and SON. The SON can account for greater than 50% of the effluent TN. The effluent SON can vary widely for different facilities and can not be manipulated by the process mechanisms in ANR processes.

2) The inorganic nitrogen is readily bioavailable for algae growth and the SON contains readily and slowly bioavailable components.

3) The rate of effluent soluble nitrogen consumption by algae steadily decreased after an initial rapid consumption of readily available compounds and as the soluble nitrogen concentration decreased. A continuous model that accounted for a decreasing SON first-order depletion rates with decreasing nitrogen concentration described the rate of nitrogen consumption accurately.

4) The effluent TP concentration consisted of a SRP component of mainly orthophosphate and SNRP component containing SAHP and SOP. At higher chemical doses used to minimize the effluent TP concentration the fraction of SNRP increased.

5) The rate of effluent soluble phosphorus consumption by algae steadily decreased after an initial rapid consumption and as the concentration decreased. A continuous model that accounted for a decreasing SP first-order depletion rate with decreasing SP concentration described the rate of SP consumption accurately.

6. The ability of a bioavailable model to describe the consumption rate of effluent soluble N and P by algae with time was demonstrated.

REFERENCES

Boudreau, B.P. and B.R. Ruddick (1991) "On a reactive continuum representation of organic matter diagenesis," *American Journal of Science* **291**, 5, 507-538.

Czerwionka, K., J. Makinia, K. Pagilla, and H.D. Stensel (2012) "Characteristics and fate of organic nitrogen in municipal biological nutrient removal wastewater treatment plants" *Water Research, 46 (7), 2057-2066.*

Gu, A.Z., Lei Liu, Annalisa Onnis-Hayden, Scott Smith, Holly Gray, Dwight Houweling, Imre Takács, (2012) "Phosphorus Fractionation And Removal In Wastewater Treatment-Implications For Minimizing Effluent Phosphorus" WERF Nutrient Removal Challenge report NUTR1R06l.

Li, B. and Brett, M.T. (2011) "Spokane Regional Wastewater Phosphorus Bio-Availability Study" WERF Nutrient Removal Challenge project NUTR4C09.

Li, B. and Brett, M.T. (2015) "The Bioavailable Phosphorus (BAP) Fraction in BNR System Effluents" WERF Nutrient Removal Challenge project NUTR1R06m.

Li, B., Fan, L., and Brett, M.T. (2015) "Mineralization Kinetics Of Soluble Phosphorus And Soluble Organic Nitrogen In Advanced Nutrient Removal Effluents" WERF Nutrient Removal Challenge project NUTR1R06p

Pagilla, K. (2007) *Presentation at Water Environment Research Foundation and Chesapeake Bay Science and Technology Advisory Committee Workshop*, Baltimore, MD, September 27/28, 2007

Pehlivanoglu, E. and D. L. Sedlak (2004) "Bioavailability of Wastewater-Derived Organic Nitrogen to the Algal Selenastrum Capricornutum," *Water Research*, **38**, 14-15, 3189-3196.

Sattayatewa, C., M. Arnaldos, and K Pagilla (2008) "Measurement of Organic Nitrogen and Phosphorus Fractions at Very Low Concentrations in Wastewater Effluents," *Water Environment Research*, **83**, 8, 675-683

Sedlak, D.L., Jeong, J., and Liu, H. (2013) "Uptake of Dissolved Organic Nitrogen From BNR Treatment Plant Effluents by Algae" WERF Nutrient Removal Challenge project NUTR1R06e.

Stensel, H.D., J.B. Neethling, D. Clark, R. Tsuchihashi, J. Sandino, A. Pramanik (2013) "Achieving Low Effluent Total Nitrogen Concentration" *87th Annual Water Environment Federation Conference Proceedings,* Chicago, Ill. October 2013.

Stensel, H.D. (2008) "Dissolved Organic Nitrogen (SON) in Biological Nutrient Removal Wastewater Treatment Processes" Compendium from WERF Nutrient Removal Challenge project NUTR1R06 (2008)

Tchobanoglous, G.T, H.D. Stensel, F.L. Burton, and R. Tsuchihashi (2014) *Wastewater Engineering: Treatment and Resource Recovery, 5th Edition,* McGraw Hill, N.Y.

Developing Attainable and Protective Permits for Nutrients

D. Clark[1*], J.B. Neethling[1], H.D. Stensel[2], J. Sandino[3], R. Tsuchihashi[4], A. Pramanik[5]
[1] HDR Inc., [2] Univ. Washington, [3] CH2M Hill, [4] AECOM, [5] WERF
*Email: dclark@hdrinc.com

ABSTRACT

Utilities work with regulators to treat wastewater to levels that protect human health and receiving water quality. Water quality criteria and permits are based on scientifically defensible and shared understanding of sources of pollutants in a watershed, as well as treatment capabilities and costs to control these in the aquatic environment. The national discussion of nutrient impacts on water quality continues to evolve – issues in high visibility water bodies such as the Chesapeake Bay, Long Island Sound, Gulf of Mexico, San Francisco Bay, and Puget Sound highlight this. The Environmental Protection Agency's (EPA) efforts to promulgate numeric nutrient standards in all states raise questions about how these standards apply to wastewater dischargers, whether they are effective, and how they affect others in the water quality arena. A Water Environment Research Foundation (WERF) report, as part of the WERF Nutrient Removal Challenge, *Nutrient Management: Regulatory Approaches to Protect Water Quality, Volume 1 Review of Existing Practices* (NUTR1R06i) provides a state-of-the art discussion of key nutrient management issues that confront point source wastewater dischargers nationwide. A second WERF report as part of the WERF Nutrient Removal Challenge, *Nutrient Management Volume II: Removal Technology Performance & Reliability* (NUTR1R06k) presents a comprehensive study of nutrient removal plants designed and operated to meet very low effluent nitrogen and phosphorus concentrations. A third report titled *"Nutrient Management Volume III: Development of Appropriate Permitting Frameworks"* focuses on nutrient discharge permitting.

KEYWORDS: WERF Nutrient Removal Challenge, bioavailability, effluent limits, limit of technology nitrogen, nutrient removal, nutrient criteria, numeric nutrient standards, NPDES permit, phosphorus, total maximum daily load, TMDL, trading, variance, water quality.

INTRODUCTION

Excessive nitrogen and phosphorus loadings to watersheds impact water quality by stimulating the growth of algae which may result in depletion of dissolved oxygen, shifts in pH, degradation of habitat, impairment of drinking water sources, and in some cases harmful algal blooms. According to the EPA, nearly every State has nutrient related pollution with impacts in over 80 estuaries/bays, and thousands of rivers, streams, and lakes. In particular, EPA cites the Gulf of Mexico and the Chesapeake Bay as examples of significant water quality impacts from 35 States that contribute to nutrient loadings.

Nutrient loadings from both point and nonpoint sources contribute to water quality impairments in the nation's waterways. Point source discharges from wastewater treatment plants can be a significant source of nitrogen and phosphorus in watersheds. Nonpoint sources contribute substantial amounts of nutrients from land use activities such as agriculture, forestry, and urban/suburban development.

Targeted nutrient levels in lakes, streams, and estuaries can be very low concentrations that are challenging to meet with treatment of point sources and application of best management practices (BMPs) to nonpoint sources. Nutrient removal treatment can substantially reduce point source discharges of nitrogen and phosphorus, however substantial investments are required to build and operate advanced wastewater treatment facilities. In some watersheds, nonpoint source nutrient loadings outweigh point sources to a degree that advanced treatment for nutrient removal, and even complete elimination of point sources, would have limited benefit to water quality. Nevertheless, point source NPDES permitted dischargers are the most directly regulated sources subject to nutrient control requirements resulting from numeric nutrient standards, total maximum daily loads (TMDLs), and water quality based permit limits.

A variety of potential approaches to establishing effluent limits for nitrogen and phosphorus are available to permit writers. The traditional permit writers' deterministic approach may result in overly restrictive effluent limits for nitrogen and phosphorus. More appropriate approaches to nutrient discharge permitting may provide greater flexibility, while at the same time arriving at limits that are protective of water quality. Depending upon site specific circumstances, combinations of these approaches to nutrient discharge permitting may provide the best overall way to meet water quality objectives in a cost effective manner.

Current Permitting Practices

Development of effluent limits typically consists of three initial technical steps: (1) development of technology-based effluent limitations (TBELs); (2) development of water quality-based effluent limitations (WQBELs); and (3) determination of final effluent limitations and conduct of an anti-backsliding analysis. Since national TBELs do not exist for nutrients, permit writers' derive effluent limitations that are protective of state water quality standards (i.e. water quality-based effluent limitations or WQBELs) as needed. The process of translating water quality standards to water quality based effluent limits consists of determining the applicable standards, characterizing the effluent and receiving waters, determining the need for limits, and finally calculating the limits.

Traditional Deterministic Permitting

The traditional deterministic approach to developing effluent limitations uses specific effluent conditions (flow and concentration) in combination with specific upstream receiving water conditions (flow and concentration) to calculate the predicted downstream concentration. The traditional deterministic approach to discharge permitting is familiar to permit writers and commonly used for water quality based effluent limits (WQBELs) for many parameters, including nutrients. Much of the guidance material used by permit writers for reference is focused on the deterministic approach. Since key reference materials emphasize protection of receiving water quality from toxics, the deterministic approach is very conservative, in that it presumes that critical conditions may never be exceeded within and at the edge of a near field effluent mixing zone in order to protect aquatic life from acute and chronic toxicity. Less reference material is available to permit writers as guidance for the development of effluent nutrient limits where broader watershed impacts, such as eutrophication, are the predominant objective for controlling nutrient discharges. Deterministic permitting may be unnecessarily restrictive because it is based upon a combination of critical conditions that are unlikely to coincide. While protection of aquatic life from potential toxicity may warrant this level of conservatism in the development of effluent limits for toxics, it is unnecessary for the control of nutrient discharges associated with enrichment and eutrophication.

Technology Based Effluent Limits

Technology-based effluent limitations (TBELs) are intended to prevent pollution by requiring a minimum level of effluent quality that is achieved using treatment technologies for reducing discharges of pollutants to surface waters. The EPA developed and promulgated "secondary treatment" regulations that are found in 40 CFR 133.102. These technology-based limits identify the minimum level of effluent quality attainable by secondary treatment in terms of five-day biochemical oxygen demand (BOD5) or five-day carbonaceous biochemical oxygen demand (CBOD5), total suspended solids (TSS), and pH. No comparable federal requirements exist for nitrogen and phosphorus, nevertheless technology based effluent limits are applied for nutrients in many situations. In the absence of national standards, technology based effluent limits are developed on a case-by-case basis.

Use of Technology Based Effluent Limits for Nutrients

Technology based effluent limits identify the performance of a wastewater treatment process by directly defining effluent phosphorus and nitrogen concentrations. In this approach, a treatment plant is required to achieve effluent quality that may be established by a variety of methods, such as selection of general treatment levels associated with a degree of technology. Examples include biological nutrient removal (BNR), or enhanced nutrient removal (ENR), or limit of technology treatment (LOT). Conventional municipal biological nutrient removal typically produces effluent total nitrogen of about 8 to 10 mg/L and total phosphorus of about 0.5 to 1 mg/L. Enhanced nutrient removal is an upgrade of the conventional nutrient removal technology to include additional reliability and performance enhancements, larger biological reactors, supplemental chemical addition, effluent filtration, etc. These processes typically produce effluent total nitrogen of about 3 to 5 mg/L and total phosphorus of about 0.1 to 0.5 mg/L. At the limits of treatment technology with the largest reactors, state-of-the-art processes, supplemental chemical addition, sidestream controls, enhanced/optimized operations, continuous monitoring, etc. to achieve the lowest effluent concentrations, effluent total nitrogen of 3 mg/L or less, and total phosphorus of 0.1 mg/L or less, may be achieved.

These generalized characterizations of levels of treatment do not address how long a stated performance can be sustained, nor the reliability and resiliency of such performance. Altered effluent nutrient speciation and bioavailability are also not addressed by these generalized characterizations. More specific and detailed information is needed, such as technology based effluent limits.

The most appropriate nutrient discharge permits will be prepared based on an understanding of both receiving water quality requirements and the capabilities of advanced nutrient removal treatment. Where either is lacking, an investment may be necessary to determine the level of nutrient management required to meet water quality objectives and link that analysis with specific objectives for effluent quality. When the relationship between nutrient loadings and water quality responses is not well defined, it is advisable to avoid overly restrictive effluent limits at the outset, since they may later prove unnecessary to meeting actual receiving water needs when they eventually become better understood. Preserving an opportunity for adaptive management approaches to guide the process of nutrient management over time may improve water quality incrementally, without overly restrictive discharge permits that result in over investment in advanced treatment. Permits structured around no net increase in existing loadings, or simple seasonal or annual loading reductions, may provide a foundation for adaptive management. In some cases, states in the process of adopting numeric nutrient criteria have used technology based effluent limits to achieve some level of point source nutrient reduction while preserving the opportunity for adaptive management approaches. This has been found to be especially important where numeric nutrient criteria are very low concentrations of nitrogen and phosphorus that have the potential to result in water quality effluent limits beyond the capabilities of advanced nutrient removal treatment.

Benefits and Limitations of Technology Based Effluent Limits

A key benefit of the application of technology based effluent limits for nutrients is the simplicity of the approach. Effluent phosphorus and nitrogen limits are specified at levels that can be inserted directly into discharge permits. The basis for the selection of technology based effluent limits for nutrients may vary and could be nominal values associated with a given level of advanced nutrient removal treatment. Alternately, effluent limits have been based on interim levels of treatment with the expectation in the long term that water quality based effluent limits will be derived, or result from a wasteload allocation in a TMDL, that eventually replaces the initial technology based effluent limits. In other circumstances, a statistical analysis of past effluent performance may be used to establish technology based effluent limits. Technology based effluent limits may be useful in achieving a degree of point source nutrient reduction while progress is made in developing all of the information necessary for the preparation of appropriate water quality based limits. In this way, technology based effluent limits may be a placeholder used with the expectation that they will later be replaced.

A potential criticism of technology based effluent limits for nutrients is that they may be selected at levels which are perceived as being too lenient to actually be protective of receiving water quality. The primary disadvantage of technology based effluent limits for nutrients is the lack of a relationship to the receiving water quality objectives. Consequently, it may not be clear whether technology based effluent limits are over-protective, or under-protective, of receiving water quality. Receiving water quality is controlled by a multiplicity of factors with complex interrelationships in the aquatic environment. Point source nutrient load reductions from wastewater treatment plants may, or may not, contribute to water quality improvements depending upon many factors. These factors include the magnitude of point sources compared to other loadings, the limiting nutrient controlling aquatic growth in receiving waters, decomposition of aquatic growth, and many receiving water characteristics related to the processing of nutrients (light penetration, scour, substrate stability, etc.). In some watersheds, nonpoint source nutrient loadings outweigh point sources to a degree that advanced treatment for nutrient removal and even complete elimination of point sources by zero discharge would have limited effect on water quality.

Technology Performance Statistics and Permitting

The understanding of advanced wastewater treatment for nutrient removal has improved substantially in recent years through operational experience, technology development and research studies, and new efforts to meet more challenging discharge permits with increasingly challenging effluent nitrogen and phosphorus limits at lower levels. Application of the additional unit processes and biological treatment modifications required to accomplish the required levels of nutrient removal has resulted in more effluent performance data from full-scale facilities, which has been studied in detail. This has provided an opportunity to define treatment performance in statistical terms based on the best designed and operated nutrient removal treatment facilities. This presents an opportunity to incorporate an improved understanding of advanced nutrient removal treatment performance in discharge permitting statistically. Technology performance statistics (TPS) describe the requirements for treatment facilities in more specific terms than common adjectives such as Advanced Treatment, Enhanced Biological Nutrient Removal, and Tertiary Treatment. Technology performance statistics can define effluent performance on average, as well as characterize the reliability expectations for the process, and the best possible effluent performance. This statistical information can be used to compare with the conditions necessary to satisfy receiving water requirements and to define the requirements of the treatment process for facility designers.

Use of Technology Performance Statistics in Permitting

Technology Performance Statistics describe the performance of a technology or process or facility under specific conditions. In this approach, the treatment plant or technology performance is tied to the statistical rank to express the probability of achieving a certain performance. Building on this statistical approach, the term Technology Performance Statistic (TPS) was used at a Water Environment Federation Technical Exhibition and Conference (WEFTEC) workshop (WEF/WERF, 2009) to assess the performance of full scale treatment plants. The TPS is determined from performance data and is linked to the operational conditions during which the data were collected (pilot, full scale, summer, winter, excess capacity available, SRT, etc). The conditions must also include external factors that impact the technology, industrial loadings, seasonality, absence of recycle streams, etc.

Benefits and Limitations of Technology Performance Statistics

A key benefit of the application of technology performance statistics to effluent discharge permitting is to define limits in terms that account directly for variability in effluent performance when pursuing lower nitrogen and phosphorus levels. This presents the opportunity to take advantage of all of the recently available information about the capabilities of nutrient removal in very specific numerical terms. Effluent limits may be based directly on technology performance statistics, such as the average or median performance for a treatment process. Probability statistics for effluent performance, such as the 95th percentile, can be linked to commonly used discharge permit structures, such as monthly limits. The advantage of using performance statistics is to define effluent limits in terms that account for realistic expectations for variability in effluent performance, as opposed to potentially overly restrictive absolute terms that presume that effluent limits will never be exceeded. In this way, discharge permit compliance issues can be avoided and unrealistic or technically infeasible permit requirements avoided.

The primary limitation of technology performance statistics is the focus on the capabilities of the nutrient removal treatment process itself, and not the linkage with receiving water quality requirements. While technology performance statistics provide an enhanced numerical description of treatment process performance, they do not define the frequency or duration of receiving water quality requirements. To do that requires receiving water quality monitoring and modeling to provide the information necessary to assess the allowable level of effluent variability.

Predictive Water Quality Models and Permitting

Predictive water quality models are tools used to estimate future receiving water conditions based on historical information and scientific relationships. A number of water quality models are available of varying complexity and capabilities for the simulation of water quality. Many of these models are based on quantitative relationships between nutrients, site-specific water quality, and ecological response indicators (dissolved oxygen, pH, algae). This information can be valuable in understanding the dynamics of a receiving water body.

Application of Water Quality Models

Water quality models are powerful tools that can provide significant insights into receiving water conditions and the impacts of wastewater discharges and other nutrient loading sources on water quality. Process-based load-response models use mathematical representations that link nutrient loads to in situ water quality and/or ecological responses. Examples are the relationships between nutrients, light, and water temperature to the growth rate of algae. Models with these capabilities are well known and include AQUATOX, CE-QUAL-W2, QUAL2K, and WASP (Bierman et.al. 2013). These models are capable of generating a significant amount of output (data) from a simulation. Since the purpose of discharge permitting is to limit the pollutants discharged to the receiving water to protect beneficial uses, the information from predictive models can be useful in developing discharge permits. Further, water quality models are tools that can be used to investigate a variety of potentially acceptable discharge permit conditions to find the most technically feasible, economical, and sustainable means of achieving compliance.

Benefits and Limitations of Water Quality Models

In some cases, the application of water quality models to inform discharge permitting is limited because modeling requires more resources (data, time, funding, expertise, etc.) than simpler permitting methods. In other circumstances, water quality modeling may be used in watershed analyses and total maximum daily loads (TMDLs) that are prepared prior to revisions in discharge permits. If permitting scenarios are not explored during a watershed analysis or TMDL, the water quality model may not be used by the permitting entity for any number of reasons including lack of budget resources, lack of modeling skills, lack of sufficient time to meet permit renewal deadlines, etc.

Applying water quality models to receiving waters and using models to inform permitting may be complex and requires adequate budget and schedule resources. Model selection, set-up, calibration, and interpretation of modeling results are potentially complex and time consuming. Each step in the predictive modeling process has hurdles to overcome, as well as the need to reach consensus among stakeholders regarding the model and interpretation of its results.

Nevertheless, these challenges may be relatively small in comparison to the implications of the capital and operating investments required of wastewater treatment facilities which are subject to the compliance requirements of effluent discharge permits. Therefore, the investment of time and resources necessary to utilize water quality models as predictive tools to inform discharge permitting may be well justified.

Probabilistic Permitting

A probabilistic approach to nutrient discharge permitting allows the variability in flows and concentrations to be recognized and the most extreme flows and concentrations placed in proper perspective with more typical conditions. A probabilistic approach to nutrient discharge permitting is advantageous because it can utilize a distribution of values for key parameters in the development of effluent limits to portray the variability that exists in effluent and receiving water flows and constituent concentrations. Deterministic permitting approaches may be overly restrictive in limiting nutrient discharges because they combine conservative assumptions in each aspect of the development of effluent limits: wastewater flow, effluent nutrient concentration, receiving water flow, and ambient water nutrient concentration. The probabilistic methodology allows for the variability of effluent and receiving water flows and nutrient concentrations to be represented by probability distributions. The potential for those variable data to coincide can then be quantified statistically and used to inform the development of discharge permits.

Probabilistic and predictive approaches may be used together to develop these distributions of potential outcomes in terms of mixed downstream conditions and can predict the frequency of their occurrence. The result is a broader perspective of whether there is reasonable potential to impact water quality, along with more information about the potential magnitude, frequency, and duration of downstream water quality conditions. This information can be compared with the receiving water requirements that have been established by numeric nutrient criteria, water quality modeling, TMDL wasteload allocation, etc. as necessary to protect receiving water quality.

Probabilistic Development of Effluent Limits

Monte Carlo analysis is a method for using the full probability distributions for each of the parameters in the mass balance approach to develop effluent limits. A Monte Carlo simulation may be used to combine the effluent and receiving water flow and concentration data and calculate the probability distribution for the downstream mixed conditions. The Monte Carlo analysis results in the probability distribution of calculated in-stream concentrations, which can then be evaluated in comparison to the in-stream target concentration. Development of effluent limits using a probabilistic approach requires calculation of the downstream conditions, followed by a comparison with the allowable frequency of exceedance. This may be followed by successive iterations with refined effluent flow and nutrient concentration values to converge on the effluent limits necessary to satisfy the downstream conditions.

The Monte Carlo analysis can test multiple combinations of parameter values based on statistical distributions. The statistical distributions for each of the four parameters (receiving water flow and concentration, effluent flow and concentration) are defined by the mean and standard deviation in log normal distributions. Minimum and maximum values are used to constrain the log normal distributions so that the tails are finite. Providing that an appropriate sample size of data are available, the mean and standard deviation for the parameters may be computed from the data time series and used in the Monte Carlo analysis to develop effluent limits. The data requirements for this approach to permitting are minimal in comparison to other modeling techniques, such as dynamic fate and transport water quality modeling. Permit writers usually have site specific receiving water flow and ambient concentration data sets available to analyze for use in traditional deterministic permit calculations.

Monte Carlo analysis has been identified as an appropriate approach to the development of effluent limits to address variability, as described in EPA's Technical Support Document for

Water Quality-based Toxics Control (TSD) (EPA 1991). In the past, the lack of readily available computing power may have limited the application of Monte Carlo analysis in discharge permitting. However, easily accessible software packages, such as the @Risk extension for spreadsheet calculators, can quickly conduct Monte Carlo simulations in minutes or seconds with multiple iterations. The software is user friendly and does not require an extensive background in statistical methods. The probabilistic approach requires little additional effort to develop the entire range of outcomes for the resulting receiving water concentration conditions. The variability in environmental factors, river flows and ambient concentrations, along with wastewater treatment plant performance, effluent flows and concentrations, are included in the Monte Carlo simulation.

Benefits and Limitations of Probabilistic Approaches

A key benefit of a probabilistic approach to nutrient discharge permitting is that it provides the ability to consider the variability in effluent and receiving water flows and concentrations. This is advantageous because the values for key parameters can be fully characterized in the development of effluent limits, rather than represented as single, extreme values. Permit development considerations are not confined to selection of the most extreme flow and concentration values commonly used in the traditional deterministic permitting process.

Probabilistic permit limits can be directly linked to satisfying receiving water quality requirements. Receiving water requirements can be expressed as either an allowable frequency of exceedance of the applicable standard, or a probabilistic representation of an acceptable downstream condition as a probability distribution rather than a single value.

Deterministic permitting approaches allow direct calculation of effluent limits. Probabilistic development of permit limits will require that permit writers use statistical characterizations of flows and concentrations and a Monte Carlo simulation to combine multiple probability distributions to calculate downstream conditions. Successive iterations of the downstream concentration calculations may be necessary to converge on an acceptable combination of effluent flows and concentrations to satisfy receiving water objectives. The data necessary for conduct of such an analysis is commonly used in deterministic permitting and includes receiving water flow data and ambient nutrient concentration data. Receiving water flow data sets are generally available covering long periods of time for most river and stream systems from sources such as USGS flow monitoring stations. Stream flow data are likely to have the greatest degree of variability and the most extreme range of minimum to maximum values of any of the data sets used in permit limit calculations. Ambient water quality data for nutrients is often available from state agency monitoring programs, watershed plans and TMDLs, and other local monitoring efforts including those conducted by wastewater utilities for receiving waters as discharge permit monitoring requirements. Computer software packages are readily available as add-ons for spreadsheet calculators to conduct the Monte Carlo simulations and can be used to conduct probabilistic permit limit calculations.

Watershed Nutrient Permitting

Watershed-based NPDES permitting is a process that emphasizes addressing all stressors within a hydrologically-defined drainage basin, rather than addressing individual pollutant sources on a discharge-by-discharge basis. Watershed-based permitting can encompass a variety of activities, ranging from synchronizing permits within a basin to developing water quality-based effluent limits using a multiple discharger modeling analysis. The type of permitting activity will vary depending on the unique characteristics of the watershed and the sources of pollution impacting it. The ultimate goal of this effort is to develop and issue NPDES permits that better protect entire watersheds (EPA, 2014).

Since nutrients are often a broad watershed scale issue in terms of water quality, consideration should be given to watershed permitting. Watershed permitting provides a structure that allows for collaboration among point source dischargers, nonpoint sources, and other stakeholders to achieve watershed nutrient management objectives. Individual discharge permit renewal schedules and other administrative factors may artificially constrain the opportunity to develop and implement watershed scale permits. Approaching nutrient management considerations from the watershed scale, as opposed to individual permits, may reveal the opportunity for watershed permits to result in effective collaborations.

Water Quality Trading

A potentially attractive tool in developing effective watershed scale nutrient management plans is nutrient trading. It is important to structure discharge permit in a manner that avoids inadvertent disincentives to nutrient trading. Combinations of both effluent concentration and mass effluent limits for nutrients may constrain the development of trades, or increase the complexity in accounting for trades.

Watershed permits formulated with trading in mind may facilitate the implementation of water quality trading. NPDES discharge permits structured with long averaging periods linked to the specific waterbody response to nutrient enrichment, such as seasonal or annual limits based on long-term average values, may present the most opportunities for trading. Effluent limits based on the total loading for the compliance period (e.g., total pounds discharged on an annual or seasonal basis) facilitate compliance and provide an opportunity for water quality offsets and trading.

CONCLUSIONS AND RECOMMENDATIONS

A great variety of approaches to establishing effluent limits for nitrogen and phosphorus have been used across the country and some have resulted in very restrictive conditions that may exceed the capabilities of advanced nutrient removal treatment. Overly restrictive effluent discharge permits generally result from the application of permitting approaches designed to protect receiving waters from short term effects in the near field effluent mixing zone, such as the approach taken to control toxics.

In other cases, effluent nutrient limits have been tailored to the site specific circumstances necessary to achieve receiving water quality objectives. This often results in more appropriate permit structures that reflect an understanding of both the impact of nutrient discharges on receiving waters and the capabilities of nutrient removal treatment. Tailoring nutrient permits to site specific circumstances often results in the need to deviate from conventional effluent limit structures applied to other parameters and combine receiving water objectives with technically feasible treatment limits.

Emphasis in nutrient discharge permitting should focus on providing the greatest amount flexibility possible in the structure of nutrient limits in order to preserve the opportunity for the most creative and economical approaches to managing nutrients. Traditional permit structures for publically owned treatment works generally include both monthly and weekly limits on both a concentration and mass basis. This may inadvertently eliminate the most effective watershed solutions to nutrient management by creating disincentives to wastewater dischargers to explore combinations of advanced wastewater treatment and other watershed management practices. It is preferable to structure discharge permits in such a way that receiving water quality objectives are met with the greatest flexibility that can be provided to the treatment processes. This is important in order to avoid unnecessarily restrictive effluent discharge conditions that result in little additional water quality protection but consume inordinate amounts of energy and chemicals that result in other deleterious environmental impacts.

Conclusions

As nutrient discharge permits become stricter (closer to zero with no margin for variability) the more challenging, expensive, and greater the environmental trade-offs become in pursuit of compliance. For these reasons and others, the traditional deterministic approach to effluent discharge permitting may result in overly restrictive limitations that accomplish little in terms of water quality improvement. More appropriate approaches for nutrients incorporate treatment technology performance statistics, predictive water quality models, and probabilistic approaches that combine technically feasible treatment limits with achievement of water quality objectives. More appropriate nutrients discharge permits may be developed when conditions include the following:

- Collaboration between permit writers and permittees to craft a flexible nutrient permits.
- Shared understanding of the frequency and duration associated with watershed nutrient management objectives.
- Shared understanding of the capabilities of advanced nutrient removal treatment.
- Recognition of the environmental trade-offs associated with nutrient removal treatment and discharge permit structures.
- Recognition of the variability in effluent characteristics and the natural environment
- Application of more sophisticated methods, water quality models, and statistical tools to arrive at permit structures that better match actual receiving water requirements.

Recommendations

Emphasis in nutrient discharge permitting should focus on providing the greatest amount flexibility possible in the structure of nutrient limits in order to preserve the opportunity for the most creative and economical approaches to managing nutrients. Traditional permit structures for publically owned treatment works generally include both monthly and weekly limits on both a concentration and mass basis. This may inadvertently eliminate the most effective watershed solutions to nutrient management by creating disincentives to wastewater dischargers to explore combinations of advanced wastewater treatment and other watershed management practices.

Water Quality Linkages

The most appropriate nutrient discharge permits will be prepared based on an understanding of both receiving water quality requirements and the capabilities of advanced nutrient removal treatment. Where either is lacking, an investment may be necessary to determine the level of nutrient management required to meet water quality objectives and link that analysis with specific objectives for effluent quality. When the relationship between nutrient loadings and water quality responses is not well defined, it is advisable to avoid overly restrictive effluent limits at the outset, since they may later prove unnecessary to meeting actual receiving water needs when they eventually become better understood. Preserving an opportunity for adaptive management approaches to guide the process of nutrient management over time may improve water quality incrementally, without overly restrictive discharge permits that result in over investment in advanced treatment. Permits structured around no net increase in existing loadings, or simple seasonal or annual loading reductions, may provide a foundation for adaptive management. In some cases, states in the process of adopting numeric nutrient criteria have used technology based effluent limits to achieve some level of point source nutrient reduction while preserving the opportunity for adaptive management approaches. This has been found to be especially important where numeric nutrient criteria are very low concentrations of nitrogen and phosphorus that have the potential to result in water quality effluent limits beyond the capabilities of advanced nutrient removal treatment.

Where the linkages with water quality requirements are less well defined, the following approaches are recommended:

- Establish a foundation for adaptive management whereby the impact of nutrient loadings on receiving water quality can be better understood over time.
- In cases where nutrient limitations are warranted, develop nutrient discharge permit limits based on no net increase in existing loadings.
  - If necessary, utilize technology based effluent limits at the basic biological nutrient removal level.
- Utilize compliance schedules in discharge permitting to provide the time necessary to develop a water quality based set of requirements for effluent limits linked with water quality response variables.

Where the linkages with water quality requirements are defined but overall watershed nutrient management and nonpoint source controls are uncertain, the following additional approaches are recommended:

- Incorporate the most basic level of nutrient limits possible in discharge permits to preserve the ability to optimize the combination of point and nonpoint source nutrient controls through adaptive management.
  - Utilize mass loading limits or technology based effluent limits at the basic biological nutrient removal level.

Technology Performance Statistics

When the linkage between water quality requirements and nutrient loadings result in the need for advanced levels of nutrient removal treatment, technology performance statistics provide a basis to define effluent performance and reliability.

Where the linkages with water quality requirements are well defined, the following approaches are recommended:

- Utilize technology performance statistics to define effluent limits based on receiving water quality requirements in terms of effluent quality and reliability.
    - Where appropriate, utilize median statistics (50th percentile) to define effluent quality such that inherent variability in treatment performance with advanced nutrient removal can be allowed.
    - Specify effluent limits in terms of average (50th), 90th, 95th, or 99th percentile statistics depending upon the reliability of treatment required for receiving water conditions.
- Establish a foundation for adaptive management whereby the impact of nutrient loadings on receiving water quality can be better understood over time.

Where the linkages with water quality requirements are well defined but water quality based effluent limits result in technically infeasible nutrient limits, the following approaches are recommended:

- Utilize the following regulatory implementation tools and define a level of feasible effluent performance for interim operation:
    - Site specific nutrient criteria.
    - Compliance schedules.
    - Variances.
    - Use attainability analysis.

Predictive Water Quality Models

When water quality models are available to simulate the water quality response to nutrient loadings, discharge permit scenarios can be simulated to develop the basis for the most flexible and sustainable permit structure possible.

Where water quality models are available to define the impact on receiving water beneficial uses in terms water quality response variables (pH, DO, algae, etc), the following approaches are recommended:

- Utilize water quality models to simulate receiving water quality responses to define effluent limits in terms of effluent quality and reliability.
- Utilized water quality models to simulate effluent discharges in alternative ways such that the critical factors affecting the response variables can be better understood, such as extended period simulations.
- Combine water quality modeling and monitoring in adaptive management approaches whereby the impact of nutrient loadings on receiving water quality can be better understood over time in pursuit of optimal watershed nutrient management.
    - Consider the changes in receiving water quality that occur following the initial reduction of point source nutrient loadings, along with each successive reduction in both point and nonpoint source loadings.
    - Select effluent nutrient limits that provide proportionate improvements in receiving water quality.

- Pursue sustainable combinations of point source nutrient removal and nonpoint source watershed nutrient management.
    - Avoid overly restrictive effluent limits that do not provide a commensurate improvement in receiving water quality, but may result in excessive use of energy and chemicals, and over production of residual biosolids.

Probabilistic Analysis

Where there is recognition that variability exists in receiving water flows and water quality, consider the application of probabilistic approaches to define levels of effluent performance to meet performance objectives and at what frequency. Extremely low receiving water flow conditions are not likely to coincide with maximum effluent discharge conditions. Likewise, aquatic life and recreational beneficial uses would generally not be thought to be impaired if a single rock, pool or riffle, or even a short reach of river had benthic algae higher than a target value. Probabilistic analysis can provide a tool to analyze the frequency at which specific conditions may occur in receiving waters based on variability in both effluent and receiving water.

Probabilistic analysis is recommended in the following circumstances:

- Where there are conditions in which there is a high degree of variability in receiving water and effluent flows and/or concentrations.
- Extremes in receiving water low flow conditions, or high ambient concentrations, are short lived or infrequent.

REFERENCES

Bierman, Victor J., Joseph V. DePinto, David W. Dilks, Penelope E. Moskus, Theodore A.D. Slawecki, Clifton F. Bell, Steven C. Chapra, Kyle F. Flynn 2013. Modeling Guidance for Developing Site-Specific Nutrient Goals. LINK1T11. WERF

Bott, C.B, and Parker, D.S. (2011), "Nutrient Management Volume II: Removal Technology Performance & Reliability" WERF Nutrient Removal Challenge project NUTR1R06k.

Clark, D.L. Hunt, G., Kasch, M.S., Lemonds, P.J., Moen, G.M., Neethling, J.B. (2010) "Nutrient Management Regulatory Approaches To Protect Water Quality - Volume Review Of Existing Practices" WERF Nutrient Removal Challenge project NUTR1R06i.

David L. Clark, D.L., Dupuis, T., Falconer, H., Hatch, L., Kasch, M.S., Lemonds, P.J., and Neethling, J.B. (2015 in press) "Nutrient Management Volume III: Development of Appropriate Permitting Frameworks" WERF Nutrient Removal Challenge Project NUTR1R06z.

EPA. 1991. Technical Support Document for Water Quality Based Toxics Control. EPA 505/2-90-001, March 1991.

EPA. 1996. U.S. EPA NPDES Permit Writers' Manual, EPA-833-B-96-003, December 1996.

EPA. 1996. *NPDES Permit Writer's Manual*, EPA 833-B-96-003.

EPA. 2003. Watershed-Based NPDES Permitting: Rethinking Permitting as Usual. EPA-833-F-03-004. May 2003.

EPA, 2010. NPDES Permit Writers' Manual. U.S. Environmental Protection Agency, Office of Wastewater Management, Water Permits Division, State and Regional Branch, EPA-833-K-10-001.

EPA. 2014. Watershed-Based NPDES Permitting, http://cfpub.epa.gov/npdes/wqbasedpermitting/wspermitting.cfm

Falk, M.W.; Neethling, J.B.; Reardon, D.J. (2011) "Striking the Balance between Nutrient Removal and Sustainability. WERF Research Project under Nutrient Challenge" WERF Nutrient Removal Challenge Project NUTR1RO6n.

Li, B. and Brett, M.T. (2011) "Spokane Regional Wastewater Phosphorus Bio-Availability Study" WERF Nutrient Removal Challenge project NUTR4C09.

Li, B. and Brett, M.T. (2015) "The Bioavailable Phosphorus (BAP) Fraction in BNR System Effluents" WERF Nutrient Removal Challenge project NUTR1R06m.

Li, B., Fan, L., and Brett, M.T. (2015) "Mineralization Kinetics Of Soluble Phosphorus And Soluble Organic Nitrogen In Advanced Nutrient Removal Effluents" WERF Nutrient Removal Challenge project NUTR1R06p.

Neethling, J.B., David Stensel, Denny Parker, Charles Bott, Sudhir Murthy, Amit Pramanik, Dave Clark (2009) "What is the Limit of Technology (LOT)? A Rational and Quantitative Approach" WEFTEC 2009.

Neethling, J.B., and H. David Stensel, 2013. Nutrient Species Implications for Technology Performance and Water Quality Impacts. WERF.

Neethling, J.B. and Stensel, H.D. "Nutrient Species Implications for Technology Performance and Water Quality Impacts," WEF/IWA Nutrient Removal and Recovery Conference, Vancouver, BC, July 28-31, 2013.

Sedlak, D.L., Jeong, J., and Liu, H. (2013) "Uptake Of Dissolved Organic Nitrogen From BNR Treatment Plant Effluents By Algae" WERF Nutrient Removal Challenge project NUTR1R06e.

Stensel, H.D. (2008). Dissolved Organic Nitrogen (DON) in Biological Nutrient Removal Wastewater Treatment Processes. Compendium from WERF Nutrient Removal Challenge project NUTR1R06 (2008)

WERF Advances the State of the Practice in the Design of Nutrient Removal Facilities

J. Sandino[1*], R. Tsuchihashi[2,] J.B. Neethling[3], Dave Stensel[4], Dave Clark[3], A. Pramanik[5]

[1*] CH2M, jsandino@ch2m.com, [2] AECOM, [3] HDR Engineering, Inc., [4] University of Washington
[5] Water Environment Research Foundation

ABSTRACT

This paper summarizes the results to date of four projects already completed under the 5-year Water Environment Research Foundation (WERF) Nutrient Removal Challenge that are deemed of particular relevance to designers of water resource recovery facilities that incorporate nutrient removal and/or recovery. The project topics include establishment of facility performance and reliability criteria, incorporation of uncertainty evaluations, membrane bioreactors design guidelines for low nutrient effluent applications, and effectively controlling simultaneous nitrification and denitrification. The emphasis of this paper is not only to present for each of the highlighted projects the knowledge gap which it is intended to breach, the corresponding study methodologies employed and the results obtained, but perhaps more importantly, the "So What" in terms of the relevance these research findings will have for the nutrient removal and recovery facility design practitioner.

KEYWORDS. Cold weather operation, External carbon, Fermenter, Membrane bioreactor, Online analyzer, Performance and reliability assessments, Simultaneous nitrification and denitrification, Uncertainty evaluation, WERF Nutrient Removal Challenge

INTRODUCTION

The Water Environment Research Foundation (WERF) Nutrient Removal Challenge is a multi-year collaborative research initiative established in 2007 to develop and provide current information about nutrient removal and recovery, their characteristics, and bioavailability in aquatic environments to help designers, operators and regulators make informed decisions. Findings from the studies focusing on nitrogen and phosphorus are reported in the papers by Stensel et al. (2013) and Neethling et al. (2013), and relevant regulatory issues on nitrogen and phosphorus removal are reported by Clark et al. (2013). The focus of this paper is on design (and to some extent operational) issues related to removal of nutrients from wastewater.

Regarding design aspects of the nutrient removal, process reliability and uncertainties associated with the process modeling have become an imminent challenge as stringent discharge limits are being imposed. In terms of treatment technologies, membrane bioreactors (MBRs) have been proven to be promising alternatives to achieve reliable nutrient removal, even though the technology is still relatively new and dissemination of the accumulated knowledge is not as widespread as conventional technologies. Another design-focused area of study that has made significant progress in recent years aims at developing a better understanding of simultaneous nitrification and denitrification.

FEATURED PROJECTS

The Nutrient Removal Challenge study areas related to design issues that are summarized herein include:

- Performance and reliability of nutrient removal and recovery facilities
- Uncertainty evaluations in model-based facility design for high level nutrient removal
- Membrane bioreactor design guide for achieving low effluent nutrient concentrations
- Simultaneous nitrification and denitrification

Performance and Reliability of Nutrient Removal and Recovery Facilities

Knowledge Gap and Aligned Project Objective. WERF cooperated with the Water Environment Federation (WEF) in a comprehensive study of nutrient removal plants designed and operated to meet very low effluent total nitrogen (TN) and total phosphorus (TP) concentrations, as well as low maximum day limits for ammonia (Bott and Parker, 2011). This effort focused on maximizing what can be learned from existing technologies in order to provide a database that helps inform key decision makers about proper choices for both technologies and rationale bases for statistical permit writing.

General Methodology. Managers of 22 facilities provided three years of operational data that were analyzed using a consistent statistical approach that considered both process reliability and the permit limits applied. A proposed set of quantitative descriptors were developed to describe the performance of facilities meeting stringent nutrient requirements in terms of effluent quality percentile statistics. Technology Performance Statistics (TPS) were defined as three separate values representing the ideal, median, and reliably achievable performance.

- The ideal TPS provides an unbiased value of the ideal performance of the technology – when it is minimally influenced by all the factors that cause statistical variability in real facilities. The ideal TPS is defined as the performance under the conditions of operation that can be sustained for a short period of time, which was adopted as being a two week (14-d) period in one year.
- The median TPS represents a measure of the concentration that was achieved on a statistical annual average basis. The project team used the median (the 50th percentile data number) in this study rather than the arithmetic average, because it is impacted less by extreme values resulting from upset events.
- The reliable Technology Performance Statistic. The reliable TPS does not represent a single percentile value for an averaging period (e.g. 95 or 99th percentile). Rather, it is a selected value depending on the technology, the averaging period used in the permit and the acceptable risk associated with frequency of violations.

Summary and Relevancy of Key Findings. This study showed that a simple statistical technique can be used to analyze treatment plant data to determine the reliability of nutrient removal process performance. Using percentiles calculated from final effluent data, the performance of the process and its associated reliability and variability can be quantified. TPS values representing the ideal performance (TPS-14d), median TPS (50%), and reliable TPS (typically

95th percentile) based on either daily or monthly data) values provide facility operators, designers and regulators a tool to determine the ability of a technology or process to meet permit limits under consideration.

A major finding of this investigation was that statistical variability is a characteristic of all the exemplary plants surveyed and that this variability should be recognized in both evaluation of technologies (e.g., stratifying them in terms of their capabilities) in an engineering environment as well as determining the appropriate effluent limits in the regulatory permit setting. This study points out that deterministic permit limits may not be appropriate for plants achieving very low nutrient limits, particularly when the limit is based on technology (concentration) rather than water quality-based (load). In addition, long averaging periods (i.e., annual average) are warranted given the inherent increase in variability of processes that must remove N and P species to concentrations approaching zero.

Uncertainty Evaluations in Model-Based Design for High Level Nutrient Removal

Knowledge Gap and Aligned Project Objective Uncertainty principles have been established outside of the water resource recovery industry that can be used with the current generation of whole plant simulators to quantify the uncertainty and variability of the model output. This WERF project, undertaken by the IWA/WEF Design and Operations Uncertainty Task Group (DOUT) as part of the WERF Nutrient Removal Challenge task of measuring and analyzing effluent variability and plant reliability in nutrient removal facilities, aims at defining how designers can use simulators to develop high-level nutrient removal and recovery systems with a desired degree of reliability. By understanding the reliability of the process, the designer can minimize the risks, costs and inefficiencies associated with either overdesigning or under-designing such systems.

General Methodology. This WERF "Issue Paper" (Belia and Johnson, 2013) introduces to stakeholders the state-of-the-art of uncertainty analysis with a brief literature review, discusses the need to extend good modeling practice in the area of model reliability and explicit uncertainty evaluations, presents the research needs for the development of a methodology that assists engineers in the incorporation of uncertainty evaluations in model-based design, and highlights the opportunities that explicit uncertainty evaluation provides to trade-off cost and probability of non-compliance associated with designing high-level nutrient removal systems.

Summary and Relevancy of Key Findings. The wastewater industry normally relies on empirical, rule-based methods for the design of nutrient removal and recovery facilities in which safety factors are applied to minimize the probability of non-compliance with effluent standards. These empirically based rules are becoming increasingly less relevant as ultra-low effluent nutrient levels are targeted. In recent years, mathematical models describing the physical, chemical, and biological processes are increasingly being applied. Indeed, the increase in the complexity of process configurations and the increased nutrient removal and recovery requirements has necessitated the use of these process-based models. However, there is no consensus on how to manage the inherent uncertainties when applying these models in the context of facility design. It also remains unclear how to translate the traditional safety factor concept to model-based design procedures.

This WERF project introduces to stakeholders the state-of-the-art of uncertainty analysis and highlights the opportunities that explicit uncertainty evaluation provides to high-level nutrient removal systems. Through the development of this Issue Paper, The WERF Nutrient Removal

Challenge aims to contribute towards the goal of promoting the development of a methodology for model-based design of water resource recovery facilities, which is especially applicable to plants trying to achieve very high nutrient removal levels. Such a methodology, as the one generally defined in the figure below, should account for uncertainty in facility design and make them more comparable and transparent with respect to performance and reliability. The main benefit of this procedure will be minimizing the over- or under-design of high nutrient removal treatment facilities by explicitly quantifying the degree of inherent safety associated with different designs.

As presented in the report, adoption of a methodological approach that evaluates uncertainty as part of model-based designs would provide significant benefits to all stakeholders:

- An owner and/or operator will be able to estimate the probability of compliance associated with different designs proposed during bid selection. They will be able to select the design that corresponds to their desired level of safety and costs.
- The designer will be able to quantitatively balance an acceptable level of process risk vs. costs of a proposed design.

Potentially a better basis for effluent guidelines may be developed and regulators may be convinced to apply probabilistic effluent limits.

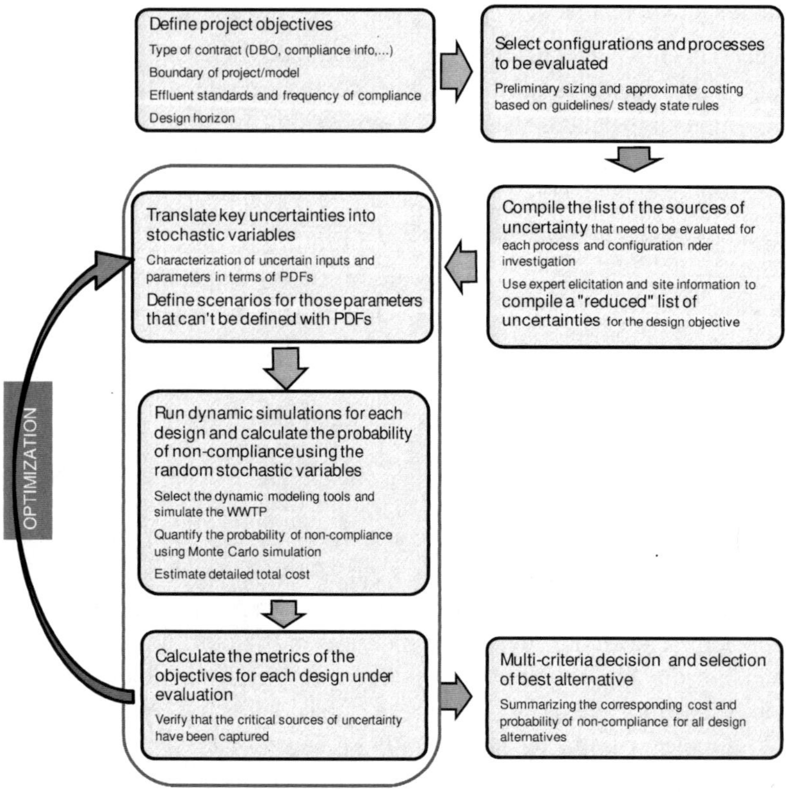

Figure 1. Proposed methodology for probabilistic plant design (Belia and Johnson, 2013). (Abbreviations: DBO = design build operate, PDF = probability density function)

Membrane Bioreactor Design Guide for Achieving Low Effluent Nutrient Concentrations

Knowledge Gap and Aligned Project Objective. Although quickly becoming a well-established water reclamation technology aimed at reuse applications, there is still a need to further understand the capabilities of membrane bioreactor (MBR) processes in achieving enhanced nutrient removal for treatment and recovery applications. Under this ongoing project, a practical guide is been developed that identifies and addresses design and operational issues that are unique to the application of MBR technology for achieving a high-level nutrient removal, i.e. TN below 3-6 mg as N/L and TP concentrations below 0.1-1.0 mg as P/L .

General Methodology. The guide begins by defining typical MBR nutrient removal (N, P, N&P) configurations, identifies advantages and limitations for some of the most common configurations, and provides a summary of a survey of full-scale facilities in the USA, based on data provided by three equipment vendors (GE, Siemens and Ovivo). In addition to this general information, data from five full scale facilities ranging in capacity from 380 m^3/d (0.1 mgd) to 68,150 m^3/d (18 mgd) achieving low nutrient levels was used to document process flow sheets, key component features and their corresponding design basis, adopted operational and control strategies, as well as their actual performance. Several key parameters such as membrane flux, permeability and cleaning requirements; mixed liquor suspended solids concentration; return activated sludge (RAS) rate and internal mixed liquor recycles; chemical addition; and control strategies are discussed in this guide.

Summary and Relevancy of Key Findings. Data from 139 MBR nutrient removal facilities provided by the three vendors was categorized according to four levels of nitrogen and phosphorus removal as shown in Table 1. Figures 2 depicts graphically the distribution of the facilities according to these treatment levels

Table 1. N and P Level of Treatment

Treatment Level	TN (mg/L)	TP (mg/L)
Level 1	< 25	< 2.0
Level 2	< 10	< 1.0
Level 3	< 6.0	< 0.5
Level 4	< 3.0	< 0.1

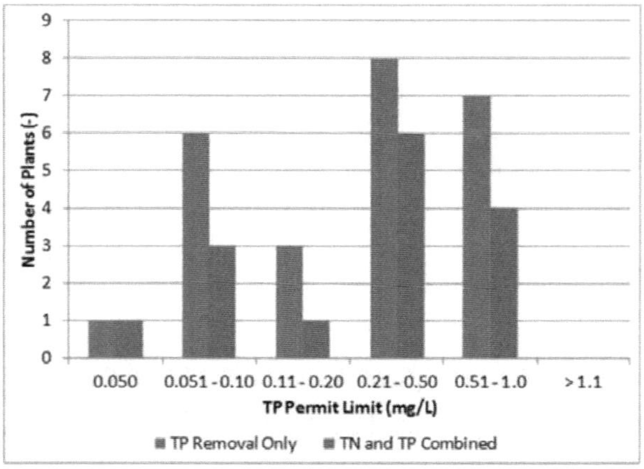

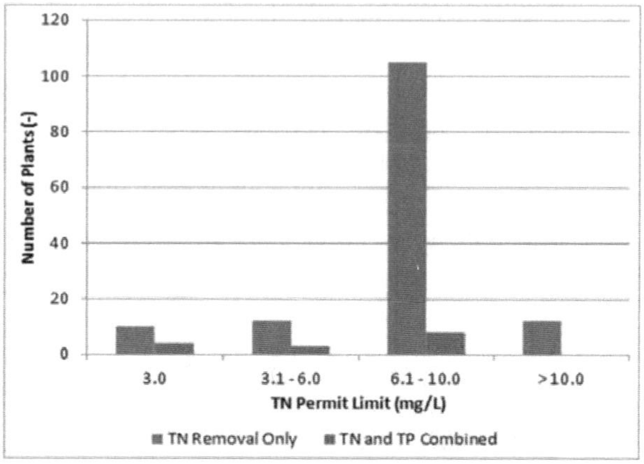

Figure 2. MBR Plants in the United States Removing TN and TP per Permit Limit Categories

Analysis of the information collected from the five case studies indicates that MBRs can reliably achieve nutrient removal to very low levels. These case studies also point out that the challenge of performance reliability is not one inherent to MBR configurations but more so to the reliability and redundancy provided for each individual component such as pumps, blowers, membranes, instruments, etc... Analysis of the historical performance and the operational approaches used at the studied facilities provides insights into key design and operational considerations for nutrient removal in MBRs:

- **MLSS Concentration**. The small footprint of MBR systems is gained in large part by operating at a high mixed liquor concentration. The primary limitation of these high suspended solids concentrations is that oxygen transfer efficiency is reduced. Additionally, high suspended solids correspond to higher return activated sludge (RAS) concentrations in the membrane tank. High RAS suspended solids concentrations are detrimental to membrane permeability. The process designer must balance the MBR footprint with energy costs associated with aeration and the required membrane area.
- **Aeration Design Consideration and DO Control**. The ability to control the dissolved oxygen (DO) levels in the aerobic zone is important to any nutrient removal design to minimize oxygen returns in mixed liquor recycles or feed into secondary anoxic and anaerobic zones. Deoxygenation (DeOx) zones provide one option for eliminating oxygen carryover from aerated to unaerated zones, which ensure stable nutrient removal. Aeration strategies for nitrogen removal are emerging to optimize energy efficiency and improve removal efficiency. Reliable online nutrient analyzers and aeration control strategies can now provide reliable signals to control the aeration process and achieve specific nitrogen objectives. Simultaneous nitrification and denitrification (SNDN) has been observed in MBR facilities and can be used to minimize or even eliminate external carbon source addition depending on the effluent total nitrogen limit. Another operational strategy to maximize energy efficiency is the use of online ammonia measurement to control blower operation. While these operational strategies are not unique to MBRs, the reduced transfer efficiencies in these systems due to the higher suspended concentrations may maximize their energy saving benefits.

- **RAS Rate and Internal Mixed Liquor Recycles.** A high recirculation rate from the membrane tanks, typically on the order of 350-400% or higher of the influent flow, is required in MBRs to keep the biomass in the membrane tank uniform to reduce membrane fouling and to maintain the biomass inventory in the aeration zones rather than the membrane tanks. High recycle rates combined with high DO in membrane tanks threaten the ability to maintain efficient anoxic or anaerobic conditions if RAS is returned in unaerated zones. There are two typical strategies to avoid interference from high DO recycle to the preanoxic zone: (1) employ a DeOx zone to reduce the DO to acceptable levels or (2) return the RAS to an aerated zone and use a separate mixed liquor recycle(s) for nutrient removal.
- **Solids Retention Time**. MBRs are typically designed at solid retention times (SRTs) above 10 days to minimize organic fouling. A good rule of thumb is to design the MBR at an SRT adequate for complete nitrification, which is temperature-dependent. Evaluation of the case study data indicate the importance for designer to balance the process requirements, treatment plant footprint, SRT, solids concentration and oxygen transfer efficiency to find an optimum point to satisfy overall performance requirements.
- **HRT – Short Circuiting.** The potential for short circuiting increases in biological zones with short retention times (smaller zones compared to the flow). This is exacerbated in MBR operational conditions: high recycle rates reduce the actual HRTs, streams with different densities move from aerated to unaerated, and unaerated to aerated zones, influent and recycle streams could be at different temperatures and solids concentrations resulting in segregation. Short-circuiting can occur even in well aerated zones. The case studies showed that short circuiting can be mitigated by installing baffles to distribute the transfer flows, by using high length to width ratios to generate good plug-flow character in the bioreactors, and by carefully evaluating the location and impact of equipment such as mixers, pumps or aerators.
- **Chemical Addition.** Chemical addition and controls are similar in conventional activated sludge (CAS) or MBR nutrient removal systems. Adequate control over chemical dosing is key to reliable performance of the system as shown in the different case studies. Well established design principles regarding mixing at the point of chemical injection, location of chemical injection points, and chemical dose control must be observed. If chemicals are added for denitrification, it is usually added in the post-anoxic zone. Chemicals for P removal are usually injected in the sludge stream ether going to the membrane tanks or going out of the membrane tanks depending on the sludge recycle pumps location (upstream or downstream of the membrane tanks).
- **Cleaning Sequence Impact.** Proper disposal of chemicals used in the cleaning of membranes (typically citric acid or chlorine) is necessary to avoid upset of performance, particularly in low effluent N and P applications.

Simultaneous Nitrification and Denitrification Process Technology

*Knowledge Gap and Aligned Project Objective***.** In simultaneous nitrification and denitrification (SND), oxygen is used for nitrification and nitrate and/or nitrite are used for denitrification in the same biological reactor (Figure 3). SND has been demonstrated in a variety of activated sludge systems with a number of design and control methodologies. This process technology is receiving increased interest for biological nutrient removal because of its apparent ability to achieve very low effluent nitrogen concentrations, save energy, and be applied in small systems with less complex tanks and recycle designs.

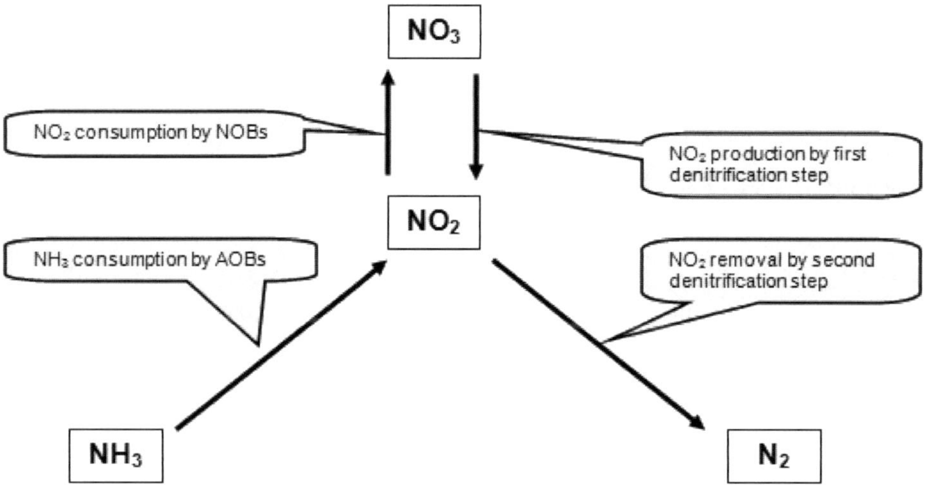

Figure 3. SND paths in terms of two-step nitrification and denitrification

General Methodology. Key information included in the report is information on SND control methods, performance, design requirements for nitrification, and denitrification and process microbiology. Of special interest are gaps in technology and information on demonstrated reliable processes for achieving low effluent nitrogen concentration. Specific topics that are covered in this publication:

- Terminology
- Definition and description of SND concept
- SND reactions and stoichiometry
- Effect on overall oxygen requirements
- Effect on carbon needed for denitrification
- Effect on aeration efficiency
- Control strategies
- Process and system design considerations
- Operational considerations
- Case studies
- Research needs

Summary and Relevancy of Key Findings. This ongoing WERF Nutrient Challenge project aims at developing a State of Technology and Research Needs summary, intended to provide up- to- date information and knowledge transfer for designers and operators that deal with activated sludge based biological nutrient removal systems, while also identifying research needs that would lead to more reliable SNDN system designs and performance.

OVERALL SUMMARY AND RELEVANCY OF KEY FINDINGS

The findings on nitrogen and phosphorus (Stensel et al., 2013; Neethling et al., 2013) through the WERF Nutrient Removal Challenge program and a number of research projects conducted in association with the program, are contributing to the fundamental understanding on the nitrogen and phosphorus removal processes, and also contributing to the better understanding of regulatory and permitting issues on nutrients (Clark et al., 2013). The studies presented in this paper are targeted for designers (and to a great extent operators) wanting to achieve technically attainable nutrient removal and recovery with increased reliability, and potentially at a lower cost. Performance reliability in nutrient removal and recovery facilities is of particular importance as the effluent discharge limits become increasingly stringent, and the need to recover some of the associated resources inherent in the water being treated gains acceptance in our industry. To this end, the WERF Nutrient Removal Challenge has over the last 5 years has significantly advanced the state of the science on this regard, and through the projects summarized herein, places this valuable information in the hands of facility designers and operators.

REFERENCES

Belia, E. and B.R. Johnson (2013) *Uncertainty Evaluations in Model-Based WWTP Design for High Level Nutrient Removal,* WERF Research Project under Nutrient Challenge, WERF Report NUTR1R06q.

Bott, C.B. and D.S. Parker (2011) *Nutrient Management Volume II: Removal Technology Performance & Reliability,* WERF Research Project under Nutrient Challenge, WERF Report NUTR1R06k.

Clark, D.L.; Neethling, J.B.; Stensel, D.H.; Sandino, J.; Tsuchihashi, R.; Pramanik, A. (2013) *Developing Attainable and Protective Permits for Nutrients*, WEFTEC 2013, Chicago, IL, October 2013.

Eleuterio, L. and J.B. Neethling (2009) *Low Phosphorus Analytical Measurement Study*, WERF Research Project under Nutrient Challenge, WERF Report NUTR1R06F.

Jimenez, J., Neethling, J.B.; Stensel, D.H (2015) *Simultaneous Nitrification and Denitrification Process Technology,* WERF Research Project under Nutrient Challenge, WERF Report (unpublished).

Liu, L., M. Bracken, D.S. Smith, D. Houweling, J.B. Neethling, H.D. Stensel, S. Murthy, A. Pramanik, and A.Z. Gu (2011) "Phosphorus Fractionation in Various Tertiary Effluents – Insights into and Implications for Advanced Phosphorus Removal," *Proceedings of Nutrient Removal and Management 2011*, Miami, FL.

Neethling, J.B., B. Bakke, M. Benisch, A. Gu, A. Stephens, H.D. Stensel, R. Moore (2005) *Factors Influencing the Reliability of Enhanced Biological Phosphorus Removal*, WERF Report 01-

CTS-3, Water Environment Research Foundation, Alexandria, VA.

Neethling, J.B. (2008) *Tertiary Phosphorus Removal*, Compendium Document by Nutrient Challenge, available online at www.werf.org/Nutrients.

Neethling, J.B.; Stensel, D.H.; Sandino, J.; Tsuchihashi, R.; Clark, D.L.; Pramanik, A. (2013) *Achieving Low Effluent Total Phosphorus Concentration*, Conference Proceedings, WEFTEC 2013, Chicago, IL, October 2013.

Pellegrin, M.L, A. Menniti, H.D. Stensel, J.B. Neethling (2013) *Membrane Bioreactor Design Guide for Achieving Low Effluent Nutrient Concentrations,* WERF Research Project under Nutrient Challenge, WERF Report (unpublished).

Stensel, D.H.; Neethling, J.B.; Clark, D.L.; Tsuchihashi, R.; Sandino, J.; Pramanik, A. (2013) *WERF's Nutrient Challenge Program Advances Fundamental Tools for the Evaluation and Optimization of Nutrient Removal Systems*, WEFTEC 2013, Chicago, IL, October 2013.

ADVANCES IN THE STATE OF THE PRACTICE FOR THE OPERATION OF NUTRIENT REMOVAL FACILITIES

R. Tsuchihashi[1]*, J. Sandino[2], J.B. Neethling[3], Dave Stensel[4], Dave Clark[3], A. Pramanik[5]

[1]AECOM, ryujiro.tsuchihashi@aecom.com
[2] CH2M
[3]HDR Engineering, Inc.
[4]University of Washington
[5]Water Environment Research Foundation

ABSTRACT

The improvements in the operation of biological and physical/chemical nutrient removal processes in recent years include better understanding of the use of external carbon for denitrification, the use of online analyzers and process control, understanding of the process control during cold and wet weather. Considerations for sustainability such as better control of aeration, alternative carbon sources, and the use of simultaneous nitrification/denitrification have also been understood by the designers and operators of nutrient removal facilities. The Water Environment Research Foundation (WERF) Nutrient Removal Challenge Program developed and supported studies to address: (1) nutrient speciation and its impact on removal, (2) uncertainty and process performance reliability, (3) improving reliability and optimizing the process, and (4) technologies for low effluent concentrations. These studies were put in the context of attainable and protective permits for nutrients. Discussion in this paper focuses primarily on the operational challenges and the results of the studies that can be reflected in the improved operation of nutrient removal facilities.

KEYWORDS

WERF Nutrient Removal Challenge, nutrient, nitrogen, phosphorus, operation, reliability, sustainability, uncertainty, external carbon, online analyzer, fermenter

INTRODUCTION

The WERF Nutrient Removal Challenge is a multi-year collaborative research initiative established in 2007 to develop and provide current information about nutrient removal at municipal wastewater treatment facilities. Initial focus was on the characteristics of nitrogen and phosphorus compounds, and bioavailability in aquatic environments to help regulators make informed decisions. Findings from the studies focusing on nitrogen and phosphorus are reported in the papers by Stensel et al. (2015), and relevant regulatory issues on nitrogen and phosphorus removal are reported by Clark et al. (2015). Design considerations for the nutrient removal facilities are reported by Sandino et al. (2015). Overall summary of the findings from the Nutrient Removal Challenge are presented by Neethling et al. (2015). Considerations for energy and greenhouse gas emissions are also recognized as important operational issues as reported by Folk et al. (2011). The focus of this paper is operational implications on findings from the studies conducted under the Nutrient Removal Challenge. In this paper, selected studies conducted under Nutrient Removal Challenge are referenced to summarize the state of the practice for the nutrient removal facilities, and key findings from these studies.

IMPROVING RELIABILITY AND OPTIMIZING THE PROCESS

One of the significant developments in the area of nitrogen removal in the past decade is increasing use and understanding of carbon addition to enhance denitrification which has helped to improve the denitrification performance. The use of primary sludge fermentation for volatile fatty acids addition has been practiced for reliable operation of Enhanced Biological Phosphorus Removal (EBPR) processes. The use of online analyzers for automated process control has become more common, even though the application tends to be limited to larger treatment facilities. Process control during the wet and cold weather has been an operational challenge for nutrient removal facilities and even to date it is relatively difficult to maintain the treatment performance during these weather conditions. The Nutrient Removal Challenge program recognized these topics as high priority research needs for improving reliability of nutrient removal facility operations and opportunities for optimizing the process. Each of these topics is discussed below based on the studies conducted and supported by the Nutrient Removal Challenge program.

Development of External Carbon Evaluation Protocol for Denitrification

As the use of external carbon source became a relatively common practice to enhance denitrification in biological nutrient removal (BNR) plants, there has been an increasing interest in exploring alternative external carbon source(s) other than methanol. With the rise in methanol price and treatment facilities' preference for using less hazardous chemicals than methanol in recent years, the development of an evaluation protocol for external carbon was identified as a priority study area.

Knowledge Gap and Aligned Project Objective. The issues often faced by the treatment plant operators when investigating a new external carbon source is that published data on various external carbon sources and their performance are limited, and sometimes it is not appropriate to compare the results from other studies directly due to different mythologies used for the evaluation of the external carbon. The main focus of the project was to establish a standardized protocol that could be used by any wastewater treatment facility so that each treatment facility can conduct the alternative carbon evaluation using the protocol and the results from testing in different treatment facilities could be compared directly (Gu and Onnis-Hayden, 2011).

General Methodology. Protocols for the laboratory scale testing of the carbon sources and protocols for the testing in the full-scale treatment facility were developed based on the various protocols being used in the past. General guidelines and approaches for the selection of potential external carbon sources and a roadmap for both laboratory-scale characterization of the carbon sources and full-scale assessment of the denitrification performance were developed (see Fig. 1).

Summary and Relevancy of Key Findings. A comprehensive literature review was conducted to summarize the kinetic rates and coefficients of various types of external carbon sources that can potentially be used for denitrification. The summary for the potential external carbon sources is useful for the facilities conducting the evaluation of external carbon. The protocol has been published as a WERF document (Gu and Onnis-Hayden, 2011). Selection of external carbon with considerations to sustainability and safety at the treatment facility is recommended in addition to the cost evaluation.

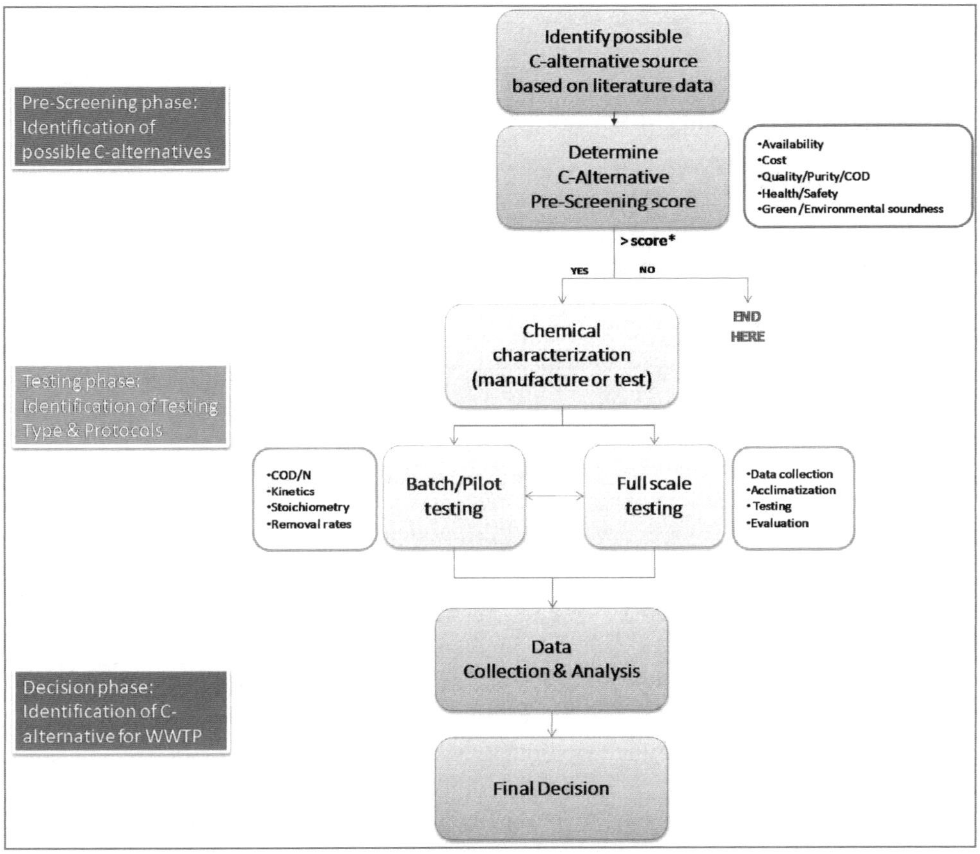

Figure 1: Roadmap for the evaluation of external carbon for denitrification (Gu and Onnis-Hayden, 2011)

Design and Operational Considerations for Tertiary Phosphorus Removal and Use of Fermenters for Biological Phosphorus Removal

Reliable and cost-effective supply of volatile fatty acid (VFA) for the enhanced biological phosphorus removal (EBPR) processes and the use of tertiary phosphorus removal processes have been considered as important design and operational challenges in phosphorous removal. Two compendium documents developed by the Nutrient Removal Challenge program (Neethling, 2008; WERF, 2011) outlined the state of knowledge and research needs in these topics.

Knowledge Gap and Aligned Project Objective. Neethling et al. (2005) reported that secondary level phosphorus removal facilities achieved effluent total phosphorus (TP) concentrations of 1 mg/L between 80 and 99 percent of the time in four out of five full-scale facilities investigated. It was also reported that the level of effluent TP can be reduced to 0.5 mg/L between 50 and 95 percent of the time, also in four facilities. However, additional tertiary processes are required to polish effluent phosphorus to achieve greater reliability and to reduce phosphorus to lower limits. Based on the findings in the previous studies, two knowledge areas in phosphorus removal were identified for compilation and dissemination of knowledge: tertiary phosphorus removal, and the use of fermenters for biological phosphorus removal.

General Methodology. A compendium document on the current practices of the use of fermenters to produce substrates for enhanced biological phosphorus removal (EBPR) processes was developed (WERF, 2011). The state of knowledge on primary sludge fermenters including configurations, advantages and disadvantages were summarized. The documentation also included secondary sludge fermenters which are a relatively new concept and the biochemical mechanisms involved are not yet well defined, but it is considered as a viable alternative to primary sludge fermentation. Limitations and suggested operational parameters were reported. Documentation on tertiary phosphorus removal by Neethling (2008) focused on factors that need to be considered to reduce phosphorus below 0.5 mg/L and, in particular, the requirements to reduce phosphorus to very low limits, below 0.05 mg/L (50 µg/L) or lower. Design considerations and future research needs were addressed. A compendium document was developed in a simple Q/A format, and made available through WERF Knowledge Area website.

Summary and Relevancy of Key Findings. In the compendium document for fermenters (WERF, 2011), five exemplary configurations for primary sludge fermentation were described. The summary included advantages and disadvantages of each fermenter configuration. Fundamentals and design and operational issues for secondary sludge fermentation were described. Design considerations for fermenters, including primary sludge pumping, grit removal, sludge grinding and screening, sludge mixing, sludge collector mechanisms, scum removal, and fermented sludge pumping were reported. The research needs identified include:

1. Development of a standardized method of expressing the short-chain volatile fatty acids (SCVFA) yield in primary sludge and secondary sludge fermenters.
2. Development of a primary sludge fermenter module capable of representing several fermenter configurations that can be incorporated into process simulators that accurately predicts the production of SCVFA through acid fermentation and its impact on the system biological phosphorus removal. Develop methods to model/simulate partial mixed fermentation vessels and solids/liquid separation through elutriation.
3. Further development and calibration of existing plant simulators to accurately model the quantities of SCVFA that can be generated through the decay and hydrolysis of secondary sludge in the anaerobic zones of the bioreactor, and its impact on the system biological phosphorus removal.
4. Development of a design manual for primary and secondary sludge fermenters and their associated odor and corrosion control systems.
5. Identification of the best method of measuring the location of the top of the sludge blanket in static fermenter/thickeners and gravity thickeners.
6. Standardization of the use of ORP measurement of fermentate to control elutriation water flow rate.

One of key issues identified in the compendium for tertiary phosphorus removal was speciation of phosphorus in wastewater. A separate compendium was developed for the measurement of low phosphorus concentrations. Nutrient Challenge also hosted studies on analytical methods for low phosphorus concentrations and phosphorus speciation for better understanding of phosphorus and phosphorus removal in wastewater (Eleuterio and Neethling, 2009; Liu et al., 2011). The processes used to convert/remove phosphorus species from wastewater were summarized (see Table 1) and research needs were addressed.

Table 1: Processes used to convert/remove phosphorus species (Neethling, 2008)

Species	Common conversion or removal process
Organic-P	Organic phosphorus can be converted to orthophosphate and polyphosphate, some organics degrade very slowly
Orthophosphate	Most abundant phosphorus species Reactive species in chemical reactions and consumed in biological growth
Polyphosphates	Condensed orthophosphates Possibly reacts with metal salts Can be used for biological growth
Chemical phosphorus	Precipitated phosphates formed by reacting orthophosphate with metal salts, or precipitates as phosphate hydroxides
Biological phosphorus	Phosphorus incorporated into the biomass for growth Excess phosphorus may accumulate under certain conditions

Use of Online Instrumentation and Automated Process Control in BNR Facilities

A comprehensive study on online instrumentation was conducted by Palmer et al. (2007), in which a comprehensive reference to understand the mechanism and performance of major nitrogen analyzers were reported. It also provided a summary of the automated process control schemes that could be adopted in BNR facilities. While useful information is becoming available, the practice of online nitrogen analyzer usage for automated control is not yet as common.

Knowledge Gap and Aligned Project Objective. The Connecticut Nitrogen Credit Advisory Board (NCAB) has been providing funding for the installation and evaluation of online analyzers for BNR process control. This effort is a major undertaking in advancing BNR process control, and also providing a unique opportunity to collectively evaluate the performance of the analyzers and investigate potential to a robust and reliable process control scheme. The use of online nitrogen analyzers and the use of online analyzers for process control at wastewater treatment plants in Connecticut are investigated under the Nutrient Removal Challenge.

The outcome of the introduction of online analyzers is evaluated in this study (Tsuchihashi et al., 2015), including the types of online analyzers installed at the treatment plants, use of the data from the analyzers, and accuracy, reliability and maintenance requirement of the analyzers. The compiled information is then used to evaluate the potential to introduce dynamic control scheme to improve BNR performance while reducing the operation and maintenance costs.

Methodology. In the first phase of the study, a survey was conducted to collect information about the use of online nitrogen analyzers, their maintenance practices, the purposes for the use of online analyzers, and considerations for the use of the analyzers for process control. In the second phase of the study, selected treatment plants are contacted, and examples of automated BNR process control and best practices for the process control are documented. Among the participated treatment facilities, each treatment plant typically had 2 or more nitrogen analyzers, and they are placed in various locations in the treatment process. All participated treatment plants responded that they calibrate and maintain the analyzers routinely, but the frequency varied. About half of the treatment plants relied on the service contract with the equipment supplier for the regular service and calibration.

Summary and Relevancy of Key Findings. An important finding from the survey is that not many treatment facilities were using the installed online nitrogen analyzers for automated process control. The use of online nitrogen analyzers by the participated treatment facilities is summarized in Table 2. As reported in Table 3, manual adjustments of the operating conditions are common, but only limited number of facilities moved on to the automated process control. The use of online nitrogen analyzers for the control of external carbon is a relatively simple operational modification that has a potential to be automated. Two facilities have started the automated control, whereas four facilities are currently adjusting carbon addition manually based on the analyzer readings.

The second phase of the study focuses on the documentation of best practices and examples of automated process control (study in progress). The outcome of the study is intended for the treatment facilities to consider moving from manual process control to automated process control for certain parts of the BNR process.

Table 2: Use of online nitrogen analyzers (multiple responses)

Use of analyzer data	Response
Monitoring trends only	5
Manual adjustment of aeration based on the reading (in addition to information from DO analyzer)	6
Manual adjustment of recycle rates	2
Manual adjustment of external carbon addition	4
Automated adjustment of aeration	1
Automated adjustment of recycle and waste rates	1
Automated adjustment of external carbon addition	2
Detect malfunctioning equipment	1

Considerations for Implementing Automated Process Control with Online Analyzers. While it is generally recognized that automated BNR process control with online analyzers can potentially improve the process performance and its reliability, implementation of automated process control is not always an easy task and in many cases online analyzers are used for a monitoring purpose or for the use with manual process control. It is also important to recognize that the automated process control with online analyzers is only as good as the reliability of the analyzer readings, and thus maintenance and calibration of the analyzers are the critical operational challenge when automated process control is implemented.

Through the survey of the nutrient removal facilities for the use of online analyzers and automated process control, it was a general consensus that automated process control works well when the control algorithm is simple and reliable analyzers are used. Examples of process control output and mechanisms for control are shown in Table 3. When an automated control is first introduced, it is generally advisable to start with a process control that is simple enough and can be monitored closely to check and adjust the control points. It is also important to note that the process control logic should include safety controls to prevent the control system from causing a major process upset. For example, the control logic should trigger an alarm when unusual conditions are encountered, such as unusually high external carbon feed flow rate, loss of analyzer signals, or unusual reading of the control parameters. The process control should be able to turn to a manual control in such conditions until the cause of the problem is identified and corrected. An example of a simple process control framework for the control of external carbon addition is shown in Figure 2.

Table 3: Process Control Output and Mechanism for Control.

Control Parameter	Input	Output	Typical Mechanism for Control
Aeration control	♦ DO (aerobic zone) ♦ Ammonia (pre-anoxic zone and aerobic zone), nitrate (aerobic zone)	♦ Blower speed ♦ Air distribution valves	♦ VFD, throttling, on/off ♦ Actuated valve that is controllable from SCADA ♦ Location(s) of the valves allow separate control? for different parts of aeration tanks
Nitrate recycle rate control	♦ Nitrate/nitrite (aerobic zone) and ammonia (pre-anoxic zone and aerobic zone)	♦ Recycle pump rates ♦ Valve opening	♦ VFD, throttling, on/off ♦ Actuated valves controllable from SCADA
RAS rate	♦ Nitrate/nitrite ♦ TSS	♦ RAS pump rates ♦ Valve opening	♦ VFD, throttling, on/off ♦ Actuated valves controllable from SCADA
External carbon dosing rate control	♦ Nitrate/nitrite (anoxic zone influent/effluent, plant effluent)	♦ External carbon dosing pump(s)	♦ Variable speed, on/off

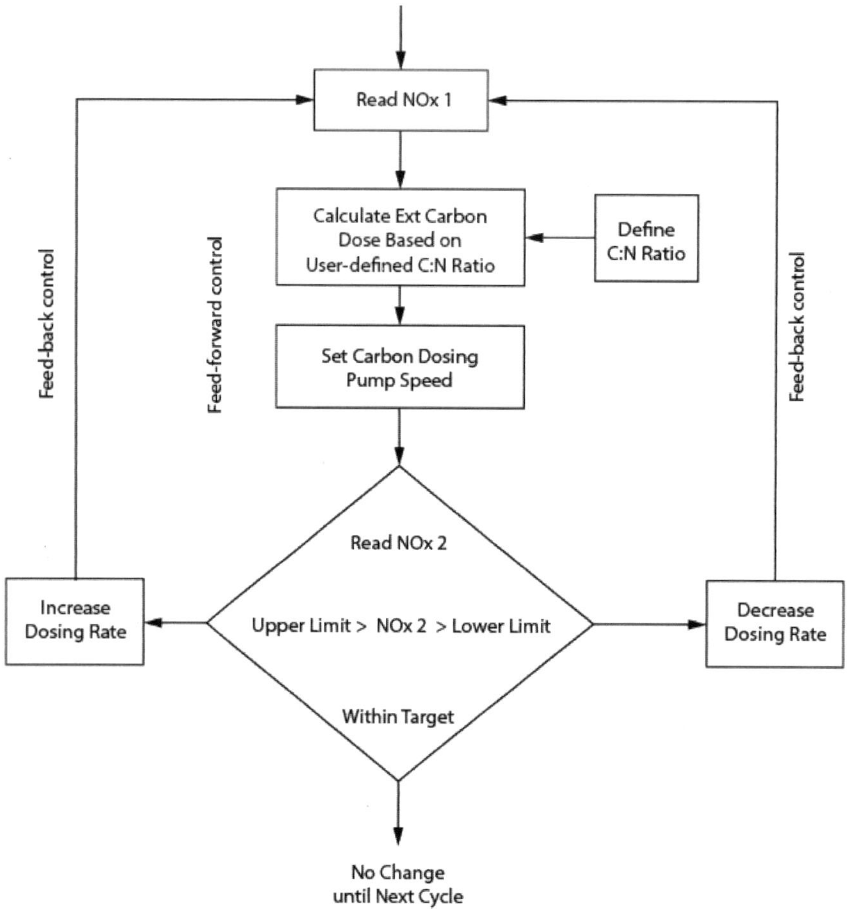

Figure 2: Example of external carbon addition control flow chart.

Operational Challenges and Approaches to Cold/Wet Weather Impacts

Process control during cold and wet weather conditions is a significant challenge for the plants with stringent discharge limits as the compromised process performance could contribute significantly to the total discharge loading. A study "Evaluating the Impact of Cold Weather Events on Biological Nutrient Removal in Wastewater Treatment Facilities" (Khunjar et al., 2014) was launched to evaluate the impacts and operational approaches to address cold and wet weather events.

Knowledge Gap and Aligned Project Objective. Because the appropriate strategies are often considered to be site-specific and depend on plant operators' experience, there has not been much documentation made available for the plant operators and engineers as a general guidance. The primary goal of this project is to assess real-world responses by BNR facilities to adverse weather events and present a compendium of strategies that have been found to be successful.

General Methodology. A two-phase study is conducted. Phase 1 of the effort is to document challenges faced by WWTPs with BNR during cold and wet weather events. Based on the outcome of the literature reviews in Phase 1, quantitative analysis of the effects of cold and wet weather events on nutrient profiles in selected WWTPs are conducted in Phase 2 (Khunjar et al. 2012).

In the first phase of the study, surveys were conducted on eleven BNR treatment facilities that experience cold and wet weather events. The information collected during the first phase included standard operating protocols, historical data. The impact of cold and wet weather events on operational parameters, and correlations between poor performance and operational parameters were evaluated. Based on the results from the first phase of the study, nutrient profile of selected treatment plants was evaluated during the cold and wet weather event by taking samples from various locations in the treatment plant. Process models are used to evaluate strategies for maintaining nitrogen and phosphorus removal performance during cold and wet weather events.

Summary and Relevancy of Key Findings. From a compilation of data from participating facilities, the most prevalent causes of nitrogen removal instability during cold/wet weather included low solids inventory, insufficient process air, insufficient carbon supply, and increased loading from sidestream during cold weather events. Similarly, the primary causes of biological phosphorus removal instability were low solids inventory in the reactor basins and insufficient carbon supply for the biological process (see Figure 3).

Potential strategies for mitigation of cold wet weather event were identified as summarized in Table 4, which represent a preliminary list of strategies employed at surveyed facilities for maintaining consistent BNR performance, or at least limiting the detrimental effects of wet weather. The study also provided a list of monitoring parameters and monitoring frequency to help anticipating the process upsets. It was recommended that treatment facilities augment existing sampling efforts to ensure that at minimum, a historical database of the critical parameters, including rainfall/snowfall data, flow rate, temperature, carbon to nutrient ratios, and others be developed. Most of the parameters are measured commonly at nutrient removal treatment facilities, and some of the parameters are monitored with online analyzers, while other parameters may not be compiled into historical database. Using the database and the strategies listed in Table 4, conceptual framework for responding to adverse weather events were proposed, as shown in Figure 4.

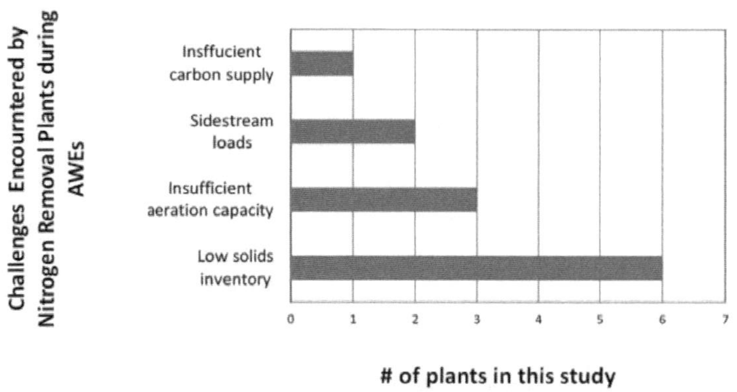

(a)

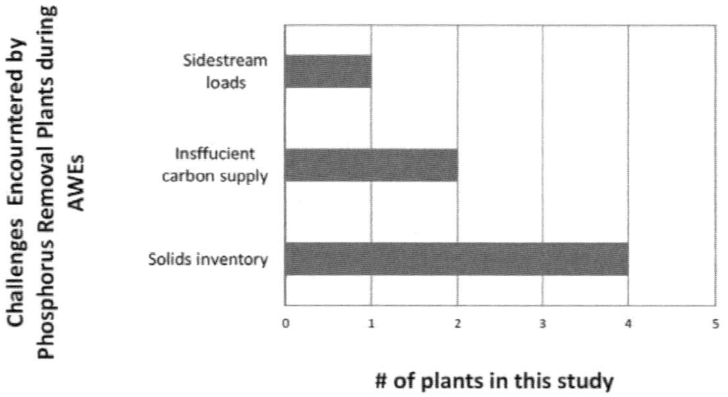

(b)

Figure 2: Challenges encountered during and after the cold wet weather events: (a) nitrogen removal, and (b) phosphorus removal (Khunjar et al. 2014).

Table 4: Potential strategies for mitigation of cold wet weather event impacts to BNR (Khunjar et al. 2014)

Strategy	Process implication/benefits
Increase MLSS	Maintain a higher MLSS concentration during cold weather conditions effectively increases the SRT, allowing retention of slow growing organisms.
Flow equalization	Equalization can act to reduce hydraulic loading to the biological process (and clarifiers), minimizing the chances of solids washout.
Step feed/contact stabilization	Step Feed / Contact Stabilization can protect sensitive biomass populations by preventing biomass washout in the system. In some cases, complete bypass of some unit processes may be necessary. Optimization of current step feed systems by operating at different flow splits during dry versus cold wet weather may also yield a significant BNR performance benefit.
Bypass/blending	For BNR plants with combined collection systems, bypassing biological treatment and blending using high-rate clarification (such as ballasted flocculation technologies) could help maintain biomass while meeting permit limits during adverse weather events. This approach may not be permitted in some regions, particularly for plants with separate sanitary sewer systems.

Biomass off-line storage/RAS equalization	Allows protection of a portion of the biomass by removing it from the treatment system during wet weather events. This strategy prevents washout and preserves important system biomass.
Increase VFA addition from fermenter	Understanding and matching VFA needs is critical to ensuring optimal biological P removal during adverse weather events when insufficient carbon may be present in the influent.
Change internal recycle rates and locations	Addition of an internal recycle stream from aerobic to pre-anoxic zone can allow maximization of use of carbon from the influent. There may also be scenarios (e.g. recovery from nitrification failure) when the internal recycle needs to be reduced/eliminated to maximize use of the aerobic volume.
Supplement biomass from solids storage	This strategy is slightly different than off-line / RAS EQ strategies. In this strategy, solids stored in the normal biosolids processing facilities can be used to bioaugment and supplement biomass in existing process units after a storm event has passed. The system may also benefit from low levels of fermentation that occurs during sludge storage.
Intermittent aeration/mixing	This strategy is aimed at preventing washout of biomass during high flow events for plants with limited storage/volume capacity. In this strategy, the plant cycles aeration and mixing for brief periods of time to maximize sludge storage within the reactor basins and also to ensure that aerobic biomass activity is maintained. The schedule of aeration cycles must be designed to allow the plants to meet weekly and monthly TN and TP limits. Metal salt addition may be needed to ensure that TP limits are met during operation of this strategy.
Aeration basin "swing" zone	An area of the aeration tanks in which the air can be turned off could prevent the biomass from being suspended, thereby reducing the risk of biomass washout. Similarly, areas that are normally unaerated could have the air turned on to ensure sufficient aerobic HRT. Metal salt addition may be needed to ensure that TP limits are met during operation of this strategy.
Sidestream treatment/equalization	Treatment of sidestream from solids digestion, thickening, and dewatering processes helps to minimize nutrient loads to biological treatment during adverse weather events, thereby reducing the risk of insufficient treatment. Sidestream storage to equalize nutrient load during adverse weather events should be done at a minimum.
Chemical addition to assist with settling and achieve P removal	This strategy focuses on using metal coagulants and/or polymer to improve settling of solids in clarifiers to prevent biomass washout. In this strategy, chemical addition is commenced prior to wet weather events in anticipation of the discharge of effluent solids from the secondary clarifier. The rate at which chemical is added should be determined through site specific jar testing.

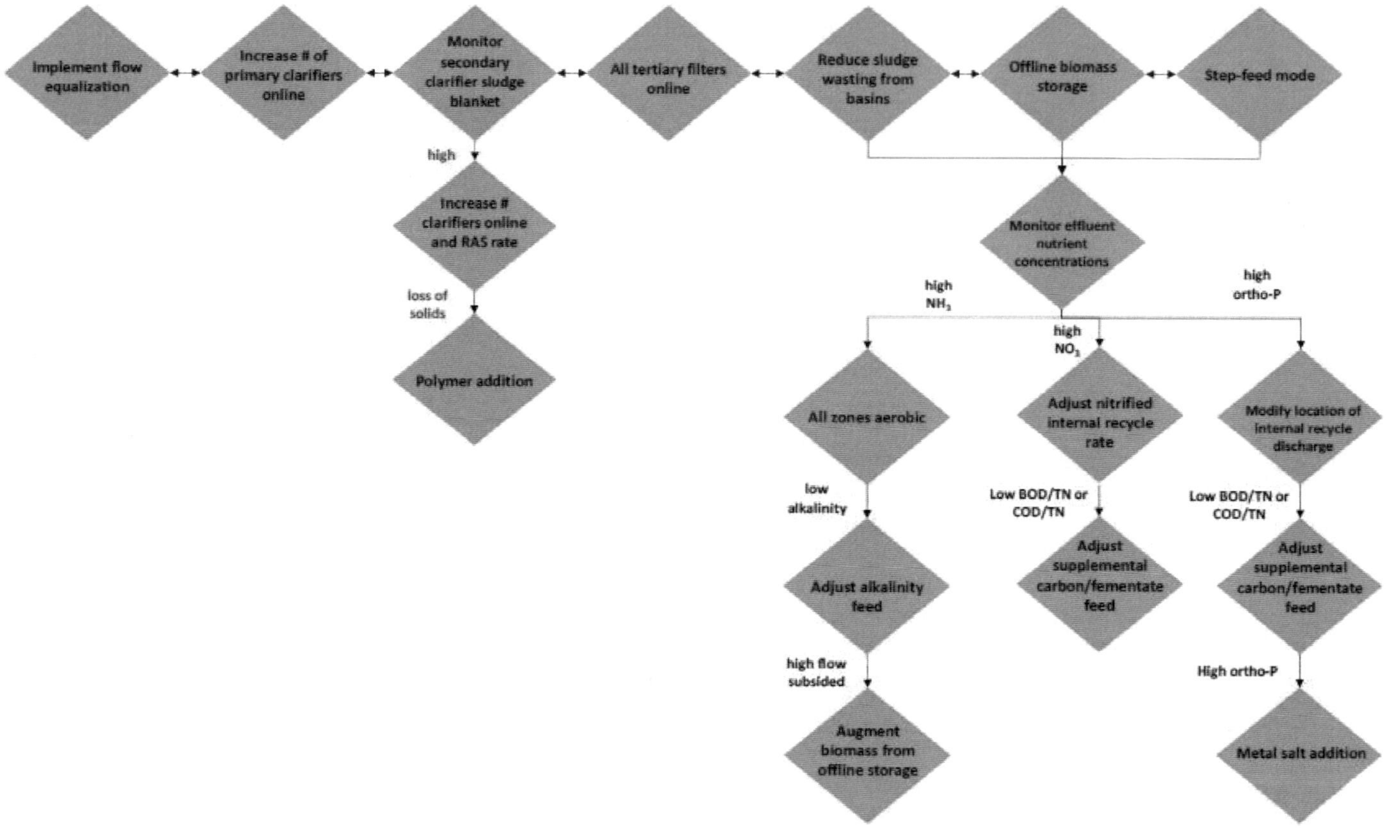

Figure 4: Conceptual Framework for Responding to Adverse Weather Events.

OVERALL SUMMARY AND RELEVANCY OF KEY FINDINGS

The findings on the characteristics of nitrogen and phosphorus (Stensel et al., 2015) through the Nutrient Removal Challenge program and a number of research projects conducted in association with the program are contributing to the fundamental understanding on the nitrogen and phosphorus removal processes. The findings are also contributing to the better understanding of regulatory and permitting issues on nutrients (Clark et al., 2015). Considerations for energy and greenhouse gas emissions are also recognized as important operational issues as reported by Folk et al. (2011). The studies presented in this paper are targeted for operators to achieve the technically attainable nutrient removal in the increased reliability, and potentially at lower costs. Even though particular findings from the studies on the characteristics of nitrogen and phosphorus were not directly reflected in the studies for operational considerations described in this paper, these findings are expected to also help operators to make well informed decision on the process control. For example, understanding of dissolved organic nitrogen that may not be removed through biological process will inform the operators the limit of effluent total nitrogen concentration the treatment facility can achieve through biological processes. Performance reliability in the nutrient removal facilities is of particular importance as the effluent discharge limits become increasingly stringent. As the treatment processes are more refined, process control with reliable analyzers becomes a crucial part of the treatment facility operations.

References

Clark, D.L.; Neethling, J.B.; Stensel, D.H.; Sandino, J.; Tsuchihashi, R.; Pramanik, A. (2015) "Developing Attainable and Protective Permits for Nutrients," WEF Nutrient Symposium 2015, San Jose, CA, July 2015.

Eleuterio, L. and J.B. Neethling (2009) *Low Phosphorus Analytical Measurement Study*, WERF Research Project under Nutrient Challenge, WERF Report NUTR1R06F.

Folk, M.W., J.B. Neethling, D.J. Reardon (2011) *Striking the Balance Between Nutrient Removal in Wastewater Treatment and Sustainability*, WERF Research Project under Nutrient Challenge, WERF Nutrient Removal Challenge Project NUTR1R06n.

Gu, A. and N. Onnis-Hayden (2011) *Protocol to Evaluate Alternative External Carbon Sources for Denitrification at Full-Scale Wastewater Treatment Plants*, WERF Research Project under Nutrient Challenge, WERF Nutrient Removal Challenge Project NUTR06b.

Khunjar, W., M. Strahota, P. Pitt, and W.J. Gellner (2014) *Evaluating the Impact of Cold Weather Events on Biological Nutrient Removal in Wastewater Treatment Facilities*, WERF Research Project under Nutrient Challenge, WERF Nutrient Removal Challenge Project NUTR1R06s.

Liu, L., M. Bracken, D.S. Smith, D. Houweling, J.B. Neethling, H.D. Stensel, S. Murthy, A. Pramanik, and A.Z. Gu (2011) "Phosphorus Fractionation in Various Tertiary Effluents – Insights into and Implications for Advanced Phosphorus Removal," *Proceedings of Nutrient Removal and Management 2011*, Miami, FL.

Neethling, J.B., B. Bakke, M. Benisch, A. Gu, A. Stephens, H.D. Stensel, R. Moore (2005) *Factors Influencing the Reliability of Enhanced Biological Phosphorus Removal*, WERF Report 01-CTS-3, Water Environment Research Foundation, Alexandria, VA.

Neethling, J.B. (2008) *Tertiary Phosphorus Removal*, Compendium Document by Nutrient Challenge, available online at www.werf.org/Nutrients.

Neethling, J.B.; Stensel, D.H.; Sandino, J.; Tsuchihashi, R.; Clark, D.L.; Pramanik, A. (2015) "WERF Addresses Challenges of Minimizing Effluent Nutrient Concentrations," WEF Nutrient Symposium 2015, San Jose, CA, July 2015.

Palmer, T.M., M.C. Ross, S.G. Nutt, S. Kharkar (2007) *On-Line Nitrogen Monitoring and Control Strategies*, WERF Report, 03-CTS-8, Water Environment Research Foundation, Alexandria, VA.

Sandino, J., Tsuchihashi, R., Neethling, J.B., Stensel, H.D., Clark, D., Pramanik, A (2015) "WERF Advances the State of the Practices in the Design of Nutrient Removal Facilities," WEF Nutrient Symposium 2015, San Jose, CA, July 2015.

Stensel, D.H., Neethling, J.B., Clark, D.L., Tsuchihashi, R., Sandino, J., Pramanik, A. (2015) "The Importance of Nutrient Species for Treatment Performance and Surface Water Impacts," WEF Nutrient Symposium 2015, San Jose, CA, July 2015.

Tsuchihashi, R., Canterbury, B., and Neethling, J.B. (2015) *BNR Process Monitoring And*

Control With Online Nitrogen Analyzers For Nitrogen Credit Exchange Program In Connecticut, WERF Nutrient Removal Challenge Project NUTR1R06y

WERF (2011) *Fermenters for Biological Phosphorus Removal Carbon Augmentation*, Compendium Document by Nutrient Challenge, B. Rabinowitz (Primary Author), J.B. Neethling (Lead Editor), available online at www.werf.org/Nutrients.

Phototrophic Membrane Bioreactor (PMBR) for the Management of Nutrients in Domestic Wastewater within the Context of Algal Biorefinery and Biofactory Concepts

Ozcan, O.[1], Jean, H.[1], Bair, R.A.[1], Prieto, A.L.[1], Pickett, M.[1], Yeh, D.H.[1]

[1]Department of Civil and Environmental Engineering, University of South Florida, Tampa 33620, FL, USA

ABSTRACT

The potential synergy of the algal biorefinery-biofactory concepts and wastewater treatment have seen increased attention in the past decade, owing to the possibility of the fusion of algae cultivation with wastewater treatment bringing down costs of biorefinery-biofactory products to competitive levels with alternative products in the market. The authors of this study investigated the viability of utilizing a Phototrophic Membrane Bioreactor (PMBR) inoculated with native species bioprospected from the primary and secondary clarifiers of a domestic wastewater treatment plant to treat pre-filtered domestic wastewater. The aims of the study were two-fold: (1) demonstrate if it is possible to treat the nutrient rich wastewater stream under low aeration conditions to save energy while meeting current surface water discharge standards, and (2) determine and demonstrate the conditions under which phototrophic cultures can raise the pH of the system over 9.25 to facilitate the stripping of excess ammonia from the system. Experiments were conducted using two identical 2 L photobioreactor columns fitted with external tubular polyvinylidene fluoride (PVDF) ultrafiltration (UF) membranes with a nominal pore size of 0.03 μm and a membrane area of 0.025 m^2 per module. Initial results of the first phase of the study indicate under low aeration conditions, the system was dominated by nitrifying species, as 59.3% of the TN loading into the system was converted to nitrate, whereas 15.8% net ammonia removal was observed after the transitional period. Under these conditions COD removal of 63.6% was reached as the process stabilized and the pH of the system decreased to 6.3. The total energy demand of the PMBR system in the first phase was 0.271 kWh/m^3. The second phase of the study is currently underway. Initial results of the second phase indicate for the PMBR system operated at 10 d SRT, the pH of both reactors remained under 9 under both high and low aeration conditions, with initial pH decreasing over time for both reactors. Low aeration conditions correlated with a consistently lower pH than high aeration conditions. The system was run at an HRT of 24 hours throughout this study. Future studies will aim to decrease HRT to competitively high rate levels after the conditions under which high rate nutrient removal using a mixed phototrophic culture are discovered and established.

KEYWORDS: algae biofuels, membrane bioreactors, algal biorefinery, domestic wastewater treatment

INTRODUCTION

Algae have been receiving increased attention in terms of their conversion to useful products within the context of their utility and synergy with wastewater treatment. Fat storing algae can be converted to biofuels and have been cited as an upcoming alternative to water and energy intensive biofuel crops from which fuel additives are currently being derived and blended with petroleum based fuels in the United States (Singh & Gu, 2010). Furthermore, algae can be processed to yield fine chemicals and nutraceutical products, as well, making them an ideal candidate for resource recovery from wastewater streams high in nutrients and water. Algal biorefineries-biofactories, or integrated renewable energy parks (IREPs) have been proposed to fuse wastewater treatment processes into the production of new materials, creating closed loop recycling systems for urban environments (Subhadra, 2010; Garcia Alba, 2011; Adarme-Vega, 2012).

There is, however, an important problem associated with the concept of marrying algal material production facilities with wastewater treatment – footprint and capital cost requirements. Especially with the rise of membrane bioreactor systems, current wastewater treatment facilities are mainly geared towards high-rate nutrient removal processes and have comparatively lower processing rates than processes that seek to incorporate algae production into main stream municipal wastewater treatment. One way to alleviate this problem and bridge the gap between the current technological trend in municipal wastewater treatment and the concept of algal material production facilities is to seek out high rate algal processes that will be attractive alternatives to current biological nutrient removal technologies.

This study aims to bridge this gap between current constraints in technology for the management of nutrients in wastewater with cultivation of phototrophic cultures for the production of new materials. For this purpose, the study was divided into two phases. The first phase of the study aims to demonstrate treatment of nutrient rich wastewater stream under low aeration conditions to be competitive in energy consumption while meeting current surface water discharge standards. The second phase was designed to demonstrate the conditions under which phototrophic cultures can raise the pH of the system over 9.25 to facilitate the stripping of excess ammonia and the precipitation of excess phosphate from the system.

METHODOLOGY

Experiments were conducted using two identical 2 L borosilicate glass photobioreactor columns fitted with external 8-mm diameter tubular polyvinylidene fluoride (PVDF) ultrafiltration (UF) membranes (Pentair, X-Flow) with a nominal pore size of 0.03 μm and a membrane area of 0.025 m^2 per module (Figure 1).

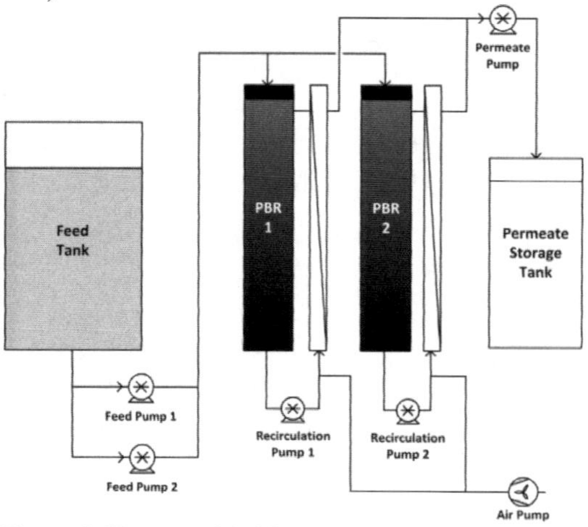

Figure 1. Phototrophic Membrane Bioreactor system

A New Kind of Nutrient Standard: Strengthening Linkages between Nutrients, Algal Blooms, and Aquatic Life

Clifton F. Bell, P.E., P.G., Brown and Caldwell, cbell@brwncald.com

ABSTRACT

This paper provides an overview of one of the most technically advanced effort to date in deriving effects-based nutrient-related criteria. The criteria are being developed for the James River estuary in Virginia, which drains approximately one-quarter of the state, including two major metropolitan areas. The Virginia Department of Environmental Quality initiated the three-year James River Chlorophyll-*a* study in 2011 to ensure that the nutrient wasteload allocations and the associated watershed implementation plans were based on strong linkages between nutrients and designated uses of the estuary. A science advisory panel oversaw a series of monitoring and research efforts that resulted in quantitative linkages between chlorophyll-*a* and various measures of aquatic life use, including the occurrence of harmful algal blooms, algal toxins, low dissolved oxygen, and high pH. As the study concludes in 2015, criteria derivation and related modeling efforts are being combined in an analysis of regulatory/management alternatives for the estuary.

INTRODUCTION

Water quality standards are the heart of the Clean Water Act, and the science behind them drives major public and private investments to protect beneficial uses. Nutrient-related water quality criteria have proven much more difficult to derive than toxics criteria, largely due to the indirect and site-specific nature of nutrient-response linkages. In many settings, simplistic nutrient criteria have lead to doubt and controversy over the cost-benefits of stringent nutrient controls. In contrast, scientifically-rigorous criteria can provide stakeholders with a great deal more confidence that nutrient controls have been set to reasonable levels. This presentation will provide an overview of what is arguably the most technically advanced effort to date in deriving effects-based nutrient-related criteria. The James River (VA) Chlorophyll-*a* Study has gone substantially beyond previous efforts in linking nutrients to beneficial uses, and could serve as a national model in managing harmful algal blooms at the watershed level.

In the mid 2000s, the Commonwealth of Virginia adopted chlorophyll-*a* criteria and nutrient loading caps for the James River estuary, which receives nutrient inputs from over 10,000 square miles of land area, two major metropolitan areas (Richmond and Hampton Roads), and over 50 permitted point sources (Figure 1). These criteria were based on a combination of scientific and attainability considerations (Virginia DEQ, 2004), but generally were not based on explicit linkages between chlorophyll-*a* and specific environmental effects. In 2010, USEPA indicated that the established nutrient loading caps were not low enough to achieve the chlorophyll-*a* criteria, and the additional nutrient controls were projected to cost more than $1 billion.

Stakeholders had major concerns over the unexpected cost escalation, especially light of model calibration problems and the semi-subjective nature of the chlorophyll-*a* criteria driving the reductions. In response, in 2011 Virginia initiated the three-year James River Chlorophyll-*a* Study with the objectives of improving both the modeling framework and the scientific basis of nutrient-related water quality criteria.

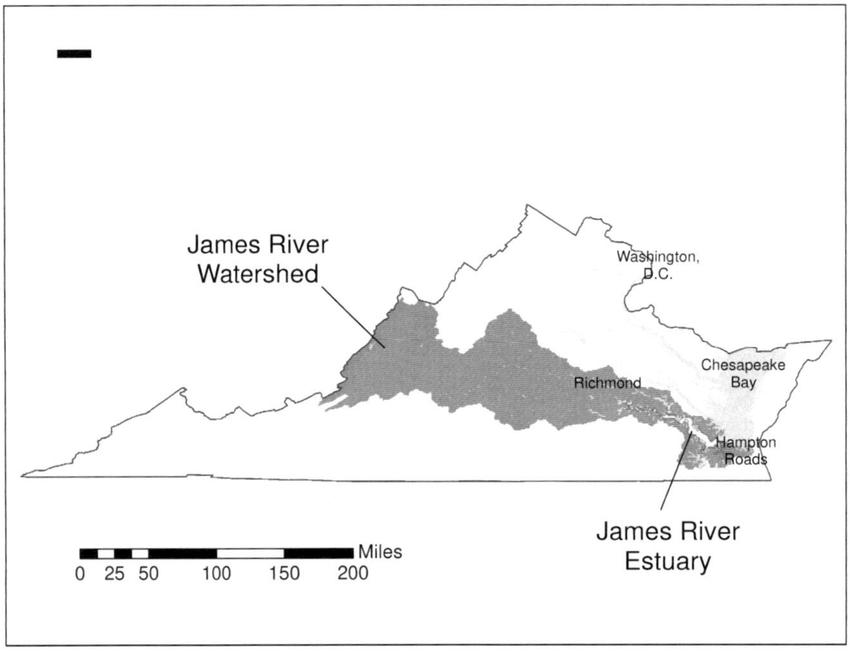

Figure 1 – Map of the James River watershed and estuary in Virginia.

METHODOLOGY

The Virginia Department of Environmental Quality (DEQ) appointed a scientific advisory panel (SAP) to lead the study, with membership from several universities (e.g., Virginia Commonwealth University, Old Dominion University, and the Virginia Institute of Marine Science), state/federal agencies, and utilities. The core of the SAP's approach has been to create a series of quantitative linkages between nutrient loads, chlorophyll- *a*, algal blooms, and living resources (Figure 2). Each of these linkages has required specialized monitoring and laboratory studies to address information gaps.

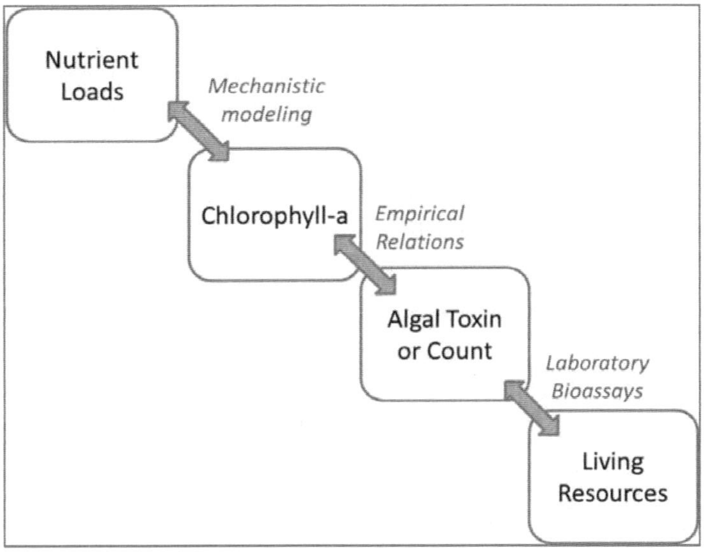

Figure 2 – Diagram of conceptual linkages between nutrients and aquatic life uses.

For example, prior to the study's initiation, it was known that the James River Estuary harbored potentially toxic cyanobacteria (*Microcystis aeruginosa*) and experienced "mahogany tides" caused by potentially harmful dinoflagellates (e.g., *Cochlodinium polykrikoides*). However, it was unknown whether these algal conditions were actually harming human health or other living resources such as fish and oysters, and if/how chlorophyll-*a* could be used to measure the risk of those effects. Study participants monitored algal toxin levels, and also performed laboratory-based bioassays to identify the densities of the algal taxa that could cause measurable harm to various categories of living resources. SAP members also conducted a comprehensive review of field-level biological data in the estuary to determine the existing status of living resources relative to frequent bloom locations (Bell, 2014).

Parallel empirical analyses related these algal densities to spatial and temporal variability in chlorophyll-*a*. This was aided by the Dataflow cruise program conducted by VIMS and the Hampton Roads Sanitation District (HRSD), which included two-dimensional mapping of chlorophyll-*a* in most of the estuary over 2005-2013. Information on the spatial and temporal chlorophyll-*a* distribution was combined with the chlorophyll-related risk of various effects (e.g., exceedance of HAB thresholds, algal toxin thresholds, or pH criteria) to calculate the expected occurrence of these effects as a function of seasonal mean chlorophyll-*a*. These risk charts were then interpreted to define the defensible range of chlorophyll-*a* criteria for each segment and season (spring or summer) of interest.

RESULTS

The current status of living resources in the James River estuary system reflects many decades of changing bottom-up and top-down effects, including water quality effects, commercial/recreational fishing, introduction of new species, habitat losses/gains, erection/removal of barriers, sediment quality, disease, and various restoration efforts. Certain living resources in the James River have "poor" absolute status that can be largely attributed to Bay-wide or even North American-wide declines (Bell, 2014). These include oysters, Atlantic sturgeon, American eel, and American shad. For oysters and sturgeon, the James River appears to actually have better status than other Virginia tributaries. Interestingly, historical zooplankton metrics in the James River were most favorable at the tidal freshwater segment, near the river's chlorophyll-*a* peak (Dauer and others, 2003). Generally speaking, the James River appears to be the most productive oyster fishery (Table 1) and finfish fishery of Virginia's major tributaries, probably due to a combination of higher primary productivity and lower rates of hypoxia. Benthic macroinvertebrates represent the single living resource category for which most of the tidal James River has poor absolute status and worse status relative to other tributaries. Previous investigations have attributed this largely to anthropogenic contamination, including toxics in sediments.

Table 1 - Comparison of Oyster Abundance by Virginia Tributary
[from Southworth and Mann (2013)]

	Average Number of Oysters per Bushel per Bar			
Tributary	**Market**	**Small**	**Spat**	**Total**
James River	40.5	507.7	888.3	1,436.4
York River	24.5	117.8	9.5	151.8
Mobjack Bay	17.0	64.5	258.8	340.3
Piankatank River	36.2	135.2	635.5	806.8
Rappahannock River	47.3	70.4	58.0	175.6
Great Wicomico River	43.7	135.2	1,277.3	1,456.2
Avg. - All Virginia Bars	*40.0*	*231.9*	*524.6*	*796.4*

The upper and lower segments of the James River estuary have very different hydrodynamic, water quality, and biological characteristics, and so results of these two regions are described separately. The upper portion of the James River estuary is characterized as tidal freshwater, and experiences a very high standing crop of phytoplankton in the summer and fall, with chlorophyll-*a* sometimes exceeding 100 ug/L (Bukaveckas and others, 2011). Although diatoms are the dominant taxon, this region also experiences relatively high cyanobacteria levels including the potential toxin former *Microcystis aeruginosa*. Study results indicated that microcystin concentrations were not high enough to impair recreational uses, cause acute aquatic life impacts, or harm human health through consumption of shellfish (Virginia DEQ, 2014).

However, cyanobacteria and microcystin levels were found to be capable of causing chronic, behavioral impacts to benthic macroinvertebrates. Specifically, microcystin was shown to significantly reduce the feeding rates of wedge clams (*Rangia cuneata*) (Wood and others, 2014). Algal activity also contributed to elevated pf (>9.0 s.u.) when chlorophyll-a exceeded 43-70 ug/L (Figure 3). In the upper estuary, the SAP discussed protective chlorophyll-*a* ranges (expressed as seasonal arithmetic means) in the 9-43 ug/L range, depending on segment and season (Bukaveckas, 2015).

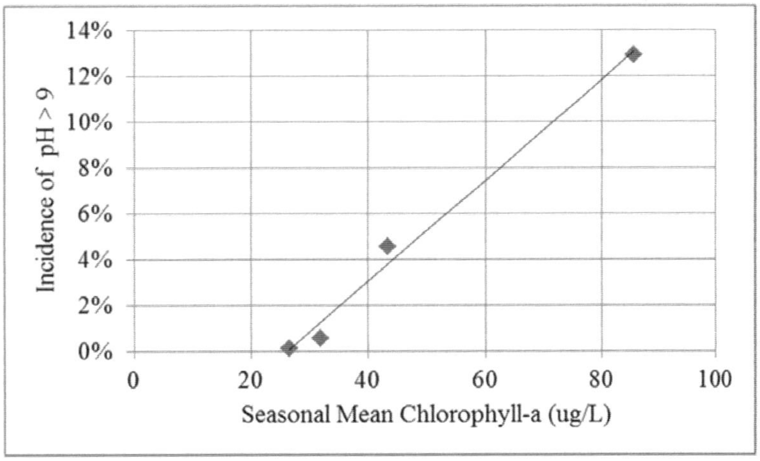

Figure 3 – Chart showing relation between the rate at which pH exceed and 9 s.u. and the seasonal mean chlorophyll-a concentrations in the lower tidal freshwater segment (summer) of the James River estuary.

In the lower estuary, the strongest linkages were between chlorophyll-*a* and the potentially harmful dinoflagellate, *Cochlodinium polykrikoides*. This "mahogany tide" taxon blooms in the late summer in response to a complex combination of stimuli that includes high temperatures and pulsed nutrient loading from runoff events (Mulholland and others, 2009). Based on local laboratory bioassays and a literature review, the SAP identified potential impacts to aquatic life when *C. polykrikoides* exceed 1,000 cell/mL (Bukaveckas, 2015). Exceedance of this algal threshold was more common when chlorophyll-*a* exceeded 30 ug/L in individual samples, or about 13 ug/L as a seasonal mean (Figure 4). Chlorophyll-*a* was also related to elevated pH in some segments. In the lower estuary, the SAP discussed protective chlorophyll-*a* ranges in the 5-21 ug/L range, depending on segment and season (Bukaveckas, 2015).

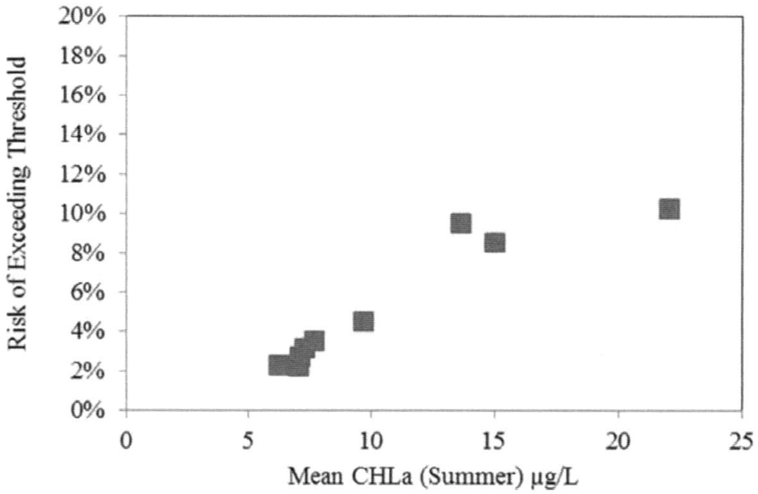

Figure 4 – Chart showing relation between the rate of exceeding 1,000 cell/mL Cochlodinium polykrikoides and seasonal mean chlorophyll-*a* in the James River mesohaline (summer) segment.

CONCLUSIONS

Based on the multiple linkages now established, chlorophyll-*a* criteria can now be based on specific levels of living resource protection, rather than on subjective judgments. As the study concludes, the new models are being employed to evaluate management alternatives, including various tiers of point and nonpoint source nutrient control. This step is supporting a direct consideration of attainability and cost-benefit in choosing the preferred watershed-wide nutrient management approach. The James River Chlorophyll-*a* study provides an example of how the additional time and cost of scientifically-rigorous criteria are well worth the investment for addressing multi billion-dollar nutrient challenges.

REFERENCES

Bell, C.F., 2014. Review of Biological Status Metrics of the Tidal James River. Report prepared for the Virginia Association of Municipal Wastewater Agencies. 24 p.

Bukaveckas, P.A. 2015. Empirical Relationships Linking Algal Blooms with Threats to Aquatic Life Designated Uses in the James River Estuary. Draft report prepared for the Virginia Department of Environmental Quality. 35 p.

Bukaveckas, P. A., L. E. Barry, M. J. Beckwith, V. David, and B. Lederer 2011. Factors determining the location of the chlorophyll maximum and the fate of algal production within the tidal freshwater James River. Estuaries and Coasts 34: 569-582.

Real Time Process Control Algorithms Combining Process Modeling and Instrumentation

Katya Bilyk[1*], Joe Rohrbacher[1], Paul Pitt[1], and David Wankmuller[1]

[1]Hazen and Sawyer
*Email: kbilyk@hazenandsawyer.com.

ABSTRACT

The purpose of this paper is to outline real-time process monitoring and control strategies that combine process modeling (in some cases) with instrumentation for the following processes: nitrification, denitrification, phosphorus removal, solids separation, and nitrogen load equalization. This paper will build upon previous work (Bilyk, et al, 2012) and includes examples of these real-time process control applications.

KEYWORDS: Modeling, real-time process control, instrumentation

INTRODUCTION

Online instrumentation is a useful and powerful tool for process control and monitoring, and hence achieving challenging wastewater treatment objectives while minimizing operational costs. Online instrumentation can include but is not limited to the following:

- Nutrient sensors
- Dissolved oxygen (DO) probes
- Oxidation reduction potential (ORP) probes
- pH meters
- Total suspended solids (TSS) monitors
- Sludge blanket level indicators
- Flow meters
- Conductivity probes

Many regions of the United States and other countries like Australia are moving towards the adoption of nutrient standards at the limits of technology (LOT), and a few are moving even beyond LOT. Online sensors measuring performance parameters such as effluent ammonia and nitrate provide real-time data that will allow automated control of process variables (i.e., DO setpoints, carbon feed) to ensure process objectives are met cost effectively.

Online instrumentation can be integrated with process modeling to provide a framework for real-time process control. Many municipalities have invested in calibrated process models and nutrient analyzers to meet stringent nitrogen and phosphorus effluent standards. The integration of modeling and operations is the next logical step in biological nutrient removal (BNR) operations, and is the subject of an ongoing WEF Plant Operations and Maintenance Committee

Task Group.

A recent paper (Bilyk, et al, 2012) outlined a framework for real-time process control and documents some of the earlier studies of online instrumentation and links to process modeling. This paper will build upon the previous work and includes some examples of real-time process control applications that do not require a calibrated process model to implement.

The purpose of this paper is to outline comprehensive real-time process monitoring algorithms that combine process modeling (in some cases) with instrumentation for all of the following processes:

- Nitrification - comparing real-time aerobic SRT to required SRT as a function of pH, DO, and temperature
- Denitrification - optimizing denitrification in real-time through pacing of internal recycle flows to nitrate leaving the first anoxic zone, and feeding supplemental carbon in real-time to match influent nitrate load
- Phosphorus removal – optimizing biological phosphorus uptake and pacing chemical addition to real-time orthophosphate concentrations
- Solids separation - providing guidance systems to manage the number of clarifiers online
- Nitrogen load equalization - utilizing equalization volume to provide a constant mass of ammonia to the biological nutrient removal (BNR) process

METHODOLOGY

This section will briefly discuss process modeling, such as that used in BNR facility design, and the key functions at BNR facilities that would benefit from real-time process control. The key functions include:

(1) Nitrification
(2) Denitrification
(3) Phosphorus removal
(4) Solids separation
(5) Nitrogen load equalization

Process Modeling

A calibrated wastewater process model has become the typical tool used to design the activated sludge process at a BNR facility. These simulators are typically dynamic to allow the simulation of varying flows and loads including wet weather. Commonly used and commercially available process simulators in the United States include but are not limited to BioWin™ and GPS-X™.

A process model is calibrated based upon a detailed characterization of the influent wastewater. Plant-specific tank volumes and configurations are known, and modeled effluent quality and process operating parameters should match observed data for the model to be considered calibrated. Using the existing plant as a starting point, modifications to the treatment process can be explored with the model. This is a common approach used by engineering firms to design new facilities or optimize existing processes, and is particularly well-suited to facilities with nutrient removal objectives because of the increased complexity of biological processes that must be considered.

The process model remains a useful tool during construction and when the facility is online. Numerous process variables can be modeled to assist operational staff with decision making that results in the optimal performance of an individual facility with regard to its specific nutrient removal objective(s). Likewise, calibrated process models can be used to develop algorithms and/or guidance to operators in real time.

RESULTS AND DISCUSSION

This section will describe the algorithms developed for specific projects to perform each function listed in the Materials and Methods section. Each system described herein was applied to a plug-flow process, and results might be slightly different for completely mixed systems.

Nitrification

The process controls that are related to nitrification performance and reliability include temperature, flow, wasting rate, solids retention time (SRT), DO concentration, airflow distribution, DO control valve positions, DO probes, and any ammonia probe(s) in the BNR basins. Assuming uninhibited growth, and proper DO control, three key parameters will determine the required SRT to maintain full nitrification—pH, DO and temperature.

The modeling/operations link that would most benefit nitrification in real-time is an ongoing comparison of actual and minimum required SRT for complete nitrification. The comparison can be made several ways depending on the operational philosophy of the plant staff and availability of instruments. For example, if TSS probes are available in the mixed liquor and on the RAS line (assuming the waste activated sludge (WAS) is at the same concentration), real-time SRT can be calculated. The corresponding minimum SRT determined by modeling can be cross-referenced at a programmed interval (operator-adjustable) with the real-time SRT. If the actual SRT drops below the required SRT, then a warning can be issued through the distributed control or SCADA system. If desired, the wasting pumps can be tied into this algorithm and their operation can be controlled to maintain a target SRT. However, in speaking with operators about this functionality, most of them preferred to be warned but not have the operating system make any changes to the wasting.

Some plants do not have online TSS probes, but most measure TSS in the mixed liquor and RAS regularly. In these cases, daily RAS and TSS data from a laboratory test could be extracted from an operations database (i.e., HACHWIMS), and combined with additional information (i.e., number of tanks in service) to estimate the actual SRT. Since RAS concentrations can vary throughout the day, it would be useful to collect and average multiple samples daily, or us a 7-day running average for RAS concentration. This data will not be in real-time, but it would likely be relevant and enhance the reliability of nitrification because it adds an additional check on nitrification. Many medium and large plants compare a target and actual aerobic SRT in daily spreadsheet reports already, but do so independently of the operating system. Automating the process could identify potential issues sooner, such as on weekends when there is less staff present. It should also be noted that wasting could occur directly from the mixed liquor and this would eliminate the need for instrumentation as the SRT is the online basin volume divided by the volume wasted.

BioWin™ process modeling was used to generate a lookup table that presents minimum aerobic SRT as a function of pH, DO and temperature for several recent projects. Online solids analyzers are used to calculate the actual SRT of the system knowing the aerobic volume online and the MLSS concentration. The corresponding minimum required aerobic SRT determined by process modeling for a specific facility is shown in Table 1. The required and actual SRTs are compared and feedback is given to the operator regarding the plant's position relative to the minimum required aerobic SRT.

Table 1. Minimum Aerobic SRT to Maintain Nitrification as a Function of Temperature and DO. This table is specific to BNR process pH values in the 6.5 – 7.0 range.

Temperature (°C)	Dissolved Oxygen Concentration (mg/L)				
	< 0.5	**0.5 - 1.0**	**1.0 - 2.0**	**2.0 - 3.0**	**> 3.0**
10	46.5	18.4	15.3	13.8	13.3
11	40.2	16.9	14.1	12.7	12.2
12	34.8	15.5	13.0	11.7	11.2
13	30.3	14.3	11.9	10.7	10.3
14	26.5	13.2	11.0	9.9	9.5
15	23.2	12.2	10.1	9.1	8.8
16	20.4	11.2	9.4	8.4	8.1
17	18.0	10.4	8.6	7.8	7.5
18	15.9	9.6	8.0	7.2	6.9
19	14.0	8.8	7.4	6.6	6.4
20	12.4	8.2	6.8	6.1	5.9
21	11.0	7.6	6.3	5.7	5.5
22	9.8	7.0	5.8	5.3	5.1
23	8.7	6.5	5.4	4.9	4.7
24	7.8	6.0	5.0	4.5	4.3
25	6.9	5.6	4.6	4.2	4.0
26	6.2	5.2	4.3	4.0	4.0
27	5.5	4.8	4.0	4.0	4.0
28	4.9	4.4	4.0	4.0	4.0
29	4.4	4.1	4.0	4.0	4.0
30	4.0	4.0	4.0	4.0	4.0

This program has been implemented at the North and South Durham Water Reclamation Facilities in Durham, North Carolina. Construction is scheduled for completion in 2017.

Denitrification

The process controls that are related to denitrification include temperature, flow, wasting rate, SRT, the amount of DO entering the anoxic zones, the ratio of biodegradable COD to total Kjeldahl nitrogen (TKN), and NOx-N and/or oxidation-reduction potential (ORP) probes in the BNR basins. Nitrate probes could be located at various locations, including but not limited to the following:

- End of the aerobic zone to measure NOx-N entering the first and second anoxic zones (if applicable).
- Secondary or plant effluent to measure NOx-N leaving the process.
- NOx-N leaving the first and/or second anoxic zones to measure the amount of nitrate removal that occurred in that zone, and to compare the effluent against a targeted setpoint. ORP probes have been used as a low cost substitute for NOx-N probes at the end of the first anoxic zone, but NOx-N probes are a direct measure of the parameter of interest.

For various reasons, a facility may elect not to locate probes in all of these places, but rather, this list represents potential locations for probes. Also, the total number of probes installed at each location will vary. For example, the number of parallel trains at the plant will determine how many instruments to purchase and is site-specific.

Carbon feed systems typically have at least three modes of operation: manual mode, flow-paced mode, and nutrient-paced mode. In manual mode, the operator sets a constant flow rate for the feed pumps at the operator workstation. In flow-paced mode, carbon feed is typically paced to influent flow with an assumption about the concentration of nitrate entering each anoxic zone.

The nutrient-paced mode is the most efficient mode as it utilizes a feedforward and feedback loop to accurately pace carbon addition to carbon demand. The nitrate concentration at the end of the aerobic zone and the flow (influent, RAS, and internal recycle rate—if carbon is being added to the first anoxic zone) are used to estimate the mass of nitrate entering both the first and second anoxic zones. Using this mass, the amount of carbon required to remove the calculated mass of nitrate is calculated and used to set the carbon feed pump rate. The mass of oxygen entering the anoxic zones can also be accounted for in setting the carbon pump feed rate. This is the feedforward portion of the algorithm. The user also sets an effluent nitrate concentration setpoint for each anoxic zone, and the anoxic zone effluent nitrate probes are used to adjust the speed of the carbon feed pumps accordingly using a proportional integral derivative (PID) loop. This constitutes the feedback loop.

It is possible to enhance the traditional flow-paced mode of addition by incorporating the modeled diurnal nitrate load into the calculation of carbon required. Therefore, instead of assuming a constant nitrate concentration entering the anoxic zone the concentration of nitrate varies throughout the day in the calculation, which potentially reduces the amount of carbon used. This is a significant enhancement that modeling makes possible, and would be advantageous for a facility that has limited or no NOx-N instrumentation.

The nutrient-probe-based carbon addition with its feedforward and feedback loops are an example of real-time process control. A few additional process enhancements that could be added are as follows:

- Develop an algorithm to modify the internal recycle rate to also maintain a nitrate setpoint leaving the 1st anoxic zone, i.e., 0.5 mg/L NOx-N. This would ensure that the first anoxic zone is not limited by nitrate load or by carbon, and retention time would be the only constraint. This also allows full utilization of the first anoxic volume and associated influent carbon. ORP probes have been used in place of NOx-N probes in this application.
- Feedback-ammonia-based DO control can also reduce energy and supplemental requirements and achieve simultaneous nitrification and denitrification. In this mode of operation an ammonia probe is located near the end of the aerobic zone. The target DO concentration is set to a low value so long as the ammonia concentration measured at the probe does not exceed a desired setpoint, say 0.5 – 1.5 mg/L, depending on the permit requirements., Denitrification can also occur under low DO conditions, which takes advantage of the remaining slowly biodegradable carbon not consumed in the first anoxic zone and also uses cell decay as a carbon source, reducing the overall need for purchased carbon. Simultaneous nitrification and denitrification (SND) operation requires higher hydraulic retention times (HRTs) and solids inventories than conventional BNR, so facilities that are not at capacity are best suited for this type of operation. It is anticipated that SND operation would occur during warm weather months. As the DO decreases, the nitrification rate in the aerobic zone decreases, therefore operating in this mode in cold weather may be problematic. Several facilities have the ability to operate or are operating in this mode including the Nansemond Treatment Plant in Suffolk, Virginia, the Neuse River WWTP in Raleigh, NC, and the North and South Durham WRFs in Durham, NC.
- An alternative to the ammonia-based DO control method is to install an ammonia probe and mixers in the last portion of the aerobic zone to allow the last DO control zone to achieve simultaneous nitrification and denitrification. The addition of the mixers would allow operation at a DO concentration that is not dependent on required minimum mixing airflows, and hence reduce the use of supplemental carbon. This differs from the ammonia-based-DO control concept in that only the last portion (typically 20 or 25 percent) of the aerobic volume is operated at low DO values. An alternative to mixers is to modify the aeration program to include short periods of no air. This strategy could be used year-round and would be suitable even for plants operating at or near capacity. The Hillsborough WWTP in Hillsborough, NC was operating in this mode for a period of time absent the ammonia probe.

Phosphorus Removal

Phosphorus removal in this section refers to biological phosphorus removal (BPR), chemical-only phosphorus removal, and a combination of BPR and chemical addition. The process controls that are related to BPR include the DO, readily biodegradable COD, volatile fatty acid, and nitrate concentrations entering the anaerobic zone; airflow rates throughout the aerobic zone; and orthophosphate probes at the reaeration channel, end of the aerobic zone, and/or end of anaerobic zone (if applicable). A phosphorus removal process can also include chemical feed pumps. For example, ferric chloride or alum can be added as needed to meet a desired orthophosphate concentration based on an algorithm that uses flow measurements and real-time probe-based orthophosphate concentrations to calculate the mass of coagulant required.

From a real-time process control perspective, an algorithm could be written to implement some operational changes when the orthophosphate concentration at the end of the aerobic zone begins to increase. The potential modifications include but are not limited to the following ideas, keeping in mind that these changes need to be viewed in the broader context of impacts to nitrogen removal as well (e.g., increasing the DO concentration can have impacts on anoxic zone performance):

- Increasing the aerobic fraction of the process to enhance uptake.
- Adding more carbon to the second anoxic zone to reduce the nitrate concentration in the RAS and make conditions more favorable for phosphorus accumulating organisms (PAOs).
- Put more anaerobic volume into service and/or reconfigure existing volumes to allocate a higher fraction of the reactor to anaerobic processes, although this may not be helpful if phosphorus uptake is an issue.
- Add coagulant at the secondary clarifier influent to adsorb and settle orthophosphate.
- Increase the airflow rate (and hence DO concentration) in the aerobic zone to enhance uptake.
- Reduce the SRT within some allowable tolerance. Excessive SRT can decrease the efficiency of BPR, and this needs to be balanced with effective nitrification.
- Increase the reaeration volume and/or DO to enhance uptake and reduce secondary release in the clarifiers.

Chemical addition can also be paced to an upstream orthophosphate analyzer reading (feed-forward control). It is best to locate the orthophosphate analyzer upstream of but as close as possible to the point of chemical addition. In this way, stoichiometric values of metal salt required to meet a desired downstream orthophosphate concentration can be estimated. One should allow the user to modify the molar ratio of metal salt to orthophosphate in order to enhance the accuracy of the process.

Solids Separation

All activated sludge plants, regardless of whether or not they have a nutrient removal objective, must achieve adequate solids separation to perform well and meet their objectives. The solids separation process is a function of the following independent variables: mixed liquor suspended solids concentration, X; hindered settling coefficient, K; settling velocity, Vo; forward flow, Q; return activated sludge (RAS) flow, Qras; and the number of clarifiers in service, represented by the total surface area, SA. Sludge volume index (SVI) is typically measured by plant operators and correlations exist to relate Vo and K to SVI. Because Vo and K are more difficult and time consuming to measure, the developed solids separation algorithm is based upon SVI as a proxy for Vo and K.

State point analysis (SPA) is commonly used to provide a simplified determination of whether or not a clarifier will fail under a given set of conditions. Applying this tool to real-time process control involves solving systems of equations with advanced mathematics and software tools.

While that is a long-term objective, a simplified empirical approach was applied on a recent project. A matrix of potential Q, Qras, X, and SVI values was modeled using SPA. The number of clarifiers in service was varied to cover the range of options. Next, each SPA was solved by adjusting X manually until the clarifier was just at the brink of failure but not failing.

A multivariable linear regression (MVLR) equation was developed to represent the required clarifier surface area as a function of variables the plant could measure in real-time: MLSS, Q, Qras, and sludge volume index (SVI), which is measured daily as is a function of Vo and K. Site specific testing was conducted at the facility to determine Vo and K.

After simulation of the matrix of alternatives, the following equation was developed using MVLR.

Required Clarifier Surface Area, ft2 = -193,090 + 981*Q + 909*SVI - 530*Qras + 34.2*MLSS

The associated R^2 for this regression is 0.87. The number of observations is 63. All of the variables were significant at the 99th percent confidence limit. This equation can easily be integrated into the SCADA system and provide real-time information as to the number of clarifiers that should be in operation (see sample screen Figure 1). In this example, the reference plant has ten 100-foot diameter secondary clarifiers and four 160-foot diameter secondary clarifiers.

Figure 1. Sample Screen for Secondary Clarifier Guidance Program

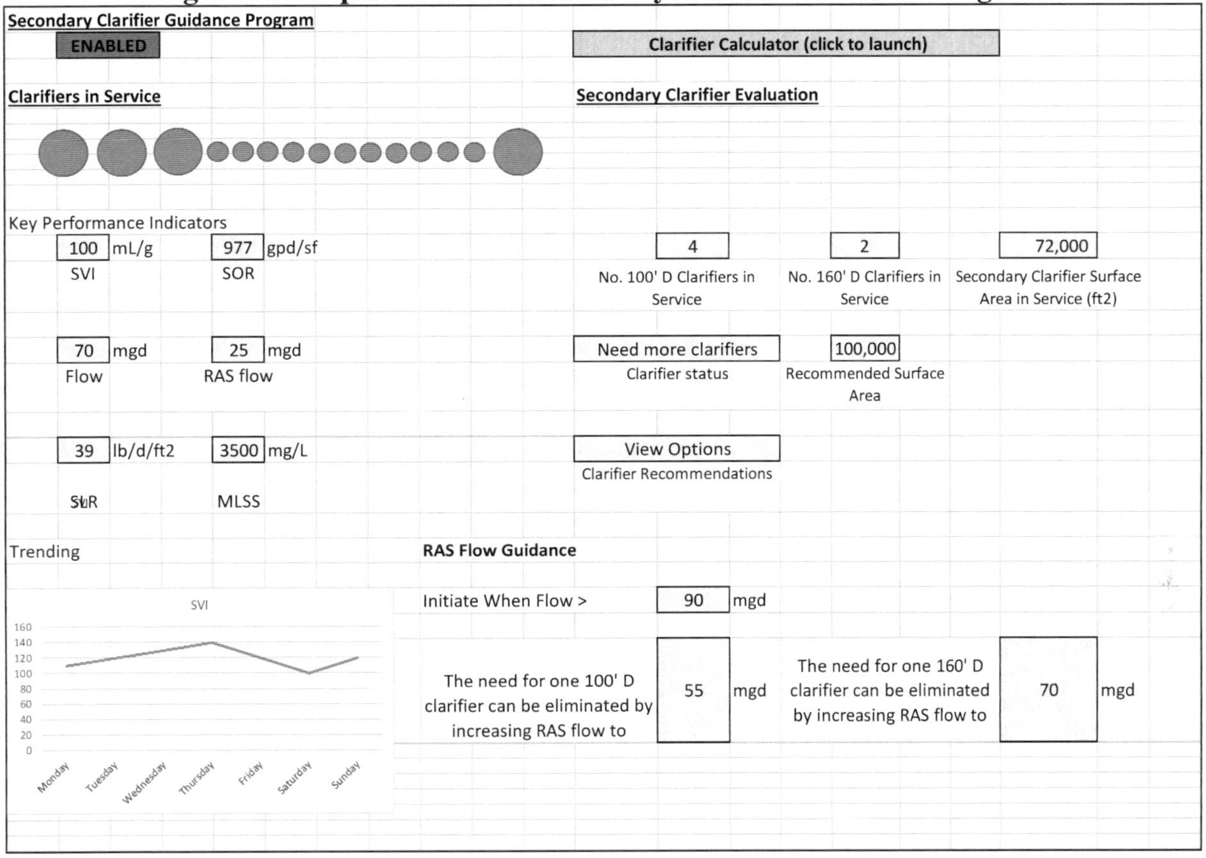

This equation will also be used to provide prompts to the user about how an increase in RAS capacity could improve settling since insufficient RAS flow can limit capacity during specific events (see lower right-hand corner of Figure 1 under "RAS flow guidance).

Another feature of this program is a built-in calculator that the operators can use to determine the number of clarifiers needed in service based upon user-entered values for Q, Qras, MLSS concentration, and SVI. This can be a useful training and planning tool.

Creating this algorithm is relatively low cost, does not require a calibrated process model, and would provide a useful second opinion to operators about whether the amount of clarifiers in service ought to be changed. This information may be particularly useful during wet weather. However, it should be emphasized and noted that this program and others like it are not intended to be a substitute for common sense. Operators should continue to walk around the plant, make observations, and use their best judgment when making decisions about taking clarifiers in and out of service.

Nitrogen Load Equalization

Plants with diurnal equalization and/or wet weather equalization may wish to control the filling and emptying of this basin volume with the objective of maintaining a consistent ammonia load to the downstream process. This can reduce the variability in required airflow, stabilizing blower

Demonstration of an up-flow biological aerated filter system for nitrogen removal from non-nitrified secondary effluent

Liyan Jin, Young J. Eum, Joon H. Min, and Hongkeun Park*

BKT Co. Ltd, Anaheim, California.
*Email: hkp@bkt21.com

ABSTRACT

BKT's BioFiltration (BBF) is an up-flow biological aerated filter (BAF) system that uses expanded polypropylene (EPP) media for maximizing biomass growth and particle filtration. In this study, a two-stage BBF pilot system was tested for nitrogen removal from a non-nitrified secondary effluent of a high purity oxygen activated sludge (HPOAS) system. The pilot was operated under various flow rates to identify the nitrification and denitrification capacity of the biofilter. Based on test results, a maximum design NH_4-N loading of 1.5 kg N/m^3/d (95 lb-N/1000ft^3/d) was recommended for achieving below 2 mg NH_4-N/L of effluent. In addition, the system was able to receive NOx-N loading rate up to 4.4 kg NO_3^--N/m^3/d (268 lb-NO_3^--N/1000ft^3/d) and achieve up to 99% nitrate removal in the denitrification stage with acetic acid as an electron donor. In addition, substantially low effluent turbidity from the BBF system would enable to eliminate or reduce the additional cost for post treatment while implemented in water reuse.

KEYWORDS: Nitrogen removal, biological aerated filter, non-nitrified secondary effluent

INTRODUCTION

The Biological Aerated Filter (BAF) is a widely used fixed-film treatment process for the removal of organic matter and nitrogen from wastewater. Not only does the BAF provide biological treatment but it also produces a high quality filtered effluent suitable for water reuse. BKT's BioFiltration (BBF) is an up-flow BAF that uses expanded polypropylene (EPP) media to support biomass growth and filtration functions with high energy-efficiency.

In September 2014, BKT Co. Ltd established a 4-gallon per minute (gpm) scale pilot demonstration using the BBF system to treat secondary effluent at the Los Angeles County Sanitation District's Joint Water Pollution Control plant (JWPCP) in California. The JWPCP uses a high-purity oxygen-activated sludge (HPOAS) system as a secondary treatment, which is typically effective at organic removal. However, its nitrification capacity is limited due to its short solid and hydraulic retention time and limited alkalinity due to CO_2 accumulation inside the HPOAS system (*Neethling et al.*, 1998). Thus, the secondary effluent at the JWPCP contained high NH_4-N with limited BOD.

With California facing a devastating drought, exploiting all available water sources including advanced treated wastewater for groundwater replenishment is essential to ensuring a stable water supply. April 1, 2015, California Governor Brown issued an executive order for first ever statewide mandatory 25% water cut due to increasing concern of the severe drought. In this severe drought situation, advanced treatment of tertiary wastewater for groundwater replenishment would potentially provide new indirect potable water source. Here, excessive nitrogen removal from HPOAS will be an essential pretreatment for the advanced water treatment process. Thus, the main objective of this study includes the evaluation of a low-energy, up-flow BBF system for nitrification and denitrification performance using the HPOAS effluent under various operating conditions, and eventually determination as one of the potential pretreatment options of advanced water treatment for indirect potable reuse.

METHODOLOGY

A pilot-scale BBF system was installed at the JWPCP consisting of two 0.48 m (19-inch) diameter acrylic columns with 1.8 m (72-inch) media depth (Figure 1 & 2). The first column in the BBF system was operated under oxic conditions to oxidize NH_4-N to NO_x-N, and the second column was operated under anoxic conditions to reduce NO_x-N to nitrogen gas with acetic acid as an electron donor.

Figure 1. BBF pilot at JWPCP

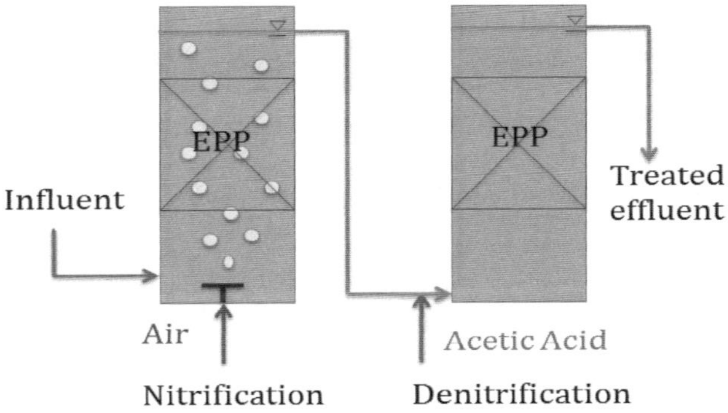

Figure 2. Conceptual Process flow diagram of the BBF pilot system

Summarized in Table 1 are characteristics of the influent water introduced to the nitrification column and denitrification column during the demonstration period. The average NH_4-N concentration in the nitrification influent was 38 mg/L with non-detectable NO_3-N and NO_2-N. The average alkalinity was ~ 333 mg $CaCO_3$/L, which was sufficient for the nitrification requirement. The average NO_3-N concentration in the denitrification influent was 31.2 mg N/L and the NO_2-N concentration was 2.7 mg N/L. The influent COD was ~48mg/L, mostly in the form of non-biodegradable organics. Thus, an external electron donor was required for denitrification.

Table 1. Feed water characteristics to the nitrification and denitrification columns

Parameters	Units	Nitrification column			Denitrification column		
		Avg.	Min	Max	Avg.	Min	Max
NH_4-N	mg N/L	38.1	34.9	40.5	4.2	1	12.4
NO_3-N	mg N/L	ND	-	-	31.2	20.8	37
NO_2-N	mg N/L	ND	-	-	2.7	0.1	8.54
pH	-	7.24	7.1	7.4	7.01	6.9	7.1
Alkalinity	mg/L as $CaCO_3$	333	218	380	118	92	130
COD	mg/L	58	43	75	48	41	56
BOD*	mg/L	3.5	3	4	-	-	-
TSS*	mg/L	10.5	9	12	5.0	3.1	7.9
Turbidity	NTU	4.4	2.3	16	1.5	0.8	2.5

*Based on average plant performance of the JWPCP from September to December, 2014

The pilot system was operated under four different flow rates, 6.4, 9.8, 12.5, and 15.1 L/min (1.7, 2.6, 3.3, and 4.0 gpm) each of which corresponded to empty bed contact times (EBCT) of 52, 34, 27, and 22 minutes, respectively. Prior to this demonstration, the nitrification column was seeded with JWPCP secondary effluent during one month of acclimation period. Operating conditions for the nitrification column are summarized in Table 2.

Table 2. Operating conditions for nitrification

Parameters	Units	1.7 gpm	2.6 gpm	3.3 gpm	4.0 gpm
EBCT	min	52	34	27	22
HSLR*	m/hr (gpm/ft^2)	2.3 (0.9)	3.3 (1.3)	4.3 (1.7)	5.0 (2.0)
NH_4-N loading rate	kg N/m^3/d (lb N/1000 ft^3/d)	1.1 (65)	1.5 (95)	2.0 (124)	2.4 (146)
SS loading rate	kg TSS/m^3/d (lb TSS/1000/ft^3/d)	0.28 (17)	0.44 (27)	0.59 (36)	0.69 (42)
Backwash trigger pressure	kPa (psi)	43.4 (6.3)	43.4 (6.3)	44.8 (6.5)	48.2 (7.0)
Pressure drop in the media layer	kPa (psi)	6.2 (0.9)	6.2 (0.9)	7.5 (1.1)	10.3 (1.5)
DO	(mg/L)	5.0~6.5			
Temperature	°C (°F)	25~33 (78~93)			

*HSLR: Hydraulic surface loading rate

Operating conditions for the denitrification column are summarized in Table 3. The NO_3-N concentration in the influent of the denitrification column was largely dependent on how well the nitrification column performed ammonium oxidation. In the JWPCP secondary effluent, the average NH_4-N concentration was 38mg/L, of which the nitrification column was able to oxidize 93.0 ± 6.6 % to NO_x-N during the operation period. The denitrification column was operated under 6.4 L/min (1.7 gpm) flow rate during the startup and optimization periods to determine the optimal electron donor dosage. After optimization, as shown in the test conditions in Table 3, the system was operated under three different flow rates, 9.8, 12.5, and 15.1 L/min (2.6, 3.3, and 4.0 gpm) with empty bed contact time (EBCT) of 34, 27, and 22 minutes, respectively. A spike test using two different NO_3-N concentrations of 56 mg/L and 68 mg/L was also conducted at the 4.0 gpm flow rate.

Table 3. Operating conditions for denitrification

Parameters	Units	Start up	Normal Operation			Challenge test @ 4.0 gpm	
		1.7 gpm	2.6 gpm	3.3 gpm	4.0 gpm	56 mg/L NO_x-N	68 mg/L NO_x-N
EBCT	min	52	34	27	22	22	22
HSLR	m/hr (gpm/ft^2)	2.3 (0.9)	3.3 (1.3)	4.3 (1.7)	5.0 (2.0)	5.0 (2.0)	5.0 (2.0)
NO_x-N loading rate	kg N/m^3/d (lb N/1000 ft^3/d)	1.1 (65)	1.4 (89)	1.6 (100)	1.9 (113)	3.3 (220)	4.4 (268)
Backwash trigger pressure	kPa (Psi)	44.8 (6.5)	48.2 (7.0)	51.7 (7.5)	58.6 (8.5)	58.6 (8.5)	58.6 (8.5)
Pressure drop in the media layer	kPa (Psi)	7.6 (1.1)	11.0 (1.6)	13.8 (2.0)	17.9 (2.6)	17.9 (2.6)	17.9 (2.6)
Temperature	°C (°F)	25~33 (78~93)					

The BBF pilot system was equipped with a supervisory control and data acquisition (SCADA) system. The flow rate, pressure, backwash volume, pH, temperature, TSS, and turbidity were monitored through online sensors via the human machine interface (HMI) shown in Figure 4. NH_4-N, NO_3-N, and NO_2-N levels in the influent and effluent were determined from grab samples twice per week, using standard method SM4500-NH_3B, SM4500-NO_3 E, and SM4500-NO_2 B respectively.

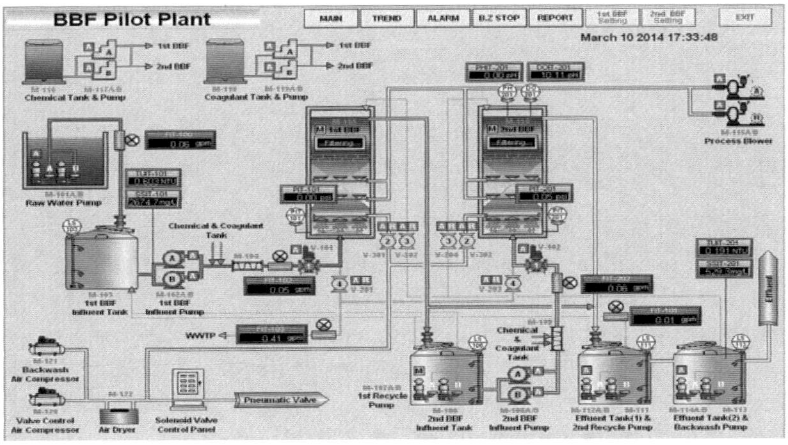

Figure 4. HMI in the BBF pilot to control and monitor the system

RESULTS

Nitrification Performance

At 52 min EBCT, BBF was able to oxidize the influent NH_4-N to non-detectable levels (Figure 5). The effluent NH_4-N increased when the EBCT was gradually reduced to 22 min. Under the condition of 22 min EBCT, the effluent NH_4-N became very sensitive to the influent NH_4-N concentration, possibly because the nitrification column had reached maximum capacity of the ammonium oxidation under such operational condition.

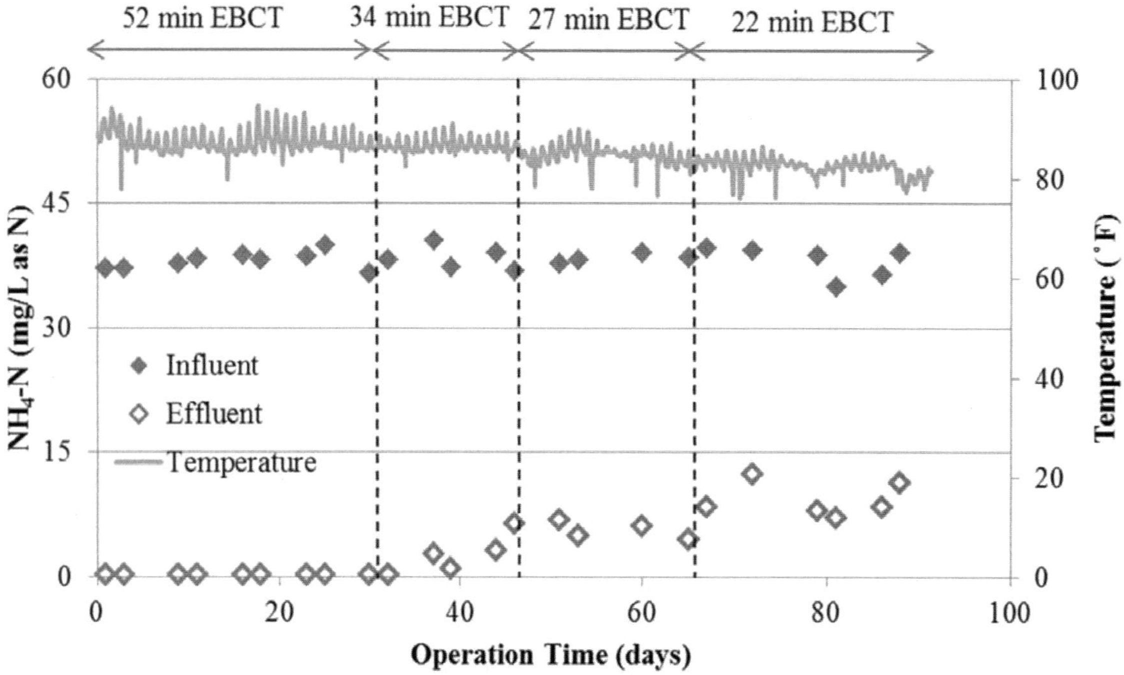

Figure 5. Influent and effluent NH_4-N with different EBCTs

The correlation between effluent NH_4-N and NH_4-N loading rates demonstrated that the effluent NH_4-N concentration remained less than 1 mg NH_4-N/L until the NH_4-N loading rate reached 1.5 kg N/m^3/d (95 lb-N/1000ft^3/d). When the NH_4-N loading rate increased above 1.5 kg N/m^3/d (95 lb-N/1000ft^3/d), a strong positive correlation (R^2=0.9002) was observed between the NH_4-N loading rate and the effluent NH_4-N. It is noteworthy that two data points shown with rather higher effluent NH_4-N than predicted by the trend line was potentially owing to limited alkalinity in the influent (218 and 264 mg $CaCO_3$/L) on the sampling days.

NUTRIENT RECOVERY AS A SUSTAINABLE NUTRIENT MANAGEMENT TOOL

Wendell Khunjar, PhD, PE[1,*]
Ronald Latimer, PE[1]
Samuel Jeyanayagam[2], PhD, PE, BCEE
Chirag Mehta, PhD[3]
Damien Batstone, PhD[3]

[1]Hazen & Sawyer, PC
[2]CH2M Hill
[3]University of Queensland, Australia

*wkhunjar@hazenandsawyer.com

ABSTRACT

This paper presents a state of the science review of nutrient recovery technologies with a special emphasis on bridging the knowledge gap currently faced by water resource recovery facilities when considering nutrient recovery for nutrient management. In this extended abstract, we suggest technology classification into accumulation-release-extraction and discuss challenges against and drivers for the implementation of these technologies. We also present a brief economic evaluation of the value of nitrogen and phosphorus products.

KEYWORDS

Nutrient Removal/Recovery, Struvite, Nitrogen, Phosphorus, Plants of the Future, Cities of the Future, Sustainable Nutrient Management, Resource Recovery

INTRODUCTION

Nitrogen (N) and phosphorus (P) are essential elements of all life forms and are extensively used in synthetic fertilizer production. These nutrients once incorporated into crops are ingested and reintroduced to the environment via liquid and solid wastestreams. To avoid the accumulation of these nutrients in the environment, we typically remove N and P prior to discharge to a water body. In this scenario, we supply energy and other non-renewable resources to constantly replenish nutrient supply for agricultural uses and then further supply energy and non-renewable resources to remove these nutrients from wastewater before discharge to the environment. This approach to nutrient management assumes a linear usage cycle which inherently posits that energy and resources are cheap and renewable. This is not the case.

Instead, recovery and reuse of nutrients provides us with an opportunity to reduce our reliance on nutrient cycling and fundamentally change the way that we view wastestreams. It should be acknowledged that nutrient recovery and reuse is not a new concept. It has been applied in different forms in the past e.g., land application of biosolids and reuse of secondary effluent for irrigation; however, extraction of a chemical nutrient product with low organic matter content, defined here as extractive nutrient recovery, has not been widely applied within the wastewater treatment industry. This paper reviews the current state of knowledge on extractive nutrient recovery. The information presented herein was generated as part of the ongoing Water Environment Research Foundation (WERF) project titled: Towards a Renewable Future: Assessing Resource Recovery as a Viable Treatment Alternative (Latimer et al., 2012a).

METHODS

A review of nutrient recovery technologies was performed by accessing data from peer reviewed journal articles as well as from responses by commercial technology providers to requests for information as part of the Water Environment Research Foundation Nutrient Recovery Challenge (Latimer et al., 2012). Historical records (40 years) pertaining to the sale of fertilizer into the agricultural industry were used to project market trends for seven recoverable products (magnesium ammonium phosphate, calcium phosphate, iron phosphate, phosphoric acid, ammonium nitrate, ammonium sulfate).

DISCUSSION

Basic Components of Extractive Nutrient Recovery

The nutrient concentration in the influent to municipal wastewater treatment plants (WWTPs) typically ranges from 10 to 50 mg N/L for N and from 1 to 10 mg P/L for P. Since the efficiency of extractive nutrient recovery technologies is lower at these concentrations, it is proposed that that extractive nutrient recovery will most likely be viable if performed within a three step framework (Figure 1):

1. Accumulation of nutrients to high concentrations,
2. Release of nutrients to a small liquid flow with low organic matter and solids content,
3. Extraction and recovery of nutrients as a chemical nutrient product.

In this approach, biological, physical, thermal, and chemical mechanisms can be used in each step to transform the nutrients present in domestic wastewater to a chemical nutrient product that has a secondary market value. One of the advantages to using this approach is that multiple options for each stage of treatment can be developed and optimized separately; thereby allowing utilities to select the most appropriate solution for their needs. It is also possible that some utilities may not need capital investment for all three processes since existing infrastructure can be reused. A complete review of state of the art options available for the accumulation, release and extraction framework is provided as part of the WERF Nutrient Recovery Challenge (Latimer 2012). This manuscript provides a brief description of the options.

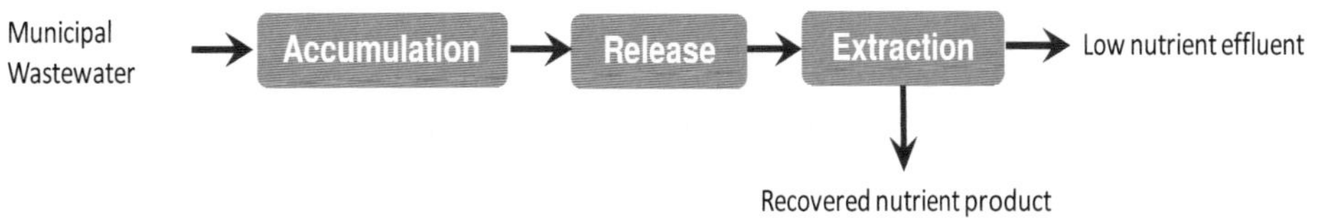

Figure 1. Components of Extractive Nutrient Recovery (Latimer et al., 2012a)

Nutrient Accumulation Options

Nutrient accumulation technologies focus on concentrating the low nutrient content of municipal wastewater (Table 1). This can be accomplished using biological (N and P), physical (N and P) and chemical (mainly P) techniques. Biological accumulation techniques center around microbial accumulation in which specially adapted microorganisms (e.g. microalgae, polyphosphate-accumulating bacteria (PAOs), purple non-sulfur bacteria (PNSB), cyanobacteria) are able to uptake (N and P) and store nutrients (P). Plants such as duckweed can also be used as part of passive nutrient treatment/accumulation strategies. Research has shown that biological systems can remove between 70 to 90 % of N and P from wastestreams and are effective for treating a wide range of nutrient concentrations including the dilute content of nutrients typically associated with municipal WWTPs. Biological processes have already been extensively applied for wastewater treatment, with over 84% of WWTPs in the USA employing some form of biological process (CWNS, 2004). Since these processes are expected to be further employed as effluent nutrient regulations become more stringent, this represents a fortuitous opportunity for the extractive nutrient recovery field, as it becomes possible to stage implementation of the extractive nutrient recovery over multiple years, with the first step initially being use of biological nutrient accumulation processes. Key requirements for using biological accumulation processes are an effective solid-liquid separation process like clarification or membranes to allow recovery of the nutrient rich biomass, as well as an appropriate release technology for subsequent processing.

Chemical accumulation using metal salt addition is another option that can be used to help accumulate nutrients (mostly P). In this process, the metal salt reacts with soluble P to form an insoluble phosphate complex, which is solid and can then be physically separated from the wastestream. Aluminum and iron solutions are often used for this purpose and can achieve greater than 85% P removal from the dilute stream, with the chemical solids being separated during clarification or filtration. One of the key challenges with using chemical accumulation techniques is that the chemically accumulated P is less useful because of the high metal salt content of the final product. This restricts its use in agricultural applications. Chemical accumulation is widely applied at domestic WWTPs. Therefore, it may be possible to accomplish extractive nutrient recovery at existing WWTPs by implementing suitable release and extraction processes to process the solids generated.

Another strategy that can be used to accumulate nutrients from the mainstream flow is adsorption and/or ion exchange. These processes can be used to remove N and P from dilute wastestreams, with removal efficiencies ranging between 50-90%. In this approach, a sorbent or ion exchange material is packed into a column. As the wastewater flows through the column, N or P (depending on the material) is either sorbed or chemically attracted to specific sites on the material. This approach has been used at pilot and full-scale for in tertiary filtration applications to help remove phosphorus. One of the biggest challenges with using adsorption and/or ion exchange for nutrient accumulation is the regeneration step which requires use of costly chemical brines and the need for replacement of spent adsorption media. Therefore, it may not currently be economically feasible to implement adsorption and/or ion exchange at larger plants. Embryonic research at the lab-scale is investigating the use of bio-regeneration as a method to help reduce costs associated with regeneration and replacement of sorbent material. It is expected that these processes will continue to become more important as WWTPs are increasingly asked to achieve effluent TP limits below 0.1 mg/L. In this scenario where adsorption and/or ion exchange is used in a tertiary filtration step, it may be possible to harvest the nutrient from this material for beneficial reuse.

Table 1. Summary of Technologies Suitable for Nutrient Accumulation at Full-Scale WWTPs (From Latimer et al., 2012a)

		Operating conditions		Pre-treatment required	Chemical Input	Commercial process	Nutrient(s) Accumulated
		Temp. (°C)	pH				
Embryonic	Microalgae	15 - 30	7.5 – 8.5	-		Lemna Technologies	N and P
	Cyanobacteria	5 - 40	6.5 - 8	-	Carbon source	-	N and P
Innovative	Adsorption/Ion exchange	NA	<8.0	Solid-liquid separation	Adsorbent, Regeneration solution	P-ROC, RECYPHOS, PHOSIEDI, RIM NUT, BIOCON	N and P
Established	EBPR	5 - 40	6.5 - 8	-	Carbon	-	P only
	Chemical	25 - 40	6.5 - 10	-	Metal salts (Al or Fe)	-	P only

embryonic – technologies that are in the developmental stage (bench/pilot scale)
innovative – developed technologies with limited full-scale application
established - commercially viable technologies with a proven history of success

Nutrient Release Options

Release technologies allow us to recover the nutrients into a low flow high nutrient content stream with minimal solids content, which can be used for extraction processes (Table 2). Release technologies typically employ some combination of biological, thermal, chemical or physical processes. Biological release is the most commonly used process that has been implemented at WWTPs. In this process, the biomass is broken down and the organic carbon, nitrogen and phosphorus are converted to carbon dioxide and methane, ammonia and soluble phosphorus, respectively. Biological release can occur under anaerobic conditions (e.g., anaerobic digestion) or under aerobic conditions (e.g., aerobic digestion), and the extent of nutrient release is dependent on the conditions employed during digestion. After biological release, the effluent streams can contain greater than 100 mg P/L and 1000 mg N/L, as well as particulate matter that must be removed. One of the biggest advantages of using an anaerobic biological release process is the opportunity to not only recover nutrients, but also the biogas. Indeed, anaerobic digestion has been extensively applied as a cost effective option for reducing the solids content of primary and waste activated sludges. Recent work performed as part of the WEF funded a national survey of anaerobic digestion and biogas use indicates that close to 25% of all WWTPs greater than 1 mgd current employ anaerobic digestion (Beecher and Qi, 2012). Also, as the industry increasingly aims to achieve energy neutrality, it is expected that anaerobic digestion processes will be increasingly implemented. In a WWTP that already employs biological and/or chemical accumulation followed by biological release using anaerobic digestion, implementation of extractive nutrient recovery would simply require the installation of the extraction step. This latter upgrade has been done at several full-scale facilities in the US and Europe.

Another option for biological release is enhanced P release processes. In this approach, phosphorus that has been biologically accumulated by PAOs is selectively released from these microorganisms (in WAS). One variation of this process is called WASStrip™ and has been patented by Clean Water Services. Enhanced P release processes can be used in combination with anaerobic digestion to help minimize the O&M requirements associated with nuisance struvite/vivianite formation that is associated with the operation of biological accumulation processes.

Another commonly used practice for releasing nutrient bound in biomass and chemical sludge is thermochemical processes coupled with chemical release. Thermochemical options can include wet oxidation, incineration, gasification or pyrolysis. In these processes, high temperature is used to destroy organic material and produce a solid product containing P, which can then be chemically released. It is important to note that N is typically lost through gaseous emissions during these processes. As a result, thermochemical processes are most suitable for extractive nutrient recovery of P. Great advances in thermochemical processes have been achieved over the past 5-10 years and they have emerged as innovative alternatives to using anaerobic digestion for managing solids at municipal WWTPs. Chemical release of nutrients can from the char, ash, biosolids (digested, dewatered activated sludge) or undigested sludge can then accomplished using concentrated acids or bases at temperatures between 100 to 200^{0}C. The liquid stream is then subjected to extraction technologies to recover the nutrients. Coupling these processes to the extractive nutrient recovery framework could allow facilities who have limited expansion capacity or are limited in disposal options for biosolids to still become resource recovery treatment plants. One of the biggest challenges associated with this release option is presence of heavy metals in the liquid stream that is generated. Post-treatment will be required to limit the heavy metal content of the chemical product. This additional treatment steps can make this option economically challenging to implement at the current market value of the chemical nutrient products that are typically recovered from these processes.

Table 2. Summary of Technologies Suitable for Nutrient Release at Full-Scale WWTPs (From Latimer et al., 2012a)

		Operating conditions		Pre-treatment required	Chemical Input	Commercial process	Nutrient(s) Released
		Temp. (°C)	pH				
Innovative	Chemical extraction	25-200	1 - 3	-	Leaching solution (sulfuric acid, hydrochloric acid, nitric acid, citric acid, oxalic acid, EDTA)	SEABORNE, STUTTGARTER VERFAHREN, LOPROX/PHOXAN, CAMBI, KREPCO, BIOCON, SEPHOS, AQUARECI, SESAL-PHOS, PASCH	N and P
	Thermochemical	150 – 1100	all	Temperature adjustment	-	MEPHREC, ASHDEC, THERMPHOS	P only
	Enhanced P Release from WAS	5 - 40	6.5 - 8	-	Carbon (volatile fatty acids)	WASStrip, PRISA	P only
Established	Anaerobic digestion	35 – 60	6.5 – 7.5	-	-	-	N and P

Nutrient Extraction Options

The next step of the extractive nutrient recovery process is the extraction and recovery of chemical nutrient products from concentrated liquid streams (Table 3). These extraction processes can be inserted downstream of accumulation or release technologies. At present, each extraction technology requires pre-treatment to reduce the solids content and/or change the temperature or pH of the liquid stream to a suitable condition for the extraction technology. One example of a commonly applied extraction technology is chemical crystallization. In this process, the soluble nutrient is precipitated and recovered as crystalline products. Products that can be generated by this process include are struvite (magnesium ammonium phosphate) and calcium phosphate (hydroxyapetite, P only). In the case of struvite formation, the pH and concentration of magnesium, phosphate and ammonium is controlled to allow the precipitation of the chemical nutrient product which is then separated from the liquid stream via gravity or mechanical separation. Further drying and processing of the product is also commonly performed.

There are multiple variations of this chemical crystallization process that have been commercialized. In each of these systems, soluble P removal efficiencies up to 90% and ammonia removal efficiencies up to 30% can be expected if struvite is the product of choice.

Addition of magnesium chloride or hydroxide as well as caustic (NaOH) is typically needed for the process to proceed. Precipitation of calcium phosphate is also possible with the addition of calcium instead of magnesium. There are over 49 full-scale installations of these processes throughout the world. For municipal WWTPs, these processes are commonly installed downstream of biological accumulation (e.g., EBPR) and biological release technologies (anaerobic digestion).

In order to recovery N only products, liquid-gas stripping of ammonia can be used. In order to extract the ammonia from the nutrient rich liquid stream, it is necessary to raise the pH above 9.3 and increase the temperature above 80^0C. Air can then be bubbled through the mixture. This forces the soluble N into the gas phase. This gas phase ammonia is then recovered by bubbling the nutrient rich gas into sulfuric or nitric acid. This process produces ammonium sulfate or ammonium nitrate, respectively. In this process, ammonia removal efficiencies up to 98% is possible; however, the relatively high cost of this method makes this option challenging for implementation in wastestreams with N content less than 2000 mg/L. As thermal hydrolysis processes like CAMBI™ and Exelys™ are increasingly implemented at municipal WWTPs, liquid-gas extraction of ammonia will become more technically feasible; however, the ultimate implementation of this process will be dependent on the cost of the products that will be recovered. While this process is established in industrial applications, it has not been extensively applied for recovery of N from municipal WWTPs.

Electrodialysis represents an embryonic extraction technology that allows for the recovery of all ions from nutrient streams at nutrient concentrations below 2000 mg/L. It represents a highly promising technology to the extractive nutrient recovery field. In this process, an electrical current is used to separate anion and cations across an ion exchange membrane.

Table 3. Summary of Technologies Suitable for Nutrient Extraction at Full-Scale WWTPs (From Latimer et al., 2012a)

		Operating conditions		Pre-treatment required	Chemical Input	Commercial process	Nutrient(s) Extracted
		Temp. (°C)	pH				
Embryonic	Electrodialysis	10 - 40	< 8.0	Solid-liquid separation	Electricity	GE Water	N and P
Innovati ve	Liquid-gas stripping	>80°C	> 9.5	pH and temperature adjustment	Caustic	ThermoEnergy Castion™	N only
Established	Struvite crystallization	25 - 40	8 – 9	Solid-liquid separation	Caustic, Magnesium or Calcium	PHOSTRIP, PRISA, DHV CRYSTALACTOR , CSIR, KURITA, PHONIX,	N and P

						OSTARA, BERLINER VERFAHEN, FIX-PHOS	

Considerations for Chemical Nutrient Products

At present, commercial technologies for extractive nutrient recovery primarily produce chemical nutrient products that are used in agricultural applications. This is because 85% of all nutrient products are associated with agronomy. Since food demand is expected to rise with an increasing global population, it is expected that demand for chemical nutrient products will also increase. This represents an opportunity for the wastewater treatment market to develop niche products that can be used in this field.

Biosolids are the primary product used to recycle nutrients from wastewater. One of the biggest challenges with biosolids is the expense associated with transporting a product with a high moisture content (~80% - 90%). Since the current value of nutrients in biosolids (~ $US 8 per tonne) is a fraction of the transport costs ($US30 per tonne to transport 50 km in USA or Australia with higher costs in Europe), nutrient recovery via biosolids can be an expensive undertaking. Even in scenarios where thermal processes are used to reduce the moisture content, the energy required (~800 kWh of energy (as gas) required to evaporate one tonne of water) is significant.

Consequently, recovery of nutrients into chemical nutrient products like struvite is the primary focus of several commercial extractive nutrient recovery technologies. In addition to struvite, other products like calcium phosphate, (hydroxyapetite), iron phosphate (vivianite), phosphoric acid, ammonium sulfate, ammonium nitrate can also be recovered depending on the nature of the wastewater as well as the secondary market being targeted. An additional advantage of recovering chemical nutrient products is the fact that some of these products have use in alternative industries (Table 4).

Table 4. Summary of Chemical Nutrient Products resulting from Extractive Nutrient Recovery Processes (From Latimer et al., 2012a)

Common Name	Chemical Name	Product Form	Uses
Struvite	Magnesium ammonium phosphate	Solid	Agricultural and ornamental crop fertilizer
Hydroxyapatite	Calcium phosphate	Solid	Agricultural and ornamental crop fertilizer. Sorbent for heavy metal contained in flue gas.
Vivianite	Iron Phosphate	Solid	Ornamental crop fertilizer. Inexpensive blue pigment for arts and crafts.
Phosphoric acid	Phosphoric acid	Liquid	Agricultural and ornamental crop fertilizer. Removal of rust, de-scaling of boilers and heat exchange tubes.
Ammonium nitrate	Ammonium nitrate	Liquid or Solid	Agricultural and ornamental crop fertilizer. Oxidizing agent in explosives.
Ammonium sulfate	Ammonium sulfate	Liquid or Solid	Agricultural and ornamental crop fertilizer. Used in flame retardant materials.

In order for chemical nutrient products to be used for agricultural purposes, they must meet some minimum requirements. For instance, all products must have consistent nutrient content and possess no/minimal odors. Solid products must have uniform size, comprise no less than 95% total solids, have less than 1% dust content and have a minimum bulk density of at least 45 pounds per cubic foot. Due to the limited mass production rate of the wastewater treatment sector, it will be challenging to compete with existing supply chains. Instead, recovered products from WWTPs should be marketed within niche markets to maximize resale. An example of this is the case of the Ostara CrystalGreen™ product, which is used as a soil amendment product.

In addition to having specific physical characteristics, chemical nutrient products must also have minimal pathogen content and low concentrations of trace organic contaminants (TOrC). To date, research has shown that chemical nutrient products resulting from extractive nutrient recovery processes have negligible pathogen or TOrC content. This is an additional benefit that these products have over biosolids.

Challenges for Implementing N and P recovery at WWTPs

Although there appears to be a general consensus that nutrient recovery can benefit the industry, there remain technical, social and economic challenges towards an industry-wide adoption of this approach (Figure 2; Guest et al. 2009). Many of these barriers largely revolve around a lack of technical and economic knowledge. For instance, although there are multiple options that can be considered for recovery, a systematic evaluation of treatment efficiencies, costs, energy balances and recovered product yields is currently absent. To address this need, the Water Environment Research Federation (WERF) is funding active research that will provide peer reviewed resources (reports/databases/tools) that can aid the technical selection process. WERF subscribers will have unlimited access to this tool.

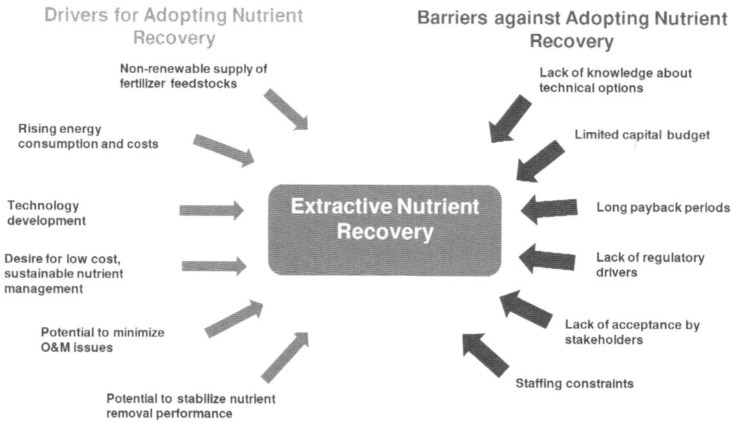

Figure 2. Drivers and Barriers for Extractive Nutrient Recovery (Latimer et al., 2012a)

CONCLUSION

Extractive nutrient recovery entails the extraction of N and P contained in municipal wastewater as a nutrient product that has value in a secondary market. Hence it can represent a revenue stream for the utility. In addition, adoption of this strategy represents a move toward more sustainable WWTP operation and can complement traditional methods of achieving N and P removal. On a global scale, the nutrients that can be recovered from human waste approximates 17 percent of N and 25 to 30 percent of P produced as fertilizer (Rogalla, 2012). It should be noted that this in itself will not offset a significant portion of the global nutrient demand and must be combined with other strategies to make a difference such as nutrient recovery from animal wastes (animal wastes represent a much larger nutrient pool than human waste) and improved agricultural practices that reduce nutrient loss.

ACKNOWLEDGEMENTS

This work was supported in part by the Water Environment Research Foundation (WERF). We thank Joseph Rohrbacher and Vivi Nguyen with Hazen and Sawyer P.C. and Todd Williams (CH2M HILL), and Ron Alexander (R. Alexander Associates Inc) for their valuable input to this project.

REFERENCES

Beecher, Ned, Qi, Y. (2012) *DRAFT Biogas Study Report – Use of Biogas at Wastewater Treatment Plants in the United States*. Water Environment Federation (WEF), 2012.

Guest, J.S., Skerlos, S.J., Barnard, J.L., Beck, M.B., Daigger, G.T., Hilger, H., Jackson, S.J., Karvazy, K., Kelly, L., Macpherson, L., Mihelcic, J.R., Pramanik, A., Raskin, L., Van Loosdrecht, M.C., Yeh, D. and Love, N.G. (2009) A new planning and design paradigm to achieve sustainable resource recovery from wastewater. Environ Sci Technol 43(16), 6126-6130.

Latimer, R., Khunjar, W.O., Jeyanayagam, S., Mehta, C., Batstone, D., Alexander, R., (2012a) Towards a Renewable Future: Assessing Resource Recovery as a Viable Treatment Alternative (NTRY1R12). Water Environment Research Foundation.

Rogalla, F. (2012). Unlocking the Energy and Nutrient Potential of Wastewater. 2012 IWA Yearbook.

BNR Process using RAS Anaerobic Zone

Gerry Stevens, AECOM

Westside Regional WWTP Operations Staff

Al Ellsworth, Mike Wyman, Angela Lambrecht

Abstract

The increasingly stringent limits imposed on wastewater treatment plant discharges demand reductions in Total Nitrogen and Total Phosphorus and improvements in process stability.

Although enhanced biological phosphorus removal (EBPR) process is considered a potentially efficient process with economic and environmental advantages compared to traditional chemical phosphorus removal alone, the benefits are often offset, in practice, by the needs to have standby chemical systems to achieve reliable and consistent performance.

Total Nitrogen removal using Anoxic/Aerobic sequencing without chemical addition has been successfully achieving effluent Total Nitrogen values of < 6 mg/L. However, when combined in an EBPR process, a competition sets up for utilization of the available carbon between the biological nitrogen and phosphorus removing processes. The use of external sources of the carbon has proven effective in stimulating de-nitrification, VFA production and Biological Phosphorous removal, however adds significantly to the treatment costs.

Side-stream RAS Anaerobic zones offers an innovative alternative to address the carbon competition, reduce the need for external carbon addition and minimize chemical usage for back-up P removal. Normally, a RAS only Anaerobic Zones requires up to 24 hour retention time, however when augmenting VFA and fermenting bacteria to this zone through the use of Primary Sludge Fermenters, the required retention time can be reduced to less than 2 hours and stabilize the Biological Phosphorus removal process. The use of a side stream Anaerobic Zone thus allows for all Primary Effluent to be directed to the denitrification stage of the process and enhance the overall Total Nitrogen removal.

Introduction

Due to the success at the Westside WWTP with a RAS Fermentation process, this site has been selected as one of six WWTP's for detailed investigation in the Water Environment Research Foundation - RAS Fermentation Study (Side-Stream EBPR Process). This study is being undertaken by staff at Northeastern University in Boston. The Study will include microbial population analyses using fluorescence in situ hybridization (FISH), Raman Spectroscopy, and PCR, phosphate release-uptake batch tests, glycogen & PHA analyses, lab-scale fermentation/hydrolysis batch tests, PHB and DAPI staining.

The study will investigate the mechanisms involved in Side-Stream EBPR Processes and utilize the information gathered during all testing to develop and/or improve process models to predict process performance when considering implementation of a side-stream hydrolysis/fermentation reactor.

AECOM has teamed up with the Westside Operating staff to optimize the RAS Anaerobic Zone, Primary Effluent By-pass, and Fermenter split flows. Detailed monitoring of all bioreactor zones is continuing in preparation for the WERF RAS Fermentation Study and site visit by N.E. University staff to take the above noted samples. This paper provides the results of the 2015 optimization period leading up to the WERF Study that commences in July 2015.

Process Configuration

A version of the Side Stream Anaerobic zone was commissioned in 2013 at the Westside Regional WWTP in West Kelowna B.C. **Figure 1 – Bioreactor #3 Train Configuration** below depicts the layout of the Pre-Anoxic, Anaerobic, Anoxic, Aerobic zones with incoming waste streams and recycles.

Figure 1 – Bioreactor #3 Train Configuration

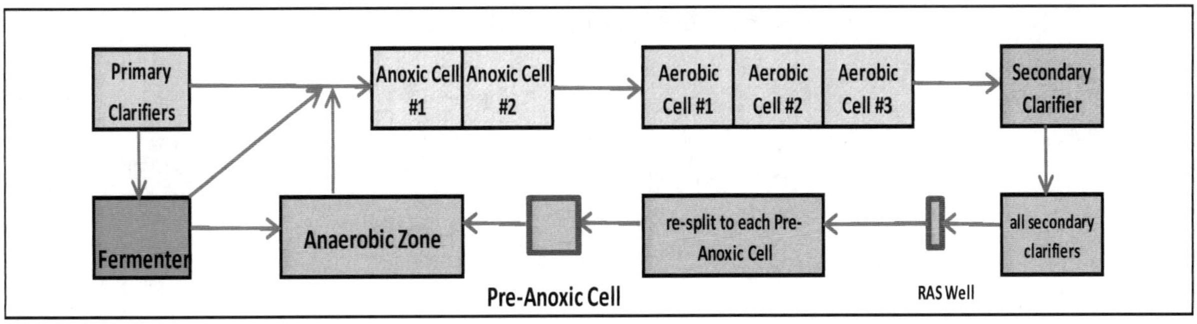

- **Primary Clarifier** – 100% of Primary Clarifier effluent flow goes to the Anoxic Zone. Primary Sludge is pumped to the Primary Sludge Fermenter.
- **Primary Sludge Fermenter** – over flow containing VFA's and fermenting bacteria is split between the Anaerobic Zone and the Anoxic Zone. Sludge is de-watered and removed from site for composting. The addition of fermenting bacteria to the anaerobic zone, accelerates the fermentation reactions and reduces the required retention time.
- **Anoxic Cells** – In Bioreactor 1&2, there are three Anoxic Cells, In Bioreactor 3, 4, 5, & 6 there are 2 Anoxic cells in series. In both cases the total volume represents 22% of the bioreactor
- **Aerobic Cells** – In all bioreactors there are three cells in series with D.O. control.
- **Secondary Clarifiers** – Each Bioreactor is dedicated to one Clarifier with individual RAS pumps, each with VFD control to maintain blanket de-nitrification.
- **RAS Collection Well** – A RAS collection well was added in the 2013 expansion program. This well collects RAS from all clarifiers. This combining event results in a significant reduction in the nitrate concentration and variability.
- **RAS Pre-Anoxic Cell** – This cell further reduces the nitrates to very low levels, i.e. < 0.5 mg/L.
- **RAS Anaerobic Zone** – This zone receives de-nitrified RAS and a portion of the Fermenter supernatant.
- **4 of 6 Bioreactors in service** – This represent a load equivalent to 110% of design capacity of 4 bioreactors.

The following **Table 1 – EBNR Bioreactor Zones** summarizes the hydraulic retention time and volume of each zone in the bioreactor. The minimum winter MLSS temperature used in this design is 10'C.

Table 1 – EBNR Bioreactor Zones

Type of Zone	Hydraulic Retention Time (hr.)	% of Bioreactor Volume
Pre-Anoxic Cell	0.3 (1)	2%
Anaerobic Cell	1.3 (1)	7%
Anoxic Cells 1, 2, 3	2.3 (2)	22%
Aerobic Cells 1, 2, 3	7.5 (2)	71%

(1) Hydraulic Retention Time based on RAS flow
(2) Hydraulic Retention Time based on Influent + RAS flow

Process Performance

The following sections describe the biological activity and performance results for each of the operating zones.

Figure 2 – Effluent Composite o-PO4 depicts the effluent results during the optimization period from January 2015 to June 2015

Figure 2 – Effluent Composite o-PO4

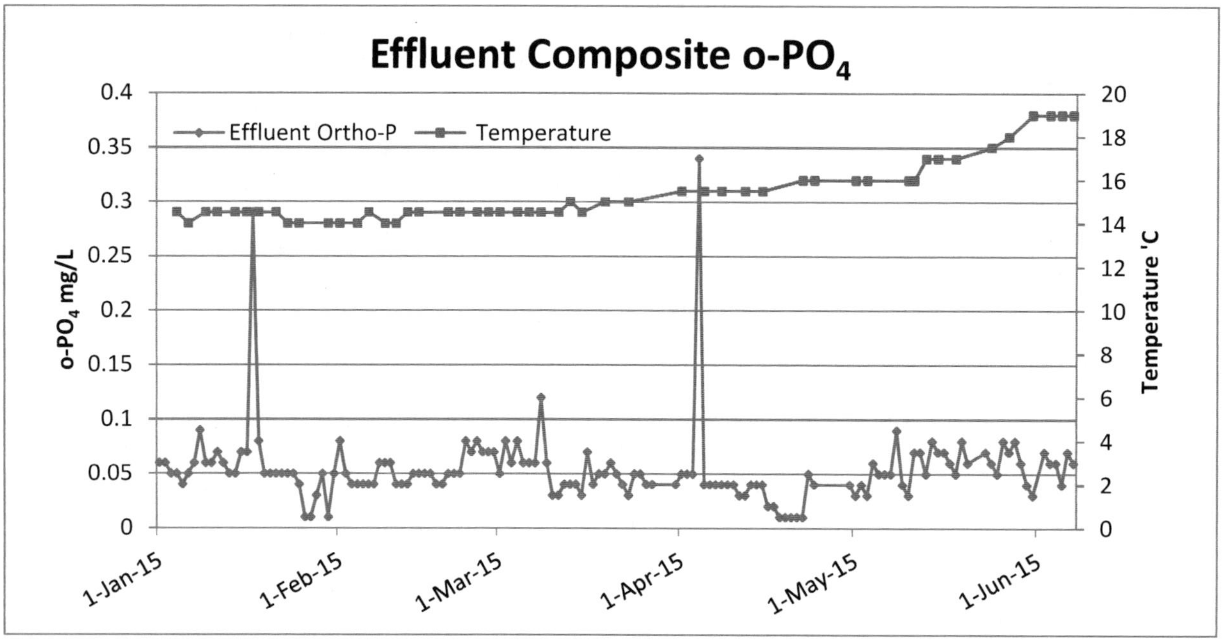

Influent & RAS Flows

Figure 3 – Influent Flow / Secondary Clarifier Flow / RAS Blanket Depth below depicts the influent flow during this period which varied from approximately 9 ML/d to a maximum of 11 ML/d and the Secondary Clarifier RAS rates which were maintained to provide a minimum sludge blanket depth of 1 foot.

The Bioreactor configuration produces a very good settling sludge which compacts readily to 2.5 to 3 times the MLSS in a very short time frame. To maintain a minimum one foot blanket depth, the RAS rates are set at approximately 60% of the inflow flow and at a constant rate 24 hours per day.

During the short de-nitrification stage in the clarifier, some fermentation begins and a small release of o-PO_4 can be measured. This release indicates the presence of some VFA which has been released during the RAS fermentation stage.

With this retention time in the Secondary Clarifier, most of the RAS Nitrates were removed. However, the clarifiers utilize a sweep helical scraper which brings the sludge to a center draw-off. When the sweep arm passes the center collection pit, a short period of lower RAS solids and higher nitrates occurs which in the previous configurations required the addition of 10% of Primary Effluent to the Pre-Anoxic cell to remove the nitrates.

In the new 2013 configuration, all the RAS is collected in a small well that mixes the RAS with high solids and no nitrates with the occasional RAS with low solids and some nitrates. In addition, the mixing also includes a small amount of fermentation by-products produced during the short retention time in the clarifier. The result is a blended RAS low in nitrates, i.e. < 0.5 mg/L which are further reduced in the

Operation of a Demonstration Scale Mixed Liquor Fermenter to Improve Total Phosphorus and Nitrogen Removal under Carbon Limitations

Rion Merlo[1*], Seppi Henneman[1], Denny Parker[1], Bob Witzgall[2], Ken Abraham[3], Steve Ramberg[4] and William Yu[4]

[1] Brown and Caldwell, Walnut Creek, California
[2] Brown and Caldwell, Elk Grove, California
[3] HDR, Elk Grove, California
[4] Sacramento Regional County Sanitation District, Elk Grove, California
* email: rmerlo@brwncald.com

ABSTRACT

A 946-m^3/d pilot biological nutrient removal (BNR) process was operated for 139 days with a mixed liquor fermenter (MLF). Throughout the testing period the highest concentration of secondary effluent (SE) ammonia (composite) was 3.7 mg-N/L. The highest concentration of SE nitrite (composite) was 0.39 mg-N/L. Intermittent acetic acid addition was necessary to meet the future average monthly nitrate concentration of 10 mg-N/L. Effectiveness of enhanced biological phosphorus removal (EBPR) was impacted by nitrate present in the return activated sludge (RAS). When RAS nitrate values approached 10 mg-N/L, an increase in SE phosphorus occurred. Acetic acid addition was necessary at times to maintain stable EBPR and denitrification. Under steady operating conditions, the MLF did not appear to have a detrimental impact to the secondary process, but its positive benefits are unclear. No volatile fatty acids (VFA) were measured in the MLF and there was not an observed increase in sCOD.

KEYWORDS: nitrogen, phosphorus, mixed liquor fermenter, biological nutrient removal

INTRODUCTION

The Sacramento Regional County Sanitation District (Regional San) must upgrade the Sacramento Regional Wastewater Treatment Plant (SRWTP) by 2021 to comply with future nitrogen discharge limitations. The future ammonia limit has a seasonal maximum daily value of 2.0 mg-N/L and an average monthly value of 1.5 mg-N/L. The nitrate limit will be set to 10 mg-N/L on an average monthly basis. Currently, the SRWTP does not provide nitrogen removal and the Regional San engaged Brown and Caldwell to design, build and operate a 946-m^3/d BNR process to demonstrate permit compliance as part of the Advanced Treatment Technology Pilot (ATTP) Project. Phase 1 testing was performed using a 4-stage BNR process.

Because Phase 1 testing showed that supplemental carbon was necessary at times to meet future nitrate limits, the design engineer (Black and Veatch [B&V]) recommended the MLF process as a way to generate carbon within the process. The MLF process was incorporated into the Cedar Creek Wastewater Treatment Plant (Kobylinski et al., 2013). The plant is an 828 m^3/hr (5.25 mgd) BNR plant and is the first such plant to use the MLF scheme as a normal operating mode, rather than on a demonstration basis. However, it should be noted that the Cedar Creek Wastewater Treatment Plant does not have primary clarifiers, whereas the pilot BNR process was treating primary effluent.

The results presented herein represent Phase 2 testing where a 5-stage Bardenpho process equipped with a MLF was tested. This configuration matches an optional operating configuration provided for in the design of full-scale facilities by B&V. In addition to ammonia and nitrate removal performance, the ability of the BNR process to provide phosphorus removal was measured during the testing period. The MLF process is not well understood nor accurately simulated using existing biological process models, and this study represents one of the first comprehensive evaluations of the MLF processes.

MATERIALS AND METHODS

The pilot-scale BNR process was designed to treat an average flow of 946 m^3/d (0.25 mgd) and a maximum flow of 1,970 m^3/d (0.52 mgd). The BNR process treated primary effluent from the SRWTP and was seeded with sludge from the full-scale SRWTP high purity oxygen activated sludge (HPOAS) process on November 12, 2013. Henneman et al. (2015) provides details of the startup procedure. The results presented herein reflect the operating period ranging from March 19, 2014 through August 4, 2014 which represents the design basis for future routine operation of the full-scale process.

The flow rate to the BNR process was variable based on a signal from the SRWTP influent flow meter such that both full-scale plant diurnal and wet weather flow variations were simulated. Figure 1 presents a process flow diagram and identifies the on-line instrumentation that was used for process monitoring and control. The MLF consisted of two unaerated tanks (termed MLF 1A and MLF 1B). It was intended that the mixed liquor would settle in the tank, to increase its residence time and thus enhance fermentation. The mixers in both MLF zones were operated for 15 minutes, twice per day for the majority of the testing period. Mixed liquor from the anaerobic zone was pumped to MLF 1A at a constant flow of 23 gpm. Primary effluent was split between the anaerobic zone and anoxic zone (Stage 1A Anoxic). Figure 2 shows a process flow diagram when the MLF was in service. Table 1 provides an overview of each compartment of the BNR process.

A classifying selector was included to mitigate biological foam formation as described by Parker et al. (2014) and Merlo et al. (2015). Nocardioform levels did not exceed 3 x 10^4 intersections/g VSS throughout the testing period (data not shown) and foam development was not an issue, consistent with Phase 1 results. The target total solids retention time (SRT) was 11 days and the internal mixed liquor return (IMLR) was limited to 2.66 times the influent flow. Acetic acid was added when necessary to target a secondary effluent nitrate concentration of 8 mg-N/L. The RAS was flow paced to half the influent flow. These operating conditions match the design conditions specified by the designer, B&V.

Process pH was controlled with the addition of caustic soda using a setpoint of 6.5 (chemical addition begins at 6.4). In general, this pH setpoint yielded secondary effluent alkalinity values ranging from 50 to 100 mg/L as $CaCO_3$ (data not shown).

The secondary clarifier was designed to produce an effluent that represents full scale so that the tertiary processes may be effectively investigated. Previous work by Parker et al. (1998) showed that, at a minimum, a 3.7-m (12-foot) diameter circular secondary clarifier was necessary to produce an effluent that reflected full-scale secondary clarification. Similar to full scale, the sludge withdrawal mechanism is a suction type (i.e., TowBro). Similarly, a flocculator centerwell was installed. The pilot plant RAS pumps were designed to mimic full-scale design parameters for future facilities (i.e., 50 percent at influent flow and a maximum of 63 percent at peak day flow). All analytical testing was performed in accordance with APHA (2012).

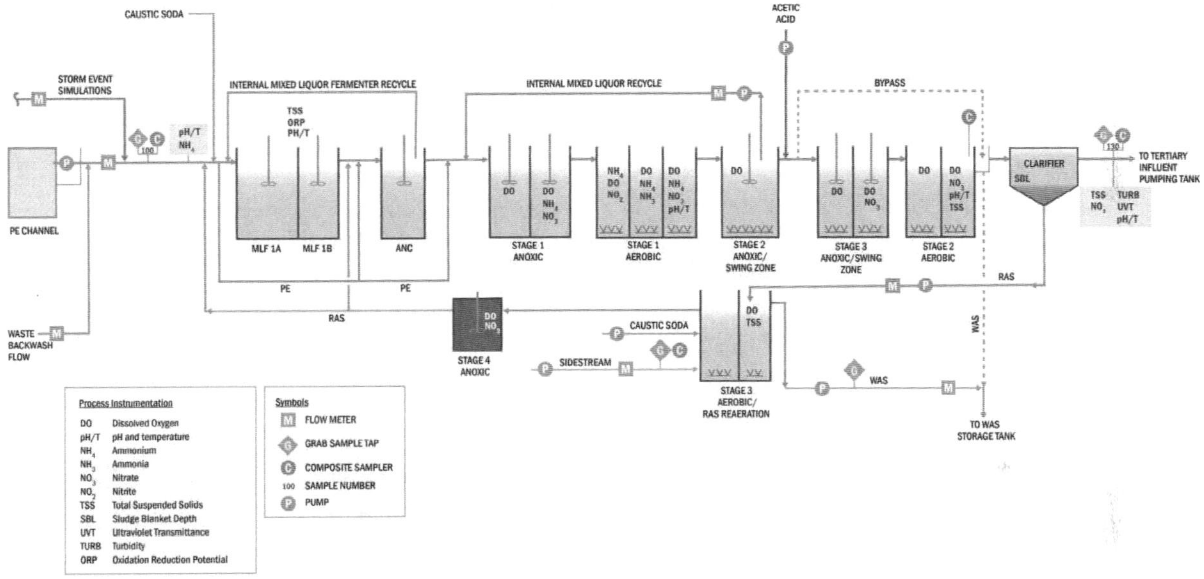

Figure 1. Process flow diagram of BNR process.

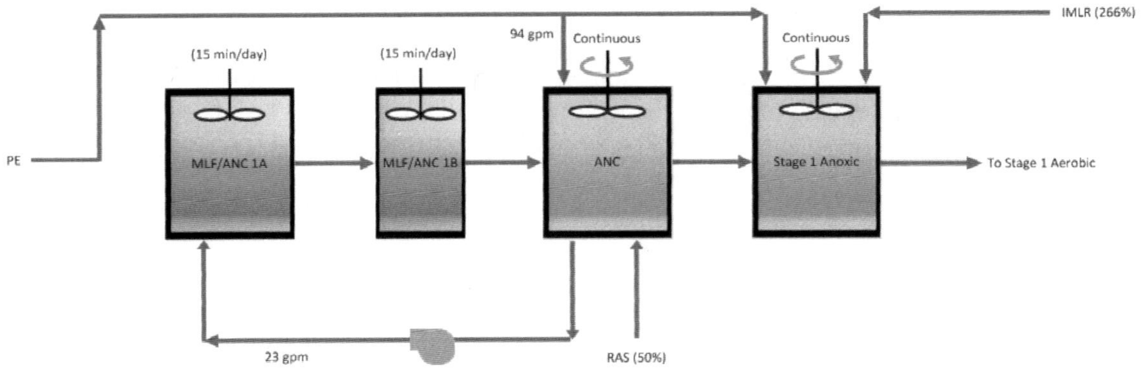

Figure 2. Process flow diagram showing operation during MLF testing.

Table 1. Overview of BNR Pilot Plant Tankage				
BNR Stage	Number of Zones	Operation	Process Objective	Tank Liquid Volume, m^3
Mixed Liquor Fermenter/ANC 1A & 1B	2	These zones were mixed once per day for 15 minutes and were otherwise not mixed. When the fermenter was not in service, these two zones served as addition anaerobic selector zones.	The fermenter was used to generate VFAs which promoted phosphorus release in the anaerobic zone as well as denitrification in the anoxic zone.	37.5
Anaerobic Selector (ANC)	1	This zone was mechanically mixed, DO should have been absent.	Promotes biological phosphorus removal as well as improved settlabililty	28.0
Stage 1A Anoxic	1	This zone was mechanically mixed, DO should be <0.1 mg/L	Promotes nitrate removal	37.1
Stage 1B Anoxic	1	This zone was mechanically mixed, DO should be <0.1 mg/L	Promotes nitrate removal	75.0
Stage 1A Aerobic Stage 1B Aerobic Stage 1C Aerobic	3	Zones 1 and 2 DO concentration was set at 3 mg/L. Zone 3 DO concentration was set at 2 mg/L using fine-bubble aeration.	Three zones mitigate short circuiting and promote nitrification. Elevated DO concentration in Zones 1 and 2 will increase nitrifier kinetics (Bratby and Parker, 2009). Reduced Zone 3 DO concentration of 2 mg/L is to prevent DO carryover into anoxic zones.	228
Stage 2 Anoxic/ Swing Zone	1	Zone is mixed with mechanical mixer. DO should be <0.1 mg/L. Alternatively, aeration can be used to maintain a DO of 2 mg/L.	During normal operation, this zone serves to "deoxygenate" mixed liquor before it flows to Stage 1 Anoxic via IMLR and Stage 3 Anoxic. If additional nitrification is necessary, the tank can be aerated.	31.2
Stage 3A Anoxic/ Swing Zone Stage 3B Anoxic/ Swing Zone	2	Zones are mixed with mechanical mixers. DO should be <0.1 mg/L. Alternatively, aeration can be used during times when additional nitrification is required.	During normal operation, this zone serves to remove nitrate. Supplemental carbon can be added to this tank if necessary. If additional nitrification is necessary, the tank can be aerated.	66.6
Stage 2A Aerobic Stage 2B Aerobic	2	Divided into two zones; both zones should be operated at DO concentration of 2 mg/L using fine-bubble aeration.	In addition to providing additional nitrification, both zones act to strip out nitrogen gas formed in Stage 3 anoxic before secondary clarification.	40.9
Stage 3 Aerobic/ RAS Reaeration	1	DO concentration was set at 2 mg/L using fine-bubble aeration.	This zone includes the classifying selector, which was used to prevent foam causing organisms from growing in the BNR	15.8
Stage 4 Anoxic	1	Zone was mixed with mechanical mixer. DO should be <0.1 mg/L.	This zone serves to "deoxygenate" return activated sludge (RAS) before it flows to ANC	13.1

RESULTS AND DISCUSSION

BNR Influent Flows and Loads

Figure 3 presents the average daily flow to the BNR process. Primary effluent (PE) flow was proportional to the flow to the SRWTP. The scaling factor used to determine the BNR process influent flow was set so that the average daily BNR process influent flow was approximately 852 m^3/d. The average daily influent flow was 880 m^3/d and ranged from 219 m^3/d 1,640 m^3/d. Figure 3 also presents the flow diverted to ANC and Stage 1A Anoxic. In general, the majority of influent flow was directed to ANC with the exception of April 1 through April 23. The high flows at the end of March and in early April represent rainfall induced peak flows. The peak flows from June 3 through June 10 are simulated wet weather testing. From June 11 to July 9, the influent scaling factor was reduced to reduce the influent loading.

Figures 4 through 7 present the BNR process influent loadings for 5-day biochemical oxygen demand (BOD), chemical oxygen demand (COD), total suspended solids (TSS), total Kjeldahl nitrogen (TKN), ammonia, total phosphorus, and ortho-phosphate. Figure 8 presents BNR influent VFA concentrations (acetic acid only). The reduction in loading that occurred from June 3 through June 10 is from the simulated wet weather testing where SE was recycled to the head of the BNR process. The reduction in loadings from June 11 to July 9 was a result of the special study to evaluate the impact of mixed liquor suspended solids (MLSS) concentration on SE turbidity by reducing the PE flow.

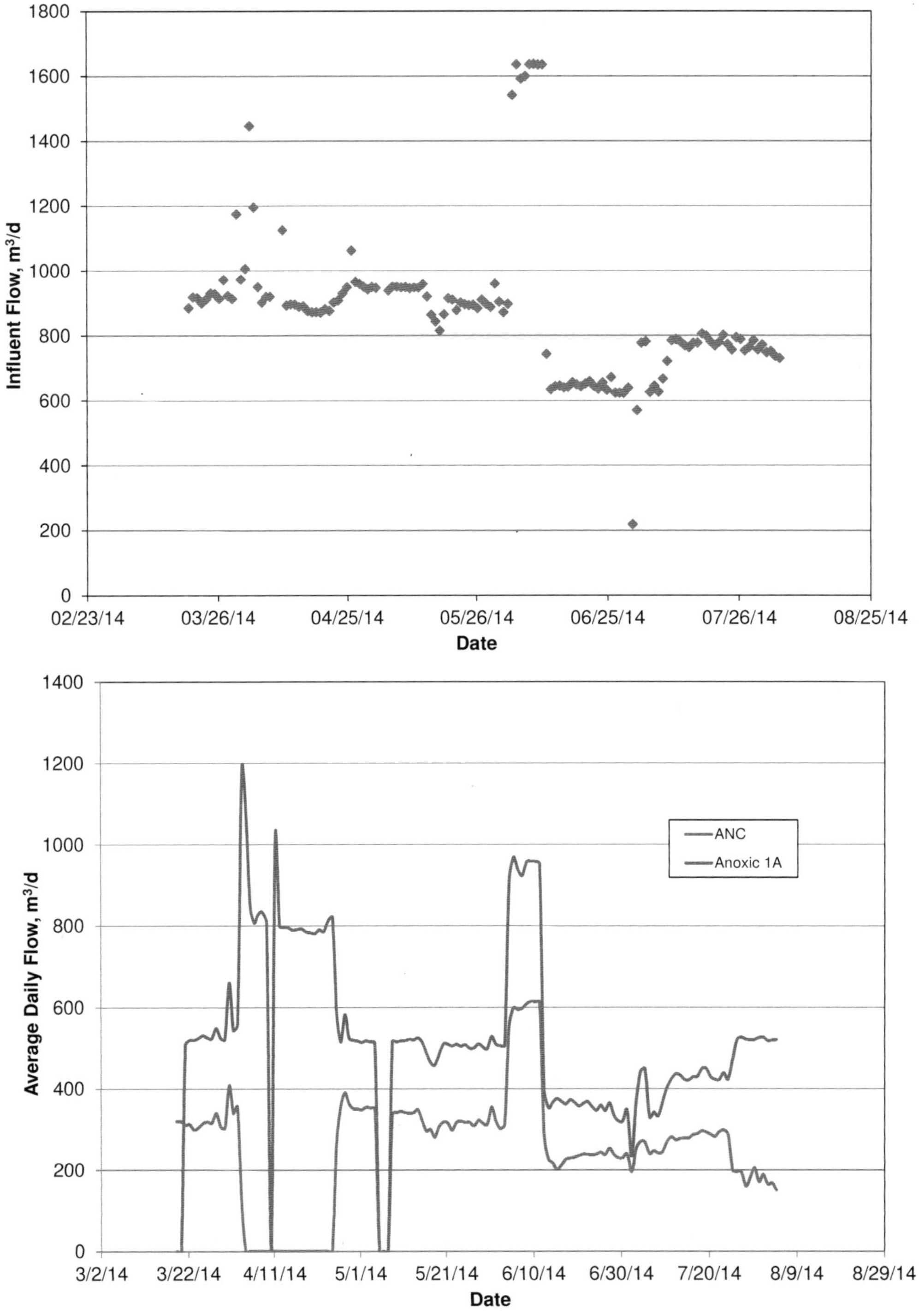

Figure 3. Average Daily Influent Flow to the BNR Process (top) and Flow to Each BNR Process Zone (bottom).

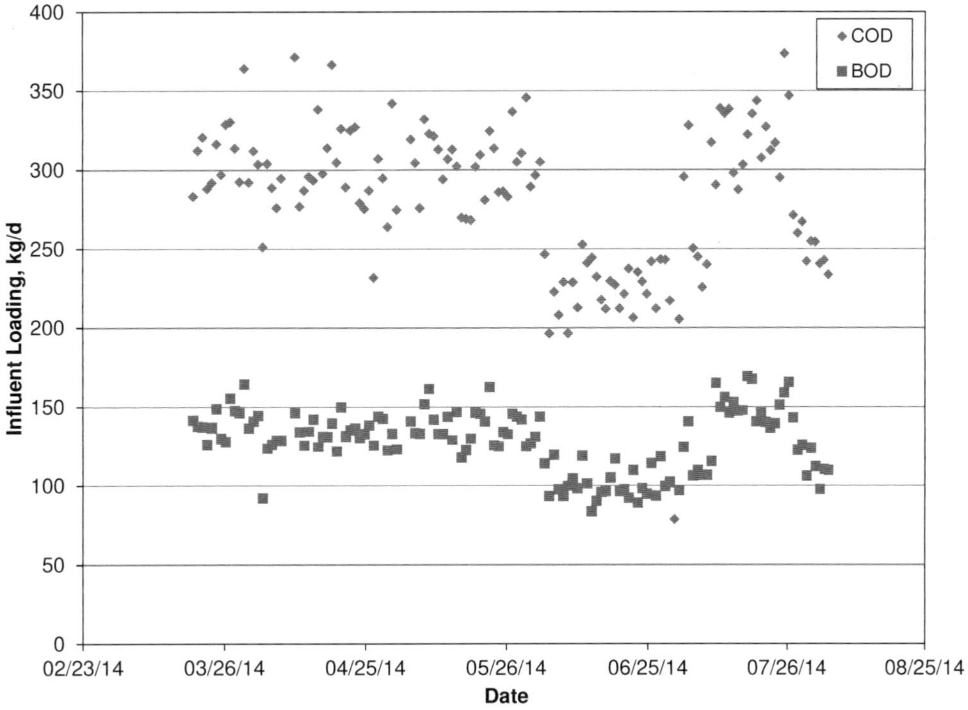

Figure 4. Pilot Influent BOD and COD Loading.

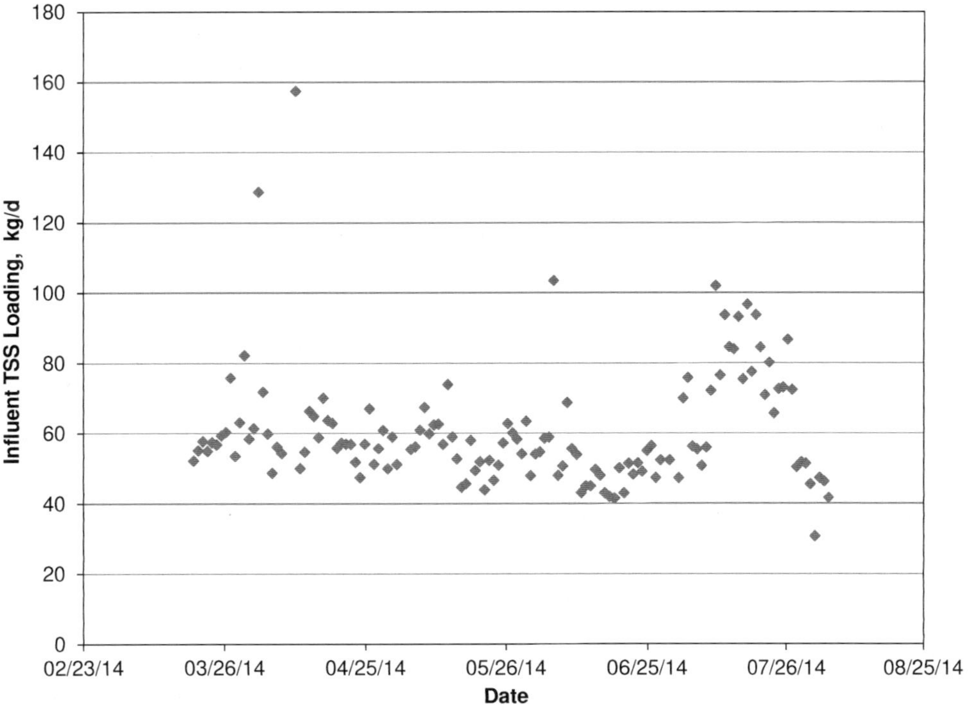

Figure 5. Pilot Influent TSS Loading.

Challenges to Achieving Stable Performance in a Site-Specific EBPR Configuration (AAnO) at the Stickney Water Reclamation Plant

Heng Zhang*, Joseph Kozak, Dongqi Qin, Yvonne Lefler, Brett Garelli, Reed Dring, Joseph Cummings, Catherine O'Connor, and Glenn Rohloff

Metropolitan Water Reclamation District of Greater Chicago
*Heng.Zhang@mwrd.org

ABSTRACT

After a site-specific enhanced biological phosphorus removal (EBPR) configuration was developed without major infrastructural change to the existing secondary treatment facility, the entire Stickney Water Reclamation Plant (WRP) was converted to this configuration in fall 2013. This site-specific EBPR configuration consists of anoxic, anaerobic, and oxic zones (AAnO) in sequence. Since the conversion, the plant treatment efficiency with respect to effluent BOD_5, suspended solids, and ammonia nitrogen has not been noticeably affected, while the effluent total phosphorus concentrations are lower. However, the EBPR performance has not reached the desired level. Occasional occurrences of the elevated dissolved oxygen (DO) and nitrate levels in the return activated sludge (RAS), shortage of carbon in the influent, and elevated chloride concentrations in the winter months pose operational challenges to this process. The potential strategies in addressing these challenges will be discussed here.

KEYWORDS: site-specific EBPR process, operational challenges, DO, nitrate, chloride

INTRODUCTION

The Stickney WRP located in Cicero, Illinois, is a municipal wastewater treatment plant, which treats domestic sewage mainly from combined sewer systems with a 2014 annual average flow of 2,760,000 cubic meters per day (m^3/d) (728 million gallons per day [mgd]) and maximum daily flow of 5,330,000 m^3/d (1408 mgd). The plant has two influent pumping stations, West Side and Southwest Side, and two primary treatment systems. Although the two influent pumping stations are linked with a tunnel, the West Side receives typically domestic sewage, while the Southwest Side receives the flows from plant recycles and deep tunnel pumpback in addition to domestic sewage. The West Side has 72 Imhoff tanks used for primary treatment and the Southwest side has 20 rectangular preliminary settling tanks. The effluents from the two primary treatment systems are sent to a centralized secondary treatment system, which was a conventional, single-stage nitrification activated sludge process. The secondary treatment system has four parallel, identical aeration batteries with a total aeration tank volume of 795,000 m^3 (210 million gallons [MG]). Each aeration battery has 8 plug flow aeration tanks with 4 passes in each tank as well as approximately 5,200 m^3 (1.4 MG) of return activated sludge (RAS) and 9,600 m^3 (2.5 MG) of influent mixing channels.

To sustainably remove phosphorus from the liquid stream and recover phosphorus from side streams as a valuable resource, a site-specific EBPR process, AAnO, was developed in one battery in May 2012. This AAnO process includes (1) anoxic, using the existing RAS and influent mixing channels, (2) anaerobic, using one eighth of an aeration tank, and (3) aerobic, using seven eighths of the aeration tank, zones as illustrated in Figure 1. Because the Stickney WRP is a nitrifying plant, it is necessary to create an anoxic zone ahead of the anaerobic zone to denitrify the nitrate being returned with the RAS. A primary factor in developing this AAnO process was to utilize only current infrastructure and push the process to maximum capacity within these confines. Mixing in the anoxic and anaerobic zones was accomplished by air through the coarse and fine bubble diffusers, and air flow was reduced to minimum, just enough to suspend the solids. The conveyance of the RAS from the secondary clarifiers back to the aeration tanks was carried out by the existing airlifts. No baffles were installed to separate zones. A secondary factor was the consideration of its unusually high TP removal efficiency, which will be briefly discussed in a later section, before the AAnO process was implemented. The details on how the AAnO process was developed and tested before it was implemented at the entire plant in fall 2013 were described by Qin et al (2014).

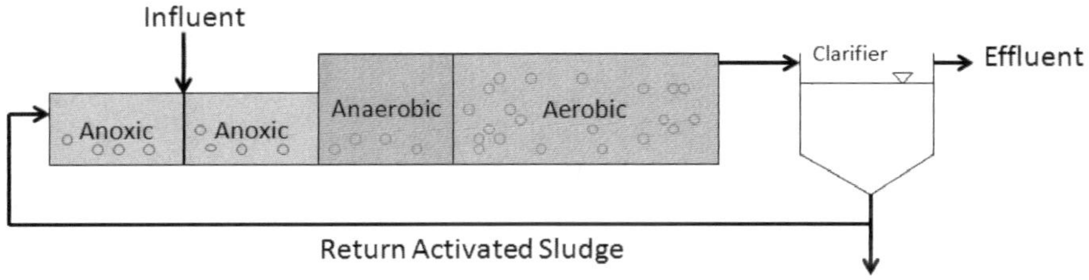

Figure 1. Flow Diagram of AAnO – a Site-Specific EBPR Process

Although this process was able to achieve an annual average total phosphorus (TP) concentration of 0.87 mg/L in the final effluent in 2014, which has been the lowest annual average TP concentration since 2002 when the daily ammonia nitrogen (NH_3-N) limits of 5 mg/L and 8 mg/L in warm and cold seasons, respectively, were first required in the NPDES permit, the Stickney WRP has not been able to achieve a monthly average TP concentration of less than 1 mg/L in the final effluent every month. The monthly average TP concentration of 1 mg/L or less is required by the newly issued NPDES permit to be achieved in 2019. Nevertheless, the organizational goals of pursuing sustainable nutrient removal and recovery require us to achieve this target as soon as possible. The challenges, as we understand at this moment, are mainly related to influent characteristics and simultaneously meeting multiple stringent effluent NPDES permit limits for a secondary wastewater treatment plant including DO, NH_3-N and total suspended solids (TSS), along with TP in the future. The AAnO process is vulnerable to occasional elevation of DO and nitrate in the RAS when there is insufficient carbon in the influent for both denitrification and EBPR. The limitation of the anoxic zone size and the fluctuation of the influent flow from the combined sewer systems also make the challenges even greater. We are exploring operational strategies to address these challenges before considering making any significant infrastructure investment. In this paper, the specific challenges and the corresponding operational strategies will be discussed.

PLANT PERFORMANCE PRIOR TO EBPR

The conventional, single-stage nitrification activated sludge process at the Stickney WRP was quite efficient with respect to removing BOD_5, NH_3-N and TSS. The effluent limits for there parameters in the NPDES permit issued to this plant are relatively stringent for a conventional secondary wastewater treatment plant, as shown in Table 1. At the plant's design average flow (DAF) of 4,540,000 m^3/d (1,200 mgd), the nominal hydraulic retention time (HRT) in the aeration tanks is only 4.2 hours. However, the median flow in 2014 was 2,570,000 m^3/d (680 mgd), which corresponds to a nominal HRT of 7.4 hours in the aeration tanks. The plant has adequate and efficient secondary clarifiers and the overflow rate at the DAF is 41 m^3/m^2/d (1,000 gal/ft^2/d).

Table 1. The NPEDS permit limits for $CBOD_5$, TSS, NH_3-N, and DO at the Stickney WRP

Parameter	Monthly Average	Weekly Average	Daily Max	Daily Min
$CBOD_5$ (mg/L)	10	15	--	--
TSS (mg/L)	12	20	--	--
NH3-N (mg/L)[1]	2.5 / 4.0	--	5.0 / 8.0	--
DO (mg/L)[2]	--	--	n/a	6.0

Note: [1]Values are for warm (April to October) and cold (November to March) seasons, respectively
[2]The value is from the 2002 NPDES permit and in the newly issued permit it was changed to 4 mg/L.

Table 2 shows the minimum and maximum monthly average effluent concentrations of BOD_5, TSS, NH_3-N and nitrate nitrogen (NO_3-N) at the Stickney WRP in each year from 2002 to 2010. The solid performance at the Stickney WRP prior to the EBPR era was partially attributed to the good settling characteristics of the mixed liquor in the secondary treatment system. The sludge volume index (SVI) varied slightly from 40 to 70 g/mL with relatively lower values in warm seasons and higher in cold seasons. The good settling characteristics at the Stickney WRP may be attributed to its plug flow type of aeration tanks with a large length to width ratio, the narrow influent mixing channel acting like an anoxic contact chamber before the aeration tanks, inefficient primary treatment and sludge gravity concentration tanks, consistent operations and other factors that have not been fully understood. In addition to maintaining the mixed liquor concentrations at a 3,000 mg/L level and sufficient SRT for nitrification, the Stickney WRP used to routinely run the DO level around 8 mg/L at the end of the aeration tanks to meet the effluent DO limit without a post-aeration facility in the plant and a little sludge blanket in the secondary clarifiers to avoid floating sludge caused by denitrification. These old operational strategies may have contributed to the good settling characteristics of the mixed liquor as well.

The heavy flocs in the mixed liquor at the Stickney WRP, which is inferred from the quick settling characteristics, benefit not only the removal of conventional pollutants like BOD_5 and TSS, but also the removal of nutrients, nitrogen (N) and phosphorus (P). Prior to the implementation of EBPR, the Stickney WRP exhibited the above-average removal of total nitrogen (TN) and TP that a conventional secondary wastewater treatment process could achieve. The average TN removal in the secondary treatment system alone was above 53% in the period of 2008 to 2010. Figure 3 shows the annual average TP removal efficiency in the secondary treatment system alone before and after the AAnO EBPR process was implemented, and Figure 4 presents the monthly average TP concentrations in the final effluent of the Stickney WRP in

2002 through 2010 prior to the implementation of the AAnO process. The level of P removed by the conventional secondary treatment system at the Stickney WRP was much greater than the stoichiometric uptake of P through the assimilation of the common heterotrophic microorganisms in the activated sludge. Lin et al. (2003) successfully developed the phosphate accumulating organisms (PAOs) sludge granules in their sequencing batch reactors. The specific gravity of the granules was much greater than the seed sludge, partially due to the presence of PAOs, and increased as the P contents in the granules increased. It was possible that some types of PAOs may have existed in the mixed liquor of the conventional, single-sludge activated sludge process at the Stickney WRP, given the fact that there was exceptional removal of P by that system. However, there was no systematic, microbiological analysis to verify the abundance of PAOs in that system. The observation of this phenomenon prompted the development of the site-specific EBPR process in belief that this process should be able to achieve the biological P removal at the level of 1 mg/L or much less as a monthly average every month after a better environment for PAOs to thrive was provided.

Table 2. Minimum and maximum monthly average concentrations of BOD_5, TSS, NH_3-N and NO_3-N in the Stickney WRP final effluent

	$CBOD_5$ (mg/L)		TSS (mg/L)		NH_3-N (mg/L)		NO_3-N (mg/L)	
Year	Min	Max	Min	Max	Min	Max	Min	Max
2002	2.3	4.3	3.7	7.3	0.18	1.16	6.80	9.90
2003	2.6	4.8	3.6	8.8	0.29	1.71	6.75	12.56
2004	2.2	4.3	4.1	6.7	0.30	1.87	6.17	10.84
2005	2.1	4.5	3.6	8.1	0.28	1.68	6.32	9.40
2006	2.3	3.7	3.7	7.6	0.11	0.91	6.03	12.71
2007	2.0	4.8	3.9	9.1	0.25	1.63	6.09	10.62
2008	2.1	5.8	3.7	9.5	0.23	1.41	5.72	11.13
2009	2.4	4.1	3.1	9.0	0.12	0.83	7.37	10.93
2010	2.0	4.9	4.1	7.3	0.31	1.38	6.14	11.40

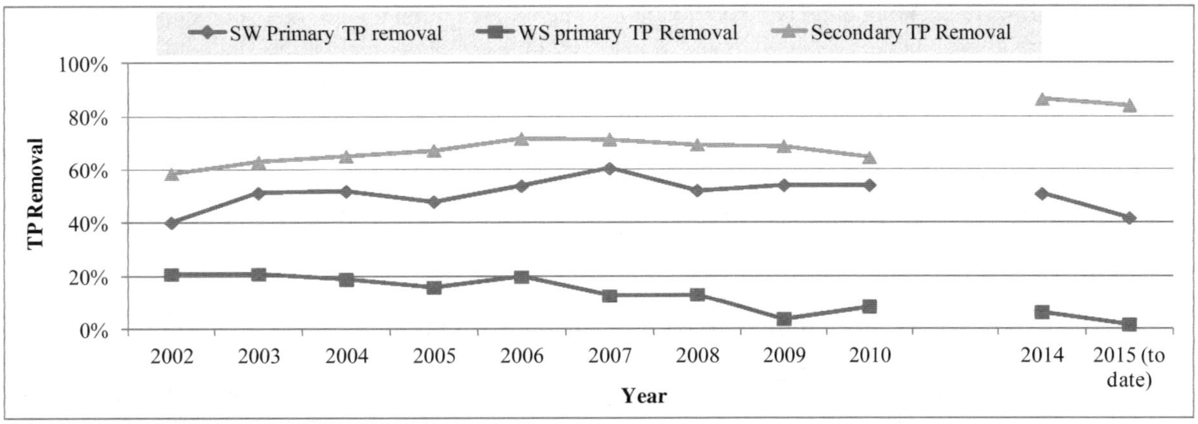

Figure 3. Annual averages of TP removal efficiency in the primary and secondary treatment systems at the Stickney WRP

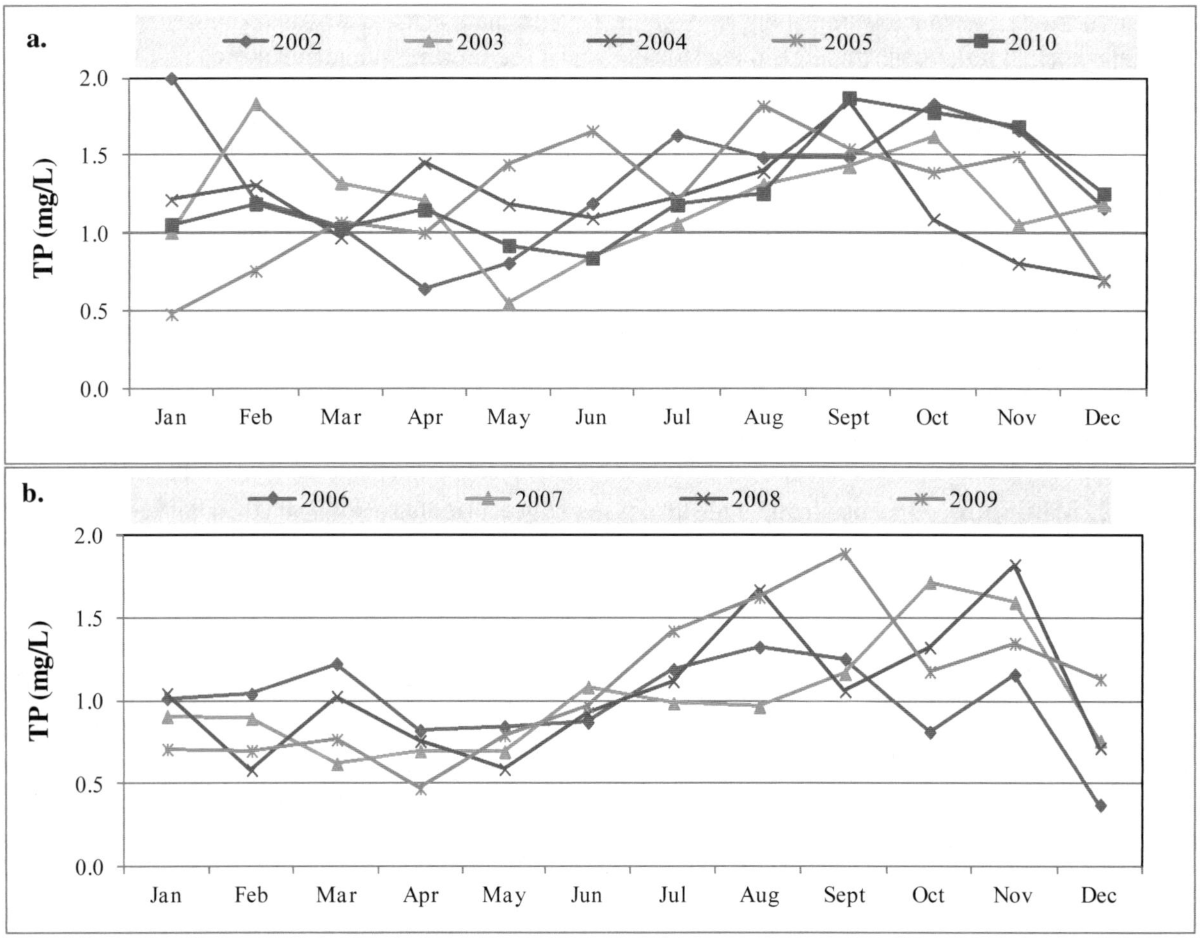

Figure 4. Monthly average TP concentrations in the final effluent of the Stickney WRP in 2002 through 2010.

LIMITATION OF AAnO EBPR PROCESS

The AAnO process was developed through maximizing the existing infrastructure and minimizing the capital investment for retrofitting the existing conventional, single-sludge activated sludge system. To create alternating anaerobic and aerobic environment for accommodating PAOs, the existing RAS and influent mixing channels were converted to an anoxic zone by reducing or completely shutting off air that was used for mixing in these channels to minimize the addition of DO. Because of the presence of nitrate in the RAS at this nitrification-required plant and relatively high DO concentrations at the effluent end of an aeration tank to facilitate the compliance with the effluent DO requirement, the role of this zone is to remove the nitrate and DO concentrations in the RAS. This is why this zone is termed as anoxic. The truly anaerobic condition can only be established after the nitrate and DO are removed (WEF, 2013). Therefore, the anaerobic zone was created by definition in the beginning of an aeration tank immediately following the anoxic zone. However, the mixing in the anaerobic zone was still accomplished by turning down the air to the minimum through the diffuse plates that were used for aeration before the AAnO process was developed. No baffle wall was installed between the anaerobic and aerobic zones with the consideration of the plug flow configuration of the aeration tank that has a length to width ratio of 50. The RAS was administrated by the air lifts, which was used for returning activated sludge from the secondary clarifies, the physical separation subsystem, to the biological subsystem of the secondary treatment system and adopted in the AAnO process without modification. The primary effluent enters the secondary treatment system at the end of RAS channel, which is about two fifths of the anoxic zone. Apparently, the readily biodegradable organic carbon including volatile fatty acids (VFAs) in the primary effluent would be used first for denitrification if nitrate was still present at the end of the RAS channel.

There are physical limitations inherited in this site-specific EBPR process, as mentioned above. However, appropriate operational strategies could overcome some of the limitations, while the others rely on influent characteristics and other means such as the modification of infrastructure if it is more sustainable than adding external carbon. The old operating strategies used for operating the conventional process, as described in the previous section, need to be modified to accommodate the requirements of PAOs. Some of the modified operational strategies have been successfully implemented, the outcome of which is evident in Figure 3, as the average P removal efficiency was higher after the AAnO EBPR process was implemented.

EBPR OPERATIONAL CHALLENGES AND POTENTIAL SOLUTIONS

As shown in Figure 5, the final effluent TP concentrations at the Stickney WRP have been fluctuated with periodic spikes since the AAnO process was implemented. These spikes were triggered by the factors, such as elevated DO and NO_3-N concentrations in the RAS consuming readily biodegradable organic carbon or being carried over to the anaerobic zones, carbon shortages for EBPR, influent flow surge due to wet weather flows from the combined sewer systems, extremely low DO concentrations in the aerobic zone due to first flush, which result in poor P uptake or even release of P to the final effluent, or possible inhibition due to the elevated chloride concentrations in the influent. Some of these factors can be addressed through the implementation of appropriate operational strategies, while the others are difficult to avoid. The main goal of developing the appropriate operational strategies is to minimize the extended elevation of effluent TP concentrations, but to allow the short-lived spikes that are triggered by unavoidable factors.

Influent Carbon

Sufficient carbon for EBPR is vital. In the literature, it has been identified that the minimum substrate-to-phosphorus requirements for EBPR are carbonaceous BOD_5 to P ratio of 25:1as a rough estimate or readily biodegradable COD (Chemical Oxygen Demand) to P ratio of 15:1 as a more accurate measure (WEF, 2010). Our experience seemed to indicate that good EBPR performance could generally be achieved in the AAnO process if the BOD_5 to TP ratio (BOD_5:TP) in the primary effluent was greater than 26:1 unless there were other influential factors negatively impacting the process. As can be seen in Figure 5, the BOD_5:TP in the primary effluent varied seasonally and was above this level most of the time in the first half of a calendar year and lower than this level in the second half. It has been observed that the effluent TP concentrations, as shown in Figure 4, followed a pattern that is complementary to BOD_5:TP, that is lower in the first half of a calendar year when carbon could be sufficient for biological P removal and higher in the secondary half when carbon would be lacking. The study conducted by Wiliński (2009) indicated that other sources of carbon, such as lactic acid and some amino acids, could be used by PAOs even though the P release rates were less than those when short-chain VFAs, such as acetic and propionic acids, were used. This widens the possibility of available carbon for biological P removal.

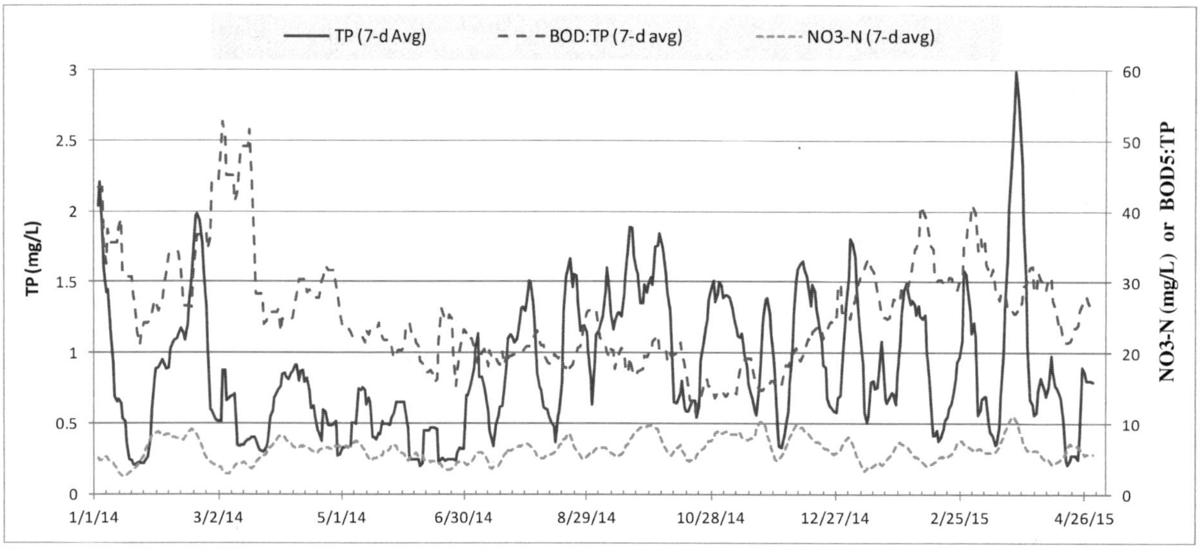

Figure 5. Final Effluent TP and NO_3-N Concentrations and Primary Effluent BOD_5:TP

Abundant influent carbon available for EBPR is even more imperative to the AAnO process because the carbon is used first by the heterotrophic microorganisms in the mixed liquor in this process if DO and nitrate are present. This requires that the DO and nitrate concentrations in the RAS should be addressed to reduce the requirements of influent carbon, which will be discussed in the next section. Meanwhile, the potential operational strategies that directly address the lack of influent carbon for the AAnO process are (1) to seek opportunity to generate readily biodegradable carbon in the anaerobic zone by encouraging mixed liquor inline fermentation, and (2) to generate carbon within the plant through primary sludge fermentation. To implement the first strategy, the air mixing in the anaerobic zone will be turned off completely to allow the solids to settle with infrequent use of the air mixing to stir up the settled solids to release the carbon. This strategy is currently being tested in two aeration tanks at the Stickney WRP.

DO and Nitrate in RAS

Due to the NPDES permit requirements, the treatment plant operators of the Stickney WRP used to raise DO concentrations in its aeration tanks for both complying with the effluent DO limit and enhancing nitrification to meet the effluent ammonia limit. This practice resulted in relatively high DO and NO_3-N concentrations in the RAS, which requires additional carbon for consuming the DO and removing the NO_3-N via biological denitrification in the anoxic zone of the AAnO process or could inhibit EBPR if they are carried over to the anaerobic zone.

Operational strategies have been developed to lower the DO and NO_3-N concentrations in the RAS including (1) maintaining a sludge blanket in the secondary clarifiers and (2) reducing the DO set points at the end of the aeration tanks. However, the stringent effluent TSS limits and flow surges due to wet weather conditions force the plant operators to give up the first strategy from time to time. The balancing point for simultaneously meeting the multiple NPDES permit limits has to be found. The strategy for handling the wet weather impact is being studied.

The second strategy is currently implemented as the opportunity for reducing DO levels in the aeration tanks occurred after the newly issued NPDES permit changed the effluent DO limit from 6 mg/L to 4 mg/L. Meanwhile, a study is being conducted to evaluate the feasibility of using an ammonia-based aeration control strategy not only to reduce the energy consumption, but also to encourage simultaneous nitrification and denitrification in the aeration tanks. The impact of this operational strategy on the EBPR process will be carefully monitored during the study.

Chloride Impact

The impact of chloride on EBPR has not been well defined in the literature. The level of chloride inhibition to PAOs varies likely depending on site-specific conditions and acclimation of microorganisms to chloride levels. As can be seen in Figure 5, the elevated effluent TP concentrations in February 2014 could not be well explained. At that time, chloride concentration was not monitored. The attention to possible chloride inhibition was brought up. Daily monitoring of chloride concentrations in the liquid phase started on November 1, 2014.

Figure 6 shows the daily concentrations of TP and chloride in the final effluent of the Stickney WRP in the winter of 2014 to 2015. It appears that each chloride spike coincides with the elevation of TP. However, data points are insufficient to derive any correlation at this time. Limited laboratory P update and release (URT) tests were conducted at the chloride levels that were detected at the plant. The results from the URT tests have been inconclusive. Monitoring of chloride concentrations in the liquid phase and more laboratory URT tests will continue in the coming winter.

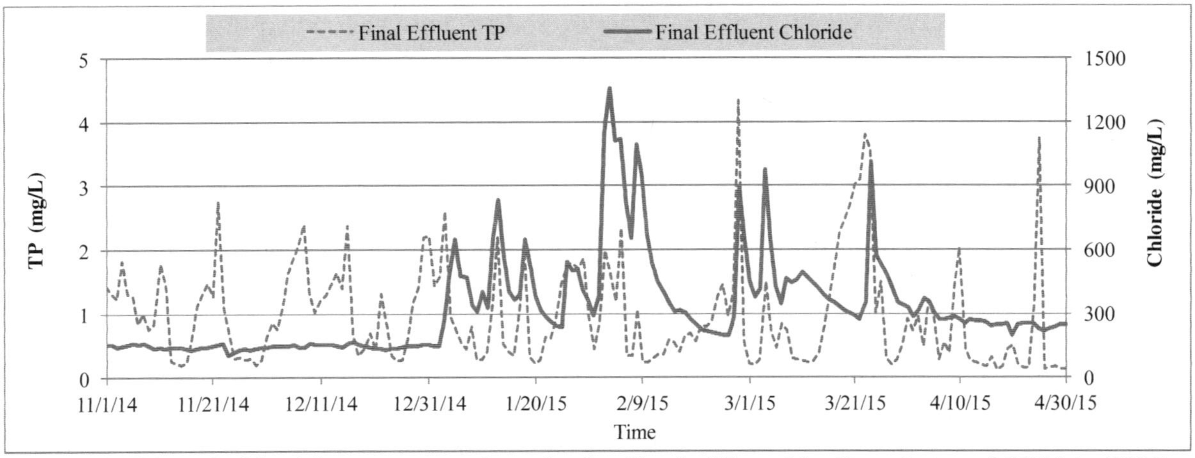

Figure 6. TP and chloride concentrations in the final effluent in the winter of 2014 to 2015

SUMMARY

Because the AAnO process is a site-specific EBPR process that is tailored to the existing infrastructure at the Stickney WRP to minimize the capital investment at this large plant and to maximize the capacity of the infrastructure and the characteristics of the mixed liquor customized to this infrastructure, the limitations of this process are acknowledged. Some of limitations have been addressed by implementing the appropriate operational strategies to overcome, while others will be addressed via further studies. Efforts are being made to achieve stable performance in this site-specific EBPR process and to explore the balance of meeting the multiple NPDES permit limits that often result in competing interests.

REFERENCES

Hua, G. and Zhang B. (2009) Pilot and full-scale evaluation of biological nutrient removal using Anoxic/Anaerobic/Aerobic process, *Proceeding of Water Environment Federation's Annual Technical Exhibition and Conference (WEFTEC),* 5912-5916.

Lin. Y.-M., Liu, Y., and Tay, J.-H. (2003) Development and characteristics of phosphorus-accumulating microbial granules in sequencing batch reactors, *Appl Microbiol Biotechnol*, **62**:430–435.

Qin, D., Lefler, M., Kozak, J. A., Zhang, H., Podczerwinski, E., Granato, T. C., Garelli, B. Cummings, J., Dring, R., Rohloff, G., and O'Connor, C. (2014) Implementation of a Site-Specific EBPR Process (AAnO) Using Existing Infrastructure at the Stickney Water Reclamation Plant, *Proceeding of Water Environment Federation's Annual Technical Exhibition and Conference (WEFTEC)*, 5586-5606.

Water Environment Federation (2010) *Nutrient Removal*, WEF Manual of Practice No. 34; Water Environment *Federation*: Alexandria, Virginia.

Water Environment Federation (2013) *Operation of Nutrient Removal Facilities*, WEF Manual of Practice No. 37; Water Environment Federation: Alexandria, Virginia.

Wiliński, Piotr Robert (2009) *Kinetics and stoichiometry of P-release with different carbon sources in the anaerobic phase of the biological phosphorus removal process in activated sludge wastewater treatment plants*, Thesis submitted to Department of Biotechnology, Chemistry and Environmental Engineering on the Section of Environmental Engineering, Aalborg University, Denmark.

The Successful Start up of a Biological Nutrient Removal Facility and Lessons Learned from Process Upsets

Seppi Henneman[1], Rion Merlo[1], Denny Parker[1], Bob Witzgall[1], Ken Abraham[2], and Steve Ramberg[3]

[1] Brown and Caldwell, Walnut Creek, California
[2] HDR, Folsom, California
[3] Sacramento Regional County Sanitation District, Elk Grove, California
* Email: SHenneman@brwncald.com.

ABSTRACT

A 946 cubic meters per day (m^3/day) biological nutrient removal (BNR) pilot facility was constructed to test a proposed treatment technology. This paper presents the methods used to successfully start this pilot facility up while gathering useful information that will assist in the development of a full-scale facility start up plan. It took approximately 2 weeks to grow sufficient nitrifying biomass to maintain low concentrations of ammonia and nitrite (<1 mg-N/L). In addition, supplemental alkalinity was necessary during the startup period. Also, in order to emulate the impacts of future BNR plant shut downs, the pilot plant was shutdown multiple times throughout the project with negative impacts to effluent water quality with regards to ammonia and turbidity. Information was gathered to evaluate the causes of upsets so that a full-scale facility shut down plan can be formulated. It was observed that deflocculation occurred when adding supplemental alkalinity. This was determined to be caused by elevated pH values in the area where the supplemental alkalinity was added, specifically in the anaerobic zone which was seen to reach pH values as high as 9.0.

KEYWORDS: Start up, nitrification, denitrification, alkalinity, caustic soda, turbidity, nitrite, deflocculation

INTRODUCTION

The Sacramento Regional County Sanitation District (Regional San) must upgrade the Sacramento Regional Wastewater Treatment Plant (SRWTP) by 2021 to comply with future nitrogen discharge limitations. The future ammonia limit has a seasonal maximum daily value of 2.0 mg-N/L and an average monthly value of 1.5 mg-N/L. The nitrate limit will be 10 mg-N/L on an average monthly basis. In addition, Regional San must meet California Title 22 recycled water standards for unrestricted reuse by 2023. In particular, the turbidity in the plant effluent must be less than 2 nephelometric turbidity units (NTU) as a daily average and less than 5 NTU as the daily 95th percentile. Currently, the SRWTP does not provide nitrogen removal and Regional San engaged Brown and Caldwell to design, build and operate a 946 cubic meters per day (m^3/d) Advanced Treatment Technology Pilot (ATTP) which included a biological nutrient removal (BNR) process to demonstrate permit compliance. The first phase of testing was performed using a 4-stage biological nitrogen removal process. After the first phase of the pilot testing, a design engineer was retained by Regional San. The design engineer modified the design basis so that the new secondary process could utilize a modified 5-stage Bardenpho process to support phosphorus removal and included a mixed liquor fermenter (MLF). The results from operation of the MLF are described by Merlo et al. 2015. This paper presents results from specific periods during the second phase of testing where a 5-stage Bardenpho process equipped with a MLF was tested. The startup procedure and associated testing are described as well as the implications to effluent quality for extended plant shutdowns due to planned plant maintenance events. A goal of this work was to determine if a nitrifying biomass could be grown from high purity oxygen activated sludge. In addition, the deflocculation effects due to increased pH (caused by alkalinity addition) are described.

MATERIALS AND METHODS

The pilot-scale BNR process was designed to treat an average flow of 946 m^3/d (0.25 mgd) and a maximum flow of 1,968 m^3/d (0.52 mgd). The BNR process treated primary effluent (PE) from the SRWTP. The flow rate to the BNR process was variable based on a signal from the SRWTP influent flow meter such that both full-scale plant diurnal and wet weather flow variations were simulated. Figure 1 presents a process flow diagram and identifies the on-line instrumentation that was used for process monitoring and control. A classifying selector was included to mitigate biological foam formation as described by Parker et al. (2014). The target total sludge residence time (SRT) was 11 days (aerobic SRT of 6 days). For reference, the full-scale facility is being designed to treat a peak flow of 14.5 cubic meters per second (m^3/s, 330 mgd) continuously for 14 consecutive days. This constant peak flow is achieved through equalization. The full-scale facility is expected to start up by 2021.

Pilot Seeding and Startup

The startup of the pilot facility was performed in such a way to provide useful information that could be used to optimize the startup of the full-scale facility in 2021. The main concern for starting up the full-scale facility is to not discharge water until the biomass has grown sufficiently to provide complete nitrification (i.e. ammonia concentrations < 1 mg-N/L). Another major concern over startup is the impact of nitrite on the full-scale chlorine disinfection system. Free chlorine reacts very rapidly with nitrite and one milligram per liter (mg/L) of nitrite as N adds an additional chlorine demand of 5 mg/L with no disinfection benefit, so the presence of nitrite in the discharge is problematic. There is always a period of high nitrite and ammonia concentrations during the startup phase when the nitrifiers and denitrifiers are still growing, so it is imperative to startup the plant in such a way to prevent nitrite from being discharged. A goal of the pilot startup was to determine how long it would take to reduce nitrite concentrations to below 1 mg-N/L to prevent excessive dosages of chlorine from having to be used.

A sampling regime was implemented which measured several constituents twice a day for the duration of the startup period. The constituent list included: Total suspended solids (TSS), volatile suspended solids (VSS), ammonia (NH_3), nitrate (NO_3), nitrite (NO_2), pH, and alkalinity. By using grab samples, each of these constituents could be measured during both the operators morning shift and during the afternoon shift (sample collection at approximately 7:30 AM and 12:30 PM, respectively). These samples were analyzed in an onsite trailer laboratory, which provided fast results by field filtering samples and using a Hach spectrophotometer with the corresponding Hach test kits for ammonia, nitrite, nitrate, and alkalinity. Receiving fast results was important so timely process modifications could be made as soon as possible, as described later in this paper. The Hach test kits are Environmental Protection Agency (EPA) approved methods. Data presented in the startup section are data produced by Hach test kits while all other analytical results are based on analyses that were performed in accordance with APHA (2012).

Process Impacts from Shutdowns

There were several occasions during the pilot study when the pilot facility had to be shut down (turned off pilot influent flow) for 21-41 hours at a time. During each of these shutdowns, the pilot influent flow was turned off, and each aeration zone was set to the minimum airflow setting specified by the manufacturer and the mixers were left on. In addition, the RAS and IMLR pumps were left on at minimum flow conditions to keep the contents of the BNR in circulation. Sometimes the pilot was shut down due to full-scale plant maintenance while other times the pilot was shutdown to test the impact of a shutdown to effluent water quality. The latter shut downs were done to simulate future shut downs in the full scale facility.

Process Impacts from Alkalinity Addition

Caustic soda was configured to be metered into the PE line. A test was performed to determine if caustic soda addition caused elevated effluent turbidity values. To determine if caustic soda addition was the cause of the turbidity spikes, caustic was metered into the pilot at the maximum pump output of 0.289 m^3/d (3.18 gph) overnight several times. Along with this, a pH profile was recorded by the operators using a portable pH probe. Effluent turbidity values were monitored during these times to determine the impact of the caustic soda addition.

Bench-Scale Testing of the Effect of pH on SE Turbidity

To emulate conditions seen in the pilot plant in an attempt to identify causes of turbidity spikes, two sets of jar tests were performed on 7/15/14. The first set of jar tests were performed by dosing three different bases (25 percent sodium hydroxide, hydrated lime, and quicklime) to reach a pH of approximately 9, mixing them, and then letting them settle for 30 minutes. Two liters of sample water (i.e., mixed liquor from ANC or thickened ANC) was poured into a square two liter beaker, followed by the addition of chemical to increase the pH to 9.0. The sample was then flocculated for 30 minutes, followed by 30 minutes of settling (to imitate flocculation/settling in a secondary clarifier). The decant water was analyzed for TSS, turbidity, ortho-phosphate, and calcium, as well as ultraviolet transmittance (UVT) at each step of a serial filtration. Initial TSS, ortho-P, and calcium were also measured for each test.

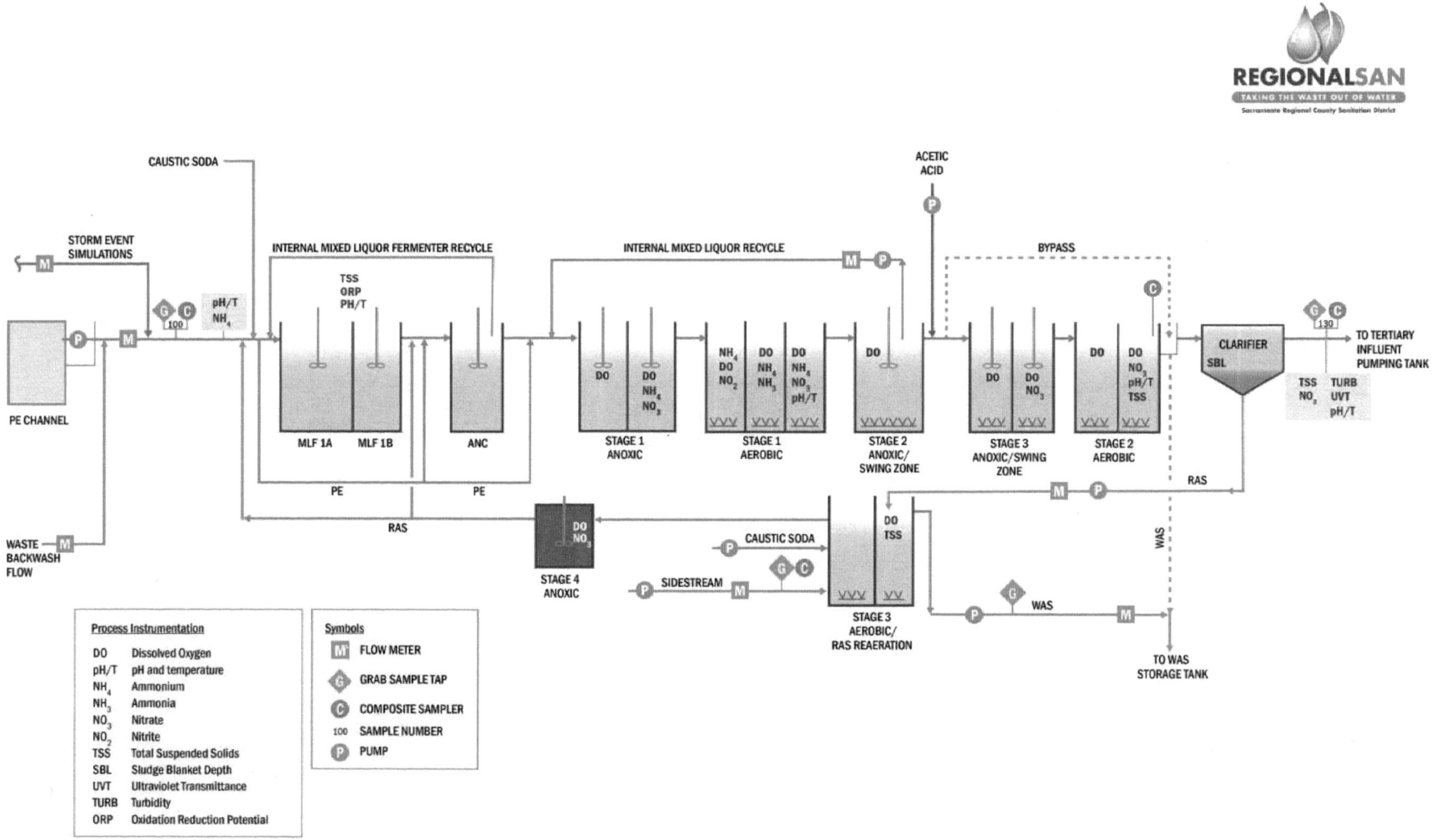

Figure 1. Process flow diagram of BNR process.

RESULTS AND DISCUSSION

Pilot Seeding and Startup

Prior to seeding, the BNR process basin tankage was partially filled with SRWTP chlorinated secondary effluent (SE), which was then dechlorinated in the pilot using dry sodium thiosulfate (approximately 16 kilograms were used in the entire BNR tankage). The pilot plant was filled with SRWTP SE because this was the most practical way of starting up the pilot facility, although this step may not be performed full-scale. The MLF, anaerobic zone, Stage 1 Anoxic, and Stage 1 Aerobic zones were filled entirely with SRWTP SE. Stage 2 Anoxic was filled half way, and Stage 3 Anoxic and Stage 2 Aerobic zones were only filled with about 0.6 meters of water depth to cover the aeration diffusers. At this point, the preparation was complete and the seeding was ready to start.

Around 9:30 AM on November 11, 2013, the control valves to all of the aerobic zones of the BNR process were set to input 0.14 cubic meters per minute per diffuser (9 inch circular fine pore diffuser) at each aerobic zone, including the swing zone. This represented the minimum airflow per diffuser per the manufacturer's recommendation. The mixers were turned on to each zone that did not have aeration. The internal mixed liquor recycle (IMLR) pump was turned on (approximately 272.5 m^3/d initially, or 50 gpm) and SRWTP high purity oxygen (HPO) return activated sludge (RAS) was added to Stage 1A Anoxic at a rate of approximately 163.5 m^3/d (30 gpm). The existing full-scale SRWTP is a high-purity oxygen activated sludge process that does not nitrify; the solids retention time is approximately 1 day. Therefore, the pilot was seeded with sludge to build up the mixed liquor suspended solids (MLSS) concentration, but was not actually seeded with any nitrifiers in order to simulate the full-scale startup. For smaller plants, it would be possible to import seed sludge from a nearby nitrifying facility to quickly startup the plant. However, Regional San's plant is too large so there is no practical way to seed the plant with any significant amount of nitrifiers from a nearby facility. Once flow started to go from Stage 1C Aerobic to Stage 2 Anoxic, the IMLR pump was turned up to approximately 2,267 m^3/d (416 gpm) to adequately circulate the contents of the BNR. Once the BNR tankage was full, the HPO RAS flow was turned off. It took approximately 24 hours to fill the pilot facility with HPO RAS. Once the HPO RAS flow was turned off, the BNR contents were allowed to circulate, through use of the IMLR pump, through the aerated and non-aerated zones to build up the MLSS concentration.

Figure 2 presents a time series plot showing the nitrogen species in Aerobic 1A and Aerobic 2B over the duration of the startup period. Note that the ammonia concentration increased immediately upon startup. The initial increase in ammonia observed immediately after startup is attributed to endogenous decay of cell material which releases ammonia. It took approximately 12 days for nitrification to stabilize to the point that both ammonia and nitrite were less than 2 mg-N/L. This is a significant finding because this means that the full-scale facility may not be able to discharge until approximately 2 weeks after startup. Nitrite values were greater than 2 mg-N/L three days after seeding and levels peaked at 38 mg-N/L and declined to less than 2 mg-N/L over three days. Although the pilot facility began discharging at a time of high nitrite (discharge began shortly after the PE pump was turned on), the full-scale plan will need to be implemented in a way that will not allow high nitrite concentrations during the discharge. One such strategy would be to bring basins online one at time, and at low flow rates (while the remainder of the flow is still be treated by the HPO system). This would dilute any nitrite to low concentrations. Even with this dilution strategy, it is important to keep the nitrite concentrations as low as possible when the basins are ready to begin discharging.

Approximately one week after seeding the pilot, a significant decrease in alkalinity was observed (Figure 3). A pH value of 6.0 was expected to significantly inhibit the growth of the nitrifiers, so alkalinity was added to the process. On November 18, 2013 at 15:30, approximately 22 kilograms of powdered lime ($Ca(OH)_2$) was added to the BNR process. The lime was spread into each zone that was in service. Even with the lime addition, the next morning (November 19, 2013) the pH had dropped to approximately 6.0 in the BNR process. To raise the pH, the primary effluent (pilot influent) pump was turned on to approximately 54.5 to 81.8 m^3/day (10-15 gpm) along with both caustic pumps on at 100 percent pump speed (25 percent sodium hydroxide solution). The PE pump had to be turned on in order to get the caustic soda into the pilot as fast as possible (the caustic soda was routed through the PE pipe). Around 12:30 on November 18, 2013, the PE flow was set to 212.5 m^3/d (39 gpm). This raised the pH and the alkalinity concentration throughout the contents of the BNR facility. At this time, the clarifier was put in service and the pilot facility began to discharge. Over the next several days, the flow was ramped up as is shown on Figures 2 and 3. The IMLR pump was set at 200 percent of the influent flow and the RAS pump was set at 60 percent of the pilot influent flow. On November 25, 2013, the pilot was operating as normal and the startup period was over.

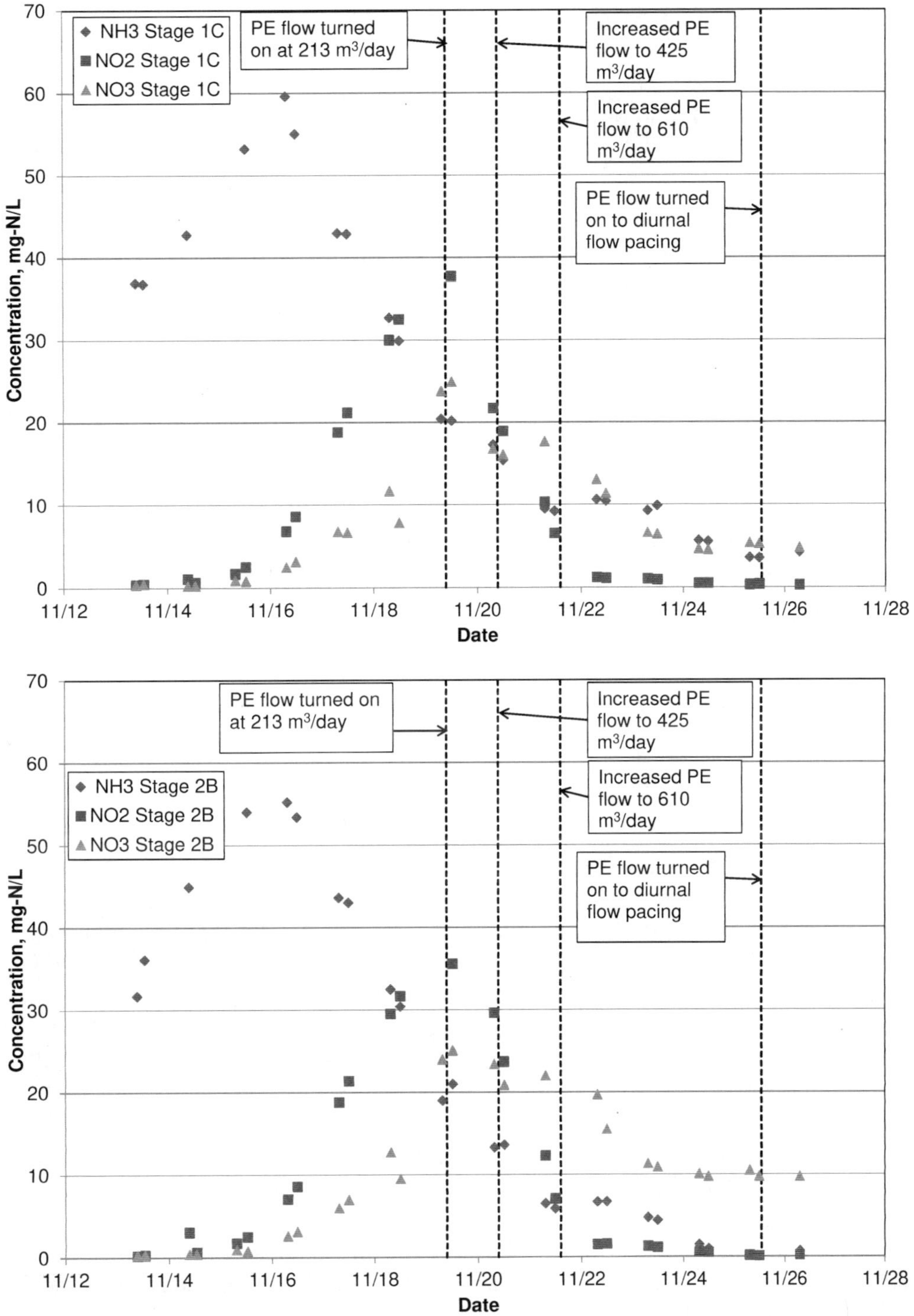

Figure 2. BNR Process Nitrogen Species during Startup for Stage 1C Aerobic (top) and Stage 2B Aerobic (bottom)

Mystic Takes (a) Charge: Startup of the BioMag Process

Megan B. Moody[1*], William C. McConnell[1], CDM Smith

[1] CDM Smith
*To whom correspondence should be addressed. Email: moodymb@cdmsmith.com

ABSTRACT

The Mystic Water Pollution Control Facility (WPCF) in Stonington, Connecticut was faced with the need to upgrade its wastewater treatment process to achieve nitrogen removal to meet the *General Permit for Nitrogen Discharges*, without constructing any new tankage. An emerging technology, the BioMag system marketed by Evoqua Water Technologies, was implemented to meet this challenge. The BioMag process utilizes a magnetite-ballasted mixed liquor process to operate at a mixed liquor suspended solids (MLSS) concentration much higher than conventional activated sludge systems. After the completion of a successful full-scale demonstration at the Mystic WPCF in 2010, a permanent system was constructed in 2014 as part of an overall plant upgrade. Since startup the facility has achieved a total nitrogen (TN) effluent concentration of less than 5 milligrams per liter (mg/L) on a consistent basis. Lessons learned at startup and during operation include foam control considerations, biological solids capture efficiency, the influence of polymer and potential impacts on the performance of downstream ultraviolet (UV) disinfection.

KEYWORDS BioMag, magnetite, nitrogen removal, ballasted biological treatment, activated sludge

INTRODUCTION

The Mystic WPCF is located on the west side of the Town of Stonington, Connecticut. The WPCF's maximum monthly permitted capacity is 3.0 megaliters per day (ML/d) [0.80 million gallons per day (MGD)]. Stonington completed long-term wastewater facilities planning in 2007 and identified an immediate need to upgrade the Mystic WPCF due to its age and condition and the need to improve treatment to meet nitrogen removal goals. TN treatment goals are based on the state-wide *General Permit for Nitrogen Discharges*, which specifies an allowable annual TN wasteload allocation for the Mystic WPCF of 12 kilograms/day (kg/d) [27 pounds per day (ppd)]. The mass allocation results in a required average annual effluent TN concentration of 5.2 mg/L based on the projected design flow rate. A desktop analysis of several processes was completed as part of the conceptual design and indicated that BioMag would enable the plant to meet its performance goals without requiring any new process tankage, and therefore was deemed the most viable and cost-effective option. The WPCF required a process that could be implemented with minimal modifications to the site. The site is constrained by the Amtrak railroad to the north, the Mystic Harbor to the west and south and a wetlands to the east. A site plan of the facility is shown in Figure 1.

The BioMag process provides a magnetite ballast to the mixed liquor in the bioreactors for dramatically enhanced settling of biological flocs. Magnetite is an inert iron oxide powder with a high specific gravity that embeds within solids to enhance secondary clarification. This allows for the secondary treatment process to operate at a MLSS concentration much higher than conventional activated sludge systems due to the enhanced setting rate of the magnetite-impregnated mixed liquor. Magnetite is recycled from the clarifiers back to the suspended growth process in the return activated sludge (RAS). Waste activated sludge (WAS) is processed for magnetite recovery and reuse using an in-line shearing mechanism to separate the sludge from the magnetite followed by a magnetic recovery drum. Recovered magnetite from the WAS is returned to the bioreactors for reuse.

Figure 1. Mystic WPCF Site Plan

A full-scale trial of BioMag at the Mystic WPCF was successfully completed in 2010 and has been previously documented (McConnell et al., 2010). The results and findings of the full-scale trial were incorporated into the final design (Moody et al., 2011).

The permanent installation was constructed in 2014 as part of the overall plant upgrade. The configuration of the upgraded treatment process includes a grit and grease removal system, primary settling tanks, bioreactors (arranged in 4-stage Bardenpho configuration), final settling tanks and UV disinfection. Solids removed from the process are thickened with a rotary drum thickener (RDT) prior to hauling off-site. The BioMag Building to the west of the bioreactors houses the magnetite storage silo, shear mill, magnetite recovery drums, ballast mix tank and ancillary systems as shown. The general arrangement of the BioMag process is shown in Figure 2. Major components of the process are described below:

- The bioreactors are equipped with baffle walls, two floating mechanical mixers per basin, fine bubble membrane diffusers, submersible internal recycle pumps, foam spray and direct surface wasting.
- The mixed liquor channel includes a surface wasting tipping tube to surface waste to a WAS wet well; the WAS pumps in the well have a recirculation discharge to mix the well with the intent of preventing solids deposition.
- The WAS is pumped to the BioMag recovery system in the building and runs through a shear mill and the magnetite recovery drum; magnetite is recycled back into the mix tank and WAS is processed by the solids thickening process. A backup drum is installed.
- The RAS wet well pumps RAS to a distribution box. The distribution box returns solids to each of the bioreactors as well as a sidestream back to the BioMag mix tank to blend virgin and recovered magnetite with the biological solids.

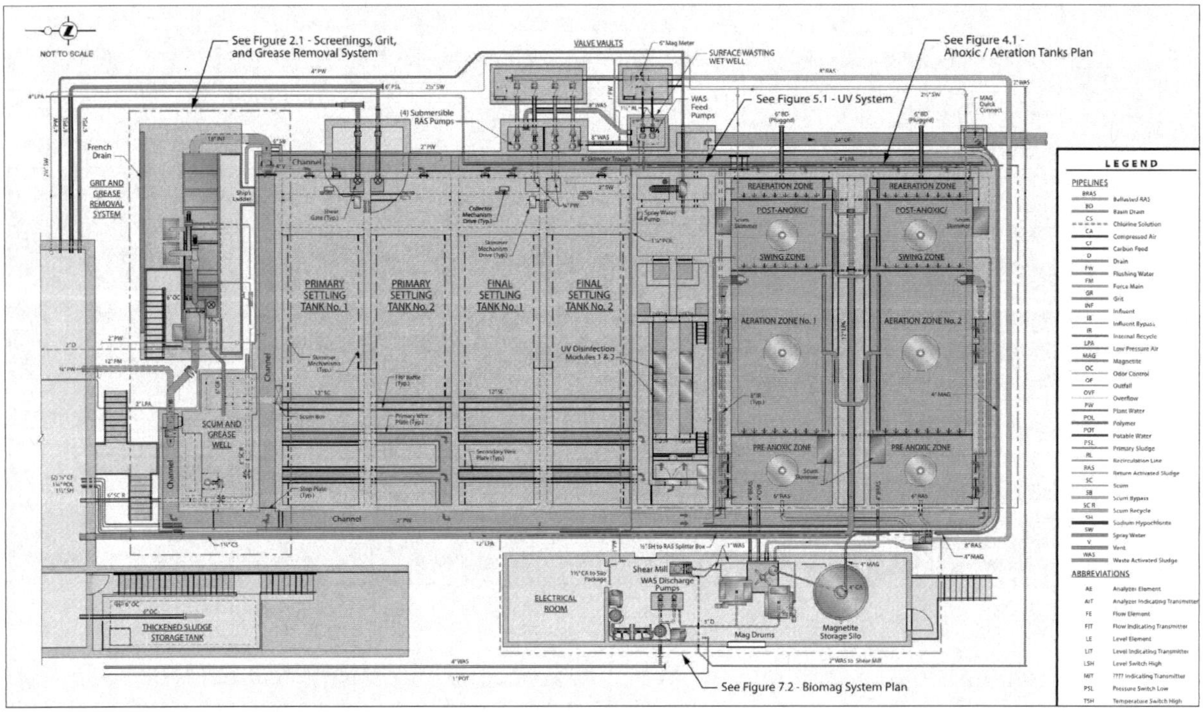

Figure 2. Process Tanks and BioMag Arrangement

Design Criteria

The upgrade to the Mystic WPCF was based on the design criteria summarized in Table 1.

Table 1.Mystic WPCF Biological System and BioMag Design Criteria

Process	Conventional activated sludge in 4-stage Bardenpho configuration; can be operated as MLE
Process Loads (30-day max.)	
- BOD_5	530 kg/day (1160 lbs/day)
- TSS	340 kg/day (740 lbs/day)
- TKN	90 kg/day (200 lbs/day)
Design Parameters	
- Design temp. (30-day min)	11 deg. C
- Aerobic SRT	9.9 days
- MLSS concentration	8,800 mg/L (biological mass only)
Aeration basins	
- Number	2
- Dimensions	6 m x 6.4 m x 4.04 m deep (52 ft x 21 ft x 13.25 deep)
- 1st-stage anoxic volume	0.180 ML (0.0475 million gallons)
- 1st-stage aerobic volume	0.416 ML (0.110 million gallons)
- 2nd-stage anoxic volume	0.180 ML (0.0475 million gallons)
- Reaeration volume	0.049 ML (0.013 million gallons)
- Total volume	0.825 ML (0.218 million gallons)
Final Settling Tanks	
- Number	2
- Dimensions	16 m x 4.0 m x 2.50 (52 ft x 13 ft x 8.2 ft deep)
- Total clarifier surface area	124.6 m^2 (1,352 ft^2)

By implementing the BioMag process along with upgrades to the aeration bioreactors, the facility has been able to reduce total nitrogen without the need to build additional tankage. In addition, the ballasted mixed liquor is able to provide better settling and reduce the SVI, significantly improving the performance of the shallow final settling tanks. At the current flowrate of 1.9 ML/d (0.5 MGD) the process is operating with a total MLSS (biological mass and magnetite) concentration typically around 11,000 mg/L and has shown that the overall goal of reducing TN to less than 5 mg/L has been successfully achieved.

Although the process is performing as intended, lessons learned at startup and during operation required further investigations, modifications to the process in the early months of operation and additional operational considerations. Modifications in the field have been made to improve the operational issues and optimize the process.

SUMMARY OF STARTUP AND OPERATIONS

Treatment Performance

The primary goal of the BioMag process at the Mystic WPCF is to produce an average annual effluent TN concentration of less than 5 mg/L. The process has performed better than expected with regards to nitrogen removal since startup of the permanent system. As shown in Figure 3, the average TN concentration to date in 2015 is 3.19 mg/L, well below the goal of 5 mg/L. Daily maximum TN concentrations have varied to more than 5 mg/L primarily due to remaining nitrate in the effluent. Though the process upgrade has provisions for supplemental carbon addition to achieve this level of treatment, to date no supplemental carbon has actually been used.

At an average 2015 flowrate of 1.9 ML/d (0.50 mgd) the process has been operating at an average biological MLSS of 5,000 mg/L. With a typical magnetite ratio of 1.1 the total MLSS concentration of the process has been operating at approximately 11,000 mg/L. The MLSS concentration variation of process operations in 2015 is shown in Figure 4.

The sludge volume index (SVI) of the final settling tanks prior to the implementation of BioMag was in the 200 to 250 mL/g range. After implementation of BioMag the final settling tanks typically operate between 60 and 80 mL/g. In May the SVI increased to just above 100 mL/g likely due to a foam outbreak in the aeration tanks with concurrent biomass inventory control issues. The SVI for the final settling tanks is shown in Figure 5.

The improved settling characteristics of the ballasted mixed liquor, and the higher concentrations this allows, has enabled the plant to successfully run with very high solids loading rates (SLR) to the shallow secondary clarifiers. The SLR is presented in Figure 6. Through the winter months, the daily SLR was typically in the 8.2-10.2 $kg/m^2/h$ (40 – 50 ppd/sf) range, and frequently peaked at 14.2 $kg/m^2/h$ (70 ppd/sf) due to the high solids inventory carried during cold temperature conditions. The SLR has reduced as the biomass inventory has been reduced with warmer temperatures.

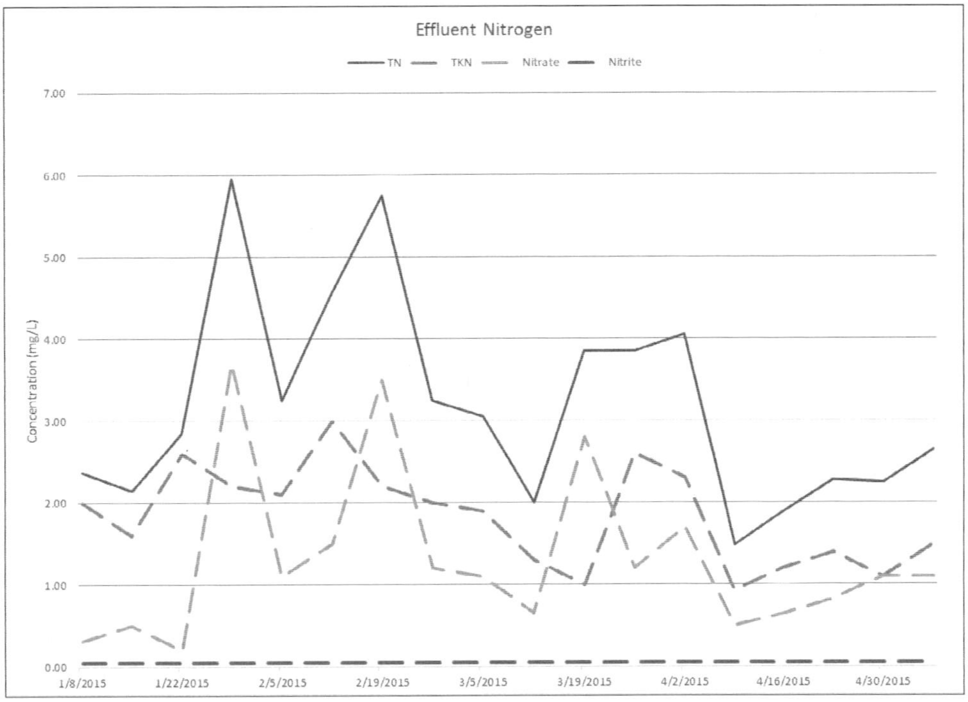

Figure 3. Effluent Nitrogen

Figure 4. MLSS Concentration

Figure 5. Sludge Volume Index

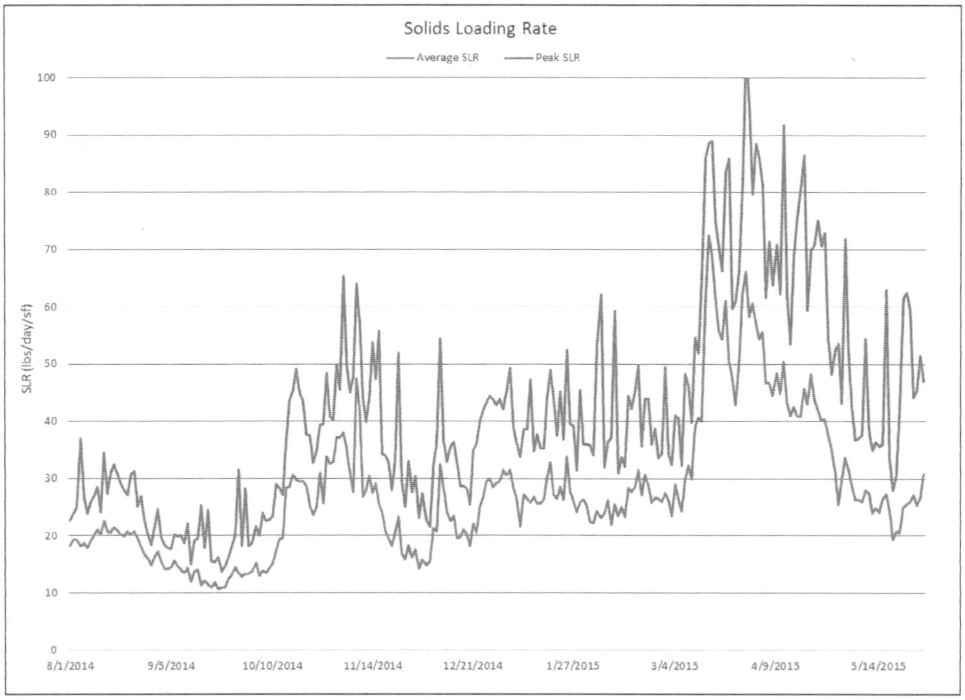

Figure 6. Solids Loading Rate

Foam Control
Excessive foam generation became problematic shortly after charging the mixed liquor with magnetite. Though the foam problem existed to some extent prior to magnetite charging, afterwards it became considerably worse. Observations and microscopic analysis of the foam did not point to a specific, identifiable culprit – filamentous bacteria of several types were present but not in numbers typically indicative of excessive foaming. It appeared as though *nocardia* was a contributor to the issue, but again not in abundance. The plant's collection system serves many restaurants due to the nature of the tourist industry in the community. The slug loads of grease to the plant, together with *nocardia,* may be a likely contributor to the foam generation. A grease removal system was installed as part of the preliminary treatment system at the plant and is capable of removing typical grease loads from the influent. However, with extreme slug loads some grease passes through into the process, contributing to foam generation.

During the full-scale demonstration, foaming was experienced to some degree (McConnell et al., 2010) and measures to mitigate this were incorporated into the design by the design team and Evoqua. Due to the limited number of online BioMag facilities, there was limited experience with foam generation and mitigation techniques. The baffle walls in the bioreactors, installed to configure the 4-stage Bardenpho process, include surface cut-outs to allow for foam to freely move to the end of the tanks with the aid of foam spray nozzles. The design also included a skimmer (tipping) trough in the mixed liquor channel prior to the final settling tanks. The intent of the skimmer trough was to surface waste foam from the channel into the WAS wet well so the magnetite could be recovered; however, the excessive foam generation during startup could not be freely moved out of the tanks and thus the problem was compounded as time passed. As a result, vactor trucks were used to physically remove the foam from the tanks. The thick mat of foam developed during the initial charging of magnetite is shown in Figure 7.

Figure 7. Aeration Tank Foam

Additional measures were implemented during startup to combat the foam and enable the plant to "keep ahead" of the foam. In addition to the foam spray and removal of foam from the effluent channel, an additional means of surface wasting directly from the bioreactors was needed. Surface wasting wells (four total) were designed by Evoqua and added to the pre-anoxic zones and the post-anoxic zones. These wells allow foam in the tanks to drop into a small wasting well in the tank, which then pump the foam out of the zone. For the pre-anoxic surface wasting well the foam is pumped to a submerged outlet within the aeration zone. The post-anoxic zone surface wasting well pumps the foam directly to the WAS wet well. A picture of the surface wasting well is shown in Figure 8. These surface wasting wells provide a means to physically move the foam out of the tanks and recover the magnetite in the foam solids.

Figure 8. Surface Wasting Well

In addition, several defoaming agents were tested to reduce the amount of foam in the bioreactors, to be used when/if needed should foam begin to re-establish. Operators have the option, at the first signs of foam generation, to add defoamer to ensure the foam does not begin to mat on the tank surface; by keeping the foam fluid means it is much more manageable. Use of defoamer on a continuous basis is not recommended, as the oxygen transfer efficiency (OTE) is decreased and the process oxygen requirement is increased when used.

Foam management has by far been the most challenging aspect of the upgraded process with BioMag. Whether due directly to an unidentified cause-and-effect from the BioMag process or not, it is strongly recommended that BioMag installations include a physical means of surface wasting foam directly from the process tanks. The solids processing impacts of continuous, or near-continuous, surface wasting need to be addressed with regards to impacts on solids processing facilities. The means for storing and pumping defoamer into the process tanks should also be considered as part of the design of a BioMag system. Several defoamers should be tested at startup to determine the best agent for the specific plant. Ample spray water should also be available to provide a means to break up foam and keep it fluid.

Magnetic Drum Biological Solids Capture Efficiency

The design and initial process operation and monitoring control did not consider or address the reality of inefficient capture of biological solids for wasting sludge from the system. While the magnetite recovery system has proven to be very efficient (with more than 95 percent of the magnetite sent through shear mill and recovery drum extracted and re-used), this recovery process also recycles a significant portion of the biological solids (the intended WAS) back to the process. The result of this less-than-anticipated biological solids capture efficiency was discovered during initial operation when the apparent WAS generation data was indicating an apparent net yield of at times more than two grams of volatile suspended solids generated per gram of biochemical oxygen demand (BOD) removed. It was discovered that the biological solids capture of the recovery drum was much less than anticipated, often in the range of 60-70 percent (i.e., a large portion of the biomass intended to be wasted from the system each day was actually not). This inefficient solids capture not only results in unintended return of WAS back to the bioreactors, but also results in "thinner" biological WAS to be processed by the plant's sludge thickening facilities with cascading impacts. When implementing a magnetite recovery system it is imperative to specify a minimum biosolids capture rate, and then to track solids inventory into and out of the recovery system.

Influence of Polymer

The BioMag process requires polymer addition to aid in the capture and settling of pin floc. Polymer is added to the mixed liquor effluent channel prior to clarification. The low flow, mid flow and high flow rates develop a dosage curve for the system. The programming and calibration of this system must be accurate to provide the correct dose. The polymer system provided by Evoqua upon startup did not provide for an adequate control of dose. Upon revisions to the program and calibration of the equipment the polymer dosage was corrected and resulted in better settling of pin floc. Incorrect dosage (either high or low) will impact the downstream UV disinfection process. Pump selection for the appropriate dose and pump selection needs to be recognized.

Clarification and Disinfection Impacts

During startup there were periods of operation where effluent from the final settling tanks deteriorated. Though the plant's effluent TSS concentration has always remained below the plant's NPDES limit of 30 mg/L, it has not always remained below its typical, but much lower threshold of performance (10 mg/L). Sporadic performance issues with the final settling tanks during startup appear to be due to several issues previously discussed, including polymer dose and location and primarily foam carryover from the bioreactors.

From a combination of bulking and foam, some solids were occasionally discharged over the final settling tanks' weirs, although effluent TSS requirements were achieved. Coincident with these conditions, the downstream UV disinfection performance became problematic in that it was not consistently achieving the plant's effluent *enterococci* limit of 35 MPN/100mL. Given the critical importance of providing adequate disinfection, the potential impacts of Biomag effluent TSS on UV disinfection were evaluated.

It has been theorized that BioMag effluent TSS particles, should they escape from the final settling tanks, may impact UV disinfection effectiveness in 3 ways: (1) by scattering or absorbing light, (2) by shading organisms, and (3) by shielding embedded organisms. These impacts are known to occur with "normal" biological TSS particles; it is reasonable to theorize that if the particles are larger, and have dense magnetite incorporated into the particle, these impacts may worsen.

An effluent sample was collected from the Mystic WPCF during a period of higher effluent TSS and analyzed for TSS particle size distribution and collimated beam testing. The results of the particle size distribution test is presented in Figure 9.

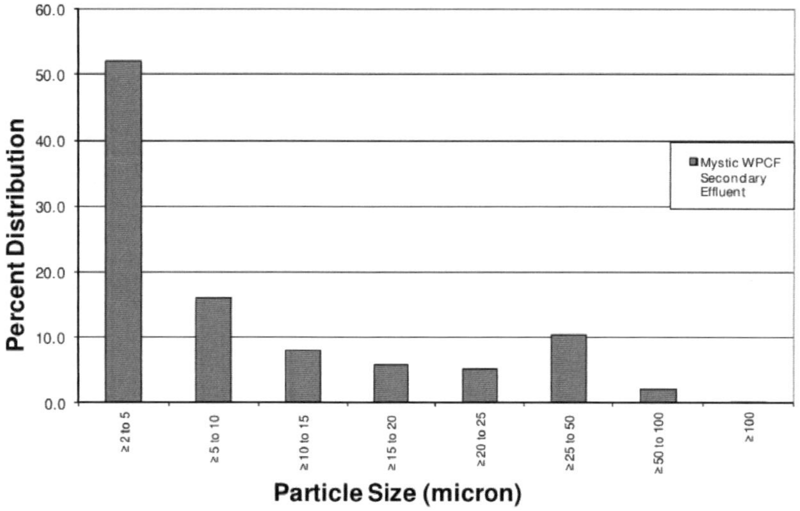

Figure 9. Particle Size Distribution of TSS from the Mystic WPCF

The results of this analysis indicate that the effluent TSS may exhibit characteristics that indeed are detrimental to the effectiveness of UV disinfection when targeting *enterococci*. The sample shows a high count of particles that are greater than 10 microns, indicating that there may be some issues with respect to disinfection because there are particle sizes that could result in bacterial shielding. It is expected that the BioMag process, which includes polymer addition to flocculate the ballasted and non-ballasted floc particles, is one of the factors that results in this particle size distribution.

The sample was tested for UV disinfection using a collimated beam. Results are presented in Table 10.

Table 10. Collimated Beam Test Results from the Mystic WPCF

Dose (mJ/cm^2)	Log Inactivation	*Enterococci* Coliform Count/100ml
0	0.000	904
10.14	0.850	128
20.27	1.394	37
30.41	1.696	18
40.54	1.628	21
50.68	1.698	18

The data presented in Table 10 indicates that there is a concentration of *enterococcus* that remains at doses above 30 mJ/cm^2 and no further reduction in the bacteria concentration is achieved even when a higher UV dose is applied. This implies a significant degree of shielding or shading may have occurred in the sample.

These data indicate that there may be an issue with respect to the impact of the effluent TSS from the BioMag system on the UV disinfection at the Mystic WPCF. The best way to provide consistent, reliable disinfection results is to limit the quantity of these particles in the UV system influent. Ongoing process-optimization is required to keep ahead of foam formation and to optimize the performance of the BioMag process, including polymer addition, recycled activated sludge (RAS) operation and clarifier performance.

Prevention of Solids Deposition

A primary design consideration was to make provisions for preventing magnetite-impregnated solids from depositing in locations that could become problematic from an operations and maintenance standpoint. The Mystic WPCF required a RAS and WAS wet well where solids potentially could settle. The few full scale BioMag facilities online, at the time, did not include wet well configurations and there was limited experience of providing a mechanism to prevent solids deposition.

Recycle systems were added to the WAS pump discharge lines to make sure that a high flow velocity is always maintained. The intent was that a percentage of the flow would discharge back into the wet well and keep the magnetite suspended. However, despite including these measures in the original design, they have proven to not adequately keep the wet well completely mixed to prevent solids deposition.

The underflow velocity into the RAS wet well was anticipated to be enough to keep the well mixed. This did not provide enough velocity and sump pumps were added to the wet well to re-suspend solids and continuously mix the wet well. Although this method did work, the pumps periodically trip and turn off. When the pumps turn off, the process would become unbalanced due to lower percent solids in the RAS and WAS. Without mixing of the wet wells the magnetite ratio is reduced and when the pumps are suspended the ratio is quickly increased.

Final modifications to the RAS and WAS wells include mixing with supplemental air available from the aeration system. The corners of each well include the ability to add air to keep the solids and magnetite suspended. When the wet wells are properly mixed the solids inventory is balanced and the process operates as intended.

SUMMARY AND CONCLUSIONS

The overall goal of implementing BioMag at the Mystic WPCF to reduce total nitrogen to below 5 mg/L was successfully achieved within the existing tankage. Providing the required treatment within the existing tankage was a crucial aspect of the project due to the site constraints. Implementation of the BioMag process saved several million dollars over the alternative processes available to provide the required level of treatment. The existing aeration basins were modified to operate in 4-stage Bardenpho configuration to achieve nitrification and denitrification at an increased MLSS concentration. The ballasted mixed liquor provided enhanced settling and decreased the SVI significantly. The BioMag process has demonstrated the ability to provide a very high level of biological treatment in a very small footprint at the Mystic WPCF, and the process should be evaluated for consideration for facilities that are facing similar challenges.

When implementing BioMag it is strongly recommended to include surface wasting and defoamer provisions in the bioreactors to remove foam. By applying defoamer and utilizing the surface wasting wells, operators can keep ahead of any foam outbreaks and the foam can be captured with the WAS to recover the magnetite. The sludge thickening process needs to account for surface wasting as it will impact operations. Sufficient spray water is also recommended for both the aeration basins and final settling tanks.

The biological solids capture of the magnetite recovery drum was in the range of 60-70 percent, much less efficient than anticipated. When implementing a magnetite recovery system it is imperative to specify a minimum biosolids capture rate and then to track solids inventory into and out of the recovery system.

Ongoing process-optimization is required to keep ahead of foam formation and to optimize the performance of the BioMag process, including polymer addition, recycled activated sludge (RAS) operation and clarifier performance in order to provide reliable disinfection results. There is potential for shielding or shading of *enterococcus* in the effluent TSS that may carry over from the clarifiers if the process is not optimized and impact disinfection.

Areas within the process that may allow for deposition of solids require provisions to keep solids suspended. A means for adequate suspension of solids in these areas are recommended and can be accomplished with recirculation pumps or mixing by air.

ACKNOWLEDGEMENTS

The implementation of the permanent installation of the BioMag process at the Mystic WPCF would not have been possible or successful without the dedication of the entire project team. A combined effort from the Stonington Water Pollution Control Authority Board and staff, the operators from United Water, the contractor (CDM Constructors Inc.), the BioMag system supplier (Evoqua) and the CDM Smith engineering team allowed for implementing modifications to the process to resolve operational issues that arose during startup.

REFERENCES

McConnell, W., Moody, M. and Woodard, S. *Full-scale BioMag Demonstration at the Mystic WPCP and establishing the Basis-of–Design for a Permanent Installation.* WEFTEC10 Conference Proceedings

Moody, M., Bishop, A., McConnell, W. *Beyond Desktop Evaluation: Key Design Criteria for Mixing and Settling of Magnetite-Impregnated Mixed Liquor.* WEFTEC11 Conference Proceedings

TECHNOLOGY INNOVATIONS THAT PROVIDE FLEXIBILITY FOR RELIABLE BIOLOGICAL NUTRIENT REMOVAL AND PHASED SITE SPACE UTILIZATION: CHALLENGES AND SOLUTIONS

Zeynep K. Erdal[1], Gary DeJesus[2], Cathi Zammit[2], Doug Berschauer[1], Brad Underwood[2] and Jay Witherspoon[1]

[1] CH2M HILL
[2] City of San Mateo

Abstract
The San Mateo Wastewater Treatment Plant (WWTP) is operated by the City of San Mateo, and treats flows from the City of San Mateo as well as the surrounding communities. The design capacity of the plant is 15.7 mgd, with planning level wet weather flows varying between 72 and 98 mgd. The facility currently consists of primary clarification, aeration basins for short SRT activated sludge operation, secondary clarifiers, pressure filtration, chlorine disinfection, and dechlorination. In addition to the current NPDES permit requirements, the facility is required to provide wet weather treatment that could result in greater than 7.5 fold flow peaking over average flow conditions. Also, the facility is located in the San Francisco Bay Area that is currently being evaluated for potential impairment with pending nutrient removal requirements to be imposed on the dischargers. The City has evaluated options to optimize the existing facility using select treatment technology options to meet three increasing levels of nutrient treatment and implement wet weather treatment in parallel to the future facilities planning.

Keywords
Nutrients, nitrification, reuse, aging infrastructure

Background and Objectives
The San Mateo Wastewater Treatment Plant (WWTP) is operated by the City of San Mateo, and treats flows from the City of San Mateo as well as the surrounding communities, including a portion of the County of San Mateo, the Town of Hillsborough, Crystal Springs County Sanitation District (CSCSD), and Foster City/Estero Municipal Improvement District (EMID). As with most wastewater collection and treatment facilities in the San Francisco Bay Area, the City's collection system and jointly owned WWTP have aging components that need improvements to continue to meet current and future flows and permit requirements, and provide functionality to support the City in administrative functions for sewer operations, maintenance, and laboratory. The facility is located in the City San Mateo in the low lying coastal area in San Francisco Bay.

The design capacity of the plant is 15.7 mgd, with planning level wet weather flows varying between 72 and 98 mgd. The facility currently consists of primary clarification, aeration basins for short SRT activated sludge operation, secondary clarifiers, pressure filtration, chlorine disinfection, and dechlorination. The current National Pollutant Discharge Elimination System (NPDES) permit for the facility was issued in 2013 and calls for seasonal effluent quality limits. In addition to the current NPDES permit requirements, the facility is required to provide wet weather treatment that could